AF443496

FUNDAMENTAL PROBLEMS IN
STATISICAL MECHANICS VII

FUNDAMENTAL PROBLEMS IN STATISTICAL MECHANICS VII

Proceedings of the Seventh International Summer School on
Fundamental Problems in Statistical Mechanics,
Altenberg, F. R. Germany, June 18–30, 1989

Editor

H. van Beijeren

State University Utrecht
The Netherlands

1990

NORTH-HOLLAND
AMSTERDAM•OXFORD•NEW YORK•TOKYO

ISBN: 0 444 88432 7

Published by:

North-Holland
Elsevier Science Publishers B.V.
P.O. Box 211
1000 AE Amsterdam
The Netherlands

Sole distributors for the U.S.A. and Canada:

Elsevier Science Publishing Company, Inc.
655 Avenue of the Americas
New York, N.Y. 10010
U.S.A.

Library of Congress Cataloging-in-Publication Data

International Summer School on Fundamental Problems in Statistical
  Mechanics (7th : 1989 : Altenburg, Leipzig, Germany)
    Fundamental problems in statistical mechanics VII : proceedings of
  the Seventh International Summer School on Fundamental Problems
  in Statistical Mechanics, Altenburg, F.R. Germany, June 18-30, 1989 /
  editor, H. van Beijeren.
        p.   cm.
    Includes bibliographical references.
    ISBN 0-444-88432-7
    1. Statistical mechanics--Congresses.   I. Beijeren, H. van (Henk)
II. Title.
QC174.7.I566   1989
530.1'3--dc20   90-32283
                                                           CIP

PRINTED IN THE NETHERLANDS

PREFACE

This book contains the lectures given at the seventh International Summer School on Fundamental Problems in Statistical Mechanics, held from June 18-30 at Haus Altenberg in Altenberg, FRG. Most of the lectures are on recent developments in statistical mechanics. The following general topics have been covered: Dynamical Systems, Phase Transitions and Critical Phenomena, Polymers and Suspensions, Spin Glasses, Kinetic Theory of Fluids and Cellular Automata and Computer Simulations in Statistical Mechanics.

The summer school was organized by the Scientific Organizing Committee with initial assistance by NUFFIC (Netherlands Foundation for International Cooperation). Financial support was received from NUFFIC in The Hague, the Bundesministerium für Forschung und Technologie in Bonn, Lorentzfonds, Leiden, The Rockefeller University, New York, Kernforschungsanlage, Jülich, Stichting Physica, Utrecht, Shell Nederland BV, The Hague and Stichting voor Fundamenteel Onderzoek der Materie (FOM), Utrecht.

The Scientific Organizing Committee consisted of:

> D. Bedeaux, State University Leiden, NL
> E.G.D. Cohen, The Rockefeller University, USA
> M.H. Ernst, State University Utrecht, NL
> B.U. Felderhof, RWTH, Aachen, FRG
> H.J. Hilhorst, Université de Paris XI, France
> H. Müller-Krumbhaar, KFA, Jülich, FRG
> H. van Beijeren, State University Utrecht, NL
> J.M.J. van Leeuwen, State University Leiden, NL

Thanks are due to Ms. L. Gerken from the KFA, Jülich for her assistance as a secretary and hostess, and to the KFA, Jülich for providing vital assistance in the local organization.

The school was attended by 103 students from 19 countries. The program consisted of 40 regular lectures, given by 13 lecturers, and 9 seminars.

It is hoped that this series, besides inspiring students and scientists interested in statistical mechanics, will document the developments in this field over the past few decades.

H. VAN BEIJEREN

TABLE OF CONTENTS

Linear Response Theory for Cellular Automata Fluids

Kinetic Theory of Dense Gases and Liquids

Computer Simulations in Statistical Physics

STATISTICAL MECHANICAL BEHAVIOR OF DYNAMICAL SYSTEMS: A ONE-DIMENSIONAL MODEL OF DIATOMIC MOLECULES

E.G.D. COHEN and A. MONGE

The Rockefeller University, New York, NY 10021

The apparent violation of the zeroth law of thermodynamics found in computer simulations of a number of simple non-integrable dynamical systems is studied on a one-dimensional model of diatomic molecules.

1. INTRODUCTION

The question we want to address here is: when does a classical dynamical system behave as a statistical mechanical system? We consider here non-integrable dynamical systems that have been studied in computer simulations and that have, from a statistical mechanical point of view, relatively few degrees of freedom.

This study is part of what is sometimes called now "experimental mathematics". We restrict ourselves here to systems consisting of particles and do not consider fields[1]. The two main questions we want to address are:

1) does the dynamical system considered approach (statistical mechanical) equilibrium, i.e. the micro-canonical ensemble? (in other words: is the zeroth law of thermodynamics[2] valid?),

2) does one then find equipartition of energy, in accordance with the (statistical mechanical) equipartition theorem?

These questions are relevant because a number of simple dynamical systems studied by computer simulations have given a negative answer to these questions.

The first system studied was that of a chain of 64 slightly anharmonic oscillators with fixed end points in one dimension by Fermi, Pasta and Ulam (FPU) in 1955[3]. Here, to their surprise, the anharmonicity did not lead to equipartition of energy over the oscillators, even after long times of several hundreds of periods of the corresponding linear system. Many subsequent studies have, in our opinion, not yet resolved this non-statistical mechanical behavior of the FPU model. Similarly Benettin, Galgani and Giorgilli (BGG)[4] studied a one-dimensional model of a gas of diatomic molecules: a number of n = 16 or 64 harmonically bound pairs of particles, with "soft" intermolecular forces, specified below. Also in this system no approach to equilibrium in physically reasonable times was found. Many other such cases have been reported in the literature[1,6].

Apart from the interest in this non-statistical mechanical behavior of these computer simulated systems for the foundations of classical statistical mechanics, there are also more practical implications. We mention as examples the problem of energy relaxation in molecules towards equipartition, in chemistry[7], and the question of the approach to and the nature of the equilibrium state of (spin) glasses[8].

Here we will restrict ourselves to a discussion of the BGG model and a closely related one, used by Erpenbeck and Cohen, the EC model[9], with hard, instead of soft, intermolecular forces. Our aim is to suggest a possible explanation of the violation of statistical mechanics in the above mentioned computer simulations.

In section 2 we discuss the BGG model, in section 3 the EC model is introduced and in section 4 we compare the results of both models and draw a number of conclusions.

2. THE BENETTIN-GALGANI-GIORGILLI MODEL

2.1. Model

We consider a one dimensional model of n identical diatomic molecules consisting of 2n atoms of mass m/2 moving on a line of length L with fixed end points. Molecule i is formed by atoms 2i-1 and 2i, bound together by a intramolecular harmonic potential. The intermolecular potential is a "soft" potential $V_s(r)$. Of the two atoms in each molecule only one interacts via the potential $V_s(r)$ with an atom in a nearest neighbor molecule, while the other atom does not interact , i.e. is transparent (cf. fig. 1).

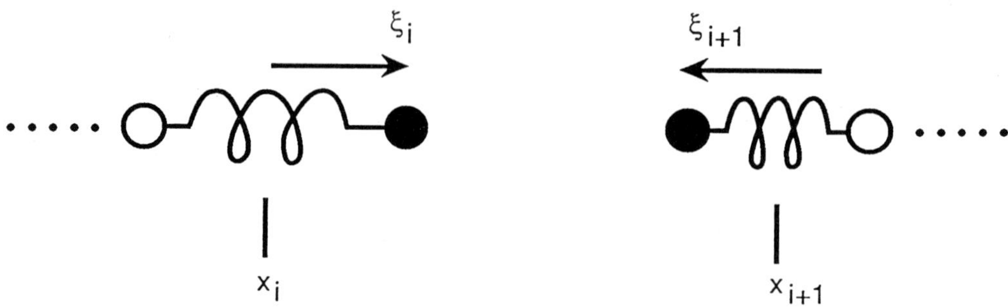

FIGURE 1

One-dimensional model of diatomic gas: in molecule i, x_i is the center of mass coordinate, and ξ_i the relative coordinate; p_i and π_i are the corresponding conjugate momenta.
Note that: 1) in the BGG model each atom has mass m/2, while in the EC model each atom has mass m; 2) in the BGG model fixed end boundary conditions are used (i.e., $x_o = 0$, $x_{n+1} = L$, $\xi_o = \xi_{n+1} = 0$), while in the EC model periodic boundary conditions are employed.

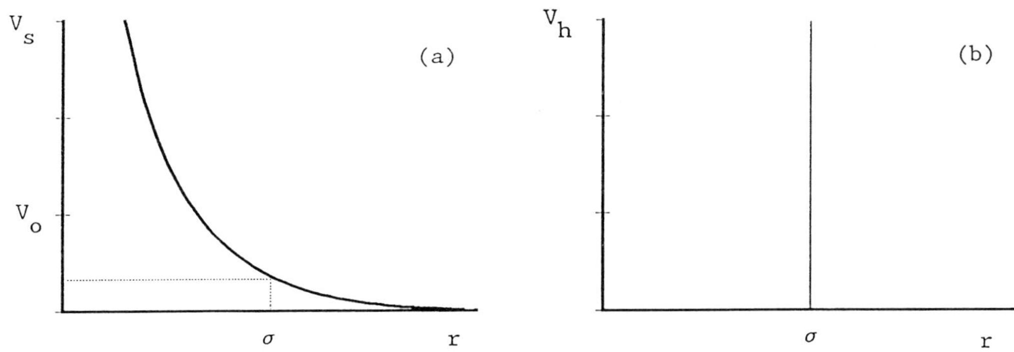

FIGURE 2

(a) "Soft" potential $V_s(r) = V_o(\sigma/r)\, e^{-(r/\sigma)^2}$.

(b) Hard core potential. The hard point potential V_h used in the EC model is obtained by taking $\sigma = 0$.

Calling the coordinates of the center of mass of molecule i, x_i, the relative coordinates of the atoms in molecule i, ξ_i, and the corresponding momenta p_i and π_i, the Hamiltonian for the system can be written as:

$$H = \frac{1}{2} \sum_{i=1}^{n} \left[\frac{p_i^2}{m} + \frac{\pi_i^2}{m} + m\,\Omega^2\,\xi_i^2\right] + \sum_{i=0}^{n} V_s\left((x_{i+1}+\xi_{i+1}) - (x_i+\xi_i)\right) \qquad (1)$$

Here Ω is the frequency of the harmonic oscillator molecules and the potential $V_s(r)$ is given by (cf. fig. 2)

$$V_s(r) = V_o\left(\frac{\sigma}{r}\right) e^{-\left(\frac{r}{\sigma}\right)^2} \qquad (2)$$

where σ is the range of the intermolecular potential.

On the right hand side of (1) , the first term represents the translational (kinetic) energy E_t of the center of mass of the molecules, the second and the third terms the kinetic E_{vt} and potential E_{vp} contributions to the vibrational energy E_v, respectively and the last term a potential energy contribution due to the intermolecular forces. Associated with the Hamiltonian H are, apart from the total energy E, a characteristic time $t_o = (m/V_o)^{1/2}\sigma$ for the duration of an intermolecular collision and a dimensionless frequency $\omega = \Omega t_o$. In addition there is a mean free time between successive collisions t_{mfp} which can be defined as $t_{mfp} = (nm/E)^{1/2}(L/n)$, so that t_{mfp} and t_o are related by $t_{mfp}/t_o = (nV_o/E)^{1/2}(L/n\sigma)$.

2.2. Computations

BGG carried out computations for n = 16 and 64, and for energies per particle E/n equal to V_o and $2V_o$. Since L was chosen to be $4(n+1)\sigma$, t_{mfp} was either $\approx 4t_o$ or $\approx 2\sqrt{2}t_o$. They computed the decay in time of the normalized vibrational energy auto-correlation function $\rho_v(t)$ defined by:

$$\rho_v(t) = \frac{\overline{E_v(T)E_v(T+t)} - \overline{E_v(T)}^2}{\overline{E_v(T)^2} - \overline{E_v(T)}^2} \tag{3}$$

Here $E_v(T) = \frac{1}{2}\sum_{i=1}^{n}\left[\frac{\pi_i^2(T)}{m} + m\,\Omega^2\xi_i^2(T)\right]$, where $\xi_i(T)$ and $\pi_i(T)$ are the relative coordinates and momenta of the atoms in molecule i at time T, respectively. The bar indicates a time average over times large compared to the characteristic time t_o. BGG used time averaging intervals several hundred times the time scale over which $\rho_v(t)$ is sensibly different from zero. The normalized auto-correlation function $\rho_v(t)$ is expected to decay from a value 1 at t = 0, to a value essentially zero in a time $t \gtrsim t_o$, the microscopic characteristic time, if statistical mechanical behavior obtains. Introducing a reduced time $\tau = t/t_o$, $\rho_v(\tau)$ was computed for a number of reduced frequencies ω, using random initial data with the condition $E_v(0) = \frac{2}{3}E$.

2.3. Results

$\rho_v(\tau)$ for n = 16 and $\omega = 8$ is plotted in fig. 3a. Although there is a decay to zero which is interpreted by BGG as an approach to equilibrium, the decay is slower than exponential, so that the decay of $\rho_v(\tau)$ was characterized by its half-life time, i.e. by that value τ_c of τ for which $\rho_v(\tau_c) = \frac{1}{2}$. This was done for two values of E/n: for E/n = V_o over values of ω ranging from about 2 to 12, and for E/n = $2V_o$, over values of ω ranging from 5 to 16. For both cases, BGG found an exponential dependence of τ_c on ω that could be represented by (cf. fig. 3b):

$$\tau_c(\omega) = A\,e^{\alpha\omega} \tag{4}$$

This implies that for the higher frequencies $\rho_v(\tau)$ will not decay to essentially zero in physical times. In fact, for $\omega = 8$, $\tau_c = 200$, much too long to be consistent with statistical mechanics (cf. section 3). BGG explain these very long times for large ω by the Nekhoroshev theorem for nearly integrable dynamical systems[10,11]. In simple terms this theorem says the following. A nearly integrable system with a Hamiltonian H = $H_o + \epsilon V$, where H_o is integrable, "looks integrable" for times t that grow exponentially with $1/\epsilon$, so that for an action variable I holds:

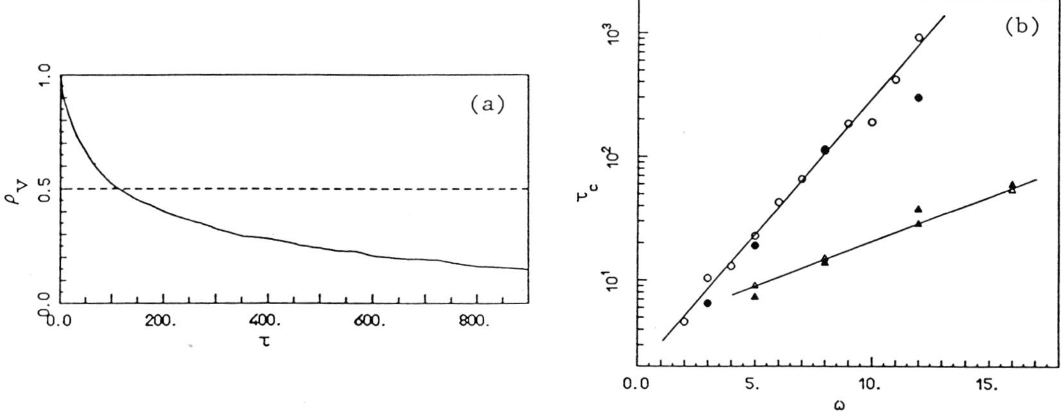

FIGURE 3

BGG model:

(a) $\rho_v(\tau)$ for $n = 16$ and $\omega = 8$ in the case $E/n = V_o$.

(b) Half-life time τ_c vs ω on semi-logarithmic scale for $E/n = V_o$ (circles) and $E/n = 2V_o$ (triangles) with $n = 64$ (full symbols) and $n = 16$ (open symbols). (from Ref. 4)

$$|I(t) - I(0)| < T\epsilon^b \tag{5a}$$

for

$$T = T_o \left(\frac{1}{\epsilon}\right) e^{\alpha\left(\frac{1}{\epsilon}\right)^a}, \tag{5b}$$

where T and T_o are a characteristic action and time, respectively and a, b and α are positive constants that depend on H_o. Thus the action variable I will change hardly at all, i.e., will be bounded by a small number $\sim\epsilon^b$ the longer, the closer the system is to an integrable system, i.e., the smaller ϵ is. Applied to the BGG system, the frequency ω corresponds to $1/\epsilon$, since the interacting diatomic molecular system is the closer to integrability, the smaller V_s is, i.e., the larger ω is. It should be remarked that the constant α in eq. (5b) depends also on the number of particles n in the system, such that $\lim_{n\to\infty} \alpha(n) = 0$ [11]. Nevertheless, BGG interpret their results as being typical for a large macroscopic system of interacting harmonic oscillators. In fact, they connect it with the old arguments of Boltzmann and Jeans to explain away the ultraviolet catastrophe in the theory of black body radiation and the anomalous specific heats of polyatomic molecules in the context of classical statistical mechanics[4,5].

3. THE ERPENBECK-COHEN MODEL

3.1. Model

This model differs from the BGG model in that the mass of each atom is m and the soft potential V_s is replaced by a hard potential V_h (cf. fig.2) [9]. Also, in this model the even numbered atoms of the molecules interact as if they were hard impenetrable points, while the odd numbered atoms do not interact at all, i.e., are transparent. The n identical diatomic molecules can now move on the infinite line, with a basic cell of length L and periodic boundary conditions. The Hamiltonian is given by eq. (1), with V_s replaced by the above mentioned hard point interaction, which, for periodic boundary conditions, can be found in the literature[9]. The only characteristic variables in this model are the characteristic time $t_{mfp} = (nm/E)^{1/2}(L/n)$ and the characteristic frequency $\omega = \Omega t_{mfp}$, since, because of the hard point interaction, the total energy E scales out via t_{mfp}.

3.2. Computations

Computations were carried out for a variety of values of n from 2 to 500, and of ω from 0.01 to 1000. We will discuss here the calculation of two quantities: 1) the same vibrational energy correlation function $\rho_v(t)$ as studied by BGG, and 2) the energy equipartition ratios, defined as follows:

$$f_\alpha = \frac{\overline{E_\alpha}^T}{<E_\alpha>_{md}} \qquad (\alpha=t, \text{ vk, vp}) \qquad (6)$$

Here $\overline{E_\alpha}^T$ and $<E_\alpha>_{md}$ are the time average and the molecular dynamics ensemble averages of the translational, vibrational kinetic and vibrational potential energies, respectively. They should all equal to 1, when there is equipartition of energy. In (6) the time average was taken over a trajectory of about 3×10^5 time-steps (t_{mfp}) and the ensemble averages equal:

$$<E_t>_{md} = E \; (n-1)/(3n-1)$$

$$<E_{vk}>_{md} = <E_{vp}>_{md} = E \; n/(3n-1) \qquad (7)$$

Here the molecular dynamics ensemble average is characterized by the parameters n, L, E and a vanishing total momentum. The factor (3n-1) is the total number of degrees of freedom of the system, while the factor (n-1) incorporates that the conservation of momentum reduces the translational degrees of freedom by 1.

3.3. Results

1) $\rho_v(\tau)$, with $\tau = t/t_{mfp}$ for n = 10 is presented in fig. 4 for the four reduced frequencies $\omega = 0.1$, 1.0, 10 and 1000, thus covering four orders of

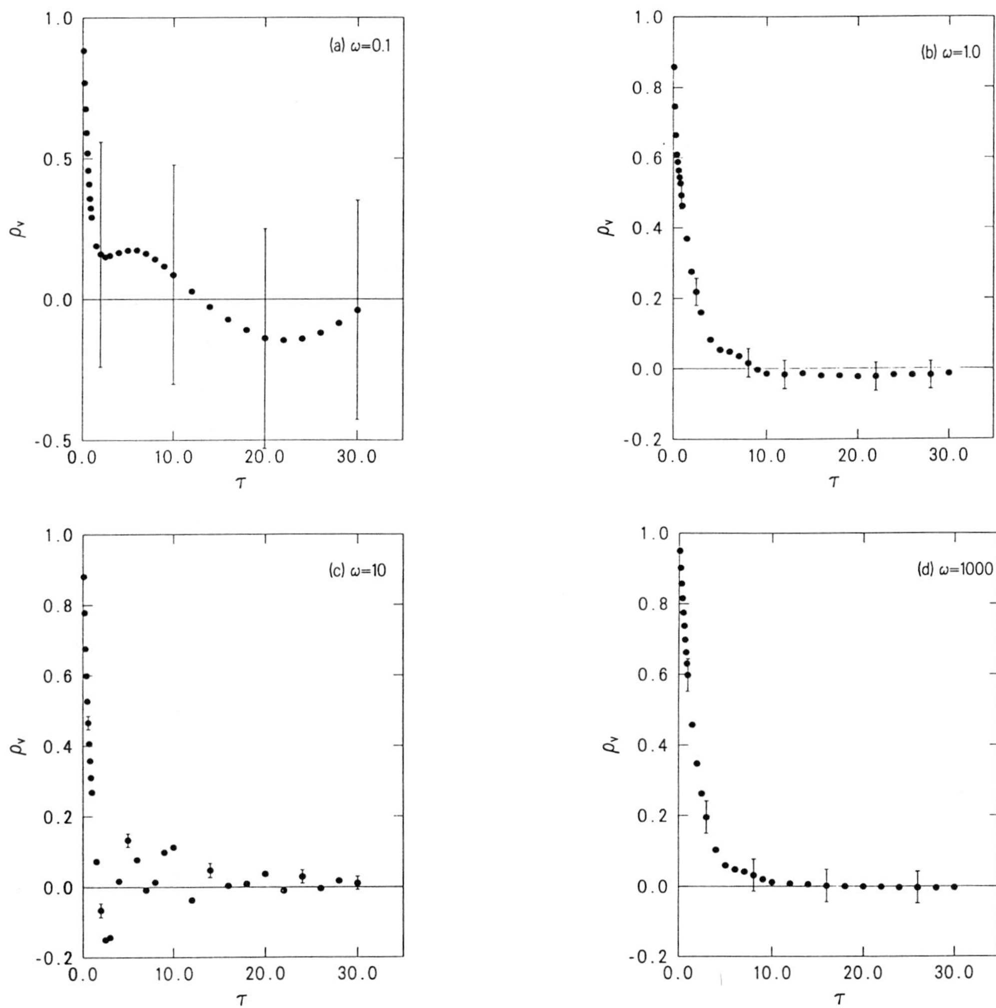

FIGURE 4

EC model:
$\rho_v(\tau)$ for systems with n = 10, and frequencies (a) ω = 0.1, (b) ω = 1,
(c) ω = 10, and (d) ω = 1000　(from Ref. 9).

magnitude. In all these cases there is a rapid, apparently exponential, decay
to zero of $\rho_v(\tau)$ in the physical time, $\tau_c \approx$ 5, independent of ω.

　2) Table I shows that the f_α's (for the maximum times considered) come
close to their equipartition values for n > 20 and ω > 0.5. Only for n $\gtrsim$ 500
and $\omega \gtrsim$ 1000 is full equipartition obtained.

　These results for the EC model indicate that for hard intermolecular forces
there is no conflict at all with the zeroth law of thermodynamics: if the
number of particles n is not too small and the frequencies are not too small
the system approaches equilibrium rapidly and behaves as a statistical

8   *E.G.D. Cohen and A. Monge*

mechanical system. The conditions on n and ω are necessary to obtain ergodic behavior (cf. ref.9). In particular, the condition on ω is required to remain sufficiently far from integrability. For, unlike the BGG system, the EC system is integrable for $\omega = 0$, since it reduces then to a system of hard points on a line. It is an interesting question, whether for small enough ω, the decay of $\rho_v(\tau)$ would occur in relaxation times $\tau_c(\omega) \sim e^{\alpha/\omega}$. In fact, although the Nekhoroshev theorem does not hold for our highly non-analytic potential V_h, one

n	ω	f_t	f_{vk}	f_{vp}
2	0.01	0.575(8)	1.012(7)	1.200(9) [a]
	0.10	0.538(1)	1.007(1)	1.224(1) [a]
	0.10	1.619(1)	0.874(1)	0.817(1) [a]
	0.10	1.537(3)	0.961(1)	0.771(1) [a]
	1.00	0.690(5)	1.036(3)	1.119(1)
	1.00	1.655(0)	1.029(0)	0.664(0)
	2.00	1.428(12)	0.872(6)	0.914(1)
	10.00	0.704(22)	1.071(6)	1.077(6)
	100.00	0.697(11)	1.073(3)	1.079(3)
	100.00	1.620(11)	0.847(3)	0.843(3)
3	0.01	0.920(7)	1.014(5)	1.039(9) [a]
	0.10	1.707(1)	1.311(0)	0.217(0) [a]
	0.10	0.941(6)	1.037(4)	1.002(7) [a]
	0.50	1.348(0)	1.323(0)	0.445(0) [a]
	0.80	1.266(0)	1.326(0)	0.497(0) [a]
	1.00	0.964(3)	1.006(4)	1.017(4)
	10.00	0.965(4)	1.012(1)	1.011(1)
4	0.10	1.068(14)	0.855(12)	1.094(22) [a]
	0.50	1.168(8)	0.941(8)	0.933(14) [a]
	1.00	1.208(2)	0.923(2)	0.921(2)
	2.00	1.209(1)	0.922(1)	0.921(0)
	5.00	1.098(2)	0.963(1)	0.964(1)
	10.00	1.165(3)	0.939(1)	0.937(1)
	10.00	1.197(2)	0.926(1)	0.926(1)
	100.00	1.074(1)	0.972(1)	0.972(1)
10	0.10	0.994(14)	1.042(13)	0.963(25) [a]
	0.10	1.056(22)	0.954(20)	0.996(40)
	0.10	1.065(14)	0.976(15)	0.966(28) [a]
	0.10	1.028(15)	0.940(15)	1.035(29) [a]
	0.50	1.042(1)	0.980(2)	0.981(2)
	1.00	1.073(1)	0.967(1)	0.967(1)
	10.00	1.076(2)	0.966(1)	0.966(1)
	10.00	1.072(1)	0.968(1)	0.968(1)
	1000.0	1.072(2)	0.968(1)	0.968(1)
20	0.10	0.986(7)	0.986(8)	1.028(15) [a]
	0.50	0.999(1)	1.002(1)	0.999(2)
	2.00	1.035(1)	0.984(1)	0.984(0)
100	0.10	0.992(4)	1.014(4)	0.994(8) [a]
	10.00	0.985(0)	1.007(0)	1.007(0)
	1000.0	1.000(1)	1.000(0)	1.000(0)
500	1000.0	1.000(1)	1.000(0)	1.000(0)

TABLE I

Equipartition ratios f_t, f_{vk}, f_{vp} for the EC model. The statistical uncertainties (given in parentheses relative to the last digit) are one standard deviation (from Ref. 9).
[a] Statistical analysis, as discussed in Ref. 9, failed in these cases.

could nevertheless, in view of the integrability of the system for $\omega = 0$, conjecture that a similar behavior would obtain near $\omega = 0$, as for a soft potential V_s near integrability for $\omega = \infty$. Of course, in the EC model the parameter ϵ in the analogue of the Nekhoroshev theorem (5) corresponds to ω.

4. DISCUSSION

1. While the hard core system behaves as a statistical mechanical system for n > 20 and ω not too small, the BGG system with a soft intermolecular potential does not appear to do so, even for n = 64. Clearly, the intermolecular potential plays an essential role here. A violation of the zeroth law would be, however, very surprising, since this system, with its repulsive intermolecular potential V_s, appears to have a perfectly well defined equilibrium state. Thus we would be confronted here with a simple system, that, although it does have an equilibrium state, it does not reach this state in a physically reasonable time. It seems to us that this behavior may be due to the following:

i) The relatively small number of particles used in the BGG model computer simulations. Statistical mechanical behavior only obtains strictly speaking, in the thermodynamic limit and it may well depend on the intermolecular potential, how many particles are needed to mimic this behavior. Thus while n = 20 suffices for this behavior for hard core interactions, a much larger number n >> 64 might be needed for V_s.

It might be useful to point out that in statistical mechanics one usually considers macroscopic systems with $n \approx 10^{20}$ particles, say and values of $\omega \approx 10^5$, interacting with potentials much closer to V_h than V_s (this value of ω follows if one takes for the vibrational frequency of infrared light $\approx 10^{14}$ and for t_{mfp} that for nitrogen or oxygen molecules in air $\approx 10^{-9}$s).

ii) The total energies E used were rather small: $E = nV_o$ and $E = 2nV_o$. Thus, for given E and n the average kinetic energy E_t available for collisions was relatively low and t_{mfp} relatively large. For large ω we expect that this will lead to relatively few collisions during an oscillation, since $\omega t_{mfp} >> 1$. In addition, each collision has relatively small effectiveness, because of the slow impact velocity and the weak force, due to the soft potential. This could then hamper the redistribution of energy and the reaching of equilibrium and equipartition.

Thus the long, unphysical time it takes for $\rho_v(\tau)$ to decay for large ω in the BGG model could be due not only to the small n and soft V_s, but also to the small values chosen for E. One would then expect that, the harder the intermolecular potential and the larger the total energy, the smaller the number of molecules is needed to approach equilibrium in physical times. This is now being investigated. If these ideas bear out, then there is no conflict for sufficiently large dynamical systems, even with weak intermolecular forces, with statistical mechanics and the observed violations are merely an artifact of the model systems used in computer simulations. We emphasize however, that the precise relationship between the various quantities n, E, ω, V_s in their effect on reaching equipartition, is not clear at present.

    2. One could wonder whether similar considerations would be relevant in understanding the violation of the zeroth law found for the FPU model.

ACKNOWLEDGMENTS

The authors are greatly indebted to Dr. J.J. Erpenbeck for a critical reading of the manuscript and to Profs. J.P. Eckmann, G. Gallavotti and S. Ruffo for stimulating discussions. One of us (EGDC) also gratefully acknowledges support of the U.S. Department of Energy under contract No. DE-FG02-88-ER13847.

REFERENCES

1) A. Patrascioiu, Phys. Rev. Lett. 50, 1879 (1983); K.R.S. Devi and A. Patrascioiu, Physica D 11, 359 (1984); A. Patrascioiu, E. Seiler, and I.O. Stamatescu, Phys. Rev. A 31, 1906 (1985); G. Benettin and L. Galgani, J. Stat. Phys. 27, 153 (1982).

2) We follow here the terminology used by G.E. Uhlenbeck and G.W. Ford in Lectures in Statistical Mechanics (AMS, Providence, 1963), chapter I (see in particular note 14, pag. 30).

3) E. Fermi, J. Pasta, and S. Ulam, Los Alamos Scientific Laboratory Report LA1940 (1955), reprinted in: Lect. Appl. Math. 15, 143 (1974).

4) G. Benettin, L. Galgani, and A. Giorgilli, Phys. Lett. A 120, 23 (1987).

5) L. Boltzmann, Nature 51, 413 (1885); J.H. Jeans, Philos. Mag. 6, 279 (1903).

6) G. Benettin, G. Lo Vecchio, and A. Tenenbaum, Phys. Rev. A 22, 1709 (1980).

7) I. Oref and B.S. Rabinovitch, Acc. Chem. Res. 12, 166 (1979).

8) D. Thirumalai, R.D. Mountain, and T.R. Kirkpatrick, Phys. Rev. A 39, 3563 (1989).

9) J.J. Erpenbeck and E.G.D. Cohen, Phys. Rev. A 38, 3058 (1988).

10) G. Benettin, L. Galgani, and A. Giorgilli, Nature 311, 444 (1984); G. Benettin, L. Galgani, and A. Giorgilli, Cel. Mech. 37, 1 (1985).

11) G. Benettin, L. Galgani, and A. Giorgilli, Comm. Math. Phys. 113, 87 (1987).

FUNDAMENTAL PROBLEMS IN STATISTICAL MECHANICS VII
H. van Beijeren, Editor
© *Elsevier Science Publishers B.V., 1990*

PATTERN FORMING INSTABILITIES AT THE CRYSTAL-MELT INTERFACE

Herman Z. Cummins

Department of Physics, City College of the City University of New York, New York, NY 10031   USA

The surface of a crystal growing into its undercooled melt is a nonlinear dynamical system of surprising complexity. It is intrinsically unstable, and rapidly undergoes a morphological transition from a planar form to a cellular or dendritic pattern. Although it has been known for many years that pattern selection is unique for a given material and given experimental conditions, the fundamental physics of the pattern selection process has only recently begun to emerge. Recent experiments providing quantitative data on the initial instability, the evolution of steady state morphology, and sidebranching will be reviewed. Other pattern forming instabilities will also be discussed.

1. Overview of pattern forming instabilities
2. Free growth into the undercooled melt: Mullins-Sekerka theory; experiments
3. The role of noise
4. Nonlinear evolution: coarsening, (self-organized criticality?)
5. Pattern selection: microscopic solvability
6. Directional solidification (subcritical bifurcation?)
7. Sidebranching

1. OVERVIEW OF PATTERN FORMING INSTABILITIES

Pattern forming instabilities in which a uniform featureless volume or a smooth interface spontaneously destabilizes and evolves to form a complex spatial pattern occur in a wide variety of systems. A partial list is shown below (for a detailed discussion of some of these systems, see Kessler et al.[1])

PATTERN FORMING INSTABILITIES

EQUILIBRIUM
    Incommensurate lattices
    Chiral liquid crystals
    Electrons on liquid helium
HYDRODYNAMIC FLOW
    Rayleigh-Bénard
    Couette-Taylor
CHEMICAL REACTIONS
    Liesegang rings
    Zhabotinsky reactions
INTERFACIAL PATTERNS
    Saffman-Taylor (viscous fingering)
    Aggregation
    Electrochemical deposition
CRYSTAL GROWTH

    Free dendritic growth
    Directional solidification

(for more examples, see <u>Random Fluctuations and Pattern Growth</u>, edited by H.E. Stanley and N. Ostrowsky, <u>Kluwer Academic Publishers, 1988</u>)

In cases such as incommensurate lattices or electrons on the surface of liquid helium, the pattern occurs at equilibrium and minimizes the free energy. In most cases, however, the instability occurs in nonequilibrium systems. The familiar convective patterns of the Rayleigh-Benard instability, the fractal patterns of viscous fingering or diffusion limited aggregation, and the dendritic patterns observed in solidification all occur in externally driven nonlinear dissipative systems far from equilibrium.

Ten years ago, Langer published a major review of instabilities and pattern formation in crystal growth which has played a major role in stimulating the research that has produced dramatic advances in this decade.[2] In these lectures I will also focus on solidification instabilities, primarily those occurring in a pure substance growing into its undercooled melt. The "global" problem we want to address includes the initial instability and growth of small fluctuations in the linear regime, evolution of the pattern through the nonlinear coarsening stage, selection of the steady state pattern (restabilization), and continuous sidebranching. Many people have contributed to the development of this field, particularly during the last few years, and I apologize to those whose contributions will not be described for lack of time and space. Also, I wish to thank my colleagues Henry Chou, X.W. Qian and Martin Muschol whose work is the basis of much of these lectures.

2. FREE GROWTH INTO THE UNDERCOOLED MELT

When a pure substance solidifies, the motion of the solid-liquid boundary is governed (in the macroscopic continuum approximation) by three equations:

    1) The thermal diffusion equation

$$D\nabla^2 T = \partial T/\partial t \tag{1}$$

    2) The Gibbs-Thomson equation (with local thermodynamic equilibrium)

$$T_I = T_M[1-(\gamma/L)\kappa] \tag{2}$$

    3) Energy conservation on the moving boundary

$$LV_n = \lambda_S \hat{n} \cdot \nabla T_S - \lambda_M \hat{n} \cdot \nabla T_M \tag{3}$$

where $T(r,t)$ is the local temperature, D is the thermal diffusion constant, γ is the surface tension, L the latent heat, κ the local curvature, λ_S and λ_M the thermal conductivity of the solid and melt, V_n the component of the growth velocity normal to the interface, and $\hat{n}$ a unit vector perpendicular to the interface pointing into the melt. These equations assume that the kinetic coefficient is infinite so that the interface is everywhere at the local melting temperature T_I [which is reduced from the bulk melting temperature by the curvature according to Eq.(2)]. The temperature of the melt far from the interface is taken to be $T_M-\Delta$ where Δ is the supercooling. Attachment kinetics, surface tension anisotropy, and mass transport are also ignored.

These equations are generally not analytically tractable because of Eq. (3) which is a boundary condition on a free surface which is itself a solution ("Stefan problem"). In his review paper in 1980 which contains a detailed discussion of the problem, J.S. Langer described the known solutions of these equations at that time.[2] He has also recently described the current state of understanding of the pattern selection mechanism based on recent theoretical advances.[3]

An exact solution to the linearized equations for a nearly planar interface (in the approximation $dT/dt = 0$) was obtained by Mullins and Sekerka in 1963.[4] In their linear stability analysis they considered an initially planar interface $\zeta=V_0 t$ moving at velocity V_0, with a small superimposed sinusoidal perturbation $\zeta_k e^{i k \cdot r} e^{\omega t}$ and found that

$$\omega(k) \approx V_0 k(1-d_0 l k^2) \tag{4}$$

or equivalently,

$$\frac{d}{dt}\zeta_k \approx V_0 k(1-d_0 l k^2)\zeta_k = \omega_k \zeta_k \tag{5}$$

In Eqs. (4) and (5), $l=2D/V_0$ is the thermal diffusion length and $d_0=\gamma T_m C_P/L^2$ is the capillary length. (In succinonitrile, with $V_0=4.5$ μm/s, $l=4.5$ cm and $d_0=27.7 \times 10^{-8}$ cm.) These equations show that the interface is unstable against all fluctuations with $k^2 \leq (d_0 l)^{-1}$ with the maximum instability at

$$k_{MAX} = (3d_0 l)^{-1/2} \tag{6}$$

There have been relatively few quantitative tests of the Mullins-Sekerka theory. Experiments by S.R. Coriell and his coworkers showed that an ice cylinder growing into undercooled water exhibits an instability of the predicted type.[5] Studies by K.A. Jackson and his coworkers in the 1960's showed that cellular and dendritic instabilities occur in solidification of various transparent molecular crystals.[6] These studies laid the groundwork for much of the later experimental studies.

In a recent preliminary experiment, we utilized a videomicroscope to study the early evolution of the instability of the planar interface of succinonitrile solidifying into its melt.[7] We found that the interface is initially roughly sinusoidal as shown in Fig. 1 and that the Fourier amplitudes ζ_k in the unstable k-range grow approximately exponentially with time as predicted by Eq. (5). Although the design of our experiment caused the melt temperature to change during solidification so that it was not possible to test the Mullins-

Sekerka theory quantitatively, its predictions were qualitatively confirmed, as shown in Fig. 2. (We believe that the same problem also occurred in the Coriell ice cylinder experiments.[5])

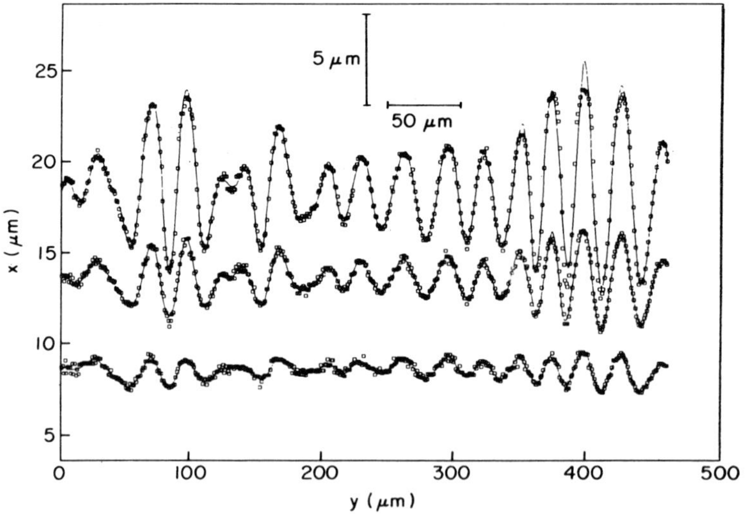

Figure 1. Experimental interface profiles of solidifying succinonitrile at 2, 3 and 4 seconds after the onset of interface motion. Points are obtained by digital interpolation of digitized images collected at a rate of six per second.

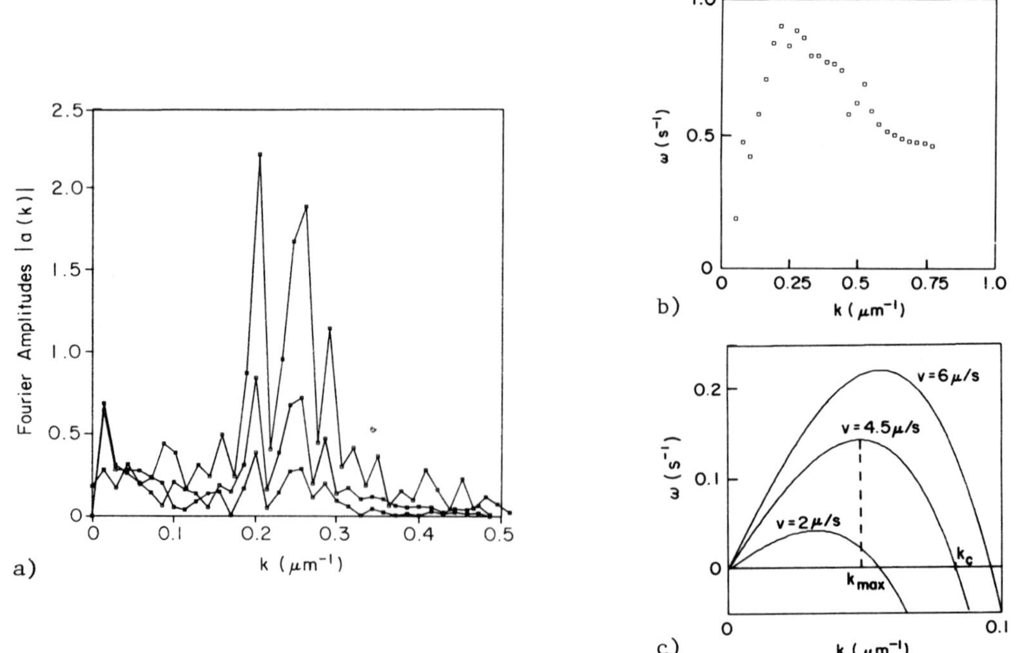

Figure 2. a) Modulus of Fourier amplitudes of the data of Fig. 1 obtained by FFT. b) Linear growth coefficients $\omega(k)$ obtained by fitting Fourier amplitudes to an exponential time dependence. c) Predictions of the Mullins-Sekerka theory for succinonitrile.

J.V. Maher and his coworkers have investigated the evolution of the Saffman-Taylor hydrodynamic instability which is governed by a set of three equations similar (but not identical) to Eqs. (1)-(3). They noted that in the case of gravity-driven flow in a closed rectangular cell (where the average interface velocity is zero), the result of a linear stability analysis is[8]

$$\frac{d}{dt}\zeta_k = C_1 k(1-C_2 k^2)$$

which has the same broad-band instability form as the Mullins-Sekerka result of Eq. (5). (The constant C_1 is proportional to the density difference between the two fluids.) By exploiting the continuously variable density difference of a binary fluid mixture near its critical temperature, DiFrancesco and Maher were able to investigate the evolving Fourier spectrum of interface fluctuations, in close analogy with our solidification study.[7] Their results show that the maximum Fourier amplitudes occur quite close to k_{MAX} of the linear stability analysis.[9] The striking similarity between the dendritic and Saffman-Taylor problems has also been noted by other authors, including Langer[3] and Kessler et al.[1]

Recently, Buka, Kertesz, Vicsek, Palffy-Muhoray and their coworkers have investigated the Saffman-Taylor instability at the interface between invading air and the liquid crystal 8CB.[10] By lowering the temperature through the isotropic-nematic phase transition at $40.5°$ C, they observed a change from dense-branching morphology to dendritic morphology demonstrating the intimate connection between anisotropy and stable dendrites, to be discussed below.

3. THE ROLE OF NOISE

The Mullins-Sekerka prediction of Eq. (5) indicates that the Fourier amplitudes ζ_k of modes with nearly identical k values should grow at nearly the same rate. The data of figure 2a, however, show wild variation in the amplitudes of adjacent Fourier components.

Note that Eq. (5) is formally solved by

$$\zeta_k(t) = \zeta_k(0)e^{\omega_k t}$$

so that even if $\omega_k>0$, ζ_k must have a non-zero value initially in order to grow. The linear stability analysis shows how the amplitude will increase, but not how it starts.

A similar problem arises in the Rayleigh-Benard convective instability. The hydrodynamic equations lead to an equation similar to Eq (5) for the amplitude A (whose square is the convective heat flux j)

$$\frac{dA}{dt} = C_1 A + C_3 A^3 + C_5 A^5 \tag{7}$$

where C_1 changes from $C_1<0$ to $C_1>0$ at the critical value of the temperature gradient ΔT_C. Meyer et al[11] converted Eq. (7) to a stochastic differential

equation by adding a random noise term f. They carried out a computer simulation with f as Gaussian white noise. The simulation was made to reproduce the
observed time evolution of j by adjusting f, which was found to be $\approx 10^4$ times
bigger than the random heat currents expected theoretically due to normal
thermodynamic fluctuations.

We recently carried out a similar analysis[12] for the data shown in Fig. 1,
starting from

$$\frac{d\zeta_k}{dt} = \omega_k \zeta_k + f_k(t) \tag{8}$$

where ω_k is given by Eq. (4).

In the computer simulation, a set of zero-mean Gaussian random numbers was
selected for the f_k at each iteration. As shown in Fig. (3), the resulting
Fourier spectrum can be made to match the general form of the experimental
data of Fig. (1) quite well. The time evolution of a single Fourier component
is shown in Fig. 4.

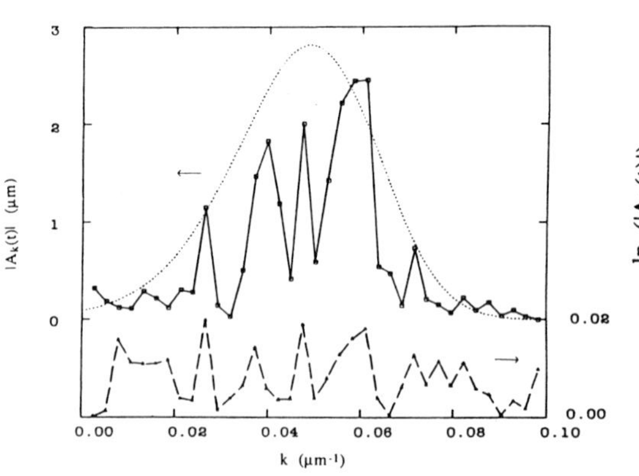

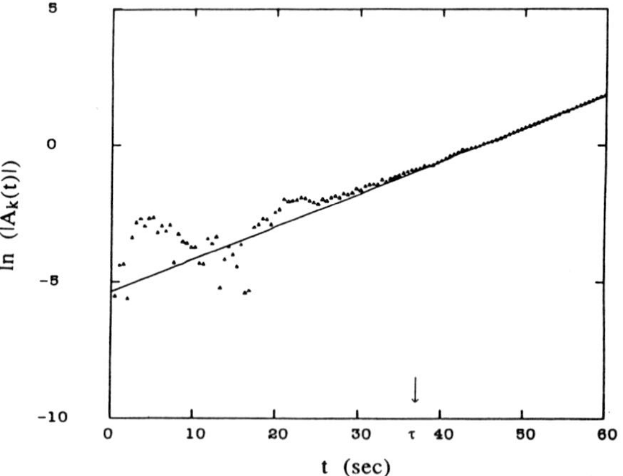

Figure 3. Modulus of Fourier coefficients of the interface from the
computer simulation. The lower and
upper data points correspond to 2.5
and 20 sec after initiation ($A_k=0$,
t=0). The smooth dotted line at the
top is the prediction of the Mullins-
Sekerka theory assuming a uniform
starting amplitude for all k of
$A_k=0.015$ μm at t=0. The rms noise
$s=2.7 \times 10^{-2}$ μm/sec$^{1/2}$. Note the
different scales in amplitude for
the two different times.

Figure 4. Logarithm of the modulus
of $A_k(t)$ for the k=0.03 μm^{-1} mode vs
time. The straight solid line is a
fit through the last 30 data points.
Its slope agrees with the prediction
of the Mullins-Sekerka theory.

The result explains the noisiness observed in the experimental data of Fig.
2(a). Each Fourier component initially follows a random walk under the influence of the random noise until it gets big enough for the deterministic growth

term to take over, as shown in Fig. 4. Those components that reach this exponential growth phase first will always be bigger than their neighbors. The size of the noise term required by the simulation also turned out to be several orders of magnitude <u>larger</u> than the fluctuating heat flux predicted by the Landau-Lifshitz microscopic fluctuation theory, in close analogy to the result of Meyer et al.[11] Although the source of this disagreement has not yet been established, it seems likely that there are some macroscopic sources of noise in these experiments which overwhelm the microscopic fluctuations. This question will be encountered again later when we discuss the sidebranching instability.

4. NONLINEAR EVOLUTION: COARSENING

As the solidification instability evolves, the Mullins-Sekerka linearized analysis becomes inappropriate once the amplitude of spatial fluctuations on the interface become comparable to their average wavelength. The system then enters a fully nonlinear regime in which dramatic competition effects occur, eventually restabilizing in a steady state dendritic pattern. The length scale of the pattern increases during the evolution (coarsening) by a factor of ~10 in succinonitrile.[7] In the next section we will review the mechanism governing steady state pattern selection. Here we offer a few tentative comments on the coarsening stage of the evolving pattern.

The phenomenon of spinodal decomposition occurs when a system with a conserved order parameter is rapidly quenched into an unstable regime where small droplets or domains of the ordered phase nucleate and then grow.[13] Numerous experiments and computer simulations indicate that as this coarsening or "Ostwald ripening" process proceeds, the average particle or domain size increases as t^n. For diffusion-controlled ripening, Lifshitz, Slyozov and Wagner showed that at large times $n = 1/3$.

Glicksman and Voorhees have reviewed the experimental and theoretical situation for coarsening of dendritic structures.[14] They note, for example, that the sidebranch spacing λ_s increases with solidification time t as $\lambda_s = At^n$ where $.25 < n < .37$, and similar results have been reported for thickening of the primary dendrite stem. Their work suggests that the theory of spinodal decomposition may provide a useful approach to understanding dendritic evolution.

Another possible approach to the nonlinear evolution stage is suggested by the form of a fully evolved dendritic pattern with primary, secondary, tertiary (and sometimes higher-order) dendrites providing a self-similar or scale invariant pattern over a limited range of scales. May and Maher have shown that the Saffman-Taylor instability in a Hele-Shaw cell asymptotically approaches a fractal dimension of 1.7 to 1.79 with increasing driving force.[15]

Similarly, the patterns formed in diffusion limited aggregation are self similar with fractal dimension 1.7. Density correlations in these systems have a power law dependence on r.[16]

In a recent series of papers, Per Bak and his coworkers have noted that many open dissipative dynamical systems evolve spontaneously into self-similar structures with long-range spatial and temporal correlations and have named such phenomena "self-organized criticality".[17] Bak has remarked that while many investigators have measured the asymptotic fractal dimension, very few have considered the dynamical processes through which such states evolve. He has shown that some computer models based on cellular automata exhibit such behaviour, and has begun (as have Hwa and Kardar[18]) to explore their dynamics. This approach, while still in an early stage of development, offers an exciting possibility for understanding how spatial patterns may actually evolve.

5. PATTERN SELECTION: MICROSCOPIC SOLVABILITY

At the end of the coarsening stage, pattern forming instabilities evolve to a steady state whose general characteristics are reproducibly determined by the relevant material properties and experimental parameters. How the steady state pattern is selected has been a major puzzle for many years. It appears to have been largely resolved in the last few years through a series of major theoretical advances leading to the microscopic solvability approach as summarized in a recent review by Kessler et al[1], although some difficult questions must still be resolved.[3]

The pre-microscopic-solvability situation, reviewed in detail in 1980 by Langer[2], can be summarized briefly as follows. In 1947, Ivanstov showed that Eqs. (1)-(3) could be solved by a steady-state parabolic needle crystal if the surface tension γ in Eq.(2) is ignored so that $T_I = T_M$ everywhere on the interface regardless of the curvature κ. Ivanstov's result, for the two-dimensional geometry, is

$$\tilde{\Delta} = e^p \sqrt{\pi p} \ \mathrm{erfc}(\sqrt{p}) \qquad\qquad (9)$$

where $\tilde{\Delta} = (T_M - T_\infty) \, C_P/L$ is the dimensionless undercooling far ahead of the interface, erfc is the complementary error function, and the Peclet number $p = \rho V/2D$ where ρ is the tip radius and V is the steady-state growth velocity. (For small p, $\Delta(p) \approx (\pi p)^{1/2}$.) A similar result occurs in the three-dimensional case. Eq. (9) predicts that for a given undercooling $\tilde{\Delta}$ there is a continuous family of parabolic solutions with constant ρV product.

Although the Ivanstov result allows infinitely sharp dendrites ($\rho \to 0$) growing at unlimited velocities, it is apparent that any nonzero surface tension will destroy such solutions by depressing the melting temperature of the highly curved tip (via Eq. 2) below that of the bulk melt, $T_\infty = T_M - \Delta$, thus stopping solidification. Various approaches to reintroducing the surface tension into the Ivanstov solution by perturbation methods produced a modified Ivanstov result with ρV=constant at large ρ, with V passing through a maximum with decreasing ρ, and then decreasing again (see Ref. 2, Fig. 11). For some years it was assumed that restabilization occurred at the maximum V, so that at its natural operating point the selected dendrite would grow at the maximum possible velocity.

Extensive studies of the free dendritic growth of succinoitrile by Glicksman and his coworkers, however, showed that this maximum velocity principle is not valid[19]; the selected velocity is actually far below the maxium V of the modified Ivantsov curve. Langer and Muller-Krumbhaar carried out a linear stability analysis on the parabolic Ivanstov dendrites and found that large-radius parabolas are unstable against tip splitting. The marginal stability point separating stable from unstable solutions for succinonitrile was shown to lie very close to the experimental results of Glicksman.[2] Langer therefore suggested that the naturally selected operating point might coincide with the marginal stability point. This marginal stability hypothesis leads to the prediction that for a particular material the collection of parameters represented by

$$\sigma = 2D d_0 / \rho^2 V \tag{10}$$

should be a constant, independent of the undercooling. Since Eqs. (9) and (10) fix ρV and $\rho^2 V$, they determine a unique (ρ, V) operating point for each undercooling. At low undercooling this result predicts that $V \propto \Delta^4$ in good agreement with several experiments.[2] For succinonitrile, σ was found experimentally to be 0.02, while the marginal stability hypothesis predicted $\sigma \sim 0.026$.

The "modern era" of pattern selection began in 1983 with the introduction of two local models of solidification which provided simplified analytically tractable versions of Eqs. (1)-(3): the geometrical model by Brower, Kessler, Koplik and Levine, and the boundary layer model by Ben-Jacob, Goldenfeld, Langer and Schon (for a review, see Kessler et al[1]). Analysis of these models led to several fundamental insights: first, that the re-introduction of surface tension to the Ivantsov continuum solutions as a perturbation is mathematically invalid since the perturbation is singular; second, that anisotropy in the surface tension plays a fundamental role, and that with no anisotropy there can be no allowed dendritic steady state; third, that inclusion of anisotropic surface tension from the beginning leads to a <u>discrete</u> set of solutions (rather than a continuum), of which only the fastest growing solution is stable.[20]

Armed with the results of these model analyses, several theorists then attacked the full nonlinear equations (1)-(3). These equations can be reduced

    H.Z. Cummins

to a single integro-differential equation for a dendrite advancing with constant velocity V along the y-direction[20,21], and the resulting equation can be studied either analytically (in limiting cases) or numerically. Here we briefly describe a numerical approach used by Kessler, Koplik and Levine.[20]

For the two-dimensional case, the integro-differential equation to be evaluated is

$$\tilde{\Delta} - \tilde{V}\kappa[1-\epsilon\cos(4\theta(x))] = \frac{1}{\pi}\int dx' K_0([x-x']^2 + (y-y')^2]^{1/2})e(y-y') \tag{11}$$

where ϵ is the crystalline anisotropy (assumed to be cubic), K_0 is a Bessel function, the rescaled velocity is $\tilde{V} = (Vd_0/2D)(T_MC_P/L)$, and lengths have been rescaled by V/2D. For zero capillarity, $\gamma = \tilde{V} = 0$, Eq. (11) is satisfied by the Ivantsov result $y_I = -x^2/2p$ with p fixed by Eq. (9). With a given nonzero value of ϵ, Eq. (11) can be solved numerically for fixed V, starting from the Ivantsov approximation, allowing nonzero slope at the tip (a cusp) to occur if needed. For each point $[x,y(x)]$ on this steady-state interface, the integral is evaluated along the entire interface $[x',y'(x')]$. The interface shape is modified at each iteration until an optimum match between the two sides of the equation at each point is obtained, which is the solution for that particular V. V is then incremented and the calculation is repeated, with the cusp magnitude at each V retained as the "result".

The physically acceptable solutions are those with zero cusp magnitude which form a discrete set. A stability analysis then shows that of this set, only the fastest is stable, which explains how a specific pattern is selected.[20] Although the integral in equation (11) extends to $\pm\infty$ along the dendrite, Levine has shown that it is sufficient to integrate over just a few tip radii to obtain the selection result.[22]

Some central results of the microscopic solvability analysis obtained analytically in the small undercooling limit $\tilde{\Delta} \ll 1$, are:[21] $\sigma^* = \sigma_0\epsilon^{7/4}$ (where σ^* is the selected value of σ in Eq. (10) and σ_0 is a constant), and (with $p \approx (\Delta/\pi)^2$), $V \propto \Delta^4\epsilon^{7/4}$ and $\rho \propto \Delta^{-2}\epsilon^{-7/4}$. In Figs. 5 and 6, results of the numerical analysis procedure of Kessler et al are illustrated. These figures were generated by X.W. Qian using a computer program generously provided by H. Levine. Figure 5 shows the dependence of V on Δ, and Fig. 6 shows the dependece of V on ϵ. Note that the computer-generated results agree well with the analytical results in the small undercooling limit.

The need for a non-zero anisotropy ϵ to produce stable dendrites results from the absence of cusp-free solutions with $V\neq0$ for $\epsilon=0$. This also explains why Saffman-Taylor fingers are not dendritic in isotropic fluids. Recent experiments by Y. Couder have shown that Saffman-Taylor dendrites can occur, however, if a tip perturbation such as a bubble is present, since a cusp then becomes physically acceptable. Experiments by Buka, Kertesz, Vicsek, Palffy-Muhoray and their coworkers have also shown that when the Saffman-Taylor instability occurs in a nematic liquid crystal, dendrites occur because the liquid phase is anisotropic.[10]

These predictions have been found to hold well for succinonitrile, but establishing their validity in general requires that other materials with different anisotropies should be carefully studied. Recent experiments by Rubinstein and Glicksman on pivalic acid, which has ten times higher anisotropy than succinonitrile, indicate serious disagreement with the theory, since they find that σ^* increases linearly with Δ rather than being a constant.[23] A departure from constant σ^* was also reported for krypton.[24] Further exploration of this question is a crucial current issue in the pattern selection field. Note that Langer has argued that these predictions will probably break down at small anisotropies where the sidebranching instability diverges.[3]

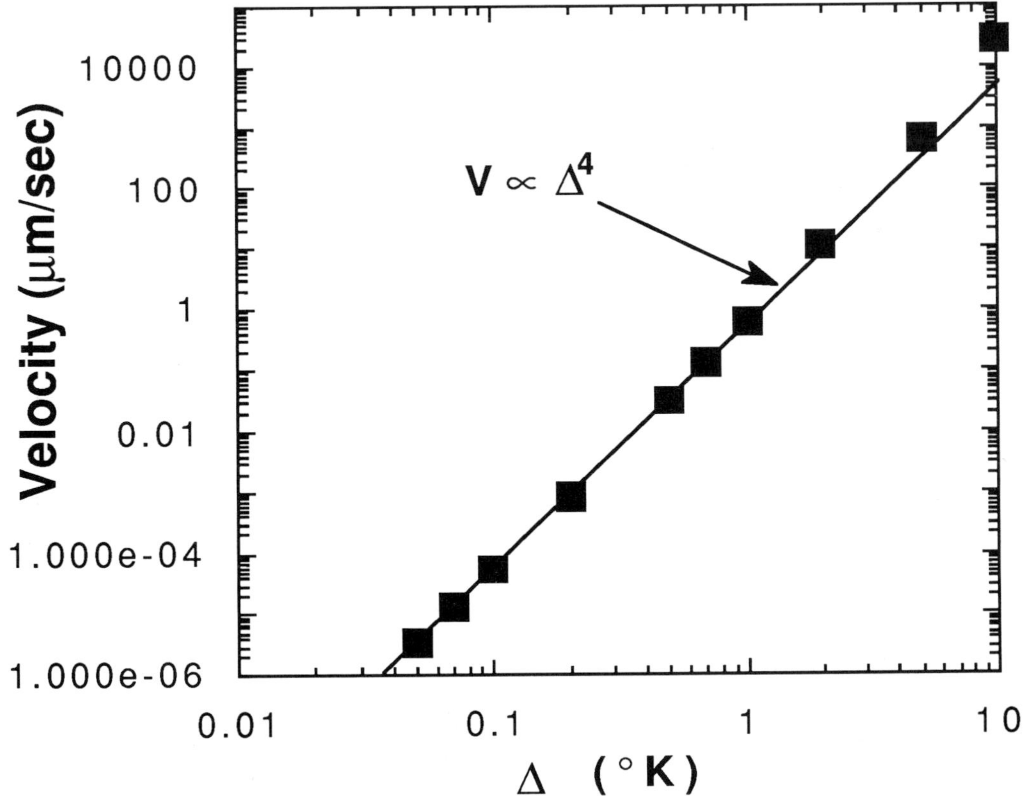

Figure 5. Dependence of tip velocity V on undercooling Δ in succinonitrile obtained by the numerical procedure of Koplik, Kessler and Levine. Note small departure from Δ^4 behaviour for undercooling above 2°K.

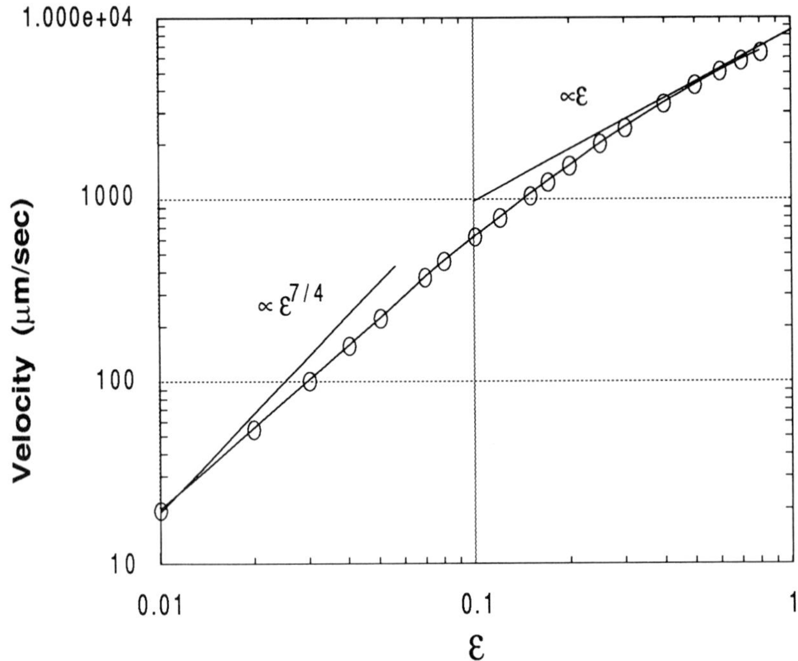

Figure 6.   Selected dendritic tip velocity vs anisotropy ε from solution of Eq. (11).   Thermodynamic parameters are those of succinonitrile.

6. DIRECTIONAL SOLIDIFICATION

If a non-facetting (molecularly rough) crystalline material is placed in a container in a temperature gradient with one end above and the other end below the melting temperature T_M, a planar crystal-melt interface will form normal to the gradient at the position where $T=T_M$. If the container is then pulled at constant velocity towards the colder end, the crystal will grow with the position of the interface remaining stationary in the laboratory, once steady state has been reached. A class of important techniques for producing high-purity single crystals, such as the Bridgman-Stockbarger method, are based on this concept.[25]

For a pure substance, the temperature gradient suppresses the Mullins-Sekerka instability and the planar interface is stable.[2] If impurities are present, however, an instability may arise because solute rejection results in a buildup of solute concentration at the interface causing freezing point depression. Ahead of the interface, where the solute concentration and freezing point depression are both smaller, the local melting temperature may be lower than the actual temperature - a situation called constitutional supercooling - and the interface therefore becomes unstable. Langer has shown that the linear stability analysis for this case is similar to the Mullins-Sekerka analysis for the one-component case considered in Section II if one replaces the dimensionless temperature field $(T-T_M)C_P/L$ by the dimensionless concentration field $\bar{\mu}/[\Delta C(\partial\mu/\partial C)]$ where μ is the chemical potential and $\bar{\mu}=\mu-\mu_{eq}$, C is the solute concentration, and ΔC is the miscibility gap. The major differences are that the instability can occur even in the presence of a positive temperature gradient, if the ratio of growth velocity to temperature gradient, V/G, exceeds a critical value, and the onset of instability is at k>0, rather than at k=0 as found in the one-component case.

Directional solidification is particularly well suited for optical microscopy studies, and was first investigated extensively with a temperature gradient microscope by Jackson and his coworkers.[6,26] They found that both cellular and dendritic interfaces can occur depending on the impurity concentration, temperature gradient, and growth speed. Glicksman and his coworkers have investigated and carefully characterized succinonitrile containing acetone as an impurity, and Trivedi and his coworkers have studied the evolution of the interface morphology, interdendritic spacing and tip radii as functions of growth rate and temperature gradient in this system.[27]

Figure 7 shows the evolution of the solidifying interface of succinonitrile containing less than 1 % acetone in a directional solidification study performed recently in our lab by X.W. Qian. The sample cell is a sealed 50μm x 1000 μm rectangular glass capillary tube, the temperature gradient was ~20°/cm, and the growth velocity was 6μm/sec. The interface is shown at: (a) 0, (b) 60, (c) 100, (d) 120, (e) 240 and (f) 300 seconds after initiation of growth. The final tip radius (f) was $\rho \sim 6\mu$m and the final interdendritic spacing ~500μm. Note how, during the coarsening phase, some dendrites are squeezed out allowing the interdendritic spacing to increase.

Although the linear stability analysis of directional solidification of a two-component system and of free solidification of a pure material give similar results, an important aspect of the directional solidification problem arises when the analysis is carried to higher order. This nonlinear analysis was first addressed by Wollkind and Segel[28] and was discussed by Langer[2], and was later approached more generally by Caroli et al.[29] As shown by Caroli et al[29], a nonlinear expansion of the dynamical equations leads to an equation of motion for the Fourier amplitudes of the interface deformation

$$\frac{d}{dt}\zeta_k = c_0(k)\zeta_k + c_1\zeta_k^3 \qquad (12)$$

With increasing V/G, the linear coefficients $c_0(k)$ becomes positive for a range of k-values so that the system becomes linearly unstable at a critical V/G value. However, near this point the cubic coefficient c_1 can be positive while c_0 is still negative, leading to a nonlinear instability. This situation is

H.Z. Cummins

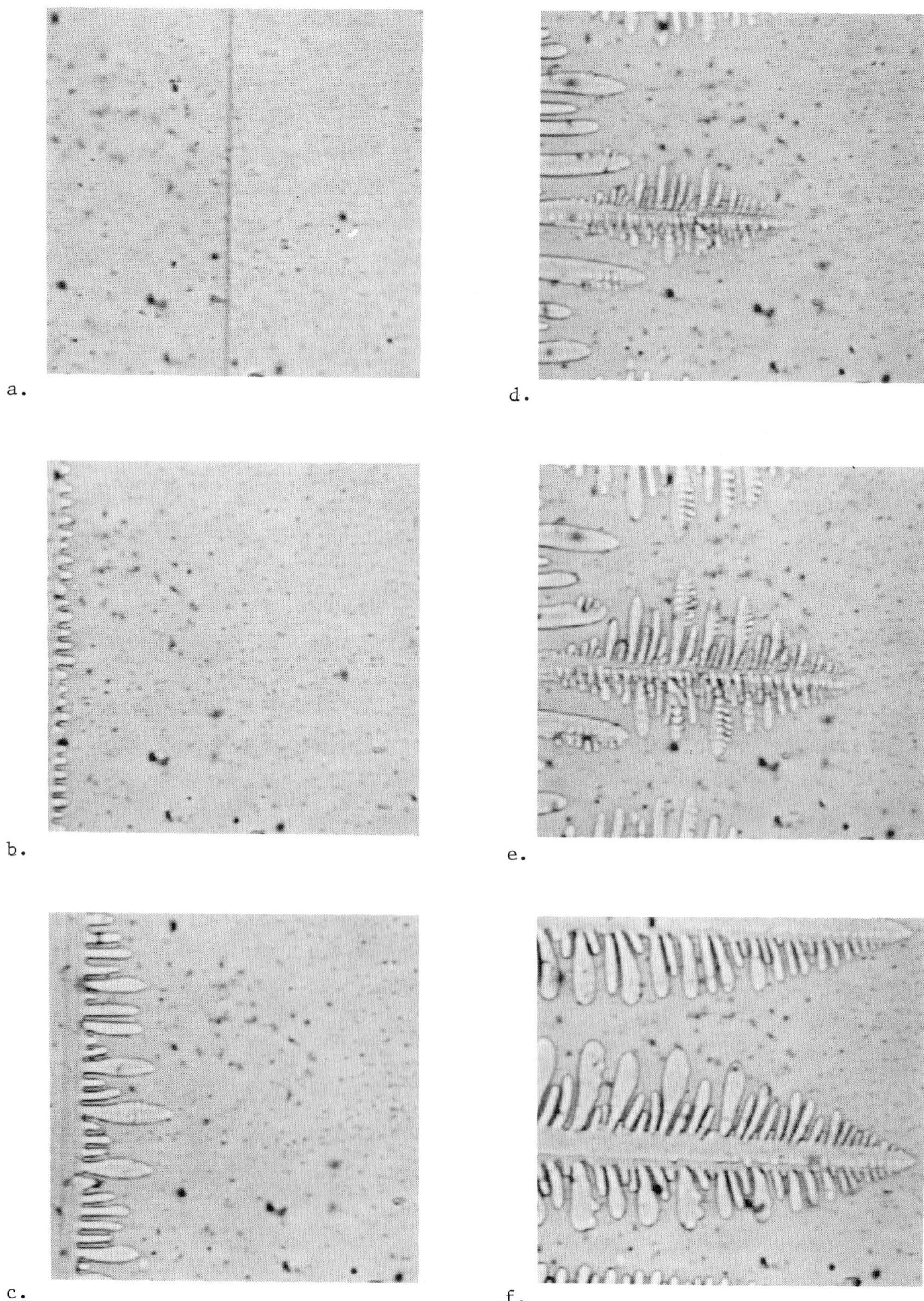

Figure 7. Evolution of the interface of solidifying succinonitrile containing 1 % acetone. Directional solidification at V = 6μm/sec.

familiar in the Landau theory of equilibrium phase transitions where a negative coefficient of the fourth order term in the Landau free energy leads to a first-order transition with hysteresis. In bifurcation theory, the equivalent situation leads to a <u>subcritical</u> bifurcation which also exhibits hysteresis. This also implies that the transition should produce an immediate jump to finite amplitude so that no truly linear regime exists! The possibility of hysteresis in these transitions and the problem of cell parameter evolution with variation on growth parameters were noted by Langer in his 1980 review.[2]

De Cheveigne et al have investigated directional solidification in CBr_4 containing Br_2 as an impurity.[30] They studied the wavelength selection process on the cellular interface and determined how the wavelength adjusts following a change in velocity. Their studies of the response to changes in growth speed show that hysteresis does occur, indicating a subcritical bifurcation in agreement with the analysis of Caroli et al. A subcritical bifurcation has also been reported in succinonitrile/acetone by Eshelman and Trivedi.[31]

Shape selection and wavelength selection in directional solidification have been studied recently by Libchaber and his coworkers who have observed the processes of wavelength adjustment following changes in growth velocity.[32] They have also observed a series of fascinating soliton-mediated instabilities at the interface between liquid crystal phases undergoing phase transitions.[33]

Because the directional solidification geometry leads to a broadband instability, the wavelength selection process is very subtle and extremely nonlinear. One type of response to a perturbation recorded recently in our lab by X.W. Qian is shown in Fig. 8. Succinonitrile containing <1 % acetone sealed in a $50\mu m$ x 1000 μm glass capillary in a temperature gradient of $20°/cm$ was undergoing directional solidification at $V = 2\mu m/sec$. The dendritic tip radius was $\rho = 12\mu m$. At t=0, the growth velocity was suddenly increased from $2\mu m/sec$ to $20\mu m/sec$. The figure shows the interface at (a) 0, (b) 39, (c) 48, (d) 51, (e) 75 and (f) 100 sec after the velocity increase. At t=100 sec, the tip radius has restabilized at $\rho \sim 4\mu m$.

Note how the interdendritic spacing has changed following the emergence of the narrow dendritic tip in (b). Tertiary dendrites, finding themselves in an "open" environment, grow rapidly and catch up with the original primary dendrite. As seen in (e) and (f), the final spacing is achieved by squeezing out some of the tertiaries, leaving an even row of equally spaced tips.

7. SIDEBRANCHING

Once a steady-state dendritic pattern has been reached, the (nearly) parabolic tips continue to grow at constant velocity V, emitting continuous trains of sidebranches which grow and coarsen as they propagate down the dendrite.

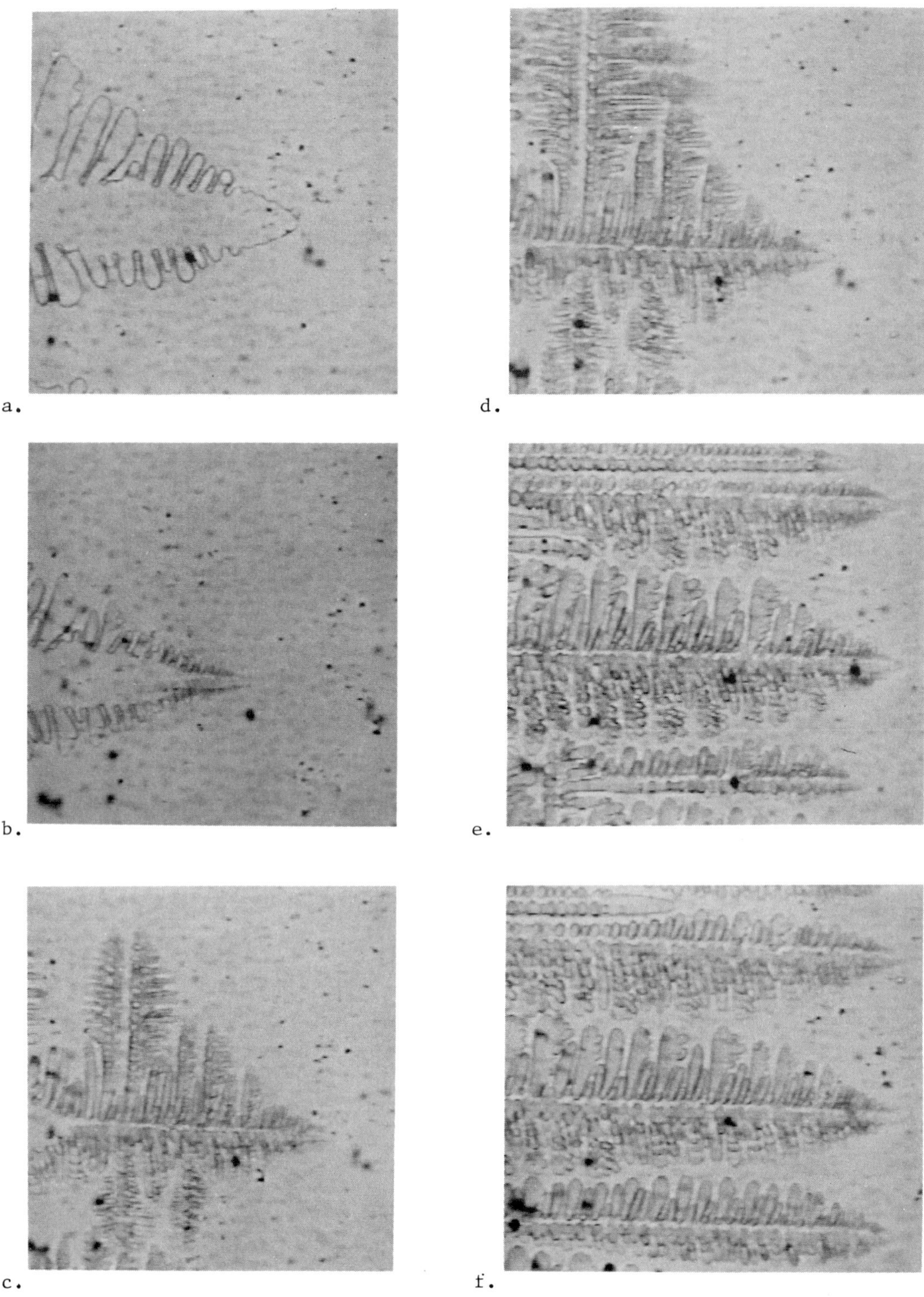

a.

b.

c.

d.

e.

f.

Figure 8.   Response to a step increase in V from 2 to 20μm/sec during
directional solidification of succinonitrile/acetone.

In the free growth geometry, these sidebranches remain nearly stationary in the laboratory frame.

Some aspects of dendritic sidebranching were studied by Glicksman and his coworkers[14,19] and by Trivedi and his coworkers.[27] Recently, Dougherty et al have performed quantitative high-resolution videomicroscopy studies of free dendritic growth of ammonium bromide from saturated aqueous solutions.[34,35] They found that $\sigma^* = 0.081\pm0.02$, in good agreement with the numerical prediction of microscopic solvability theory $\sigma^* = 0.065\pm0.02$. These experiments revealed three major attributes of the sidebranching process:

 (i)   The sidebranching is only roughly periodic, even quite close to the tip.

 (ii) There is little correlation between sidebranches on opposite sides of the tip.

(iii) The rms sidebranch amplitude is an exponential function of distance from the tip, with no apparent onset distance.

The theory of sidebranch formation has been discussed by several groups in the last few years, including Nittman and Stanley[36], and Langer, Pieters and their coworkers.[3,27] A puzzle arose in this analysis in that the dendritic tip turns out to be linearly stable. However, the Langer-Pieters analysis, following methods developed by Zeldovich for flame fronts, revealed that localized wave packets generated by noise pulses near the tip grow in amplitude and stretch as they move down the dendrite, although at any point at a fixed distance down the tip, the disturbance dies out after the wave packet induced by the pulse has passed. This is the sense in which the front is stable even though the sidebranches continue to grow.

The growth of the sidebranch amplitude A(s) with distance s from the dendrite tip was found to be[3]

$$A(s) \approx \exp\left[\frac{0.647}{(\sigma^*)^{1/2}} \left(\frac{s}{\rho}\right)^{1/4}\right] \tag{13}$$

Eq. (13) also explains, as noted by Langer, that at small anisotropy ϵ (where $\sigma^* \to 0$), the microscopic solvability theory will break down since the growth of the sidebranching instability will overwhelm the steady state solution.

This theory suggests that the sidebranching instability results from selective amplification of noise at the dendrite tip. Whether or not thermal fluctuations alone are sufficiently large to account for the observed sidebranches has not been conclusively established, although Langer's estimate indicates that they probably are not.[3] In a recent numerical simulation of dendritic growth, Saito et al have shown that numerical noise present in the simulation is sufficient to produce sidebranching.[38]

Martin and Goldenfeld[39] have carried out calculations in the 2-d boundary
layer model of the sidebranching susceptibility to periodic heat pulses. They
find that some of these questions could be addressed by imposing a pulsed heat
perturbation at the tip, an experiment which we are currently setting up.
Rabaud et al have performed a related (nonlocal) experiment by modulating the
air pressure in a Saffman-Taylor fingering experiment.[40] They found that such
periodic forcing does generate a train of coherent sidebranches.

LOOKING AHEAD: SOME OPEN QUESTIONS

Having reviewed the progress made during the 1980's in understanding pattern forming instabilities at the crystal-melt interface, we can see that
while tremendous progress has been made on the theoretical level, very few
experimental tests exist and experiments should be vigorously pursued in order
to test as many of the theoretical predictions as possible. I conclude with a
list of some of the challenges and questions which various experimental
groups, including ours, will try to explore during the next few years:

1) What is the source of the noise that initiates the Mullins-Sekerka instability and the steady state sidebranching instability?

2) How will the sidebranch formation process respond to a periodic pulse perturbation? Will the sidebranches then be spatially coherent?

3) Can we find an appropriate group of materials with different anisotropies
   and characterize them adequately to thoroughly test the microscopic solvability predictions?

4) To what extent is it legitimate to ignore mass transport, convection and
   the microscopic details of crystal growth (including attachment kinetics)
   in carrying out theoretical analyses? If they are significant, can they be
   correctly included?

5) Since the "selected" dendritic tip geometry is known to differ only
   slightly from the zero-order parabolic approximation (for succinonitrile,
   typically 10^{-4} to 10^{-5}), is it really permissible to ignore the sidebranching perturbations near the tip in analyzing shape selection?

6) Can the global evolution from initial linear instability to final steady-
   state dendritic pattern be understood as a single global process? Does it
   constitute an example of self-organized criticality in which the selected
   pattern constitutes a fractal attractor with a global basin of attraction?

7) Is the cellular instability in a directional solidification of alloys generally a subcritical bifurcation?

REFERENCES

1) D. Kessler, J. Koplik and H. Levine, Adv. Phys. $\underline{37}$, 255 (1988).

2) J.S. Langer, Rev. Mod. Phys. $\underline{52}$, 1 (1980).

3) J.S. Langer, Physica $\underline{140A}$, 44 (1986); Science $\underline{243}$, 1150 (1989).

4) W.W. Mullins and R.F. Sekerka, J. Appl. Phys. $\underline{34}$ 323 (1963); $\underline{35}$, 444 (1964).

5) c.f., S.C. Hardy and S.R. Coriell, J. Cryst. Growth $\underline{3}$, $\underline{4}$, 569 (1968); S.R. Coriell and R.L. Parker, J. Appl. Phys. $\underline{36}$, 632 (1965).

6) c.f., K.A. Jackson and J.D. Hunt, Acta Metallurgica $\underline{13}$, 1212 (1965); K.A. Jackson, Mat. Sci. Res. $\underline{4}$, 229 (1969).

7) H. Chou and H.Z. Cummins, Phys. Rev. Lett. $\underline{61}$, 173 (1988).

8) M.W. DiFrancesco, S.N. Rauseo and J.V. Maher, Superlattices and Microsctructures $\underline{3}$, 617 (1987).

9) M.W. DiFrancesco and J.V. Maher, Phys. Rev. A $\underline{39}$, 4709 (1989); Phys. Rev. A (to appear - July, 1989).

10) A. Buka, J. Kertesz and T. Vicsek, Nature $\underline{323}$, 424 (1986); A. Buka and P. Palffy-Muhoray, Phys. Rev. A $\underline{36}$, 1527 (1987); S. Arora, A. Buka, P. Palffy-Muhoray, Z. Racz and R. Vora, Europhys. Lett. $\underline{7}$, 43 (1988).

11) C.W. Meyer, G. Ahlers and D.S. Cannell, Phys. Rev. Lett. $\underline{59}$, 1577 (1987).

12) X.W. Qian, H. Chou, M. Muschol and H.Z. Cummins, Phys. Rev. B $\underline{39}$, 2529 (1989)

13) c.f., J.D. Gunton, M. San Miguel and P.S. Sahni, in: <u>Phase Transitions and Critical Phenomena</u>, vol. 8, edited by C. Domb and J.L. Lebowitz (Academic, London, 1983); D.A. Huse, Phys. Rev. B $\underline{34}$, 7845 (1980); C. Roland and M. Grant, Phys. Rev. Lett. $\underline{60}$, 2657 (1988).

14) M.E. Glicksman, R.J. Schaefer and J.D. Ayers, Metall. Trans. $\underline{7A}$, 1747 (1976); M.E. Glicksman and P.W. Voorhees, Metall. Trans. $\underline{15A}$, 995, 1081 (1984).

15) S.E. May and J.V. Maher, The Fractal Dimension of Radial Fingering Patterns (to be published).

16) c.f., T.A. Witten, Jr. and L.M. Sander, Phys. Rev. Lett. $\underline{47}$, 1400 (1981).

17) c.f., C. Tang and P. Bak, Phys. Rev. Lett. $\underline{60}$, 2347 (1988);

18) T. Hwa and M. Kardar, Phys. Rev. Lett. $\underline{62}$, 1813 (1989).

19) c.f., S.C. Huang and M.E. Glicksman, Acta Metall. $\underline{29}$, 701, 717 (1981).

20) D.A. Kessler and H. Levine, Phys. Rev. Lett. $\underline{24}$, 3069 (1986); Phys. Rev. B $\underline{33}$, 7867 (1986); D.A. Kessler, J. Koplik and H. Levine, Phys. Rev. A $\underline{33}$, 3352 (1986).

21) A. Barbieri, D.C. Hong and J.S. Langer, Phys. Rev. A $\underline{35}$, 1802 (1987).

22) H. Levine (private communication).

23) E.R. Rubinstein and M.E. Glicksman (preprint).

24) J.H. Bilgram, M. Firmann and W. Kanzig, Phys. Rev. B 37, 685 (1988).

25) c.f., G.J. Sloan and A.R. McGhie, Techniques of Melt Crystallization (John Wiley and Sons, New York, 1988).

26) K.A. Jackson, Materials Science Research 4, 229 (1969).

27) R. Trivedi and K. Somboonsuk, Mat. Sci. Engin. 65, 65 (1984); K. Somboonsuk, J.T. Mason and R. Trivedi, Metall. Trans. $15A$, 967 (1984); R. Trivedi, Metall. Trans. $15A$, 977 (1984).

28) D. Wollkind and L. Segel, Phil. Trans. R. Soc. London, 268, 351 (1970).

29) B. Caroli, C. Caroli and B. Roulet, J. Phys. (Paris) 43, 1767 (1982).

30) S. de Cheveigne, C. Guthmann and M.M. Lebrun, J. Cryst. Growth 73, 242 (1985); J. Phys. (Paris) 47, 2095 (1986).

31) M.A. Eshelman and R. Trivedi, Acta. Metall. 35, 2443 (1987).

32) J. Bechoefer and A. Libchaber, Phys. Rev. B 35, 1393 (1987).

33) P. Oswald, J. Bechhoefer and A. Libchaber, Phys. Rev. Lett. 58, 2318 (1987); J. Bechhoefer, P. Oswald, A. Libchaber and C. Germain, Phys. Rev. A 37, 1691 (1988); A.J. Simon, J. Bechhoefer and A. Libchaber, Phys. Rev. Lett. 61, 2574 (1988).

34) A. Dougherty, P.D. Kaplan and J.P. Gollub, Phys. Rev. Lett. 58, 1652 (1987).

35) A. Dougherty and J.P. Gollub, Phys. Rev. A 38, 3043 (1988)

36) J. Nittmann and H.E. Stanley, J. Phys. A 20, L891, L1185 (1987).

37) R. Pieters and J.S. Langer, Phys. Rev. Lett. 56, 1948 (1986); J.S. Langer, Phys. Rev. A 36, 3350 (1987); M.N. Barber, A. Barbieri and J.S. Langer, Phys. Rev. A 36, 3340 (1987); R. Pieters, Phys. Rev. A 37, 3126 (1988).

38) Y. Saito, G. Goldbeck-Wood and H. Muller-Krumbhaar, Phys. Rev. A 38, 2148 (1988).

39) O. Martin and N. Goldenfeld (unpublished).

40) M. Rabaud, Y. Couder and N. Gerard, Phys. Rev. A 37, 935 (1988).

FUNDAMENTAL PROBLEMS IN STATISTICAL MECHANICS VII
H. van Beijeren, Editor
© *Elsevier Science Publishers B.V., 1990*

SCALING BEHAVIOR & THERMODYNAMICS[1]

J.J.P. Veerman[2]
M.J. Feigenbaum

The Rockefeller University
1230 York Avenue
New York, New York 10021

I   INTRODUCTION

It is our purpose in this review to explain the thermodynamic formalism, in particular its application to dynamical systems where scaling behavior is present. While intended for students in statistical physics, it is our hope that the material will also be accessible for a somewhat wider audience. We have therefore included some calculations that are standard in statistical physics.

The general setting of the problem is the following: We are given a Cantor set S (in one dimension) generated by some underlying dynamical procedure. In some generality, this means that set S is generated in successive approximations in a level-by-level fashion similar to the way in which one constructs a middle third Cantor set. In sections 2 and 3, we will discuss these recursive procedures in some detail. Those who appreciate a very informal introduction to the notions of Cantor set and dimension can find it in Mandelbrot (1982).

Since most of the sets so constructed have zero measure, their Hausdorff dimension (section 4) is of interest. However, the central issue here is: what are the characteristic properties of the recursive procedure by which the set is naturally generated?

This is of importance for the following reason. In doing an experiment (numerical or otherwise) that creates the Cantor set from an otherwise unknown dynamical system, one has two pieces of information. The first is an approximation of the Cantor set, the second, more important one is the behavior of successive approximations. More detailed information may be hard or impossible to obtain. While this information is not sufficient to completely characterize the Cantor set, it gives a decidedly more detailed description of the Cantor set than just the Hausdorff dimension (see section 5).

It will turn out that the Cantor set is approximated by an ever-growing number of smaller and smaller intervals. The ratios of the length of an interval in one level to that of an interval in the

---

[1] Based on the 1989 NUFFIC Summer School lectures by M.J. Feigenbaum
[2] Current address: Institute for the Mathematical Sciences
        SUNY at Stony Brook, New York 11794

previous level are called scalings. The partial characterization we seek is the precise analysis of statistics on the sizes of the scalings (sections 6 and 7). This analysis turns out to be the same as thermodynamics in classical statistical mechanics.

To illustrate the use of this, we will in sections 8, 9 and 10 treat a rather complicated Cantor set which plays an important role in the study of one dimensional dynamical systems. We will set up and solve for an approximate thermodynamics associated with the recursive scheme that generates the set. The notion of phase transition will be discussed with this context as an example.

II   TREES

As a first example, we will construct the so-called two scale Cantor set without memory. Throughout this work we will denote Cantor sets by S.

Start with an interval $\Delta \subset \mathbb{R}$, normalized to have length one. So Δ may be taken to be the interval $[0, 1]$. Split this interval in two pieces $\Delta(0)$ and $\Delta(1)$, each of which contains one of the endpoints of Δ. Denote the length of an interval I by $|I|$. The ratios

$$\sigma(\varepsilon_1) = \frac{|\Delta(\varepsilon_1)|}{|\Delta|} , \quad \varepsilon_1 \in \{0,1\}$$

are called "scalings." Clearly,

$$\sigma(0) + \sigma(1) \leq 1 .$$

At the next level of resolution, each of the intervals $\Delta(\varepsilon_1)$ is split in two. We obtain intervals $\Delta(\varepsilon_2\varepsilon_1)$ with the following properties. The four endpoints of $\Delta(\varepsilon_1)$ are also endpoints of the four intervals $\Delta(\varepsilon_2\varepsilon_1)$. For the lengths of the intervals one has

$$|\Delta(\varepsilon_2\varepsilon_1)| = \sigma(\varepsilon_2)|\Delta(\varepsilon_1)| . \tag{2.1}$$

If one continues the construction of the intervals in the n-th approximation, their lengths satisfy

$$|\Delta(\varepsilon_n...\varepsilon_1)| = \sigma(\varepsilon_n)\sigma(\varepsilon_{n-1})...\sigma(\varepsilon_1)|\Delta| .$$

The set consisting of the 2^n intervals at the n-th level will be called S_n. The Cantor set S consists of those points which are elements of S_n for each n:

$$S = \bigcap_{n>0} S_n .$$

A convenient way of keeping track of the lengths of the intervals $\Delta(\varepsilon_n ... \varepsilon_1)$ and their ordering in S_n is to draw a binary tree T. Each node of T has a "0" $-$ and a "1" $-$ branch emanating from it. The nodes to which one assigns the lengths of the intervals $\Delta(\varepsilon_n \ldots \varepsilon_1)$ have been labeled in such a way that the zeros and the ones appear in the order $\varepsilon_n \ldots \varepsilon_1$ when one travels <u>from the node to the top of the tree</u> (see figure 2.1). Notice that T completely determines S.

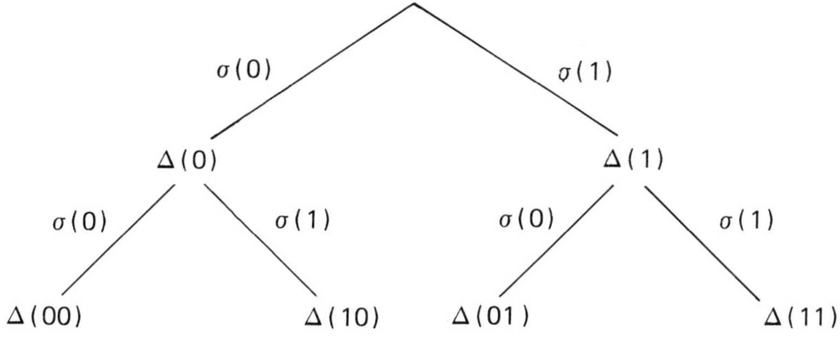

figure 2.1

There are obvious generalizations of the above construction. (First of all, we could have split each interval into $k > 2$ pieces.) We would then get a k-nary tree (each node has k branches). One could also consider sets described by "pruned" k-nary trees. For example if in figure 2.1, consecutive "1" branches are forbidden, one obtains figure 2.2.

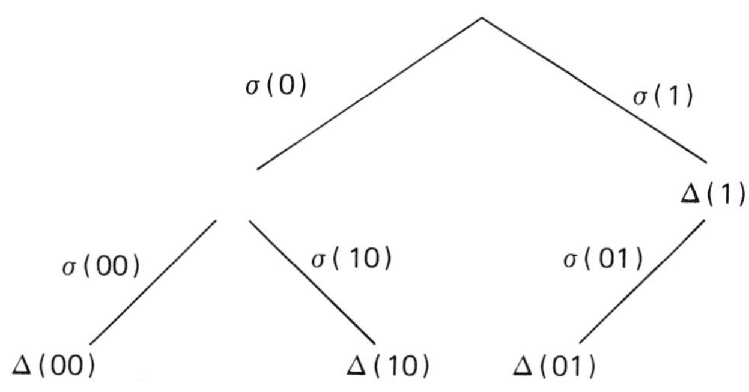

figure 2.2

While scalings without memory serve the useful purpose of giving insight into the combinatorics associated with the trees, for the interesting sets occurring in dynamical systems one has to define scalings more generally:

$$\sigma(\varepsilon_n \ldots \varepsilon_1) = \frac{|\,\Delta(\varepsilon_n \ldots \varepsilon_1)\,|}{|\,\Delta(\varepsilon_{n-1} \ldots \varepsilon_1)\,|} \tag{2.3}$$

The scaling now depends on the entire path to the node rather than only on the last node. In figure 2.3, we have drawn the tree for the case that the scalings depend on the last two branches leading to the node. However, we will adhere to the convention that the combinatorics of the tree is without memory: the way new branches emanating from a certain branch are labelled depends only on the label of the current branch. This means in particular that there are only four binary trees of this type, because after the level nodes on the first level the choices are exhausted. Notice that these choices determine the spatial ordering on the line of all intervals $\Delta(\varepsilon_n \ldots \varepsilon_1)$. Furthermore, it implies that two nodes are <u>nearby</u> on the tree, when the <u>low epsilons agree</u> (i.e., $\varepsilon_1, \varepsilon_2$, etc.). It is clear, then, that the scalings plus the spatial ordering of $\Delta(\varepsilon_n \ldots \varepsilon_1)$ determine the Cantor set.

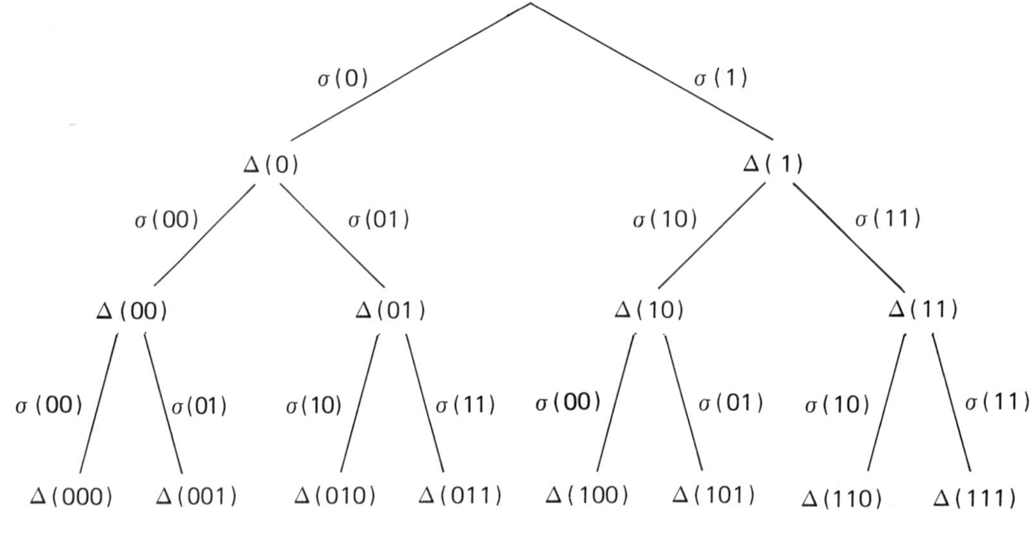

figure 2.3

For the scalings we have the result that they are asymptotically invariant under smooth coordinate transformations. The latter is easily understood. A smooth invertible coordinate transformation restricted to a very small interval is asymptotically affine (i.e., has constant derivative). A high level scaling is the ratio of the lengths of a very small interval and a subinterval of it. Thus, the asymptotic geometry of the Cantor set is completely determined by invariant quantities that are the limit points of the scalings (together with the ordering). The set of limit points is called the scaling function $\sigma(x)$ and can be defined as follows. Let a be the branching ratio of the tree. Then

$$x = \sum_0^\infty \frac{\varepsilon_{n-1}}{a^{i+1}} \;\Rightarrow\; \sigma(x) = \lim_n \sigma(\varepsilon_n \varepsilon_{n-1} \ldots \varepsilon_1).$$

The purpose of this labeling scheme is to do it in such a way that the dependence of the scaling upon lower ε's falls off fast. We will argue later that this decay is exponential. The scaling function is then a well-defined notion (the limit exists). Moreover, the fast decay means that the scaling function is very well approximated by $\sigma(\varepsilon_n \ldots \varepsilon_{n-k})$ for some finite k. Hence the importance of figure 2.3.

In order to get the ε-dependence to fall off fast, one may have to change the ordering properties of the intervals. The way it is done in figure 2.4 will be important throughout this work. The first two branches are as usual. The rest of the tree can be constructed recursively by noting that the subtree starting at the node "1" and the entire tree are each others' mirror image.

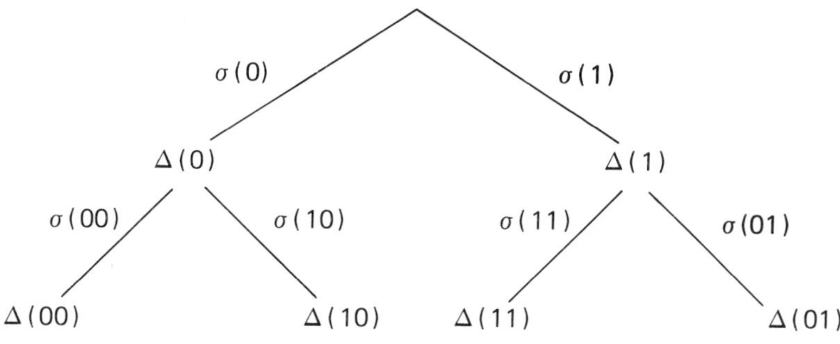

figure 2.4

As remarked before, the tree determines the Cantor set. The converse is not true. One makes the observation that the set determined by the tree of figure 2.4 could also have been generated by the ternary tree of figure 2.5. The first level of this tree consists of the nodes labeled 0, 11, and 01 of figure 2.4. One skips the nodes that end on 10. The trees themselves are equivalent to the set of scalings plus the ordering properties of the intervals.

So, if the Cantor set does not determine a unique set of scalings, what is the reason we are interested in scalings? This is easily answered. The sets we study come from an underlying dynamical system. This provides us with extra information. Imagine, for example, that there is an orbit that slowly fills out the entire Cantor set. Points on that orbit have a temporal ordering (as well as a spatial ordering).

A natural way to select a tree (and one that is often used) is to insure that each node at level n of the tree corresponds to an interval that is visited equally often (namely with frequency $1/N_n$) by orbits that trace out the Cantor set. It will now be clear that the trees of figure 2.4 and 2.5 cannot both satisfy that requirement, because an interval in level n of figure 2.5 may correspond to an interval in different levels in figure 2.4. Of the class of trees with equal probability in each interval one selects the simplest.

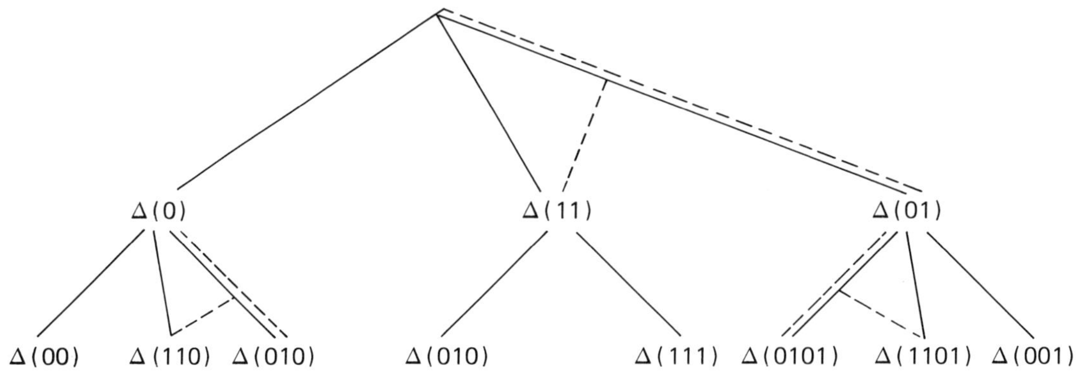

figure 2.5

In conclusion: it is crucial to realize that the sets we are interested in are associated with a dynamical process. The dynamical process selects a tree.

III   PRESENTATION FUNCTIONS

There is a more revealing way to present the tree structure of the Cantor sets discussed in the previous section. To this end we construct a map F_ε for each direction in which the tree can branch:

$$F_\varepsilon(\Delta(\varepsilon_n \ldots \varepsilon_2)) = \Delta(\varepsilon_n \ldots \varepsilon_2\varepsilon) \tag{3.1.a}$$

$$\Delta(\varepsilon_n \ldots \varepsilon_1) = F_{\varepsilon_1} \circ F_{\varepsilon_2} \circ \ldots \circ F_{\varepsilon_n}(\Delta) . \tag{3.1.b}$$

Let γ and γ^1 be distinct ε-sequences of the same length with $\gamma > \gamma^1$ in the ordering discussed in section 2. Then, for a given tree, the ordering of $\gamma\varepsilon$ and $\gamma^1\varepsilon$ depends only on ε. Therefore the F_ε's are monotone functions from Δ into itself. As a consequence the inverses F_ε^{-1} of the functions F_ε form branches of a single function from Δ into itself. We will name this function the 'presentation function' (Feigenbaum, 1988).

The case we will most thoroughly study is the one drawn in figure 3.1 with F_1 orientation reversing and F_0 orientation preserving. This corresponds to the tree exhibited in figure 2.4.

[Note: strictly speaking, we have defined F_ε only on the endpoints of the intervals $\Delta(\varepsilon_n \ldots \varepsilon_1)$. These points are dense in the Cantor set S. Since two nearby nodes x and x' in the tree are mapped to nearby nodes $\alpha\varepsilon$ and $\beta\varepsilon$ by F_ε, one may extend F_ε to continuous functions on S. The details of F_ε outside the Cantor set will play no role in the analysis to come, and one may just interpolate to get functions defined on Δ.]

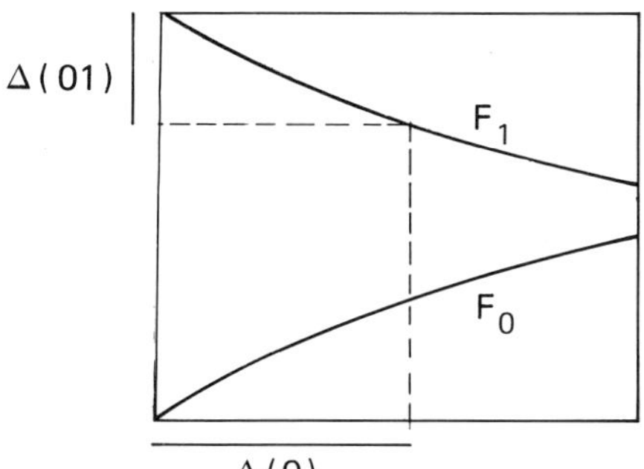

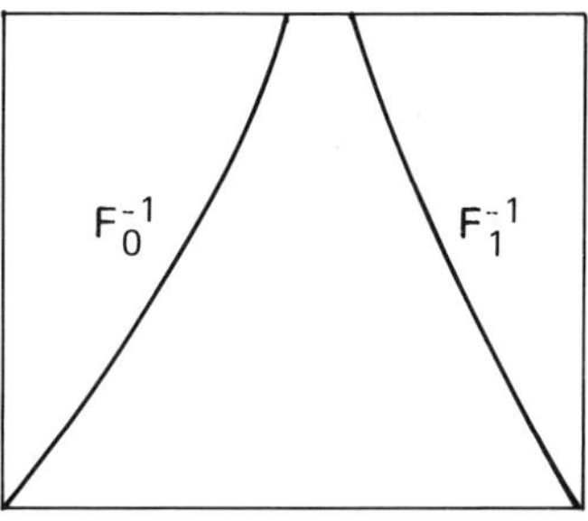

figure 3.1

Putting together equations (2.3) and (3.1), we obtain:

$$\sigma(\varepsilon_n \cdots \varepsilon_1) = \frac{|\, F_{\varepsilon_1} \circ \cdots F_{\varepsilon_{n-1}}(F_{\varepsilon_n}(\Delta))\,|}{|\, F_{\varepsilon_1} \circ \cdots F_{\varepsilon_{n-1}}(\Delta)\,|} \tag{3.2}$$

It is also clear from (3.1) that the constructed intervals have the ordering of the nodes on the tree. So the presentation function and the tree are equivalent notions (given one of them, one constructs the other).

We will now show that the exponential decay of the dependence of σ on the ε's is implied by the smoothness of the F_ε (Sullivan, 1988). Since this decay is an assumption that enters at every stage of the theory, it is of paramount importance to understand the smoothness of F_ε.

Assume that the F_ε are contractions and the derivatives are α Holder continuous for some α, that is: there are α, λ, C with

$$0 < 1 - \lambda \le F_\varepsilon'(x) \le \lambda \tag{3.3}$$

and

$$0 < \frac{F_\varepsilon'(\xi) - F_\varepsilon'(\eta)}{|\xi - \eta|^\alpha} < C, \quad 0 < \alpha < 1 \tag{3.4}$$

for all x, ξ and η in Δ. From equation (3.2) one obtains

$$\sigma(\varepsilon_n \cdots \varepsilon_1) = \sigma(\varepsilon_n \cdots \varepsilon_{k+1}) \frac{F_{\varepsilon_1} \cdots F_{\varepsilon_k}(F_{\varepsilon_{k+1}} \cdots F_{\varepsilon_n}\Delta)}{F_{\varepsilon_{k+1}} \cdots F_{\varepsilon_n}\Delta} \frac{F_{\varepsilon_{k+1}} \cdots F_{\varepsilon_{n-1}}\Delta}{F_{\varepsilon_1} \cdots F_{\varepsilon_k}(F_{\varepsilon_{k+1}} \cdots F_{\varepsilon_{n+1}}\Delta)}.$$

Divide by $\sigma(\varepsilon_n \ldots \varepsilon_{k+1})$. The right hand side now consists of two terms, which by the mean value theorem are equal to derivatives in some points $\xi_1 \in \Delta(\varepsilon_n \ldots \varepsilon_{k+1})$ resp. $\eta_1 \in \Delta(\varepsilon_{n-1} \ldots \varepsilon_{k+1})$. Subtracting 1 from each side, one gets:

$$\frac{\sigma(\varepsilon_n \ldots \varepsilon_1) - \sigma(\varepsilon_n \ldots \varepsilon_{k+1})}{\sigma(\varepsilon_n \ldots \varepsilon_{k+1})} = \frac{\frac{d}{dx}\{F_{\varepsilon_1} \ldots F_{\varepsilon_k}(\xi_1)\} - \frac{d}{dx}\{F_{\varepsilon_1} \ldots F_{\varepsilon_k}(\eta_1)\}}{\frac{d}{dx}\{F_{\varepsilon_1} \ldots F_{\varepsilon_k}(\eta_1)\}} .$$

Now set

$$\xi_{i+2} = F_{\varepsilon_{k-i}}(\xi_{i+1})$$
$$\eta_{i+2} = F_{\varepsilon_{k-i}}(\eta_{i+1}),$$

and use the chain rule to obtain

$$\frac{\sigma(\varepsilon_n \ldots \varepsilon_1) - \sigma(\varepsilon_n \ldots \varepsilon_{k+1})}{\sigma(\varepsilon_n \ldots \varepsilon_{k+1})} = \frac{\prod_{i=0}^{k-1} F'_{\varepsilon_{k-i}}(\xi_i) - \prod_{i=0}^{k-1} F'_{\varepsilon_{k-i}}(\eta_i)}{\prod_{i=0}^{k-1} F'_{\varepsilon_{k-i}}(\eta_i)} .$$

Since ξ_i and η_i both are in $\Delta(\varepsilon_n \ldots \varepsilon_{k-i})$, by using (3.3) we have

$$F'_{\varepsilon_{k-i}}(\xi_i) = F'_{\varepsilon_{k-i}}(\eta_i) + k_i \lambda^{\alpha(n-k+i)}$$

where $|k_i|$ is bounded by C. Therefore,

$$\left| \frac{\sigma(\varepsilon_n \ldots \varepsilon_1) - \sigma(\varepsilon_n \ldots \varepsilon_{k+1})}{\sigma(\varepsilon_n \ldots \varepsilon_{k+1})} \right| \leq \frac{K\lambda^{\alpha(n-k)}}{1 - \lambda} = K\,\beta^{n-k} , \tag{3.5}$$

for some $\beta < 1$.

IV   HAUSDORFF DIMENSION

The sets we consider often have the property that at each level a definite fraction of the length is taken out of the interval. The total length of the remaining intervals at level n then approaches zero. These sets have zero Lebesgue measure. There is a property that can distinguish such sets, namely the Hausdorff dimension.

At this point we should warn the reader that there are many different notions of dimensionality. These notions are certainly not equivalent in general. In spite of its somewhat cumbersome definition, the Hausdorff dimension has in the context of this work convenient properties. For other purposes (e.g., numerical work) other notions may be more fit.

We follow the definitions of Falconer (1985). A δ-cover of a set S is a (countable) collection of intervals $U_i(\delta)$ with length $|U_i(\delta)| \leq \delta$. The β-measure $H^\beta(\delta)$ of S is given by

$$H^\beta(S) = \lim_{\delta \to 0} \inf_{\delta-\text{covers}} \sum_{i=0}^{\infty} |U_i(\delta)|^\beta \ . \tag{4.1}$$

Here, the infimum is taken over all possible countable collections $\{U_i(\delta)\}_{i=1}^{\infty}$ that cover S. According to this definition, one first covers the set S in an optimal way with intervals that are not bigger than δ. Then one takes a limit as $\delta \to 0$. The Hausdorff dimension is an invariant of the Cantor set (under smooth coordinate transformations).

The β-measure has the following property. There is a number d (the Hausdorff dimension of the set S) such that

$$0 \leq \beta < d \ \Rightarrow \ H^\beta(S) = \infty$$
$$\beta > d \ \Rightarrow \ H^\beta(S) = 0 \ . \tag{4.2}$$

To illustrate these concepts, we will now calculate the Hausdorff dimension of the two scale Cantor set S (figure 2.1). The Hausdorff dimension will be denoted by HD.

We can get an upper estimate of $H^\beta(S)$ easily. Cover S with any convenient cover, such as the intervals $\Delta(\varepsilon_n \ldots \varepsilon_1)$ of the n-th level of construction of the set S. These form a δ-cover of S where

$$\delta = \max_{\{\varepsilon\}_n} | \Delta(\varepsilon_n \ldots \varepsilon_1) | \ .$$

(In fact, these sets are a closed, not open, cover of S. But by replacing them with open sets $\tilde{\Delta}(\varepsilon_n \ldots \varepsilon_1)$ ever so slightly larger, the calculation would not be changed.) Thus

$$H^\beta(S) \leq \lim_{n \to \infty} \sum_{\{\varepsilon\}_n} | \Delta(\varepsilon_n \ldots \varepsilon_1) |^\beta \ = \ \lim_{n \to \infty} \sum_{\{\varepsilon\}_n} [\sigma(\varepsilon_1)\ldots\sigma(\varepsilon_1)]^\beta \ = \ \lim_{n \to \infty} \sum_{\varepsilon} [\sigma(\varepsilon)^\beta]^n \ .$$

Therefore, if d satisfies

$$\sum_{\varepsilon} \sigma(\varepsilon)^d \ = \ 1 \ , \tag{4.3}$$

then

$$HD(S) \ \geq \ d.$$

To show that HD(S) is equal to d, one would have to show that for some $C > 0$

$$H^d(S) \geq C \lim_{n \to \infty} \sum_{\{\varepsilon\}_n} | \Delta(\varepsilon_n \ldots \varepsilon_1) |^{HD(S)} \ .$$

Equation (4.3) then implies that

$$HD(S) = d. \tag{4.4}$$

This is much harder and we will not pursue this any further. A more general statement can be found in Ruelle (1988): if the F_ε satisfy (3.3) and (3.4) and if T is a finitely pruned k-nary tree, then the HD(S) is given by:

$$HD(S) = \lim_{n \to \infty} d_n$$

where

$$\sum_{\{\varepsilon\}_n} |\Delta(\varepsilon_n \ldots \varepsilon_1)|^{d_n} = 1.$$

(Finitely pruned means: the pruning of a branch depends on finitely many, say N, previous branches.)

For the two scale Cantor set with equal scaling one obtains

$$2\sigma^{HD(S)} = 1 \implies HD(S) = -\frac{\ln 2}{\ln \sigma}.$$

V   A TANTALIZING EXAMPLE

In this section we will study the statistics of the scalings on the n-th level of construction of the Cantor set. We will do so from first principles by means of an example, the two scale Cantor set. The general theory will be set forth in the next section. Recall from the first two chapters that the scalings and therefore all the statistics associated with them are characteristics of the presentation function rather than of the set itself.

For the Cantor set S one has (figure 2.1)

$$|\Delta(\varepsilon_{2n} \ldots \varepsilon_1)| = \sigma(0)^{n-i} \sigma(1)^{n+i}, \tag{5.1}$$

where $n - i$ of the ε's are zero, and the rest are ones. The set of arguments of Δ consists of all binary words of length $2n$. So:

$$\sum |\Delta(\varepsilon_{2n} \ldots \varepsilon_1)|^\beta = \sum_{i=-n}^{n} \binom{2n}{n+i} \sigma(0)^{(n-i)\beta} \sigma(1)^{(n+i)\beta} \tag{5.2}$$

or

$$\sum |\Delta(\varepsilon_{2n} \ldots \varepsilon_1)|^\beta = (\sigma(0)^\beta + \sigma(1)^\beta)^{2n}. \tag{5.3}$$

The ratio of the $(i+1)$-st and the i-th term in (5.2) is given by

$$\frac{C_{i+1}}{C_i} = \frac{n-i}{n+i+1}\left[\frac{\sigma(1)}{\sigma(0)}\right]^{\beta} \cong \frac{1-i/n}{1+i/n}\left[\frac{\sigma(1)}{\sigma(0)}\right]^{\beta} .$$

It is easy to see that the contribution to the sum in (5.2) is maximized when this ratio equals one, or

$$\frac{i}{n} = \frac{\left[\dfrac{\sigma(1)}{\sigma(0)}\right]^{\beta} - 1}{\left[\dfrac{\sigma(1)}{\sigma(0)}\right]^{\beta} + 1} . \tag{5.4}$$

Now assume n to be very large. It is clear that for n large enough the contributions of $\frac{i}{n} \in \{[-1, -1+\varepsilon] \cup [1-\varepsilon, 1]\}$ are negligible if ε is small. In fact, the contribution from those intervals is drowned in the contribution of the terms $\frac{i}{n} = -1+\varepsilon$ and $\frac{i}{n} = 1-\varepsilon$, as one concludes readily from (5.2).

Of course, there are many terms for which i/n is approximately constant. It seems handy to replace i by a new variable $\mu = i/n$ (or some affine function of it). Since the factorials are not very convenient for this, we apply Stirling's formula first. (Notice that this is justified by our remark that contributions of the tail are vanishingly small.)

Stirling's formula gives:

$$\sum_{\{\varepsilon\}_{2n}} |\Delta(\varepsilon_{2n} \cdots \varepsilon_1)|^{\beta}) = \sum_{i=-n(1-\varepsilon)}^{n(1-\varepsilon)} \frac{(2n)^{1/2}}{(2\pi)^{1/2}(n+i)^{1/2}(n-i)^{1/2}} \exp[2n \ln 2n - (n+i) \ln(n+i)$$
$$- (n-i) \ln (n-i)] \cdot \sigma(0)^{(n-i)\beta} \sigma(1)^{(n+i)\beta}$$

where we have neglected terms of order $\frac{1}{\varepsilon n}$ for fixed ε. With the new variable μ this becomes

$$\cdots = \sum_{\mu=-1+\varepsilon}^{1-\varepsilon} \left[\frac{1}{\pi n(1-\mu)(1+\mu)}\right]^{1/2} \exp\left[2n \left\{\ln 2 - \tfrac{1}{2}(1+\mu) \ln(1+\mu) - \tfrac{1}{2}(1-\mu) \ln (1-\mu)\right.\right.$$
$$\left.\left. + \tfrac{1}{2}\beta \ln \sigma(0)\sigma(1) + \tfrac{1}{2}\mu\beta \ln \frac{\sigma(1)}{\sigma(0)}\right\}\right] . \tag{5.5}$$

With obvious changes of notation this will be rewritten as

$$\cdots = \sum_{\mu=-1+\varepsilon}^{1-\varepsilon} C(\mu)\, e^{2n\{s(\mu)-\beta(c_1\mu + c_2)\}} . \tag{5.6}$$

Since μ increments in steps of $\frac{1}{n}$, this can be written as an integral (neglecting order $\frac{1}{n}$ terms):

$$\cdots = \int n\, d\mu\, C(\mu)\, e^{2n\{s(\mu)-\beta(c_1\mu + c_2)\}} . \tag{5.7}$$

As shown before,

$$-X(\mu, \beta) = s(\mu) - \beta(c_1\mu + c_2)$$

has a unique maximum (see (5.4)) at $\mu = \bar{\mu}$. Further:

$$-\frac{d^2}{d\mu^2} X\Big|_{\mu=\bar{\mu}} = \frac{-1}{(1+\bar{\mu})(1-\bar{\mu})} < 0, \tag{5.8}$$

so that the maximum is non degenerate. This implies that one may use steepest descent methods to evaluate the integral (5.7).

$$\int n\, d\mu\, C(\mu)\, e^{-2nX(\mu,\beta)} = n\, C(\bar{\mu})\, e^{-2nX(\bar{\mu},\beta)} \int d\mu\, e^{-n(\mu-\bar{\mu})^2\, \partial^2 X}$$

$$= n\, C(\bar{\mu}) \left[\frac{\pi}{n\partial_\mu^2 X(\bar{\mu},\beta)} \right]^{1/2} e^{-2nX(\bar{\mu},\beta)}. \tag{5.9}$$

In the last term we have again (in applying the steepest descent method) neglected $\frac{1}{n}$ terms. Comparison with (5.8), (5.4) and the definition of $C(\mu)$ shows that all prefactors cancel. One finally obtains (replacing $2n$ by n)

$$\sum |\Delta(\varepsilon_n \dots \varepsilon_1)|^\beta) = e^{-nX(\bar{\mu},\beta)}. \tag{5.10}$$

(As an exercise, the readers should convince themselves that when one works out what $X(\bar{\mu}, \beta)$ is using equations (5.4) and (5.5), one gets equation (5.3) back.)

This calculation yields

$$X(\bar{\mu}, \beta = HD) = 0,$$

if one combines (5.10) and the results in section 4. This equation then determines the Hausdorff dimension of the set S.

Here then is the first step of the statistics of the problem. The largest single contribution to the sum in (5.2) comes from the terms with $\frac{i}{n} = \bar{\mu}$. There are $C(\bar{\mu})e^{ns(\bar{\mu})}$ of these intervals and each has length $e^{-n(c_1\bar{\mu} + c_2)}$ (i.e., average scaling is $e^{-(c_1\bar{\mu} + c_2)}$). If we add up nearby ($\frac{i}{n} \cong \bar{\mu}$) contributions, they come from intervals with smaller length or have lengths that occur less abundantly. From (5.10) one concludes that, on the average, one can say that the sum is dominated by $e^{ns(\bar{\mu})}$ intervals with average scaling $e^{-(c_1\bar{\mu} + c_2)}$.

Suppose we were interested in the set $S_n(\mu_0, \delta) \subset S_n$ of intervals which have an average scaling of $e^{-(c_1\bar{\mu} + c_2)}$ where $\mu \in [\mu_0 - \delta, \mu_0 + \delta]$ for some positive, small δ. We will denote the sum over these intervals by $\sum_0$. It is worth pointing out that $|\mu_0|$ should not be taken too close to one, since that would preclude our applying Stirling's formula in what follows: the whole reasoning described below fails miserably if the number of pieces $\Delta(\varepsilon_{2n} \ldots \varepsilon_1)$ with scaling μ_0 is not large. This is important to bear in mind when doing numerical or real life experiments.

As before, Stirling's approximation gives equation (5.5) with the summation now running over $[\mu_0 - \delta, \mu_0 + \delta]$. Expressing this as an integral as in (5.7) yields

$$\sum\nolimits_0 |\Delta(\varepsilon_n \ldots \varepsilon_1)|^\beta = \int_{\mu_0-\delta}^{\mu_0+\delta} n\, d\mu\, C(\mu)\, e^{ns(\mu)-n\beta(c_1\mu+c_2)}. \tag{5.11}$$

Clearly, then, $S_n(\mu_0, \delta)$ dominates the above sum if β is chosen to be equal to β_0 with

$$\mu_0 = \bar{\mu}(\beta_0),$$

in which case one obtains, in agreement with (5.10),

$$\sum\nolimits_0 |\Delta(\varepsilon_n \ldots \varepsilon_1)|^{\beta_0} = e^{-nX(\bar{\mu},\beta_0)}.$$

For $\beta \neq \beta_0$ one clearly has

$$\mu_0 \neq \bar{\mu}(\beta)$$

That is: for small enough δ the maximum of $s(\mu) - \beta(c_1\mu + c_2)$ lies outside the interval of integration of (5.11). To evaluate the integral in this case, set $\mu = \mu_0$ and integrate:

$$\sum\nolimits_0 |\Delta(\varepsilon_n \ldots \varepsilon_1)|^\beta = 2\delta n\, C(\mu_0)\, e^{ns(\mu_0)-n\beta(c_1\mu_0+c_2)+O(n\delta)}. \tag{5.12}$$

The exponential error here comes from the fact that

$$\frac{1}{2\delta e^{-nx}} \int_{x-\delta}^{x+\delta} e^{-n\mu}\, d\mu = \frac{e^{+n\delta} - e^{-n\delta}}{2\delta} = e^{O(n\delta)}.$$

In analogy with the definition of the Hausdorff dimension, we now define the exponent f as follows:

$$\text{for all } \beta < f \quad \lim_{n\to\infty} \sum\nolimits_0 |\Delta(\varepsilon_n \ldots \varepsilon_1)|^\beta = \infty$$

$$\text{for all } \beta > f \quad \lim_{n\to\infty} \sum\nolimits_0 |\Delta(\varepsilon_n \ldots \varepsilon_1)|^\beta = 0. \tag{5.13}$$

Then, of course, (5.12) implies

$$s(\mu_0) - f(c_1\mu_0 + c_2) = 0$$

where

$$\mu_0 = \bar{\mu}(\beta_0)$$

Thus

$$f = \frac{s(\bar{\mu}(\beta_0))}{c_1\bar{\mu}(\beta_0) + c_2}. \tag{5.14}$$

This formula gives the relation between the exponent β_0 such that the μ_0 scalings dominate the sum $\sum |\Delta|^{\beta_0}$ and the exponent f for which the sum $\sum |\Delta|^f$ crosses over as defined by (5.13).

Recall that each elementary interval has a 'probability' N_n^{-1} associated with it. The length of the 'average' interval that dominates $\sum |\Delta|^{\beta_0}$ is $N_n^{-(c_1\bar{\mu} + c_2)}$. Define the exponent α as:

$$N_n^{-1/\alpha} = N_n^{-(c_1\bar{\mu} + c_2)}$$

or

$$\alpha = \frac{1}{c_1\bar{\mu} + c_2}. \tag{5.15}$$

Then f can be expressed as a function of α (provided $c_1 \neq 0$):

$$f(\alpha) = \alpha \cdot s\left(\frac{\alpha^{-1} - c_2}{c_1}\right).$$

The definition of s in (5.6) gives:

$$f(\alpha) = \alpha \left\{ \ln 2 - \frac{1}{2}(1 + \frac{\alpha^{-1} - c_2}{c_1}) \ln(1 + \frac{\alpha^{-1} - c_2}{c_1}) - \frac{1}{2}(1 - \frac{\alpha^{-1} - c_2}{c_1}) \ln(1 - \frac{\alpha^{-1} - c_2}{c_1}) \right\}.$$

Note the word "analogy" in the definition (5.13) of f. If certain conditions are met, one can prove (Bohr and Rand, 1987) that

$$S(\mu_0) = \{ \text{all } x \in \Delta \mid X \text{ has average scaling } e^{-(c_1\bar{\mu} + c_2)} \}$$

has Hausdorff dimension $f(\alpha)$ if $\alpha = (c_1\mu_0 + c_2)^{-1}$. Notice that $S(\mu_0)$ is a strange set. Since every point in X can be arbitrarily well approximated by points in $S(\mu_0)$, the closure of $S(\mu_0)$ is S. Furthermore, there are points in S that do not have an average scaling (that is, the limit does not converge). So $S \setminus \bigcup_{\mu} S(\mu) \neq \emptyset$.

VI THERMODYNAMICS
(CANONICAL AND MICROCANONICAL ENSEMBLES)

The reader with some training in statistical physics will undoubtedly have recognized the last section as a treatment of the microcanonical ensemble. We will now set forth the general principles behind this.

While we are in the position that we can apply mathematical notions borrowed from statistical physics, the actual 'physics' cannot be transplanted. The difference with the statistical physics of spin systems is that there the 'tangible' objects are the actual spins. Here the spins (or scalings) arise from the presentation function which in turn is generated by an underlying dynamical process. Thus the 'tangible' object, the Cantor set, does not itself determine the scalings (see section 2).

With this in mind we may call $X(\bar{\mu},\beta)$ of equation (5.10) the "free energy" although it is not an energy in the physical sense of the word. To render the description of the theory transparent we will write $F(\beta)$ for that quantity. Similarly we will replace $c_1\mu + c_2$ of section 5 by μ.

Write the number of pieces $\Delta(\varepsilon_n \dots \varepsilon_1)$ at n-th level of construction of S as

$$N_n \cong a^n \tag{6.1}$$

where
$$a = \lim_{n\to\infty} \frac{1}{n} \ln N_n .$$

Let
$$|\Delta(\varepsilon_n \dots \varepsilon_1)| = N_n^{-h(\varepsilon_n \dots \varepsilon_1)} . \tag{6.2}$$

Then
$$\sum_{\{\varepsilon\}} |\Delta(\varepsilon_n \dots \varepsilon_1)|^\beta = \sum_{\{\varepsilon\}} N_n^{-\beta h(\varepsilon_n \dots \varepsilon_1)}$$

$$= \sum_{\mu>0} N_n^{-\beta\mu} \sum_{\substack{\{\varepsilon\} \\ h(\varepsilon_n \dots \varepsilon_1)=\mu}} 1 = \sum_{\mu>0} N_n^{-\beta\mu+s(\mu)} . \tag{6.3}$$

This equation defines μ as the value of $h(\varepsilon_n \dots \varepsilon_1)$ or the "microcanonical energy." The quantity $e^{ns(\mu)}$ equals the cardinality of the configurations with $h = \mu$, that is to say: $s(\mu)$ is the "entropy."

In the previous section we showed for the two scale Cantor set $s(\mu)$ has negative second derivative. (This can be generalized to all finitely pruned k-nary trees satisfying (3.3) and (3.4), see Ruelle (1978). Thus $s(\mu) - \mu\beta$ has a non-degenerate maximum at $\mu = \bar{\mu}$ As in the previous section, the contribution of the maximum dominates the sum if n is very large:

$$\frac{\partial}{\partial\mu}\bigg|_{\mu=\bar{\mu}} \{s(\mu) - \mu\beta\} = 0$$

$$\frac{\partial^2}{\partial\mu^2}\bigg|_{\mu=\bar{\mu}} \{s(\mu) - \mu\beta\} < 0 \tag{6.4}$$

$$\sum_{\{\varepsilon\}} |\Delta(\varepsilon_n \ldots \varepsilon_1)|^\beta \cong a^{n(s(\bar{\mu})-\beta(\mu))} \tag{6.5}$$

(note that $\bar{\mu}$ depends on β). We can also regard

$$\sum_{\{\varepsilon\}} |\Delta(\varepsilon_n \ldots \varepsilon_1)|^\beta = \sum_{\{\varepsilon\}} a^{-\beta n h(\varepsilon_n \ldots \varepsilon_1)} = a^{-nF(\beta)} \tag{6.6}$$

as a canonical ensemble with $F(\beta)$ the "free energy per particle" (sometimes called "pressure" in the mathematics literature). Therefore, we have

$$F(\beta) = \beta \bar{\mu}(\beta) - s(\bar{\mu}(\beta)) \tag{6.7}$$

which is the equivalent of (5.10).

We note that by (6.4) $F(\beta)$ is precisely the Legendre transform of the entropy. Therefore $F(\beta)$ satisfies

$$F''(\beta) < 0$$
$$\tag{6.8}$$
$$F'(\beta) = \bar{\mu}(\beta) > 0.$$

We will use the same definitions for f and α as we used before in (5.13) and (5.15). So

$$f = s(\bar{\mu}) / \bar{\mu}$$

$$\alpha = 1/\bar{\mu} \tag{6.9}$$

By equation (6.8) $F(\beta)$ is strictly monotone and therefore invertible. We will call its inverse b(F). We can now express f as a function of α. Since

$$f = \beta - \frac{\beta\bar{\mu} - s(\bar{\mu})}{\bar{\mu}} = \beta - \alpha(\beta\bar{\mu} - s(\bar{\mu}))$$

and

$$\frac{d\beta}{dF}\left\{\frac{d}{d\beta}[\beta - \alpha(\beta\mu - s(\mu))]\right\} = 0,$$

one finds that $-f(\alpha)$ is the Legendre transform of b(F):

$$f(\alpha) = -\min_F \{\alpha F - b(F)\}. \tag{6.10}$$

Some papers take $f(\alpha)$, a quantity we derive from $F(\beta)$, as their starting point. They give a direct definition of $f(\alpha)$ in terms of the intervals Δ. We reproduce that statement as a corollary of the above.

Consider the intervals of S_n. As explained in section 2, one can define a probability measure p such that each of these intervals has an equal probability (see (6.1)) $p_n = a^{-n}$. According to (6.5), the 'average' length l_n of the intervals that dominate the sum is $a^{-n\bar\mu}$. So

$$p_n = l_n^{1/\bar\mu} = l_n^{\alpha}. \tag{6.11}$$

The cardinality of the intervals that dominate the sum is (see (6.5)) $a^{ns(\bar\mu)}$. One obtains

$$a^{ns(\bar\mu)} = l_n^{-s(\bar\mu)/\bar\mu} = l_n^{-f} \tag{6.12}$$

The convenient property of $f(\alpha)$ that is often used is the following: Equation (6.10) implies that $f(\alpha)$ has a unique maximum, and moreover at the maximum

$$f'(\alpha_{max}) = F(\beta = HD) = 0 \tag{6.13}$$

In that case, as discussed in section 5, f is precisely the Hausdorff dimension of S.

We emphasize that if we had constructed the same set S by means of a different tree, then $F(\beta)$ would have been different and so would $f(\alpha)$. It is left to the reader to show that the tree of figure 2.5 has associated with it:

$$F(\beta) = -\ln\{\sigma(0)^{\beta} + \sigma(0)^{\beta}(\sigma(0)^{\beta} + \sigma(1)^{\beta})\}.$$

To exhibit more clearly the connection between classical statistical mechanics and successive approximation of Cantor sets, we will give the "Hamiltonian" h (see(6.2)) in a few simple cases. For instance, for the two scale Cantor set one concludes from (5.2) that

$$h(\varepsilon_n \ldots \varepsilon_1) = \tfrac{1}{2}\ln\sigma(0)\,\sigma(1) + \tfrac{1}{2}\tfrac{i}{n}\ln\frac{\sigma(1)}{\sigma(0)}$$

$$= \ln\sigma(0) + \ln\frac{\sigma(1)}{\sigma(0)} \cdot \frac{1}{n}\sum\varepsilon_i\,. \tag{6.14}$$

This corresponds with a system of n non-interacting spins that assume the values 0 or 1 (Ising) in an external field of strength $\ln\dfrac{\sigma(1)}{\sigma(0)}$.

It is straightforward to derive the Hamiltonian for the binary Cantor set with 1 level of memory (figure 2.3). Note that

$$\sigma(\varepsilon, \varepsilon') = \sigma(00) \left[\frac{\sigma(10)}{\sigma(00)}\right]^{\varepsilon} \left[\frac{\sigma(01)}{\sigma(00)}\right]^{\varepsilon'} \left[\frac{\sigma(11)\sigma(00)}{\sigma(10)\sigma(01)}\right]^{\varepsilon\varepsilon'} .$$

We have

$$|\Delta(\varepsilon_n \ldots \varepsilon_1)| = \sigma(\varepsilon_n \, \varepsilon_{n-1}) \, \sigma(\varepsilon_{n-1} \, \varepsilon_{n-2}) \ldots \sigma(\varepsilon_2 \, \varepsilon_1).$$

Thus

$$h(\varepsilon_n \ldots \varepsilon_1) = \ln \sigma(00) + \ln \frac{\sigma(10)}{\sigma(00)} \sum_2^n \frac{\varepsilon_i}{n} + \ln \frac{\sigma(01)}{\sigma(00)} \sum_1^{n-1} \frac{\varepsilon_i}{n} + \ln \frac{\sigma(11)\,\sigma(00)}{\sigma(10)\sigma(01)} \sum_1^{n-1} \frac{\varepsilon_i\varepsilon_{i-1}}{n}$$

$$\cong \ln \sigma(00) + \ln \frac{\sigma(10)\,\sigma(01)}{\sigma(00)^2} \frac{1}{n}\sum_1^n \varepsilon_i + \ln \frac{\sigma(11)\,\sigma(00)}{\sigma(10)\sigma(01)} \frac{1}{n}\sum_1^{n-1} \varepsilon_i\varepsilon_{i-1} . \qquad (6.15)$$

This now corresponds with an Ising model with nearest neighbor interaction. In general, if we allow k levels of memory, we will get a k-nearest neighbor Ising model. Notice, however, that under assumptions (3.3) and (3.4) these interactions will decay exponentially with distance (because the scalings do so, see (3.5)).

We finally remark that from (6.14) it is immediately apparent that in determining thermodynamic quantities we discard information (Feigenbaum, 1987). In (6.15) only $\{\sigma(00),$ $\sigma(10)\,\sigma(01),\ \sigma(11)\sigma(00)\}$ is needed as input. The tree, however, is given by 4 scalings plus ordering properties.

VII MARKOV GRAPHS
(GRAND CANONICAL ENSEMBLE)

In the previous section we have described the statistics associated with equation (6.6). This equation says that the β-sum over the lengths of the intervals behaves as a pure exponential if we go down deep enough in the tree. In order to investigate how accurate this statement is for finite, but large n, we have to take into account fluctuations in F. It is natural to follow tradition here and construct a grand canonical ensemble (Feigenbaum 1987).

Suppose

$$\sum_{\{\varepsilon\}_n} |\Delta(\varepsilon_n \ldots \varepsilon_1)|^{\beta} = a^{-nF_n(\beta)} \ , \quad \lim_{n\to\infty} F_n(\beta) = F(\beta)$$

and define the "Gibbs potential" G

$$e^{-G} = \sum_n z^n \sum_{\{\varepsilon\}_n} |\Delta(\varepsilon_n \ldots \varepsilon_1)|^{\beta} . \qquad (7.1)$$

Since the scalings depend exponentially weakly on the low ε's (see (3.5)), we can use equation (2.3) to approximate higher level Δ's very accurately with the ones that are one level lower. Suppose we approximate with k levels of memory:

$$e^{-G} = \sum_n z^n \sum_{\{\varepsilon\}_n} [\sigma(\varepsilon_n \cdots \varepsilon_{n-k})]^\beta \, |\Delta(\varepsilon_{n-1} \cdots \varepsilon_1)|^\beta$$

$$= \sum_n z^n \sum_{\{\eta_n \cdots \eta_{n-k+1}\}} \sum_{\{\varepsilon\}_n} T_{(\eta_n \cdots \eta_{n-k+1}),(\varepsilon_{n-1} \cdots \varepsilon_{n-k})} |\Delta(\varepsilon_{n-1} \cdots \varepsilon_1)|^\beta , \qquad (7.2)$$

where T is a transition matrix of rank 2^k with

$$T_{(\eta_n \cdots \eta_{n-k+1}),(\varepsilon_{n-1} \cdots \varepsilon_{n-k})} = 0$$

<u>except</u>

$$T_{(\varepsilon_n \cdots \varepsilon_{n-k+1}),(\varepsilon_{n-1} \cdots \varepsilon_{n-k})} = [\sigma(\varepsilon_n \cdots \varepsilon_{n-k})]^\beta .$$

If we now think of

$$\sum_{\{\varepsilon_{n-k-1} \cdots \varepsilon_1\}} |\Delta(\varepsilon_{n-1} \cdots \varepsilon_1)|^\beta = \Psi^{(n-1)}_{(\varepsilon_{n-1} \cdots \varepsilon_{n-k})}$$

as a vector $\Psi^{(n-1)}_\varepsilon$, then (7.2) becomes

$$e^{-G} = \sum_{\{\eta\}} \sum_n z^n (T \Psi^{(n-1)})_\eta = \sum_{\{\eta\}} \left(\sum_n z^n T^n \Psi^{(0)} \right)_\eta \qquad (7.3)$$

for some appropriate $\Psi^{(0)}$. So

$$= \sum_{\{\eta\}} ([1 - zT]^{-1} \Psi^{(0)})_\eta = \sum_{\{\varepsilon\}\{\eta\}} \frac{\mathrm{cof}(\{\eta\}, \{\varepsilon\}) \cdot \Psi^{(0)}_\varepsilon}{\det(1 - zT)} , \qquad (7.4)$$

where $\mathrm{cof}(\{\eta\}, \{\varepsilon\})$ stands for the cofactor of the matrix $(1 - zT)$ centered on the element $(\{\eta\}, \{\varepsilon\})$.

The asymptotic behavior of T (which has positive entries and therefore its largest eigenvalue is real and positive) is dominated by its largest eigenvalue. It is clear from (7.4) that e^{-G} has a pole when z^{-1} equals an eigenvalue of T. We therefore look for

$$\lambda(\beta) = \text{largest eigenvalue of } T = \frac{1}{\text{smallest pole of } e^{-G}} . \qquad (7.5)$$

Then if λ_2 is the second largest eigenvalue

$$e^{-G} = \sum_{\{\varepsilon\}} z^n \, \lambda(\beta)^n \, \Psi^{(0)}\left(1 + O\left(\left[\frac{|\lambda_2|}{\lambda}\right]^n\right)\right) .$$

Thus for some α between 0 and 1 (7.1) implies

$$F_n(\beta) = - \frac{\ln\,[\lambda(\beta)(1 + O(\alpha^n))]^{1/n}}{\ln\,a} \;\Rightarrow\; \lim_{n\to\infty} F_n(\beta) = - \frac{\ln\,\lambda(\beta)}{\ln\,a} . \qquad (7.6)$$

The reason we went through this formalism is that T is a sparse matrix and hence one may calculate $[\det\,(1 - zT)]^{-1}$ by means of Markov graphs. We will now see how one achieves this.

One draws a graph on which the nodes are labelled with $(\varepsilon_n \ldots \varepsilon_{n-k+1})$ for all allowed ε-words (in figure 2.2, for instance, "11" is not allowed). One then draws arrows from $(\varepsilon_{n-1} \ldots \varepsilon_{n-k})$ to $(\varepsilon_n \ldots \varepsilon_{n-k+1})$ respresenting $z[\sigma(\varepsilon_n \ldots \varepsilon_{n-k})]^\beta$. According to (7.2), e^{-G} is obtained by summing over the contributions of all possible paths through the graph. We are now interested in the poles of the expression thus obtained (see (7.4)).

Before we continue with examples, let us write up the rules by which the summation can be done easily. The following rules must be applied in the order in which they are given here.

1) Links that are in series are multiplied

2) Links that are parallel (i.e., start at node a, end at node b, and $a \neq b$) are added up

3) Links that form loops (i.e., starting and ending on the same node) form a geometric sequence $s \to \Sigma\, s^n = \dfrac{1}{1 - s}$.

These rules can be inferred easily by comparing (7.1) and (7.2) with the definition of a tree. If $a = \frac{1}{2}\log N_n$ (see (6.1)) is unknown, one derives from (7.1):

4) $\Sigma\,|\,\Delta(\varepsilon_n \ldots \varepsilon_1)\,|^0 = N_n = a^n$.

Thus for $\beta = 0$, the smallest zero of $\det\,(1 - zT)$ is equal to a^{-1}. Finally, from (6.12)

5) If $\beta = HD(S)$ then $z = 1$ is a zero.

Here are the examples. The Markov diagram drawn in figure 7.n corresponds to the tree drawn in figure 2.n.

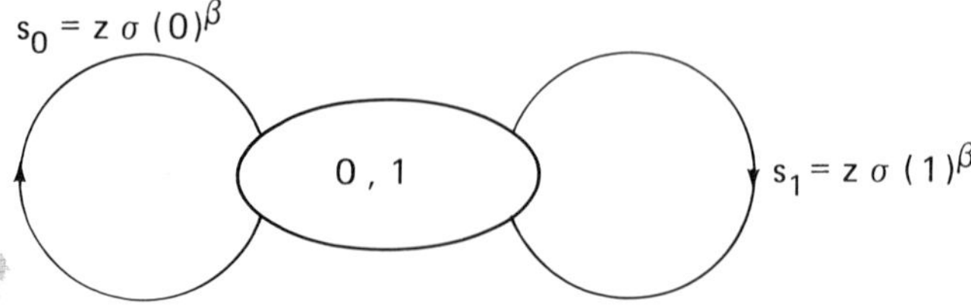

figure 7.1

To sum the contributions in figure 7.1, apply rule 2 first and then rule 3. One obtains

$$e^{-G} = \sum_n (s_0 + s_1)^n = \frac{1}{1 - z(\sigma(0)^\beta + \sigma(1)^\beta)} .$$

As before

$$\sigma(0)^{HD} + \sigma(1)^{HD} = 1$$

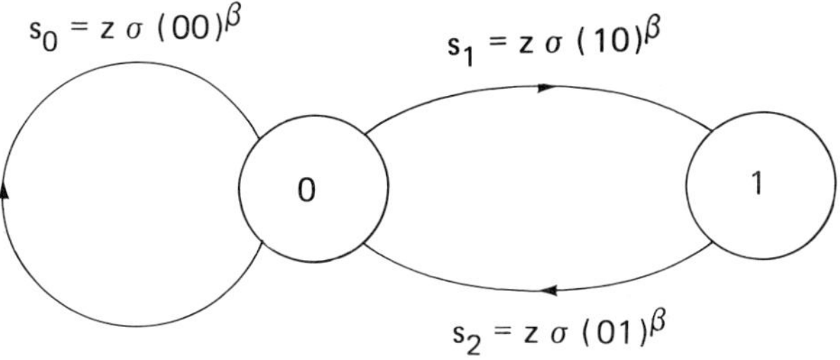

figure 7.2

To sum the contributions, add s_1 and s_2 first. Then figure 7.1 results with s_1 replaced by $s_1\,s_2$.

$$e^{-G} = \frac{c}{1 - z\sigma(00)^\beta - z^2\sigma(10)^\beta\sigma(01)^\beta} .$$

The consistency requirements 4 and 5 give

$$1 - a - a^2 = 0 \implies a = \frac{1 + \sqrt{5}}{2}$$

and

$$1 - \sigma(00)^{HD} - \sigma(10)^{HD}\,\sigma(01)^{HD} = 0.$$

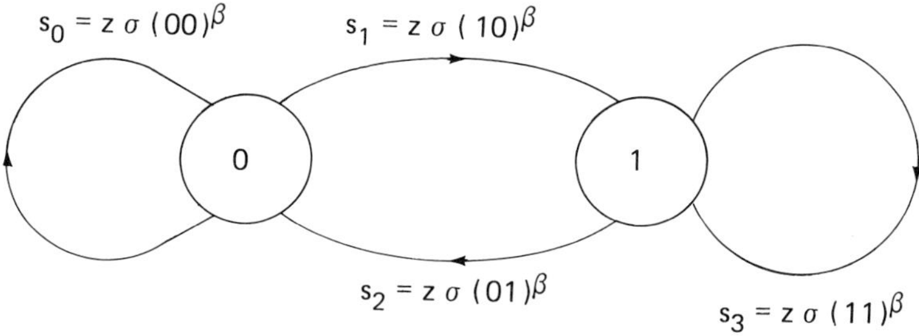

figure 7.3   figure 7.4

As stated before, thermodynamics "forgets" the ordering properties, therefore figure 7.3 and 7.4 are identical. To sum the graphs, first sum the part to the left of node 1. This gives

$$s' = \frac{z^2\,\sigma(01)^\beta\,\sigma(10)^\beta}{1 - z\,\sigma(11)^\beta}\,.$$

Now the figure is reduced to figure 7.1. Thus

$$e^{-G} = \frac{1 - z\,\sigma(11)^\beta}{1 - z(\sigma(00)^\beta + \sigma(11)^\beta) - z^2(\sigma(01)^\beta\,\sigma(10)^\beta - \sigma(00)^\beta\,\sigma(11)^\beta)}\,.$$

One can check that a = 2.

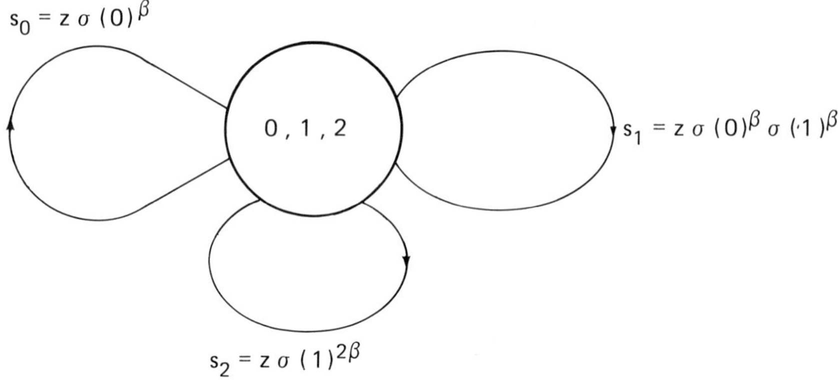

figure 7.5

As in figure 7.1

$$e^{-G} = \frac{1}{1 - z[\sigma(0)^\beta + \sigma(0)^\beta((\sigma(0)^\beta + \sigma(1)^\beta)])}\,.$$

VIII   CIRCLE MAPS

In the remaining sections we will discuss an application of the foregoing ideas. This section will provide the reader with some background material needed to appreciate the example. No attempts to give demonstrations have been made in this section.

A circle map is a map that takes points of the circle to points of the circle. One can also describe this as a map f of the real line to itself with the additional property that

$$f(x + 1) = f(x) + 1\,.$$

By "forgetting" the integer of the map one obtains the map of the circle to itself. A commonly studied example is

$$x_{i+1} = (x_i + \omega - \frac{k}{2\pi} \sin 2\pi x_i) \bmod 1 \tag{8.1}$$

This is a degree one map, meaning that it wraps around the circle only once. Maps of degree one have the property that if they are monotone their dynamics is fairly simple. In figure 8.1 we have drawn the map (8.1) for $0 \le k < 1$ when it is strictly monotone, for $k = 1$ when it is critical (i.e., it is invertible but there is a point x_0 such that $f'(x_0) = 0$), and for $k > 1$ when it is non-invertible.

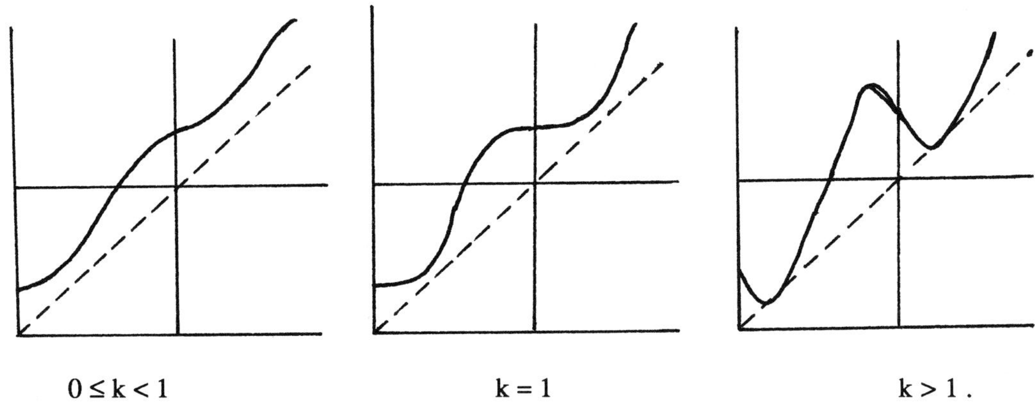

figure 8.1

The rotation number is defined as the average (angular) velocity along an orbit. More precisely, at a point x we define

$$\rho_f(x) = \lim_{n \to \infty} \frac{f^n(x)}{n} \quad \text{(if the limit exists)} \tag{8.2}$$

(Here f is regarded as a map on the real line.) The rotation number is one of the most important tools in studying the dynamics of circle maps. It describes the average behavior of an orbit.

In the case that $k > 1$ the map may or may not exhibit sensitive dependence on initial conditions ("chaotic behavior"). For $0 \le k \le 1$ there is no sensitive dependence on initial conditions. The case $k = 1$ is sometimes called "critical."

For the purposes of this paper it is sufficient to consider monotone $(0 \le k \le 1)$ maps. These maps have the property that for each point x the limit in (8.2) exists and moreover that its value does not depend on x. However, the rotation number is not constant as we let f vary.

Consider a one-parameter family of circle maps, for constant k, let f_ω be defined as in (8.1) where now $\omega \in [0, 1]$. The question we will address is: how does the rotation number vary as a function of ω. In other words: where do qualitative differences in behavior (namely change of rotation number) occur as we let ω vary through the parameter space? The function that gives, for constant k, the rotation number as a function of ω will be denoted by $\rho(\omega)$. In particular, let f be twice differentiable and monotone of degree one. Note that $f_\omega(x)$ is also a continuous monotone function of ω. The function $\rho(\omega)$ has the following properties (see for example Devaney 1985):

    i) $\rho(\omega)$ is monotone increasing and continuous

    ii)   The inverse image of a rational number $\frac{p}{q}$ is an interval, denoted by $I_{p/q}$.

    iii)   The inverse image of an irrational number α is a point.

Such a graph (see figure 8.2) is sometimes called the "devil's staircase."

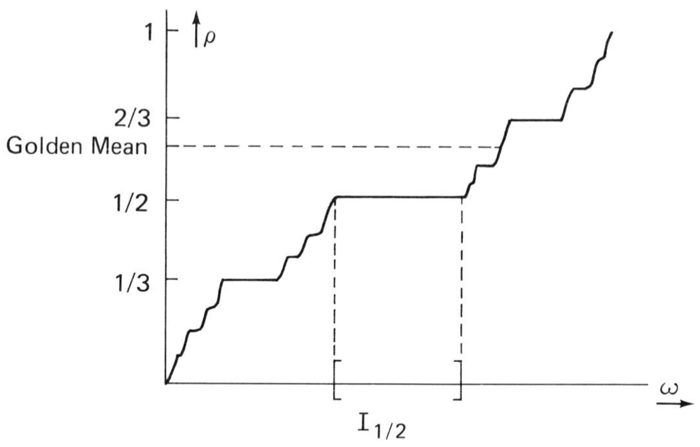

figure 8.2

The object is to study the properties of the Cantor set A defined as follows

$$A = [0, 1] \setminus \left\{ \bigcup_{p/q} \overset{\circ}{I}_{p/q} \right\}, \tag{8.2}$$

where $\overset{\circ}{I}$ means the interior of I. All one has to do to apply the theory in the previous sections is to find a natural way to construct that Cantor set by means of a tree.

The simplest example of the set A is obtained when we let k go to zero. The intervals $I_{p/q}$ now collapse to zero length, i.e., to a point. In this case,

$$\rho(\omega) = \omega.$$

To apply the theory we must define a "formal Cantor set" as follows. Replace in the unit interval each rational point $\frac{p}{q}$ by two rational points $\frac{p}{q}+$ and $\frac{p}{q}-$ (the boundary of $I_{p/q}$ if $k > 0$), without changing the distances between any two points. The interval thus modified will henceforth be denoted by A.

To generate the Cantor set, we first generate the rational numbers by means of a tree. The natural way to do that is to use the so-called Farey tree (which also plays an important role in the theory of renormalization of circle maps). The first few levels of the tree are drawn in figure 8.3.

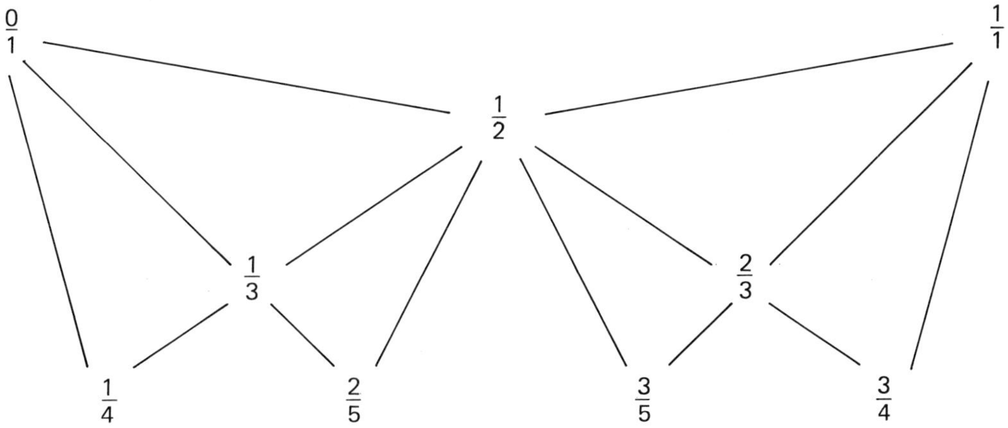

figure 8.3

Without further explanation, we give the rules here (see Hardy and Wright, 1924). The number $\frac{1}{2}$ is the rational between 0 and 1 with the smallest denominator. It is a neighbor of 0 and of 1 in the tree. Its level in the tree is 1. The numbers on level 2 are obtained similarly: between every two neighbors we find the number with the smallest possible denominator. The new pairs of numbers are 0 and $\frac{1}{3}$, $\frac{1}{3}$ and $\frac{1}{2}$, $\frac{1}{2}$ and $\frac{2}{3}$, $\frac{2}{3}$ and 1. One constructs the tree level by level defining the number with the smallest denominator between each pair of neighbors. Notice that each number in the tree now has a "direct parent" and an "indirect parent." In the sequel we will draw only the direct parents, so that a binary tree results (see figure 10.2).

More formally, one can show that $\frac{p}{q}$ on level n and $\frac{r}{s}$ on level $n - \ell$ are Farey neighbors if and only if

$$p \cdot s - q \cdot r = \pm 1. \tag{8.2}$$

Their "child" at level $n + 1$ is given by

$$\text{child} = \frac{p + r}{q + s}. \tag{8.3}$$

The tree enumerates all rational numbers.

The tree that generates the Cantor set can now be defined as follows. Let level 1 consist of

$$\Delta(0) = [0, \tfrac{1}{2}]$$

$$\Delta(1) = [\tfrac{1}{2}, 1] .$$

Level 2 consists of the lengths of

$$\Delta(00) = [\tfrac{0}{1}, \tfrac{1}{3}]$$
$$\Delta(10) = [\tfrac{1}{3}, \tfrac{1}{2}]$$
$$\Delta(11) = [\tfrac{1}{2}, \tfrac{2}{3}]$$
$$\Delta(01) = [\tfrac{2}{3}, \tfrac{1}{1}]$$

We have chosen the labelling so that the topology of the tree is the same as in figure 2.4. The presentation function will be similar to the one drawn in figure 3.1. The reason for this choice is that it is the most natural in terms of the continued fraction expansions, as will be discussed in the next section. We will denote this tree as the "Farey tree of the rational intervals."

IX   THERMODYNAMICS OF THE FAREY TREE

In this section the thermodynamics of the Farey tree of rational intervals is discussed. As we have seen in the previous section, this corresponds to the thermodynamics of the set A for the family f_ω given in (8.1) with $k = 0$ (rotations).

There are various reasons to treat this simple case. First of all, the problem can to a large extent be treated analytically and therefore yields insight into the more general case. Further, numerical calculations indicate that the scaling function for $k = 1$ resembles the scaling function for $k = 0$ very closely. Therefore one would expect the thermodynamics to be very similar. The third reason lies a little beyond the scope of this article. Summarized, it is the following. If k is smaller than 1 and if $\rho(\omega)$ assumes an irrational value (a Diophantine number), then (renormalized) high iterates of f_ω will more and more resemble (renormalized) high iterates of rotation by ρ. This suggests that scalings should be the same locally around those irrational values of $\rho(\omega)$. The case $k = 0$ has therefore more general information than one would suppose at first sight.

We first need to discuss some elementary but crucial number theory. Consider a real $\rho \in (0, 1)$ and define the continued fraction coefficients c_i of ρ as follows (see Hardy and Wright, 1924):

$$\rho = \cfrac{1}{c_1 + \cfrac{1}{c_2 + \cfrac{1}{\ddots}}} \overset{D}{=} [c_1, c_2, ...]..$$
(9.1)

The sequence $\{c_i\}$ is finite if and only if ρ is rational. The continued fraction approximate of ρ are given by

$$\frac{p_k}{q_k} \overset{D}{=} \cfrac{1}{c_1 + \cfrac{1}{c_2 + \ddots + \cfrac{1}{c_k}}} = [c_1, c_2, ..., c_k].$$
(9.2)

Note that without loss of generality we take $c_k \geq 2$. In matrix form this becomes

$$M \overset{D}{=} \begin{pmatrix} p_{k-1} & p_k \\ q_{k-1} & q_k \end{pmatrix} = \begin{pmatrix} 0 & 1 \\ 1 & c_1 \end{pmatrix} \cdots \begin{pmatrix} 0 & 1 \\ 1 & c_k \end{pmatrix},$$
(9.3)

as one can check by multiplying out the c_k's in equation (9.2).

The continued fractions are very "natural" in the following sense. If we consider

$$R_\rho(x) = (x + \rho) \bmod 1$$

then the sequence

$$\{R_\rho^{q_k}(x)\}_{k=1}^{\infty}$$

forms a sequence of <u>closest returns</u> to x. (That is, they approximate x better than $R_\rho^q(x)$ with $q < q_k$.)

We can now define a generating function for the Farey tree (the inverses will turn out to be the presentation function for the Farey tree of the rational intervals)

$$\begin{aligned} F_0 &: [c_1\, c_2 \, ... \,] \rightarrow [1 + c_1, c_2, ...] \\ F_1 &: [c_1\, c_2 \, ... \,] \rightarrow [1, c_1, c_2 \, ... \,] \end{aligned}$$
(9.4a)

By using the definition of c_k, one finds

$$F_0(x) = \frac{x}{1 + x}$$

$$F_1(x) = \frac{1}{1 + x}.$$
(9.4b)

As in section 8, we say that the number $\frac{1}{2}$ is in level 1 of the Farey tree; the numbers $\frac{1}{3}$ and $\frac{2}{3}$ form level 2 and so on. We will show that the n-th level of the Farey tree consists of the numbers

$$\bigcup_{\{\varepsilon_1\cdots\varepsilon_n\}} F_{\varepsilon_1} \ldots F_{\varepsilon_n}(1) = \text{level } n . \tag{9.5}$$

To start the reasoning one shows that p_{k-1}/q_{k-1} and p_k/q_k are Farey neighbors. This comes about because the determinant of the matrix M defined in (9.3) has determinant $(-1)^k$, which is equivalent to criterion (8.2) for being neighbors. To find out who the child of these two numbers is, use (8.3):

$$\text{child} = \frac{p_{k-1}+p_k}{q_{k-1}+q_k} = M\begin{pmatrix}1\\1\end{pmatrix} = M\begin{pmatrix}1 & 1\\0 & 1\end{pmatrix}\begin{pmatrix}0\\1\end{pmatrix}$$

$$= \begin{pmatrix}0 & 1\\1 & c_1\end{pmatrix}\cdots\begin{pmatrix}0 & 1\\1 & c_k\end{pmatrix}\begin{pmatrix}1 & 1\\0 & 1\end{pmatrix}\begin{pmatrix}0\\1\end{pmatrix}$$

$$= \begin{pmatrix}0 & 1\\1 & c_1\end{pmatrix}\cdots\begin{pmatrix}0 & 1\\1 & c_k+1\end{pmatrix}\begin{pmatrix}0\\1\end{pmatrix} = [\, c_1, \ldots, c_{k-1}, c_k+1] . \tag{9.6}$$

It follows from this that the "direct" parent (see section 8) of $[c_1 \ldots c_k]$ is $r/s = [c_1 \ldots, c_k-1]$. If $c_k > 2$, we have that

$$r/s \neq p_{k-1}/q_{k-1}.$$

Therefore, the second child of p_k/q_k is:

$$\text{child} = \frac{r+p_k}{s+q_k} = \begin{pmatrix}0 & 1\\1 & c_1\end{pmatrix}\cdots\begin{pmatrix}0 & 1\\1 & c_{k-1}\end{pmatrix}\left\{\begin{pmatrix}0 & 1\\1 & c_k-1\end{pmatrix}+\begin{pmatrix}0 & 1\\1 & c_k\end{pmatrix}\right\}\begin{pmatrix}0\\1\end{pmatrix}$$

$$= \begin{pmatrix}0 & 1\\1 & c_1\end{pmatrix}\cdots\begin{pmatrix}0 & 1\\1 & c_{k-1}\end{pmatrix}\begin{pmatrix}0 & 1\\1 & c_k-1\end{pmatrix}\begin{pmatrix}0 & 1\\1 & 2\end{pmatrix}\begin{pmatrix}0\\1\end{pmatrix} = [c_1 \ldots c_k-1, 2]. \tag{9.7}$$

In the case that $c_k = 2$ (it is not smaller than 2 according to (9.2)), one finds the two parents of $[c_1, \ldots, c_{k-1}, 2]$ and then calculates their children. Thus the above implies that the sum of the continued fraction coefficients increases by one as one goes from level n to level $n+1$. Since by (9.4a) we have

$$[c_1 \ldots c_k] = F_0^{c_1-1} F_1\, F_0^{c_2-1}F_1 \ldots F_0^{c_k-1}F_1(1), \tag{9.8}$$

equation (9.5) follows.

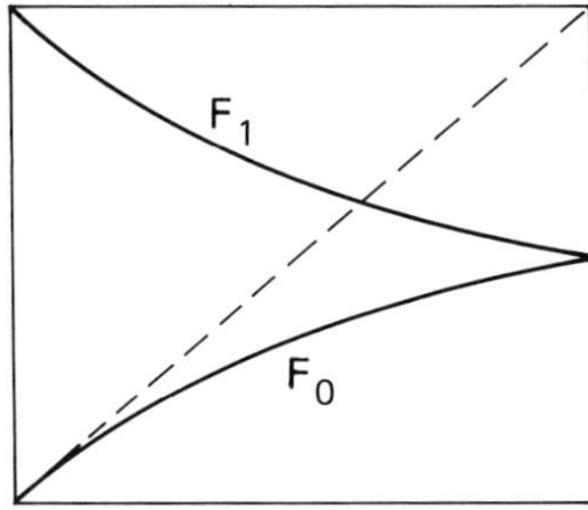

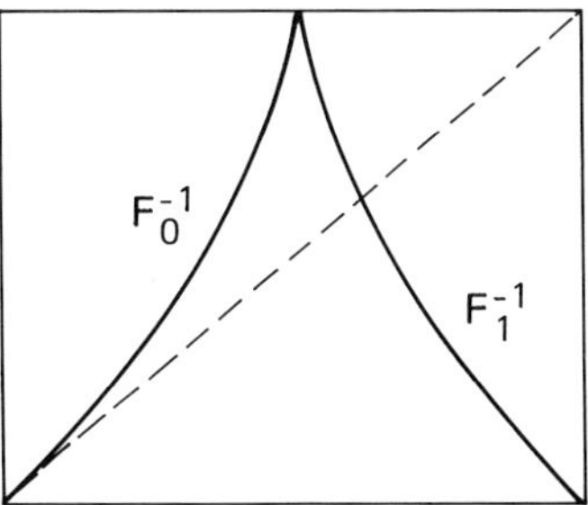

figure 9.1

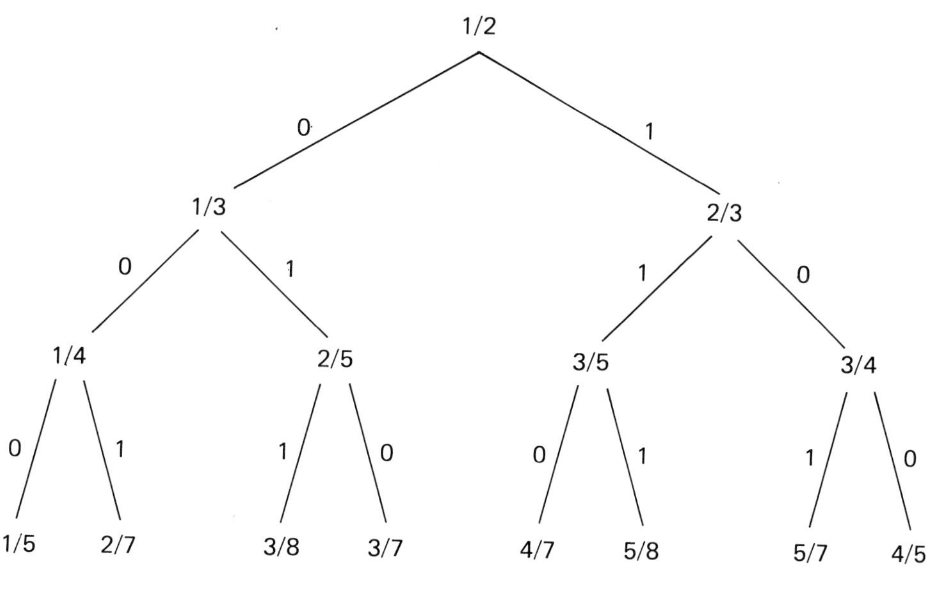

figure 9.2

In figures 9.1 and 9.2 we have drawn the presentation function and the Farey tree. Note that it is of the type of figure 2.4. In figure 9.3 the tree is drawn in its continued fraction representation.

Now all we have to do is to determine the Farey tree T of the rational intervals as defined in (8.4). However, it is not hard to see that the inverses of F_0 and F_1 are precisely the presentation function for T in the way this was described in section 3. (The endpoints of the $\Delta(\varepsilon_n \ldots \varepsilon_1)$ are the rationals which are generated by the F_ε.)

The intervals $\Delta(0\varepsilon_{n-1} \ldots \varepsilon_1)$ and $\Delta(1\varepsilon_{n-1} \ldots \varepsilon_1)$ on level n of T are bounded by the Farey neighbors and thus they must have lengths

     J.J.P. Veerman and M.J. Feigenbaum

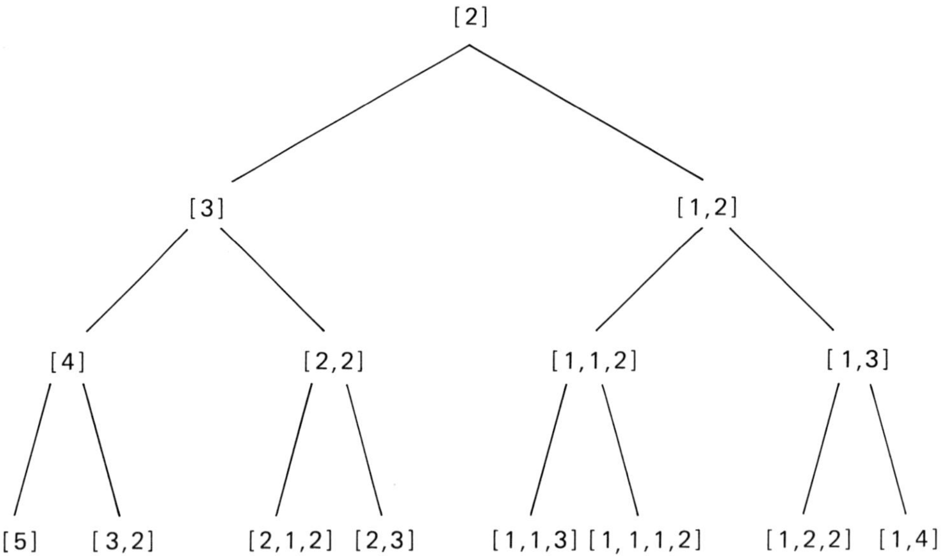

figure 9.3

$$A = |\,[c_1 \ldots c_k - 1] - [c_1 \ldots c_k]\,|$$

$$B = |\,[c_1 \ldots c_{k-1}] - [c_1 \ldots c_k]\,| \tag{9.9}$$

where $[c_1 \ldots c_k] = F_{\varepsilon_1} \ldots F_{\varepsilon_{n-1}}(1)$. (The expressions in (9.9) are valid even if $c_k = 2$ or $c_{k-1} = 1$.)

To calculate the length A and B, use (8.2) to see that the distance between two Farey neighbors equals the reciprocal of their two denominators. It follows from (9.3) that if q_k is the denominator of $[c_1 \ldots c_k]$ then

$$\prod_{i=1}^{k} c_i \leq q_k \leq \prod_{i=1}^{k} (c_i + 1) . \tag{9.10}$$

Using (9.9) one obtains:

$$\{c_k(c_k + 1) \prod_{i=1}^{k-1} (c_i + 1)^2\}^{-1} \leq A \leq \{(c_k - 1)\, c_k \prod_{i=1}^{k-1} c_i^2\,\}^{-1}$$

$$\{(c_k + 1) \prod_{i=1}^{k-1} (c_i + 1)\}^{-1} \leq B \leq \{c_k \prod_{i=1}^{k-1} c_i^2\,\}^{-1}$$

$$\Rightarrow \{\prod_{i=1}^{k} (c_i + 1)^{2\beta}\} \leq A^\beta + B^\beta \leq \{\prod_{i=1}^{k} c_i^{2\beta}\}^{-1} \cdot c_k^\beta ,$$

Forming the β-sum as in equation (6.6) one obtains

$$\sum_{\Sigma c=n} \prod_{i=1}^{k} (c_i + 1)^{-2\beta} \leq 2^{-nF(\beta)} \leq \sum_{\Sigma c=n} c_k \prod_{i=1}^{k} c_i^{-2\beta} \tag{9.11}$$

In the next section we will investigate the approximate thermodynamics.

We close this section with the important observation that our F_ε's do not satisfy (3.4), i.e., they are smooth, but they are not contractions. Our F_ε's have slopes smaller or equal to 1 (in absolute value) and equality is attained for both branches. It is not known how to generalize the proofs of the results of sections 4 to 7 for this case. Numerical results, however, consistently indicate that the thermodynamic formalism still holds. We will therefore proceed with the calculations.

X   APPROXIMATE THERMODYNAMICS

In order to solve for the thermodynamics approximately we will define an approximate thermodynamics. The new model then is (compare with (9.11)):

$$2^{-nF(\beta)} = \sum_{\substack{\{c\} \\ \Sigma c=n}}^{n} \prod_{i=1}^{k} \mu^{-\beta} c_i^{-2\beta} \tag{10.1}$$

for some fixed $\mu > 1$. Note that equation (6.2) now implies

$$h(\varepsilon_n \dots \varepsilon_1) = \sum_{i=1}^{k} (\ln \mu + \ln c_i)$$

where $\Sigma c = n$. If k is of order n, then $h(\varepsilon_n \dots \varepsilon_1)$ ("the energy") of the corresponding interval is big and therefore its contribution to the sum in (10.1) is small. One can compare this model with a model in which k particles have a distance c_i/n to each other. Each particle carries an energy $\ln \mu$ with it and furthermore the particles repel each other with a potential energy $\ln c_i$. (Repelling means that the "state" in which they are equidistant with small distance in between contributes very little to the sum for $\beta > 0$.)

Following equation (7.1), we define the grand canonical ensemble

$$e^{-G} = \sum_{\Sigma c=2}^{\infty} z^{(\Sigma c)} \sum_{\{c\}} \prod_{i=1}^{k} \mu^{-\beta} c_i^{-2\beta} \ . \tag{10.2}$$

Therefore

$$e^{-G} = -\frac{z}{\mu^\beta} + \sum_{\Sigma c=1}^{\infty} \sum_{\{c\}} \prod_{i=1}^{k} z^{c_i} \mu^{-\beta} c_i^{-2\beta} . \qquad (10.3)$$

Because of (10.1) for

$$| z \cdot 2^{-F(\beta)} | < 1 ,$$

the above series converges absolutely and we may therefore change the order of summation in (10.3). Summing first the terms of order $\mu^{-\beta}$ then $\mu^{-2\beta}$ and so on, one obtains

$$e^{-G} = -z\mu^{-\beta} + \sum_{k=1}^{\infty} \left\{ \sum_{n=1}^{\infty} z^n \mu^{-\beta} n^{-2\beta} \right\}$$

$$= \sum_{k=1}^{\infty} \left\{ \Lambda_\mu(z, \beta) \right\}^k - z\,\mu^{-\beta} = \frac{\Lambda}{1 - \Lambda} - z\,\mu^{-\beta} \qquad (10.4)$$

where

$$\Lambda_\mu = \sum_{n=1}^{\infty} \frac{z^n}{\mu^\beta n^{2\beta}} .$$

According to (7.4) and (7.5), the free energy is determined by the pole $[\lambda(\beta)]^{-1}$ of e^{-G}:

$$F(\beta) = -\ln \lambda(\beta) / \ln 2 , \quad \Lambda(\lambda(\beta), \beta) = 1 \qquad (10.5)$$

Since we know that the Hausdorff dimension of the set whose thermodynamics we are trying to calculate equals one (see section 9), we have the constraint:

$$\lambda(1) = 1 \ \Rightarrow \ \Lambda(1, 1) = \mu^{-1} \sum_{i=1}^{\infty} \frac{1}{n^2} = 1 .$$

Denote the Riemann zeta function $\sum_{i=1}^{\infty} n^{-s}$ by $\zeta(s)$, then this implies that

$$\mu = \zeta(2) = \pi^2/6.$$

Thus from (10.5) one concludes that the (approximate) thermodynamics is determined by

$$\sum_{n=1}^{\infty} [\lambda(\beta)]^{-n} n^{-2\beta} [\zeta(2)]^{-\beta} = 1. \qquad (10.6)$$

To investigate this thermodynamics, let us first consider the behavior of $\lambda(\beta)$ as β goes to $-\infty$. We claim that the asymptotic behavior of $\lambda(\beta)$ is given by:

$$\lambda(\beta) = \left[2 \sqrt{\zeta(2)} \right]^{-\beta} = [2.565...]^{-\beta}. \tag{10.7}$$

To verify this, one first observes that (10.7) implies that the n-th term C_n of the series is given by

$$C_n = \left[2 \sqrt{\zeta(2)} \right]^{n\beta} n^{-2\beta} [\zeta(2)]^{-\beta} = [2^{-n} n^2 \zeta(2)^{(1-n/2)}]^{|\beta|}.$$

One now checks that asymptotically as $\beta \to -\infty$

$$C_n \leq \delta^{n|\beta|}, \quad \text{for some } \delta < 1, \ \underline{\text{except}} \ C_2 = 1.$$

Using (10.5), one deduces that asymptotically

$$\sum_{\{\varepsilon\}_n} |\Delta(\varepsilon_n \ ... \ \varepsilon_1)|^\beta = 2^{-nF(\beta)} = [2.565...]^{-n\beta}.$$

If one reasons along the line of the remark below (10.1), it is clear that for the original model described in section 9, we expect the golden mean to dominate. That is, asymptotically as $\beta \to \infty$

$$\sum |\Delta(\varepsilon_n \ ... \ \varepsilon_1)|^\beta \sim [\sigma(11 \ ...)]^{n\beta}.$$

Here $\sigma(11 \ ...) = \dfrac{2}{3 + \sqrt{5}} \approx \dfrac{1}{2.618}$ is the derivative of F_1 at its fixed point. This differs less than 2% from the prediction of the approximate model.

We will now determine the nature of the free energy when β is very close to one. This requires some algebra.

$$\int_0^\infty \frac{x^{2\beta-1}}{e^{x-F} - 1} dx = \sum_1^\infty e^{nF} \int_0^\infty x^{2\beta-1} e^{-nx} \, dx \ . \tag{10.8}$$

By substituting $u = nx$

$$= \sum_1^\infty \frac{e^{nF}}{n^{2\beta}} \Gamma(2\beta),$$

which according to equation (10.6) $= \zeta(2)^{\beta} \, \Gamma(2\beta)$.   (10.9)

Now set $2\beta = 2 - \varepsilon.$

Note that the left hand side of (10.8) and (10.9) are infinitely differentiable with respect to β for $\beta > 0$. One thus (asymptotically as ε goes to zero) solves F from:

$$\int_0^{\infty} \frac{x}{e^{x-F} - 1} \, dx + \varepsilon C_1 \; = \; \zeta(2) \, \Gamma(2) + \varepsilon C_2 \, .$$

From (10.8) and (10.9) one then finds that

$$\int_0^{\infty} \left[\frac{x}{e^{x-F} - 1} - \frac{x}{e^x - 1} \right] dx \; = \; \varepsilon C.$$

Writing $e^x = 1 + u$, rationalizing the integrand and splitting the integral in two pieces:

$$\int_0^b \frac{\ln(1 + u)}{u} \, \frac{(1 - e^{+F})(1 + u)}{(1 + u)(e^{-F} - 1)} \, du + \int_b^{\infty} \ldots \, du \; = \; \varepsilon C.$$

Taking $-F \ll b \ll 1$, we see that the second integral does not depend on F, and its outcome is a constant K. In the remaining integral the terms $\dfrac{\ln(1 + u)}{u} (1 + u)$ cancel approximately. The result is an elementary integral which yields

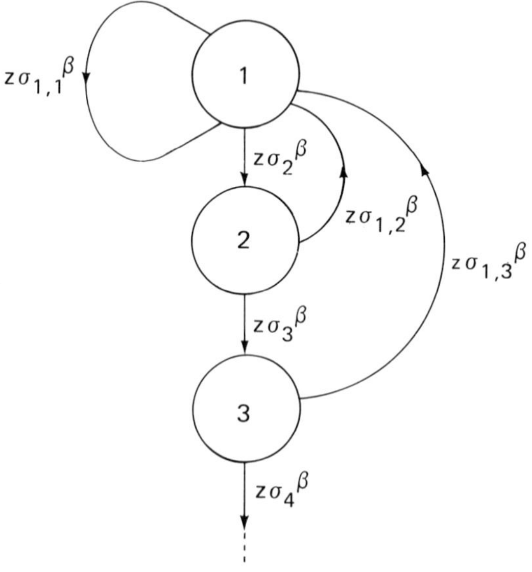

figure 10.1

$$(1 - e^F) \ln[(1 + u)e^{-F} - 1] \Big|_0^b + K = \varepsilon C.$$

It is straightforward (neglecting F^2 terms) to show that this implies that asymptotically

$$F \ln (-F) = (\beta - 1)C - K. \tag{10.10}$$

Therefore $F(\beta)$ has a branch point at $\beta = 1$.

It is also possible to draw the Markov diagram associated with (10.3). While we have not formally written down a transition matrix, it is not hard to see how the Δ's change from one level to the next. From (10.1) we have:

$$\sigma_n = \frac{|\Delta(c_k = n + 1)|}{|\Delta(c_k = n)|} = \left(\frac{n}{n + 1}\right)^2, \quad n \geq 2. \tag{10.11a}$$

When $c_{k+1} = 2$, one has

$$\sigma_{1,n} = \frac{|\Delta(c_k = n - 1, c_{k+1} = 2)|}{|\Delta(c_k = n)|} = \frac{n^2}{(n - 1) 2^2 \mu}. \tag{10.11b}$$

These describe all the possible transitions in this approximate model. The Markov diagram is given in figure 10.1. Note that the diagram is infinite. As remarked before, at $\beta = 1$ the term that contributes most is $c_k \to \infty$. This is reminiscent of Bose condensation in statistical physics.

To end this section, a remark about numerical work with circle maps. From the theoretical point of view taken in this work, it is appropriate for us to think about the δ's as iterates under the

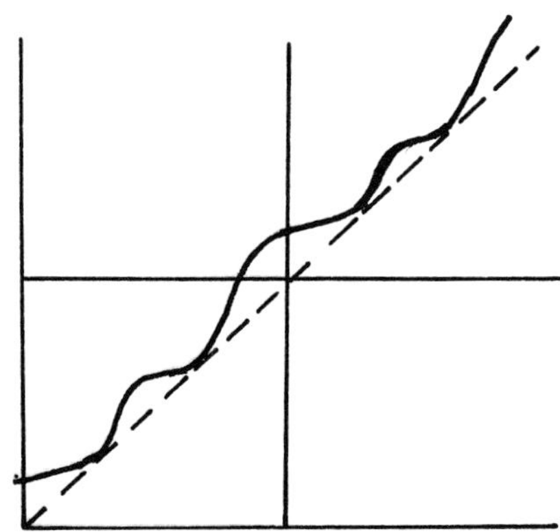

figure 10.2

presentation function as is done throughout this text. Since this presentation function in general is not given this would mean that one has to find the boundaries of the rotation intervals.

Numerically this is not a good scheme. The reason is that the boundaries are very hard to locate with good accuracy due to the marginally unstable character of the dynamics at such parameter values (see figure 10.2). To illustrate this point we have drawn a circle map at the upper boundary of the rotation interval $I_{1/2}$.

There is a much stabler method at hand. This is to find the midpoints $\omega(p/q)$ of the rotation interval $I_{p/q}$. This is dynamically very stable (derivative zero for critical maps). One can now set up the thermodynamics using only intervals $\delta(\epsilon_n \ldots \epsilon_1)$ whose endpoints are the midpoints $\omega(p/q)$ (see figure 10.3).

In this scheme, somewhat different from the foregoing, one takes out <u>only</u> one level of δ's (that is: the previous level δ's are first restored). The thermodynamics obtained from such a scheme is identical to the one discussed above. We leave as an ambitious exercise for the readers to convince themselves of this.

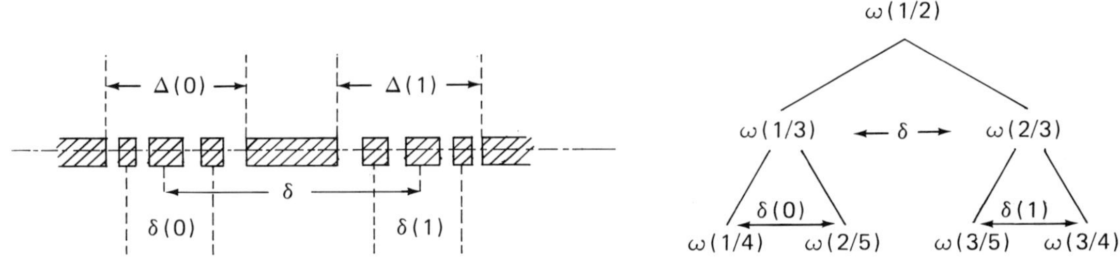

figure 10.3

REFERENCES

T. Bohr, D. Rand, *Physica* **25D**, 387-398, 1987.

R.L. Devaney, *An Introduction to Chaotic Dynamical Systems*, Benjamin/Cummings, 1986.

K.M. Falconer, *The Geometry of Fractal Sets,* Cambridge Univ. Press, 1985.

M.J. Feigenbaum, *J. Stat. Phys.*, **46**:5/6, 919-924, 1987.

M.J. Feigenbaum, *J. Stat. Phys.*, **46**:5/6, 925-932, 1987.

M.J. Feigenbaum, *J. Stat. Phys.*, **52**:3/4, 527-569, 1988.

G.M. Hardy, E.M. Wright, *An Introduction to the Theory of Numbers*, Oxford Univ. Press, 1960.

B.B. Mandelbrot, *The Fractal Geometry of Nature*, W.H. Freeman, 1982.

D. Ruelle, Thermodynamic Formalism, *Encyclopedia of Mathematics and its Applications*, **5**, Addison-Wesley, 1978.

D. Sullivan, *Differentiable Structures on Fractal Like Sets*, preprint CUNY, 1988.

ACKNOWLEDGMENTS

Both authors express their gratitude to the organizers of the Summer school for their hospitality. One of the authors (JJPV) is especially indebted to Eddie Cohen for his encouragement to participate in the Summer school and to write this article. He also acknowledges useful discussions with Donald Spear. While writing this work he was partially supported by DOE grant DE-AC-02-83-ER-13044. Finally, JJPV also expresses his admiration for the excellent research atmosphere in the Physics Department of Rockefeller University, of which he was a part for two years.

FUNDAMENTAL PROBLEMS IN STATISTICAL MECHANICS VII
H. van Beijeren, Editor
© Elsevier Science Publishers B.V., 1990

FRENKEL-KONTOROVA MODELS OF COMMENSURATE-INCOMMENSURATE PHASE TRANSITIONS

Robert B. Griffiths

Physics Department, Carnegie-Mellon University, Pittsburgh, PA 15213, U.S.A.

TABLE OF CONTENTS

INTRODUCTION

The material in this article was originally prepared for a series of lectures providing an introduction to Frenkel–Kontorova models, in the context of commensurate-incommensurate phase transitions, for advanced students. In writing up the notes I have tried to retain some of the freedom and informality which are possible in such a setting. Thus while not hesitating to replace what I actually said by what I think I should have said, I have not tried to write a review article. There is a heavy bias in Ch. 3, and to some extent in Ch. 4, on topics with which I and various collaborators have been heavily involved. Their inclusion, and the omission of many other topics equally pertinent to the subject matter, is not to be taken as an indication of what is and is not important in the field, but merely the fact that I can best serve students (and others) by discussing the things I understand the best.

Chapter 5, whose subsections have been numbered to correspond with the chapter numbers in the main text, indicates review articles and other literature giving further details on various topics taken up in this article. It can be considered a guide to the list of references at the end of the paper, and consequently I have been sparing in my use of references in the main text.

1. COMMENSURATE AND INCOMMENSURATE PHASES

1.1. Modulated Phases

An ordinary crystal is a structure which is periodic in space; in particular, it is invariant under translation by any one of three mutually independent primitive lattice vectors $\mathbf{a}_1$, $\mathbf{a}_2$, and $\mathbf{a}_3$. This means the Fourier transform has delta function peaks at wave vectors $\mathbf{q}$ of the form

$$\mathbf{q} = n_1\mathbf{b}_1 + n_2\mathbf{b}_2 + n_3\mathbf{b}_3 \qquad (1.1.1)$$

where the $\mathbf{b}_j$ are reciprocal lattice vectors and the n_j are integers. These wave vectors are the positions of the peaks observed in X-ray and neutron scattering experiments.

In some crystals diffraction peaks are observed not only at wave vectors of the form (1.1.1), but additional peaks occur at positions

$$\mathbf{q} = n_1\mathbf{b}_1 + n_2\mathbf{b}_2 + n_3\mathbf{b}_3 + m\mathbf{c} \ , \qquad (1.1.2)$$

where m is an integer and $\mathbf{c}$ is a modulation wave vector whose relationship to the $\mathbf{b}_j$ depends on temperature, pressure, etc. In such a modulated crystal or <u>modulated phase</u> there is an additional periodicity "competing" with the ordinary crystal lattice. If there are integers m_j, not all of which are zero, such that

$$m_1\mathbf{b}_1 + m_2\mathbf{b}_2 + m_3\mathbf{b}_3 + m_4\mathbf{c} = 0 \; , \qquad (1.1.3)$$

the modulated phase is called <u>commensurate</u>; otherwise it is <u>incommensurate</u>.

To be sure, there is no distinction in principle between a commensurate modulated phase and an ordinary crystal, since whenever (1.1.3) is satisfied one can choose a new set of linearly independent reciprocal lattice vectors $\mathbf{b}_1'$, $\mathbf{b}_2'$, $\mathbf{b}_3'$ such that any $\mathbf{q}$ of the form (1.1.2) can be written as a linear combination of the $\mathbf{b}_j'$ with integer coefficients. However, if $\mathbf{c}$ varies with temperature and pressure, the choice of the $\mathbf{b}_i'$ will also vary, and thus (1.1.2) can provide a more economical physical description of a series of commensurate phases in cases in which $\mathbf{c}$ varies smoothly, or in small steps, as a function of temperature or pressure.

More complicated modulated structures are possible in which more than four reciprocal lattice vectors are needed in order to give the position of diffraction peaks by a formula of the type (1.1.2). An analogous situation arises in quasicrystals; for example, six reciprocal lattice vectors are needed for an icosahedral quasicrystal. However, the relationship among these vectors does not vary with temperature and pressure, in contrast to modulated structures.

The simplest modulated structures are those in which $\mathbf{c}$ is simply a scalar multiple of one of the $\mathbf{b}_j$, say

$$\mathbf{c} = \beta\mathbf{b}_3 \; . \qquad (1.1.4)$$

In this case the modulation can be thought of as arising in the following way: some quantity which is constant in each plane of atoms perpendicular to $\mathbf{b}_3$ varies from plane to plane with a period $1/\beta$ in units of $\mathbf{a}_3$. An example is provided by the magnetic structure of various rare-earth metals which form hexagonal close-packed crystals.[1,2] Neutron scattering shows that the magnetization in a plane perpendicular to the hexagonal axis is constant (i.e., the elementary magnetic moments of all the atoms in a given plane are parallel to each other), but varies from plane to plane in a manner which depends on the temperature. The corresponding β in (1.1.4) is found to depend continuously on the temperature in some cases, whereas in other cases it remains constant, equal to some rational number, through some range of temperatures before jumping discontinuously to another rational value at a transition temperature.

If β in (1.1.4) is a rational number, one has a commensurate phase, while if it is irrational, the phase is incommensurate. Thus if β varies continuously there must be an infinitely rapid transition between commensurate and incommensurate phases. However, the fact that β is sometimes observed to "stick" at rational values over some range of temperatures suggests that there may be a fundamental difference between commensurate and incommensurate phases, and perhaps even in those cases in which it seems to vary continuously there might be a similar sticking phenomena which could be seen if the experimental resolution (etc.) were improved sufficiently. Could it be that β takes on only rational values in thermodynamic equilibrium? That is, can there exist true incommensurate phases? It is questions of this sort which make the study of modulated phases a fascinating and very challenging subject for both experiment and theory.

1.2. Statistical Models of Modulated Phases

Substances which exhibit modulated phases are usually rather complicated, and the various interactions which give rise to the observed structures are, at best, imperfectly understood. Hence it is not surprising that a lot of theoretical effort has been devoted to the study of simplified models. The hope is that by focusing on a small number of physical parameters and working out their consequences for modulated structures, one may gain some insight into what is and is not important in real materials.

This is a well-trodden path in statistical mechanics: e.g., the insights into phase transitions in real magnets obtained by studying the two-dimensional Ising model are well known. But, as this example suggests, even relatively simple models can pose formidable mathematical problems in terms of obtaining exact or even approximate solutions!

One of the best-known and most studied models of modulated phases is the axial next-nearest neighbor Ising (ANNNI) model.[9,10] For a three-dimensional cubic lattice the interactions have the following form. Chains of spins parallel to the z axis are coupled through a ferromagnetic nearest-neighbor interaction $J_1 > 0$ and an antiferromagnetic next-nearest-neighbor interaction $J_2 < 0$. In addition, there are ferromagnetic nearest-neighbor interactions J_0 in planes perpendicular to the z axis. The model has not been solved exactly, but quite a bit of information about the phase diagram has been obtained through series expansions, Monte-Carlo simulations, a mean-field approximation, and the like. Although many questions remain unanswered, it seems clear that this model can exhibit a wide range of commensurate and incommensurate phases, as a function of the temperature and the ratio $-J_2/J_1$, which resemble those in the rare earth metals in the sense that the average magnetization is constant in each plane perpendicular to the z axis, but varies from plane to plane. Along with these phases there

are, of course, phase transitions of various sorts, and the overall phase
diagram is extremely complicated.

Another model of modulated phases is the chiral clock model[9,10] in which
the Ising spin is replaced by a "clock" variable taking any one of $p \geq 3$
equally spaced directions in a two-dimensional spin space. On a cubic lattice
these "spins" are coupled ferromagnetically (so they tend to be parallel) in
planes perpendicular to the z axis, while the coupling between spins in
adjacent planes is "chiral": it favors a twist in the spin direction upon
going from one plane to the next. Unlike the ANNNI model, there are no next-
nearest-neighbor interactions in chains parallel to the z axis. Once again,
various methods have been used to investigate the phases and phase diagrams,
for various values of p, with results indicating similar phonomena to those
found in the ANNNI model.

1.3. Phenomenological Models of Modulated Phases

Given the complexity of even the simplest statistical models of modulated
phases, it seems worthwhile looking at phenomenological models of the
phenomena constructed in the same spirit as the well-known Landau approach
to second-order phase transitions.[3] Thus given the experimental evidence
that the magnetization is constant in planes perpendicular to the hexagonal
axis of a rare-earth metal, one might treat the average magnetizaion M_j in
the j'th plane as an independent variable, and suppose that there is a
phenomenological free energy of the form

$$F = \sum_j \left[\Phi_1(M_j) + \Phi_2(M_j, M_{j+1}) + \Phi_3(M_j, M_{j+1}, M_{j+2}) + \ldots \right] \quad (1.3.1)$$

where Φ_1, Φ_2, etc. are phenomenological free energies which depend in some
way on the temperature, pressure, etc.

Note that M_j is the sum of the magnetic moments of a very large number of
atoms lying in a particular plane. Hence it is plausible that one can treat
it as a classical (non-quantum) vector with well-defined magnitude and
direction. And as the main thermal fluctuations have already been taken
into account in calculating the M_j, it is plausible that the equilibrium
state of the crystal, so far as the magnetization of the different planes is
concerned, can be obtained by finding the global minimum of F over all
possible configurations $\{M_j\}$.

To be sure, if our goal is that of predicting the magnetic structure of a
particular rare earth metal at a particular temperature, (1.3.1) represents
no progress, for we do not know the phenomenological functions Φ_1, Φ_2, etc.
However, in a phenomenological approach the strategy is very different. We
look for simple functional forms for the Φ_j in (1.3.1), containing one or two
parameters, with the expectation that we can carry out the minimization

explictly as a function of these parameters. We can then "watch" how the
modulated phases depend on these parameters, and hope to see at least some
of the features exhibited by real materials in the laboratory, such as a
"locking" to a rational period when a parameter is varied. If this
approach is successful, we may gain some insight even if we do not know the
relationship between some model parameter and the temperature of (say)
erbium.

The task of minimizing an expression like (1.3.1), even numerically with
a computer, turns out to be a non-trivial problem, as we shall see. Hence
there is an advantage in looking at the simplest possibilities before going
on to more complicated cases. Instead of a three-component vector, let us
suppose that M_j is a real scalar. And since further neighbor interactions
only make the problem more complicated, let us suppose that Φ_j is zero for
all $j \geq 3$. (If, in addition, we set $\Phi_2 = 0$, the model is no longer
interesting, because minimizing F will yield an M_j independent of j, and
thus not a modulated phase.) We have then arrived at what might be called
the "Frenkel-Kontorova" class of models: a scalar variable for each j, and
an "interaction" which does not extend beyond neighboring planes. As we
shall see in the next section, the usual Frenkel-Kontorova model is of this
type (although the scalar variable cannot always be thought of as a
"magnetization").

The mean-field approximations to the ANNNI and chiral clock models
results in a minimization problem of the form (1.3.1). For the ANNNI model,
M_j is a scalar, but Φ_3 is non-zero. The same is true of the phenomenological frustrated ϕ^4 model of Janssen and Tjon.[5] For the chiral clock
model, Φ_3 and all further neighbor terms are zero, but M_j is a two-component
vector rather than a scalar.

Thus it appears that the Frenkel-Kontorova model is the simplest
phenomenological model of the type (1.3.1) capable of exhibiting both
commensurate and incommensurate phases, along with various transitions which
take place between them. And, not surprisingly, it is the model for which
the most detailed results are available, whether analytical or numerical.

2.   FRENKEL-KONTOROVA MODELS

   2.1. Notation and Examples

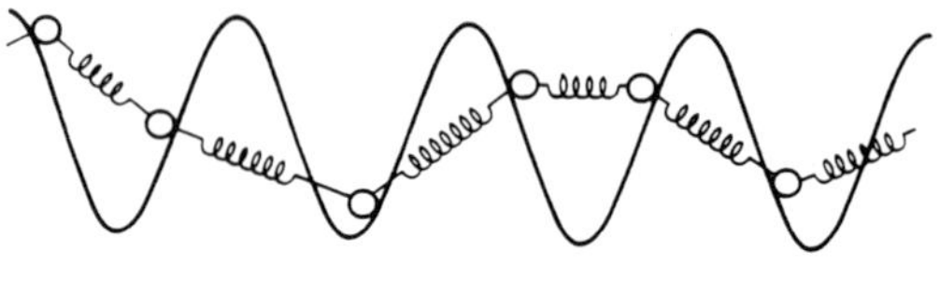

            FIGURE 2.1

A Frenkel-Kontorova model can be visualized as a set of particles or atoms in one dimension placed in a periodic potential and connected by springs, see Fig. 2.1. The total potential energy is thus of the form

$$H = \sum_n \left[V(u_n) + W(u_{n+1} - u_n) \right] \tag{2.1.1}$$

where u_n is the position of the n'th atom, $V(u)$ is the periodic potential, which we assume for convenience has period one,

$$V(1+u) = V(u) \ , \tag{2.1.2}$$

and $W(\Delta u)$ is the spring potential depending on the separation of neighboring atoms.

In what I shall call the <u>standard</u> <u>model</u>, because of its connection with the "standard map" §2.4 below), V and W are given by:

$$V(u) = K(1-\cos 2\pi u)/(2\pi)^2 \ , \tag{2.1.3}$$

$$W(\Delta u) = \frac{1}{2}(\Delta u)^2 - \sigma \Delta u \ , \tag{2.1.4}$$

where $K \geq 0$ is a measure of the "depth" of the periodic potential (the factors of 2π in (2.1.3) are chosen so that d^2V/du^2 is equal to K at the minimum), and σ is the "unstretched length of the spring," the value of Δu at which W is a minimum.

A second example is the chiral XY model in a magnetic field,[23] for which $V(u)$ is the same as (2.1.3), but W has the form

$$W(\Delta u) = [1-\cos 2\pi(\Delta u - \gamma)]/(2\pi)^2 \ . \tag{2.1.5}$$

As both V and W are periodic, one can think of $2\pi u_n$ as the angle (in radians) defining the value of the XY "spin" at the site n in a one-dimensional chain. A non-zero γ in (2.1.5) tends to favor a rotation of the angle going from one site to the next, whereas K in (2.1.3) is like a magnetic field which favors (in terms of minimum energy) an angle of zero.

A third example[25,26] arising in the study of magnetoelastic instabilities has

$$V(u) = Ku^2/2 \tag{2.1.6}$$

$$W(\Delta u) = -\frac{1}{2}(\Delta u - \gamma)^2 + \frac{1}{4}(\Delta u - \gamma)^4 \ . \tag{2.1.7}$$

Note that $V(u)$ is no longer periodic, and that $W(\Delta u)$, the potential for a very non-linear spring, has a double well, in contrast to the single well of (2.1.4).

The energy (2.1.1) is a particular case of the general nearest-neighbor interaction,

$$H = \sum_n K(u_{n+1}, u_n) \tag{2.1.8}$$

with

$$K(u_{n+1}, u_n) = V(u_n) + W(u_{n+1} - u_n) \ . \tag{2.1.9}$$

Note that alternative choices for K are possible, e.g., we could replace
$V(u_n)$ by $V(u_{n+1})$ on the right side of (2.1.9). More generally, (2.1.8) is
formally the same, except for "end effects," if K is replaced by

$$\bar{K}(u_{n+1}, u_n) = K(u_{n+1}, u_n) + f(u_{n+1}) - f(u_n) \tag{2.1.10}$$

where f is an arbitrary function. If V has the property (2.1.2), then K is
also periodic in the sense that

$$K(u, u') = K(1+u, 1+u') \ . \tag{2.1.11}$$

We shall also think of (2.1.8) as defining a (generalized) Frenkel-Kontorova
model.

We shall refer to the collection $\{u_n\}$ of positions of atoms (or angles in
example 2) for $-\infty < n < \infty$ as a <u>configuration</u>. A configuration is <u>periodic</u>
with period Q provided there are integers Q > 0 and P such that

$$u_{n+Q} = u_n + P \tag{2.1.12}$$

is valid for all n. The <u>average spacing</u> or "winding number" (the term is
used because of its association with cylinder maps, §2.4 below) of a
configuration is

$$\omega = \langle u_{n+1} - u_n \rangle = \lim_{(q-p) \to \infty} (u_q - u_p) / (q-p) \tag{2.1.13}$$

provided the limit exists for any sequence of pairs of integers p < q for
which (q-p) tends to infinity. Of course for a periodic configuration
(2.1.12), ω is equal to P/Q.

The presence of an integer P on the right side of (2.1.12) is rather
natural as long as (2.1.2) holds. However, it should be set equal to 0 for
the corresponding definition when applied to the second and third examples
above. In addition, ω is equal to zero in the second and third examples (at
least for the configurations of physical interest), though it is possible to
introduce other quantities which can play an analogous role.

2.2. Minimum Energy and Enthalpy Configurations; Ground States

In what follows we shall concentrate on the ground states of Frenkel-
Kontorova models, because it is these which correspond to the thermodynamic
equilibrium states of minimum free energy for the phenomenological models
introduced above in §1.3. The argument presented there in effect maps the
problem of a modulated phase at a <u>finite</u> temperature in a three-dimensional
system onto the <u>zero</u> temperature ground state of a one dimensional system,
as the free energy F in (1.3.1) corresponds to the Frenkel-Kontorova energy H

of (2.1.1) or (2.1.8). Similarly, a variation of temperature in the three dimensional model corresponds to changing the parameters (such as K and σ in (2.1.3) and (2.1.4))in the corresponding one-dimensional system at zero temperature. (To be sure, one can study the problem of a Frenkel-Kontorova model at a finite temperature by using a Boltzmann weight of $e^{-\beta H}$, but that is outside the scope of the present discussion.)

While the meaning of "ground state" or "minimum energy state" is unambiguous for finite systems, the same is not the case for a configuration $\{u_n\}$, $-\infty < n < \infty$, in a Frenkel-Kontorova model. The problem is that (2.1.1) and (2.1.8) do not, in general, make sense as infinite sums. However, they can still be used, in an obvious way, to define energy <u>changes</u> if only a <u>finite</u> number of the u_n are altered. Thus, following Aubry,[14,16] we can define a <u>minimum energy configuration</u> $\{u_n\}$ as one in which the energy change is always positive, or at least not negative, if any finite set of u's is altered. An equivalent definition is that, given any integers p < q, the quantity

$$H_{pq} = \sum_{j=p}^{q-1} K(u_{j+1}, u_j) \qquad (2.2.1)$$

is an absolute (not just a local) minimum as a function of the variables u_{p+1}, u_{p+2}, $\cdots$ u_{q-1}, with u_p and u_q held at the original values.

The energy per particle ε of a configuration $\{u_n\}$ is defined with the help of (2.2.1) by

$$\varepsilon = \lim_{(q-p) \to \infty} (q-p)^{-1} H_{pq} , \qquad (2.2.2)$$

that is, by evaluating the energy per particle for longer and longer segments, and taking the limit as the length of the segment becomes infinite. For ε to be defined, this limit must exist no matter which sequence of pairs of integers p < q is used to evaluate the right hand side. It is easy to construct examples of configurations for which ε is not defined.

While it can be shown that ε is defined for a minimum energy configuration under suitable restriction on K(u,u') in (2.1.8), this value need <u>not</u> be the minimum possible value of ε, as shown by the following example. If $\sigma = 0$ in (2.1.4), then both V in (2.1.3) and W are non-negative, and the minimum value of ε over all configurations is 0, which is its value when $u_n = 0$ for all n. On the other hand the configuration

$$u_n = n \qquad (2.2.3)$$

has $\varepsilon = 1/2$, and nonetheless it is a minimum energy configuration, as one can verify using (2.2.1). The reason is that while the spring energy W for two adjacent atoms can be lowered by bringing them together, doing so will inevitably elongate other springs, as long as only a finite number of atoms can be displaced, and so there is a net energy increase.

Since a minimum energy configuration need not have the minimum possible energy per particle, it is convenient to introduce an alternative minimum enthalpy condition.[20] In thermodynamics the enthalpy is the energy plus pressure-times-volume. For a one dimensional system we can write

$$\text{Enthalpy} = \text{Energy} - \sigma L, \qquad\qquad (2.2.4)$$

where σ is the stress or tension (and thus minus the "pressure"), and L is the length. In a finite system of N particles with $u_1 \leq u_2 \leq \ldots u_N$, L is $u_N - u_1$, and hence the last term in (2.2.4) is

$$-\sigma(u_N - u_1) = -\sigma\sum_n (u_{n+1} - u_n) \ . \qquad\qquad (2.2.5)$$

Thus to go from energy to enthalpy requires only a simple alteration in $W(\Delta u)$ in (2.1.1); note that this alteration already appears in (2.1.4), so that for the case of the standard model, (2.1.1) as we have written it can already be called an "enthalpy".

Assuming that (2.1.2) or, more generally, (2.1.11) holds, and that the stress term (2.2.5) has been incorporated into $K(u,u')$, we shall say that $\{u_n\}$ is a <u>minimum</u> <u>enthalpy</u> configuration provided H_{pq} in (2.2.1) will not decrease when u_{p+1}, u_{p+2}, $\ldots u_{q-1}$ are altered in any way <u>and</u> u_q <u>is</u> <u>changed</u> <u>by</u> <u>adding</u> <u>an</u> <u>arbitrary</u> <u>integer</u>. Altering u_q by an integer can be thought of, intuitively, as a process in which every u_k for $k > q$ is altered by the same integer as u_q. Because K is periodic, (2.1.11), the energy of this "tail" of the configuration $\{u_n\}$ remains unaltered. (We could also, of course, allow u_p to change by an integer amount; the resulting definition is obviously the same.)

The enthalpy per particle, η, is defined by (2.2.2), assuming the limit exists, in the case where the stress term (2.2.5) has been incorporated into $K(u,u')$. It is the same as ε when σ is zero, and more generally the two are related through the formula

$$\eta = \varepsilon - \omega \sigma \qquad\qquad (2.2.6)$$

We are now ready to comment on a major difference between Aubry's approach to the study of Frenkel-Kontorova models and that presented in Chapter 3 below. Aubry treats ω as an independent variable: the minimum energy configuration depends on ω, and so does its energy per particle, $\varepsilon(\omega)$, with "energy" defined by (2.2.2) with $\sigma = 0$. On the other hand, I use σ as an independent variable, and calculate $\eta(\sigma)$, the enthalpy per particle of the corresponding minimum enthalpy configuration. (Roughly speaking, the two approaches correspond to using the constant volume and constant pressure ensembles, respectively, in statistical mechanics.) The functions $\varepsilon(\omega)$ and $\eta(\sigma)$ are related to each other by Legendre transformations in a manner analogous to energy and enthalpy in thermodynamics; in particular the formulas

$$\eta(\sigma) = \min_{\omega} \; [\varepsilon(\omega) - \omega\sigma] \qquad (2.2.7)$$

$$\varepsilon(\omega) = \max_{\sigma} \; [\eta(\sigma) + \omega\sigma] \qquad (2.2.8)$$

make explicit what is meant by (2.2.6), and show that

$$\eta = d\varepsilon/d\omega , \quad \omega = -d\eta/d\sigma \qquad (2.2.9)$$

provided the derivatives exist.

While the distinction between energy and enthalpy is useful for the first example of §2.1, it is irrelevant for the second and third examples, since $\omega = 0$.

Aubry defines a <u>ground</u> <u>state</u> to be a <u>recurrent</u> minimum energy configuration.[14,16] One can define a recurrent configuration $\{u_n\}$ to be one with the property that given any integer p and any number $\varepsilon > 0$, there are integers m and q > 0 such that both inequalities

$$\begin{aligned}
\left| u_{p+q} - u_p - m \right| &< \varepsilon \\
\left| u_{p+q+1} - u_{p+1} - m \right| &< \varepsilon
\end{aligned} \qquad (2.2.10)$$

are satisfied. Intuitively, the idea is that any part of a configuration will reappear again, within an arbitrarily close approximation, later in the same configuration. The use of an integer m in these inequalities reflects the fact that V or K is periodic, (2.1.2) or (2.1.11). Aubry's own definition of recurrence makes use of a cylinder map (§2.4 below). It is obvious that a periodic minimum energy configuration, see (2.1.12), is recurrent and is therefore a ground state.

A simple illustration of the difference between a ground state and a general minimum energy configuration is provided by a ferromagnetic Ising chain in which each u_n can take on only the values +1 and −1, and

$$K(u_{n+1},u_n) = -Ju_n u_{n+1} \qquad (2.2.11)$$

with J > 0. Configurations (a) and (b) in Fig. 2.2 are minimum energy configurations, and as they are periodic, they are ground states. Configuration (c), on the other hand, is a minimum energy configuration which is <u>not</u> recurrent. Finally, configuration (d) is not a minimum energy configuration, since changing the − to + lowers the energy.

(a) $+ + + + + + + + +$

(b) $- - - - - - - - - -$

(c) $+ + + + + - - - - -$

(d) $+ + + + - + + + + +$

FIGURE 2.2
Ferromagnetic Ising chain: (a), (b),
(c) are minimum energy configurations;
(a) and (b) are ground states.

2.3. Convex W: Aubry's Theorems

The results presented below hold for cases in which H is of the form
(2.1.1), V is periodic, (2.1.2), V and W and their first and second
derivatives are continuous, and W is strictly convex in the sense that there
is a constant c such that

$$W''(\Delta u) \geq c > 0 . \tag{2.3.1}$$

Alternatively, in the more general case (2.1.8), K must satisfy (2.1.11), be
twice continuously differentiable, bounded below, and satisfy the condition

$$\partial^2 K(u,u')/\partial u \partial u' \leq - c < 0 . \tag{2.3.2}$$

(Some of these conditions can be relaxed to some extent; we refer the reader
to Aubry's papers for details.)

Using these conditions, Aubry has established a number of mathematical
theorems.[14-16] The results which are most important for our purposes are the
following:

A1. Every minimum energy configuration has a well-defined average spacing
or "winding number" ω; that is, the limit (2.1.13) exists.

A2. For every ω there is at least one ground state (recurrent minimum
energy configuration).

A3. The quantity ω is a continuous function of σ. (It is also a
continuous function of certain other parameters on which $K(u,u')$ depends
continuously. I am, unfortunately, unable to give a statement which is
at the same time precise and concise.)

A4. If ω is a rational number P/Q, where Q > 0 and P are integers with no
common factors, any ground state configuration is periodic, i.e., (2.1.12)
is satisfied, with period Q.

A5. If ω is <u>irrational</u>, there is an <u>envelope</u> or <u>hull</u> <u>function</u> $f_\omega(x)$ with
the following properties.

i) The function f_ω is monotone strictly increasing

ii) It satisfies the relation

$$f_\omega(1+x) = 1 + f_\omega(x) ; \tag{2.3.3}$$

equivalently, $f_\omega(x)-x$ is a periodic function of x with period 1.

iii) If $f_\omega(x)$ is continuous, any ground state configuration $\{u_n\}$ with average spacing ω is of the form

$$u_n = f_\omega(n\,\omega + \alpha) \tag{2.3.4}$$

for some choice of α. Conversely, for any choice of α, (2.3.4) defines a ground state configuration.

iv) If $f_\omega(x)$ is discontinuous at some point x_o, it is also discontinuous at all values of x of the form

$$x = x_o + p + \omega\,q \tag{2.3.5}$$

where p and q are any integers. Let f_ω^+ and f_ω^- be the right-continuous and left-continuous versions of f_ω (see below). Then every ground state configuration is either of the form

$$u_n = f_\omega^+(n\,\omega + \alpha) \tag{2.3.6}$$

or of the form

$$u_n = f_\omega^-(n\,\omega + \alpha) \tag{2.3.7}$$

for some α. Conversely, for any choice of α, (2.3.6) defines a ground state configuration, and the same is true of (2.3.7).

v) The function f_ω is unique in the following sense: given any $\hat{f}_\omega$ with the same properties, there is a real number β such that

$$\hat{f}_\omega(x) = f_\omega(x+\beta) \ . \tag{2.3.8}$$

(See below for comments on this in the case where f_ω is discontinuous.)

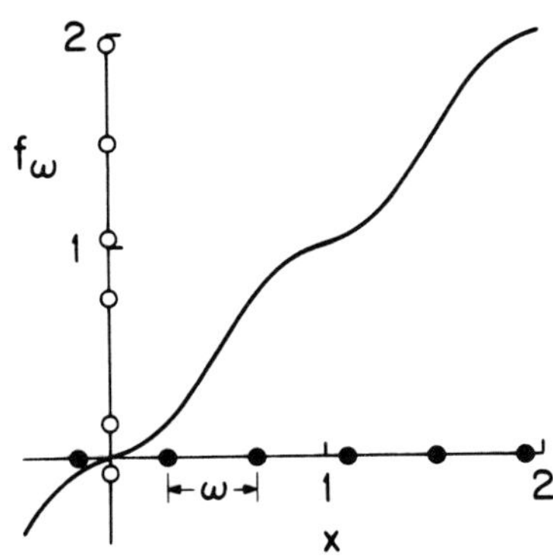

FIGURE 2.3
A continuous hull function (schematic).
The open circles are the values of f_ω
corresponding to the values of x
indicated by the filled circles.

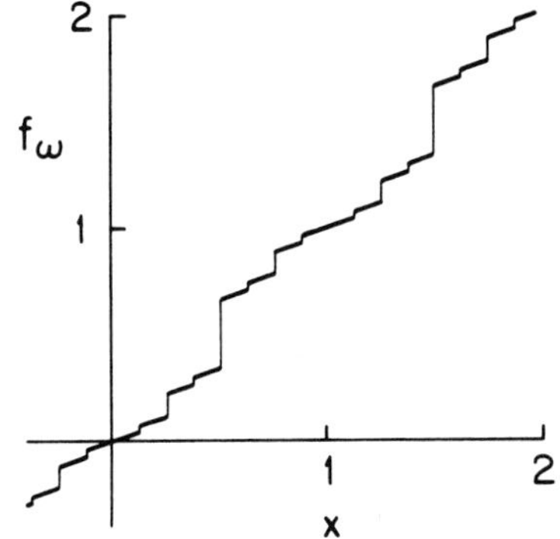

FIGURE 2.4
A discontinuous hull function
(schematic). Only a few of the
discontinuities are large enough
to be visible.

While all of these results are important for the study of Frenkel-

Kontorova models, it is A5 which is in some ways the most interesting, but also the hardest to understand. Thus it is worthwhile exploring what it means in a bit of detail.

To begin with, let us suppose that f_ω is continuous, as illustrated schematically in Fig. 2.3. Here a series of solid dots have been placed on the x axis at a constant separation ω. The corresponding values of $f_\omega(x)$ are denoted by open circles on the f_ω axis; these give the positions of successive atoms in the ground state.

The case in which f_ω is discontinuous is harder to visualize because the places where it is discontinuous form a dense set on the x axis, since ω is irrational. That is, arbitrarily close to any point in the x axis, f has a discontinuity. I have tried to represent this schematically in Fig. 2.4 where vertical lines have been drawn at the points where f ashows large discontinuities (most of the discontinuities are, of course, very small).

Wherever a discontinuity occurs, the definition of f_ω is somewhat ambiguous. Mathematically a function can only be assigned a single value at this point, and f_ω^+ and f_ω^- are defined by taking the extreme possibilities. In particular, f_ω^+ takes the value at the upper end at each point of discontinuity, the solid circle in Fig. 2.5, and f_ω^- the value at the lower end, the open circle in this figure. Equivalently, $f_\omega^+(x)$ is the limit of $f_\omega(x')$ as x' decreases to x, and $f_\omega^-(x)$ the limit of $f_\omega(x')$ as x' increases to x.

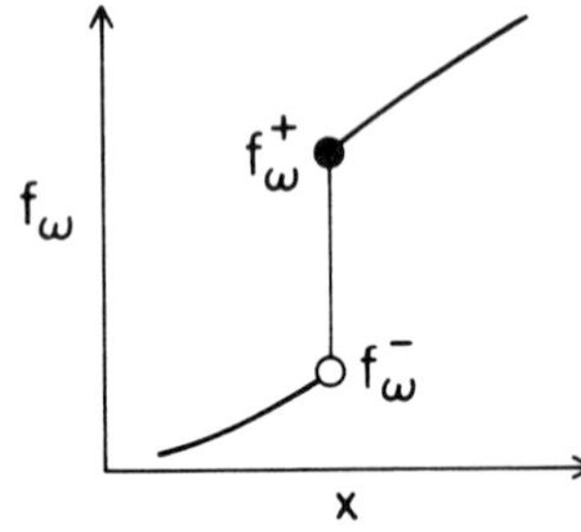

FIGURE 2.5
The functions f_ω^+ and f_ω^- are defined as shown here at points where f_ω is discontinuous.

Note that while f_ω is discontinuous at a countably infinite set of points, it is continuous, and $f_\omega^+(x)$ is equal to $f_\omega^-(x)$, at an uncountably infinite, and thus much larger set of points. In addition, both f_ω^+ and f_ω^- can be constructed, via the limiting process just described, with x' restricted to the points where f_ω is continuous. Hence, even though f_ω^+ and f_ω^- are distinct functions, it is best to think of them as the right- and left-continuous versions of the "same" function f_ω , and it is in this sense that A5(v) applies in the discontinuous case: one can place a superscript + (or a superscript -) on both sides of (2.3.8).

The calculation of the function f_ω is, in general, a difficult numerical problem. The usual procedure is to compute a succession of periodic configurations whose winding numbers form a better and better approximation to the desired ω, and use these configurations to approximate f_ω. Nonetheless, the very existence of this function has non-trivial consequences. One of these is that in a given unit interval of the u axis, say from one peak to the next in Fig. 2.1, there are either $\text{Int}(1/\omega)$ or $1 + \text{Int}(1/\omega)$ atoms in a ground state configuration, where $\text{Int}(y)$ is the largest integer not exceeding y. Another consequence is that given two distinct ground states $\{u_n\}$ and $\{v_n\}$ with the same winding number, either $u_n < v_n$ for all n, or else $v_n < u_n$ for all n.

2.4. Convex W: Area Preserving Twist Maps

We assume that H is of the form (2.1.8) with K given by (2.1.9). An _equilibrium configuration_ $\{u_n\}$ (which can be stable or unstable) is defined by the property that the force on each atom is zero:

$$\frac{\partial H}{\partial u_n} = \frac{\partial K(u_{n+1}, u_n)}{\partial u_n} + \frac{\partial K(u_n, u_{n-1})}{\partial u_n} = 0 \ . \tag{2.4.1}$$

Let the quantity p_n be defined by

$$p_n = \frac{\partial K(u_n, u_{n-1})}{\partial u_n} = W'(u_n - u_{n-1}) \ . \tag{2.4.2}$$

It is the tension in the spring pulling atom n to the left, Fig. 2.6. In order that the net force on this atom be zero, it is necessary that the tension p_{n+1} in the spring pulling to the right compensate for p_n and for the force due to the periodic potential V, and thus that

$$p_{n+1} = p_n + V'(u_n) \ , \tag{2.4.3}$$

an equation which is equivalent to (2.4.1). The position u_{n+1} of the next atom must, therefore, be chosen to yield the spring tension (2.4.3); that is, it is determined by solving

$$p_{n+1} = W'(u_{n+1} - u_n) \ , \tag{2.4.4}$$

see (2.4.2), for u_{n+1}.

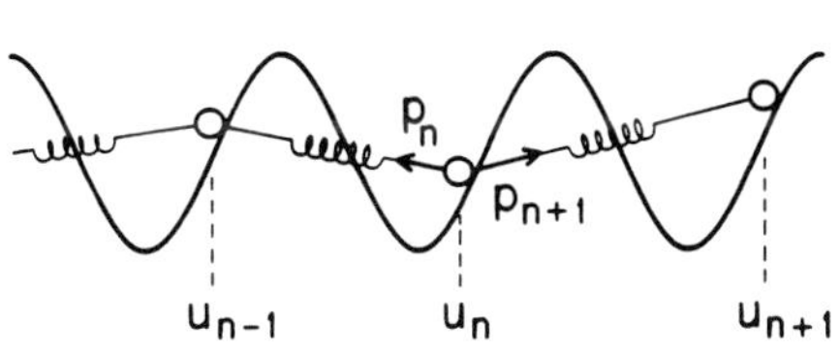

FIGURE 2.6
Given the position u_n of the n'th atom and the tension p_n of the spring to its left, the balance of forces determines p_{n+1} and thus the position u_{n+1} of the next atom.

Thus we see that if u_n and p_n are given, the condition that the net force on atom n be zero specifies a unique u_{n+1} and p_{n+1}, assuming that (2.4.4) has a solution and that this solution is unique. The existence and uniqueness of this solution is ensured by the strict convexity condition (2.3.1) or the analogous (2.3.2). Thus there is a well defined mapping

$$(u_{n+1}, p_{n+1}) = T(u_n, p_n) \tag{2.4.5}$$

of the real plane $\mathcal{R}^2$ onto itself. It can be shown that if V and W (or K) are continuously differentiable an appropriate number of times, that T is continuous (and continuously differentiable), has a continuous inverse, preserves areas, and satisfies the "twist condition":

$$(\partial u_{n+1}/\partial p_n)_{u_n} > 0 \ . \tag{2.4.6}$$

The property of preserving area follows from the fact that the Jacobian of (2.4.5), $\partial(u_{n+1}, p_{n+1})/\partial(u_n, p_n)$, is equal to one.

If $V(u)$ is periodic, (2.1.2), or, more generally, if (2.1.11) is satisfied, one can introduce a quantity

$$\tilde{u} = u \,(\mathrm{mod}\ 1), \ 0 \leq \tilde{u} < 1 \ , \tag{2.4.7}$$

which is best thought of as lying on a circle S^1 (the unit interval with periodic boundary conditions), and a map

$$(\tilde{u}_{n+1}, p_{n+1}) = \tilde{T}(\tilde{u}_n, p_n) \tag{2.4.8}$$

which carries the cylinder $S^1 \times \mathcal{R}$, Fig. 2.7, onto itself. One can think of $\tilde{T}$ as equivalent to T when the real plane $\mathcal{R}^2$ is "rolled up" onto the cylinder $S^1 \times \mathcal{R}$. The twist condition (2.4.6) now has the geometrical interpretation that under the action of $\tilde{T}$ points further up on the cylinder, on a given vertical line, move further to the right (in the direction of increasing $\tilde{u}$) than do those further down.

Area preserving twist maps have been extensively studied as examples of

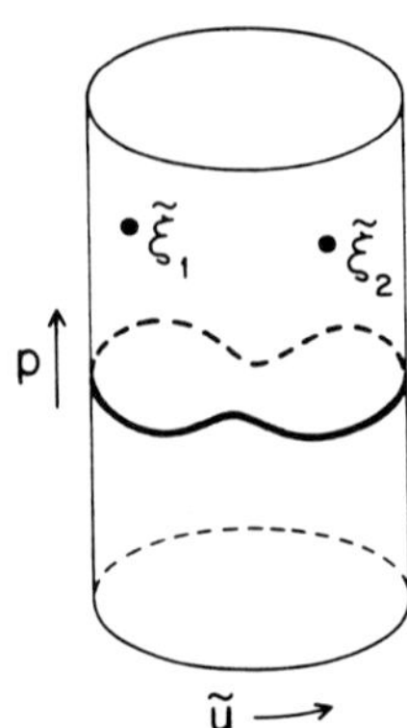

FIGURE 2.7

A portion of the cylinder $(\tilde{u}, p)$ which is mapped by $\tilde{T}$ onto itself. The points $\tilde{\xi}_1$, $\tilde{\xi}_2$ form an orbit of period 2. The curve encircling the cylinder is an invariant "circle" (KAM curve) mapped by $\tilde{T}$ onto itself.

simple non-linear dynamical systems. In particular, there have been many studies of the "standard map"

$$u_{n+1} = u_n + p_n + (K/2\pi)\sin(2\pi u_n) \tag{2.4.9}$$

$$p_{n+1} = p_n + (K/2\pi)\sin(2\pi u_n) \tag{2.4.10}$$

corresponding to equilibrium configurations of the "standard" Frenkel-Kontorova model defined by (2.1.3) and (2.1.4). Note that the stress σ in (2.1.4) cancels out of the equilibrium conditions (2.4.1), so that it is absent from the map T or $\tilde{T}$.

Given any point $\xi_O = (u_O, p_O)$ in the real plane, successive applications of T and T^{-1} produce an <u>orbit</u> $\{\xi_n\}$ of the map with the property that

$$\xi_{n+1} = T\xi_n \tag{2.4.11}$$

for every n. We shall denote the corresponding orbit on the cylinder by $\{\tilde{\xi}_n\}$. If $\{\xi_n\}$ is known, the corresponding equilibrium configuration $\{u_n\}$ is also known. Conversely, if we know $\{u_n\}$ and the function $K(u,u')$, $\{\xi_n\}$ can be reconstructed using (2.4.2).

An orbit on the cylinder is <u>periodic</u> with period $Q > 0$ provided

$$\tilde{\xi}_{n+Q} = \tilde{\xi}_n \tag{2.4.12}$$

for every n. It is <u>recurrent</u> if for every n and every $\varepsilon > 0$ there is some $P > 0$ (which can depend on n and ε) such that

$$\|\tilde{\xi}_{n+P} - \tilde{\xi}_n\| < \varepsilon \ , \tag{2.4.13}$$

where

$$\|(\tilde{u}, p)\| = [v^2 + p^2]^{1/2} \tag{2.4.14}$$

with v the minimum of $\tilde{u}$ and $1 - \tilde{u}$.

Since minimum energy configurations are particular examples of equilibrium states, they can be thought of as orbits of T or $\tilde{T}$. In particular, a periodic configuration in the sense of (2.1.12) corresponds to a periodic orbit of $\tilde{T}$, and the definition of a recurrent configuration in (2.2.10) is chosen to coincide with that of a recurrent orbit of $\tilde{T}$.

However, it should be kept in mind that minimum energy configurations (including ground states) represent only a fraction of the orbits which exist on the cylinder, and (depending on the form of V and W) this is sometimes a very small fraction of the total. Also note that ground states of all different winding numbers ω occur simultaneously on the cylinder. Because of the twist condition, ground states with larger ω tend to lie higher up (a larger values of p) than ground states with smaller ω. This is intuitively reasonable, as a larger average spacing will mean a larger tension in a typical spring.

    2.5. Commensurate and Incommensurate Phases

    We shall call the ground state of a Frenkel-Kontorova model "commensurate"
if it is periodic in the sense of (2.1.2), and "incommensurate" if it is not
periodic. Let us now examine whether (or to what extent) these definitions
coincide with those in Sec. 1.1 based upon the Fourier transform of the
structure.

    In terms of the discussion in Sec. 1.3, one could regard u_n as
representing the average magnetization in the n'th plane of atoms in a
magnetic crystal, or as some analogous quantity in some other type of
modulated crystal. Such an identification is problematical in the first
example of §2.1 because the u_n will, in a typical case, tend to infinity with
n, so for this example one needs to make the corresponding identification
using $\tilde{u}_n$, see (2.4.7), which lies between 0 and 1.

    Then the Fourier transform of interest will have the form (where we omit
various irrelevant factors)

$$\phi(k) = \sum_n \Lambda_n e^{ikna} \tag{2.5.1}$$

where $k = q_3$ in the notation of Sec. 1.1, $\underline{a}$ is the (average) separation of
neighboring planes of atoms, and

$$\Lambda_n = \Lambda(u_n) \tag{2.5.2}$$

is some function of u_n, which for the first example of §2.1 should be
assumed to be periodic,

$$\Lambda(u+1) = \Lambda(u) . \tag{2.5.3}$$

    Let us first consider the case in which the configuration $\{u_n\}$ is
periodic with period Q, (2.1.12), in which case the Λ_n are also periodic,

$$\Lambda_{n+Q} = \Lambda_n . \tag{2.5.4}$$

As a consequence, $\phi(k)$ in (2.5.1) consists of a set of δ functions located at

$$k = (2\pi/a)m/Q \tag{2.5.5}$$

where m denotes any integer. With

$$|\mathbf{b}_3| = 2\pi/a , \tag{2.5.6}$$

we can set β in (1.1.4) equal to $1/Q$ or, more generally, to P/Q, where P is
an integer, and P and Q are relatively prime.

    Next assume that $\{u_n\}$ is not periodic, but is given by a hull function
(2.3.4) for some irrational ω. Formally, (2.3.4) can be thought of as
defining u_n for all real values of n and not just integers, and consequently

Λ_n is defined for all real n using (2.5.2). In view of (2.3.3) and (2.5.3), this extended function Λ_n is periodic in n with period $1/\omega$, and hence can be expanded as a Fourier series:

$$\Lambda_n = \sum_\ell \lambda_\ell e^{i2\pi\ell n\omega} \tag{2.5.7}$$

(where of course the the λ_ℓ depend on ω). One comes to the same conclusion for the second and third examples of §2.1 in cases in which the u_n are given by (2.3.4) with f_ω a periodic function of period 1.

Inserting (2.5.7) in (2.5.1) yields a $\phi(k)$ consisting of a set of δ functions located at

$$k = (2\pi/a)(m+\ell\omega) \tag{2.5.8}$$

where m and ℓ are any two integers. Consequently we may identify β in (1.1.4) with ω, and as ω is irrational, the corresponding structure is incommensurate in the sense employed in Sec. 1.1.

To be sure, a hull function (2.3.4) has been proved to exist only under rather specific conditions, and this leaves open the possiblity that there may be non-periodic ground states which are not of this form. What Aubry's theorems (§2.3) show is that there are at least some models of the Frenkel-Kontorova type (those with convex W) which exhibit both commensurate and incommensurate phases in the sense of Sec. 1.1, and transitions from one to the other as σ as varied.

3. MINIMIZATION EIGENVALUE METHOD

3.1. Methods for Finding Ground States

In general it is a difficult task to find the ground state of a Frenkel-Kontorova model (2.1.1). Only in special cases have solutions been obtained in closed form, and thus numerical procedures are usually necessary. The usual strategy is to study periodic ground states, and to approximage non-periodic states (if they exist) with periodic states.

One approach is to look for periodic orbits of the $\tilde{T}$ map (Sec. 2.4) corresponding to equilibrium configuration. This method is limited to cases in which W is convex (so that a map exists). Its main disadvantage is that there may be several different orbits with a given winding number ω, and there is no "signature" which can unambiguously identify one of these as the ground state.

A second approach is to start with an approximate ground state and then refine it. For a periodic state the number of equilibrium equations of the form (2.4.1) is finite, and numerical techniques are available for finding a better solution once an approximate solution is available. The difficulty

with such methods is always that of finding a good first approximation. One
strategy is to slowly vary a parameter (such as K in the standard model),
starting at a value for which the ground state is already known, and using
the previous ground state as a first approximation every time the parameter
is altered. If the parameter changes are sufficiently small, one can hope to
"track" the ground state in this manner, but if, as sometimes happens, there
is a "phase transition" in which the ground state changes discontinuously as
the parameter varies, the results can be misleading. In brief, the problem
with refining an approximate ground state is that numerical procedures may
well yield a metastable state in place of the time ground state, and there is
no unambiguous "signature" distinguishing the two.

The minimization eigenvalue approach described below has the advantage
that unlike the mapping method it can be applied to non-convex as well as
convex W, and unlike both approaches discussed previously, it is guaranteed
to yield the ground state (or at least $\underline{a}$ ground state in cases of degeneracy)
rather than a metastable state. To be sure, the minimization eigenvalue
method has its own shortcomings, which will emerge in the course of
discussion. Nonetheless, it has by now been demonstrated to be a fairly
effective numerical procedure for both convex and non-convex W, and further
improvements are possible.

3.2. Minimization Eigenvalue Equation

To find the ground state of the Frenkel-Kontorova Model (2.1.1), we look
for a solution to the following non-linear eigenvalue equation (whose origin
is discussed in §3.3 below):

$$R(u) + \eta = V(u) + \min_{u'} [W(u-u')+R(u')] , \qquad (3.2.1)$$

where η and $R(u)$ are the (unknown) eigenvalue and eigenfunction, respectively.
For the generalized model (2.1.8), the corresponding equation is

$$R(u) + \eta = \min_{u'} [K(u,u')+R(u')] . \qquad (3.2.2)$$

If V and W (or K) are continuous and the real variables u and u' belong to
some compact interval J, it can be shown that (3.2.1) (or (3.2.2)) has a
solution with a unique eigenvalue η and a continuous eigenfunction R. This
is the situation for the second and third examples of Sec. 2.1. (In the
third example the fact that V(u) tends to infinity as u goes to $+\infty$ or $-\infty$
means that the u_n in a ground state configuration are always confined to a
compact interval.) If V is periodic, (2.1.1), or K satisfies the
corresponding (2.1.11), one must impose the "boundary condition" that R has
the same period,

$$R(1+u) = R(u) . \qquad (3.2.3)$$

It turns out that η is the ground state energy per particle, or the ground state enthalpy per particle in situations in which energy and enthalpy are distinct (see Sec. 2.2). For the latter, the stress σ must be included in W (as, for example, in (2.1.4)) or in K.

To find the ground state configuration requires an additional step. Given a solution to (3.2.1) or (3.2.2), let

$$\tau(u) = u' \qquad (3.2.4)$$

be the (possibly multiple-valued) function which for each u gives the value of u' where the minimum is achieved. Then any infinite configuration $\{u_n\}$ with the property that

$$\tau(u_n) = u_{n-1} \qquad (3.2.5)$$

is a minimum enthalpy configuration, and if this configuration is recurrent, (2.2.10), it is a ground state. Furthermore, starting with an arbitrary u, the sequence u, $\tau(u)$, $\tau(\tau(u))$, etc. converges towards points which lie on a minimal enthalpy configuration. Since in currently-available numerical implementations of the method, u and u' are confined to a finite grid of points, there is always at least one periodic ground state, and finding it by iteration of the τ map is quite simple. (Once again, "enthalpy" can be replaced by "energy" in those situations where the two are not distinct.)

When V is periodic and R satisfies (3.2.3), (3.2.1) can be rewritten as

$$R(u) + \eta = V(u) + \min_{u'} [W^*(u-u') + R(u')] \qquad (3.2.6)$$

where

$$W^*(\Delta u) = \min_{n} [W(\Delta u - n)] , \qquad (3.2.7)$$

with the minimum over all integers. A geometrical construction for W* is shown in Fig. 3.1: displace the graph of W horizontally by integer amounts, and take the lower envelope of the resulting curves. The advantage of (3.2.6) is that both u and u' can be confined to the interval [0,1), which is an obvious advantage if the equation is solved numerically. (Such a restriction without replacing W by W* leads to errors.)

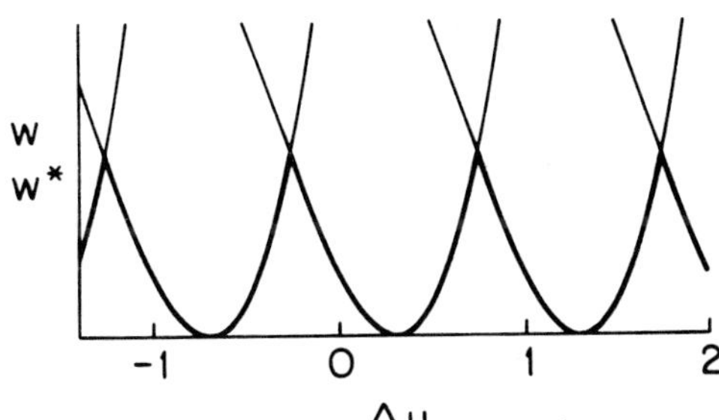

FIGURE 3.1
The curve with its minimum at Δu = 0.3 is $W(\Delta u)$, and the other curves are $W(\Delta u - n)$ for $n = \pm 1, \pm 2$. The heavy curve forming the lower envelope is $W^*(\Delta u)$.

The previous remarks on existence of solutions also apply to the left eigenvalue problem

$$L(u') + \eta = \min_{u} [L(u) + K(u,u')] \ . \tag{3.2.8}$$

In addition, the (unique) eigenvalue η is the same for (3.2.2) and (3.2.8). The eigenfunctions R and L, on the other hand, are in general different, and need not be unique. There is a "trivial" non-uniqueness which arises from the fact that adding a constant to L or to R yields another eigenfunction. It is often convenient to adopt a "normalization"

$$\min_{u} [R(u)] = 0 \tag{3.2.9}$$

in order to fix the value of this additive constant. But sometimes one observes a non-trivial degeneracy in which two or more solutions to (3.2.2) do not simply differ by a constant, leading to two or more distinct τ maps. This situation arises if there are two or more distinct ground states. Of course the same comments apply to (3.2.8).

In cases in which R and L are non-degenerate (apart from adding a constant), we define their sum to be

$$F(u) = L(u) + R(u) \ . \tag{3.2.10}$$

The function F has the property that it achieves its minimum value at points corresponding to a minimum enthalpy configuration.

3.3. "Derivation" of the Minimization Eigenvalue Equation

The two lines of argument presented in this section are not intended as derivations or justifications of (3.2.1) or (3.2.2) in a mathematical sense. Indeed, formal proofs of the properties of such an equation and the relationship of its solutions to the ground state problem can be based on the equation itself. Nonetheless, informal "derivations" of the sort presented here can add an intuitive understanding which formal arguments do not necessarily provide.

The first approach is based upon the zero-temperature limit of a transfer operator. Given a one-dimensional energy of the form (2.1.8), a natural procedure in statistical physics is to introduce a transfer operator (or matrix, if u and u' take on a finite number of values)

$$T(u,u') = e^{-\beta K(u,u')} \ , \tag{3.3.1}$$

where β is the reciprocal temeprature. The partition function, and thus the thermodynamic properties of the system, are determined by the largest eigenvalue of T, regarded as a linear operator. This eigenvalue is real and positive, and the corresponding right eigenvector can be chosen to be real and positive, so they may be written in the form $\exp-\beta\eta$ and $\exp-\beta R(u)$, respectively. The eigenvalue equation then takes the form:

$$\int du' e^{-\beta K(u,u')} e^{-\beta R(u')} = e^{-\beta \eta} e^{-\beta R(u)} \ . \qquad (3.3.2)$$

As we are interested in the limit as $\beta \to \infty$ (temperature goes to zero), it
is plausible that we can approximate the left side of (3.3.2) by the maximum
value of the integrand multiplied by some quantity proportional to a power
of β. Thus the dominant term on the left side is

$$\max_{u'} \exp\{-\beta[K(u,u') + R(u')]\} \ , \qquad (3.3.3)$$

and equating this with the right side yields (3.2.2). Of course a similar
approximation using the left eigenvector of T will yield (3.2.8).

In the second approach we consider the ground state problem for a finite
number of atoms. Let $\tilde{R}_N(u)$ be the minimal enthalpy (i.e., as given by
(2.1.1), but with a σ contribution included in W, as in (2.1.4)) of a chain
of N atoms, u_1, u_2, ... u_N, with the constraint that $u_N = u$; i.e., the N'th
atom is at some fixed poisition, but the other atoms are free to rearrange
themselves in an optimal way so as to minimize the total enthalpy. Thus

$$\tilde{R}_1(u) = V(u) \qquad (3.3.4)$$

$$\tilde{R}_2(u) = V(u) + \min_{u'} [W(u-u') + \tilde{R}_1(u')] \qquad (3.3.5)$$

and, in general,

$$\tilde{R}_{N+1}(u) = V(u) + \min_{u'} [W(u-u') + \tilde{R}_N(u)] \ . \qquad (3.3.6)$$

Next let us suppose that for large N, $\tilde{R}_N(u)$ approaches some function R(u)
plus a constant proportional to N,

$$\tilde{R}_N(u) \simeq R(u) + N\eta \ . \qquad (3.3.7)$$

Then in the large N limit, (3.3.6) is the same as (3.2.1), and in addition,
we have an intuitive interpretation for R(u): it is the effective potential
determining the force on the right-most atom of a semi-infinite chain, Fig.
3.2(a), when that atom is held fixed at a position u while the other atoms
are free to move so as to achieve a state of minimum enthalpy.

FIGURE 3.2
The effective potentials R(u), L(u) and
F(u) refer to atoms at the right edge
(a), left edge (b) of semi-infinite
chains, and the center (c) of a doubly
infinite chain, respectively. From
Ref. 20.

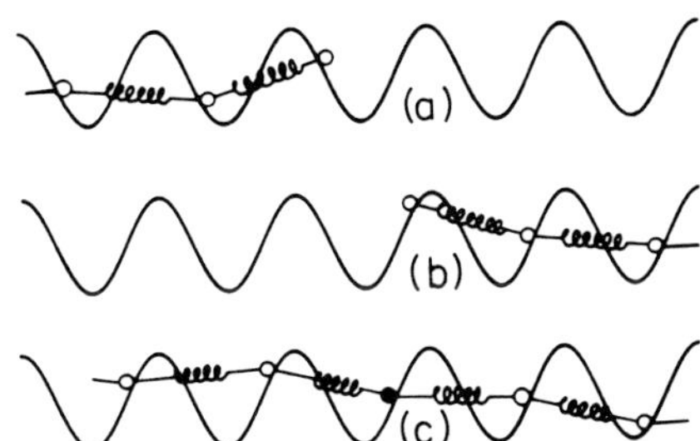

There is a slight difficulty in that numerical studies show that (3.3.7) is often not satisfied! In order to salvage the argument, let us suppose that (3.3.4) is replaced by

$$\tilde{R}_1(u) = R(u) , \qquad (3.3.8)$$

which is to say the left-most atom of any finite chain is subject to a special boundary condition: it "sees" the potential R(u) in place of V(u). In this case (3.3.7) is satisfied exactly. Hence our previous intuitive picture is correct, subject to an appropriate "boundary condition at minus infinity," since a semi-infinite chain has no left-most atom.

This picture also suggests why it is that iterating τ yields a sequence of points which eventually approach a minimum enthalpy configuration (§3.2). If the right-most atom in a semi-infinite chain is at u, the atom to its immediate left will be at $\tau(u)$, the one to the left of that at $\tau(\tau(u))$, etc. Hence by iterating a sufficient number of times one finds positions of atoms far from the right edge of the chain, and therefore less perturbed by the existence of that edge.

There is a similar interpretation for L(u), (3.2.8), as an effective potential which determines the force on the left-most atom in a semi-infinite chain extending to the right, when this atom is held at a position u and the other atoms take positions which minimize the enthalpy; see Fig. 3.2(b). (Note that the precise interpretation of L, and also of R, depends on how one expresses K in terms of V and W; see the remarks following (2.1.9).) Finally F(u), the sum of R and L, (3.2.10), can be thought of as the effective potential for an atom in the center of a doubly infinite chain whose position is constrained to be equal to u, while all the other atoms are allowed to move so as to minimize the enthalpy; see Fig. 3.2(c).

3.4. Numerical Procedures

The basic numerical problem is to solve (3.2.1) or (3.2.2) for η and the effective potential R, or (3.2.8) for η and L. Let us focus on (3.2.6), where u and u' are confined to the unit interval, as it illustrates all of the essential ideas. It will be convenient to denote the right hand side by $\mathcal{K}R$, where $\mathcal{K}$ is a non-linear operator, so that the equation has the form

$$R + \eta = \mathcal{K}R . \qquad (3.4.1)$$

The first step is to replace the continuous variables u and u' by discrete variables (which can be denoted by the same symbols) taking on a total of N values uniformly spaced on the unit interval (or circle). Typically we use a value of N between 10^2 and 10^4. Obviously a larger N yields a better approximation to the original problem, but it also takes more time and more computer memory. By using this discretization, we approximate the original eigenvalue problem by a corresponding finite matrix problem.

Two procedures have been employed to solve the discretized version of (3.4.1). The first is a simple iteration. Let $R^{(0)}(u)$ denote any function whatsoever (typically we set $R^{(0)} = V$), and calculate $R^{(n)}$ iteratively by means of the formula

$$R^{(n)} = \left[R^{(n-1)} + \mathcal{K}R^{(n-1)} - \eta_n\right]/2 \qquad (3.4.2)$$

where η_n is a constant chosen so that the minimum value of $R^{(n)}$ is zero, that is, this function is "normalized" in the sense of (3.2.9).

The iteration is continued until

$$\Delta_n = \max_u \; |R^{(n)}(u) - R^{(n-1)}(u)| \qquad (3.4.3)$$

is less than some predetermined error limit ε, at which point the iteration stops, and $R^{(n)}$ and η_n are considered adequate approximations to the desired eigenfunction and eigenvalue, respectively.

It is, by the way, essential to average $\mathcal{K}R^{(n-1)}$ with $R^{(n-1)}$; replacing the right side of (3.4.2) with $\mathcal{K}R^{(n-1)} - \eta_n$ yields a procedure which does not, in general, converge.

The number of iterations required depends both on the choice of ε and on the parameters which enter $\mathcal{K}$, such as σ and K in the standard model, (2.1.4) and (2.1.3). In practice one finds that 20 or 30 iterations are quite sufficient to obtain "reasonable" precision; that is, the remaining errors are considered insignificant in comparison with the effects of a discrete grid. As the basic numerical operation is minimizing a function of u' for a fixed value of u, the time required per iteration of (3.4.2), and thus the total time to find a solution, varies as N^2. Various tricks can be used to shorten the iteration time, although they require more complicated programming.

The second procedure for solving (3.4.1) involves searching directly for the ground state of the discretized problem. Consider a <u>cycle</u> C of values $u_1, u_2, \ldots u_p, u_{p+1} = u_1$ of the variable u, and define the <u>cyclic average</u> of the matrix $K(u,u')$ over this cycle by means of the formula

$$<K>_C = p^{-1} \sum_{j=1}^{p} K(u_{j+1}, u_j) \; . \qquad (3.4.4)$$

Here p, the length of the cycle, may be any integer greater than zero, but for a discretized grid of N points it suffices to consider cycles of length $\leq N$. It is clear that $<K>_C$ is the enthalpy per particle for a periodic configuration, and hence the enthalpy per particle of the ground state of the discretized problem is given by the minimum cyclic average,

$$\eta = \min_{C} \, \langle K \rangle_C \; , \qquad\qquad (3.4.5)$$

where the minimum is over all cycles.

There are various procedures for finding the minimum cyclic average and the minimizing cycle in a finite number of steps. Once the minimizing cycle is known, the eigenfunction R (or L) can also be constructed in a finite number of steps. The method which we have employed[27] is based on ideas of Karp[28] and von Golitschek,[29] and is too complicated to describe here. Suffice it to say that the memory requirements and the speed are comparable to those used in the iteration procedure described earlier; in particular, the time required, while depending somewhat sensitively on the parameters entering V and W, is roughly proportional to N^2.

The procedure based on the minimum cyclic average yields the exact answer (to within numerical roundoff) in a finite number of steps, while the iteration procedure described previously never (apart from very special cases) converges exactly. But this distinction is not all that important, since the practical limit on precision in both cases is provided by the discreteness of the grid: the fact that N is finite. Even the exact ground state on a discrete grid can differ in significant ways from its counterpart when u and u' are continuous. In particular, the former is always periodic (or, to be more precise, there is always a periodic ground state), whereas we know that in the continuous case, non-periodic ground states with irrational ω are possible, at least when W is strictly convex: see Sec. 2.3.

Thus it is the discreteness of the grid which is the main limitation to numerical studies of ground states by means of the minimization eigenvalue method. It is, of course, possible to use the results on a discrete grid as an approximate ground state for the continuous system, which can then be further refined as indicated in §3.1. This should work provided the ground state in the continuous case has the same period as in the discrete case. However, there is always the danger that refinement will yield a metastable state with an energy slightly above that of the true ground state. Could one make better use of the discrete grid by, for example, introducing some interpolation procedure to obtain R between grid points? I think this is not impossible, despite the fact that in typical cases R has "kink points" where the first derivative decreases discontinuously (see Fig. 3.3 below).

Despite the limitation just noted, the minimization eigenvalue method has shown itself to be a very effective tool for the construction of phase diagrams for Frenkel-Kontorova models. Once it has provided an overall "picture" of what is going on, further refinements are sometimes possible using other procedures. We now turn to some examples.

   3.5 Selected Results

   We shall begin with some results[20] for the standard model, (2.1.3) and
(2.1.4). Figures 3.3 and 3.4 show the effective potentials R(u) and F(u)
for the case K = 3 and σ = 0.36. The corresponding map τ(u) is in Fig. 3.5.
In this case the ground state has period 3 with ω = 1/3; the atoms lie at
positions indicated by the minima of the F function. Note that R and F,

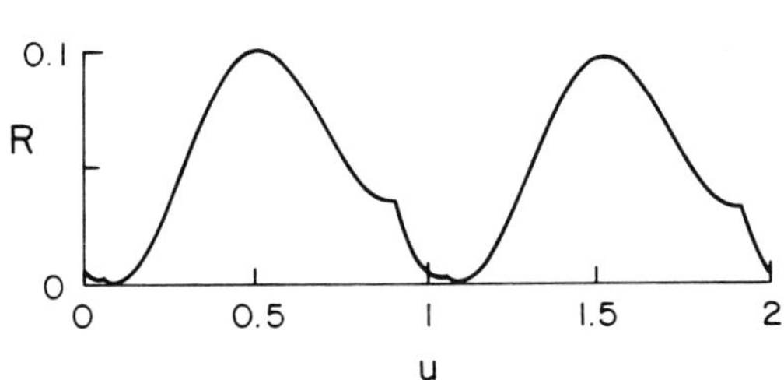

FIGURE 3.3
The effective potential R(u) for the
standard model with K = 3, σ = 0.36.
From Ref. 20.

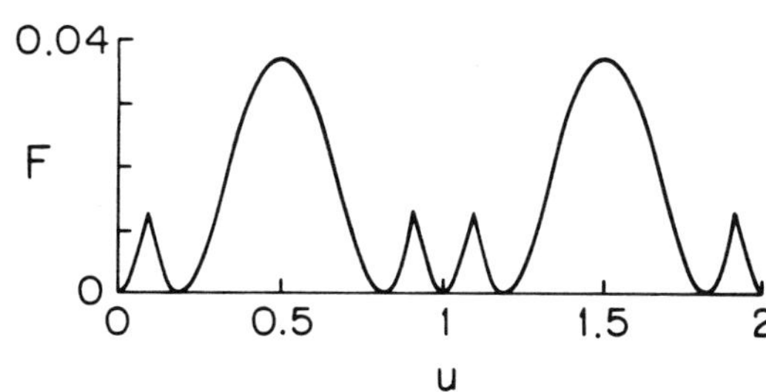

FIGURE 3.4
The effective potential F(u) for the
standard model with K = 3, σ = 0.36.

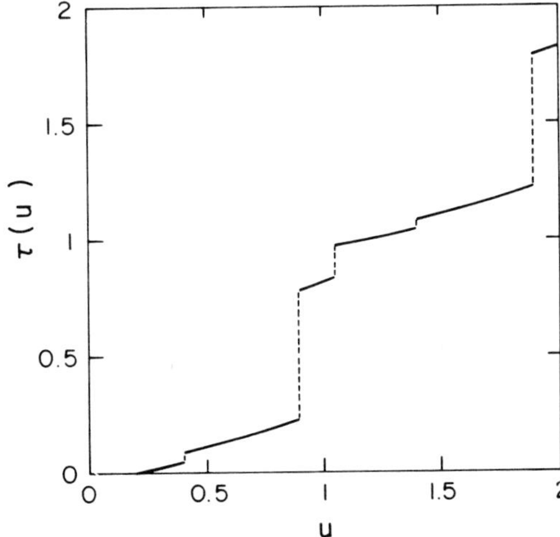

FIGURE 3.5
The map τ(u) for the standard model
with K = 3, σ = 0.36. From Ref. 20.

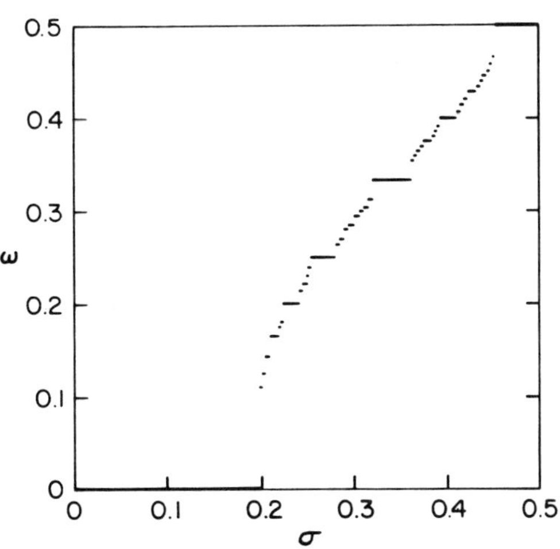

FIGURE 3.6
The average spacing or winding number
ω for the standard model with K = 1,
as a function of σ. From Ref. 20.

while continuous, have discontinuous derivatives. In the present case, R has
three such "kink" points in every period, one at each place where τ(u) is
discontinuous. (One can show that as long as W and V are smooth functions,
any discontinuity in the derivative of R, or L or F, is always of the form

shown here: a discontinuous decrease, rather than a discontinuous increase, as u increases.)

Figure 3.6 shows the winding number or average spacing ω in the ground state as a function of σ with K = 1. Numerically one finds a series of "steps" or, to be more precise, "plateaus" on which ω is constant as a function of σ. These occur at rational values $\omega = P/Q$, and the steps tend to be narrower as Q increases. The numerical evidence is consistent with the expectation, based on a model which Aubry has solved exactly, that there is a plateau for every rational value of ω. Such a function is often called a "devil's staircase." Also note that one of Aubry's theorems, A3 in Sec. 2.3, implies that $\omega(\sigma)$ is a continuous function, contrary to what one might guess from simply glancing at Fig. 3.6!

The positions of four of these plateaus, with ω = 0, 1/4, 1/3, and 1/2, are shown as a function of K in the phase diagram in Fig. 3.7. In this diagram the region occupied by each "phase" with a given ω appears as a "tongue". Between those tongues which are shown in the figure there are a large (presumably infinite) number of other tongues which are not shown. In addition, because ω is a continuous function of σ, there must be lines corresponding to irrational values of ω lying between the tongues for rational values. These irrational values, as noted earlier, are not directly accessible by present numerical techniques.

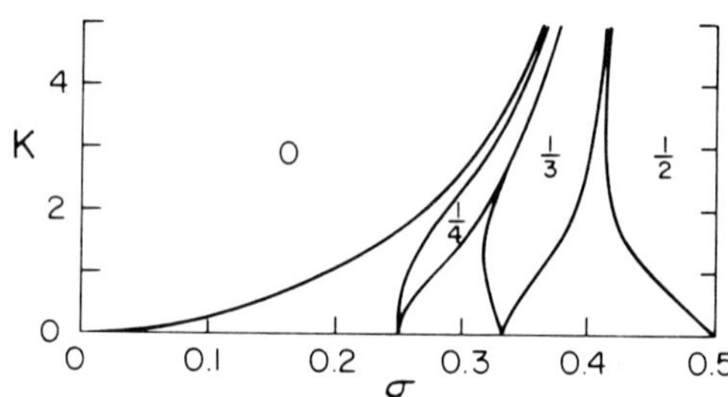

FIGURE 3.7
Phase diagram for the standard model showing the ω = 0, 1/4, 1/3, and 1/2 "tongues" for $0 \leq \sigma \leq 0.5$. There are a multitude of other "tongues" lying in the spaces between those which are shown here. From Ref. 20.

The (relatively!) simple phase diagram with tongues bounded by (apparently) smooth curves can be modified in an important way by adding small amounts of higher harmonics to the simple cosine in (2.1.3), resulting in a potential

$$V(u) = \frac{K}{(2\pi)^2} \sum_k \varepsilon_k [1-\cos 2\pi ku] , \qquad (3.5.1)$$

with $\varepsilon_1 = 1$. Figure 3.8 shows a phase diagram with a small admixture of second harmonic: $\varepsilon_2 = 0.2$, and $\varepsilon_k = 0$ for k > 2. (The portion of the diagram with σ less than about 0.23 resembles Fig. 3.7, and has been omitted.)

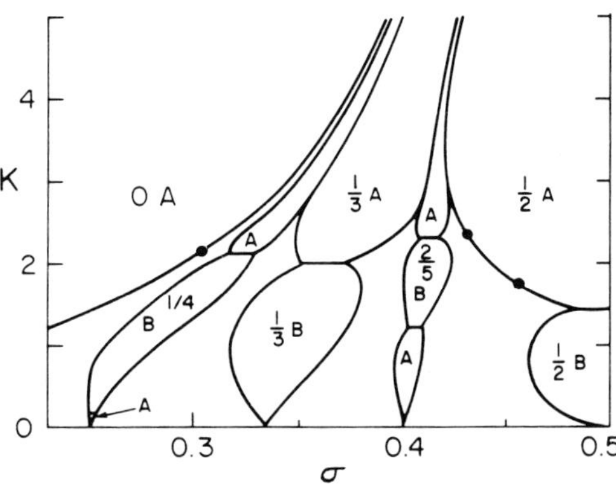

FIGURE 3.8
Phase diagram for a modified standard model with $\varepsilon_2 = 0.2$ in (3.5.1). The dots indicate some of the accumulation points of lines separating A and B phases. From Ref. 20.

The new feature in Fig. 3.8 is the appearance of additional phase transitions as K is varied, for all periodic phases with period Q > 1. At these transitions, which are indicated by horizontal lines, the configuration changes abruptly from one which we call type A to another, type B. The two types are shown schematically in Fig. 3.9 for the case $\omega = 1/2$. In type B there is always an atom on the top of the potential hill.

Both A and B type phases possess lines of reflection symmetry. However, a potential of the form (3.5.1) can also give rise to ground states of type C in which there is no reflection symmetry at all. Of course, given a ground state C_1 which breaks the reflection symmetry of (3.5.1), its mirror image C_2 is also a ground state. See Fig. 3.9 for an illustration with $\omega = 1/2$. A phase diagram[24] for $\varepsilon_2 = 1/4$ and $\varepsilon_3 = 1/6$ is shown in Fig. 3.10. Here the dashed horizontal lines represent continuous (second order) phase transitions, and the solid horizontal lines discontinuous (first order) transitions. Once again this is only a partial phase diagram, as one expects an infinite number of other "tongues", with equally complicated behavior, lying between those shown in the figure.

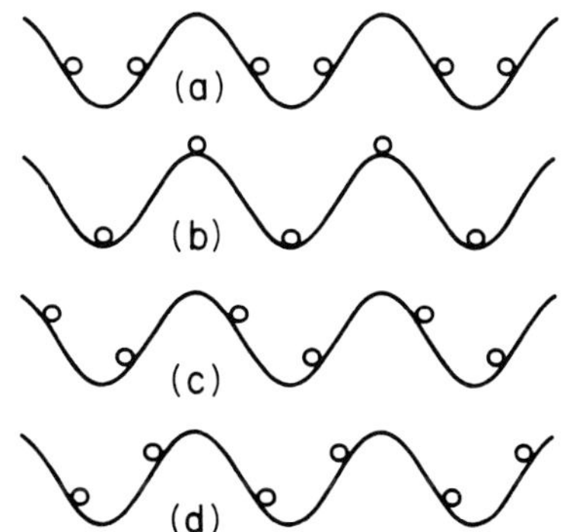

FIGURE 3.9
Phases of type (a) A, (b) B, (c) C_1 and (d) its mirror image C_2 for $\omega = 1/2$ (schematic).

Numerical studies show no indication of phases other than the A type in

the standard model with a pure cosine potential; that is, $\varepsilon_k = 0$ for all
$k \geq 2$. Nonetheless, B and C type phases can occur with arbitrarily small
mixtures of higher harmonics. In this sense the (relative) simplicity of
the standard model may be somewhat misleading.

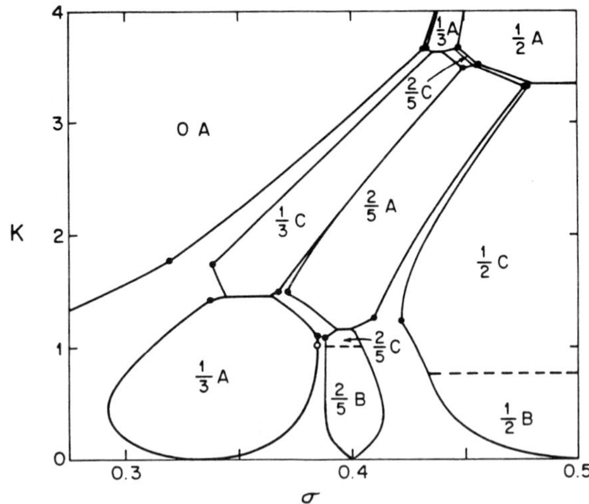

FIGURE 3.10
Phase diagram for a modified
standard model with $\varepsilon_2 = 1/4$
and $\varepsilon_3 = 1/6$ in (3.5.1).
From Ref. 24.

Next let us consider some cases in which W is not convex. We begin with
the chiral XY model in a magnetic field,[23] (2.1.5), whose phase diagram is
shown in Fig. 3.11. For small values of K, this model closely resembles the
standard model. It is possible to define the analog of ω, and as a function
of γ (which plays a similar role to σ) at fixed K, ω exhibits a "devil's
staircase" structure. However, for large values of K the situation is very
different, and one finds only a small number of phases separated by
discontinuous (first order) or continuous (second order) transitions.

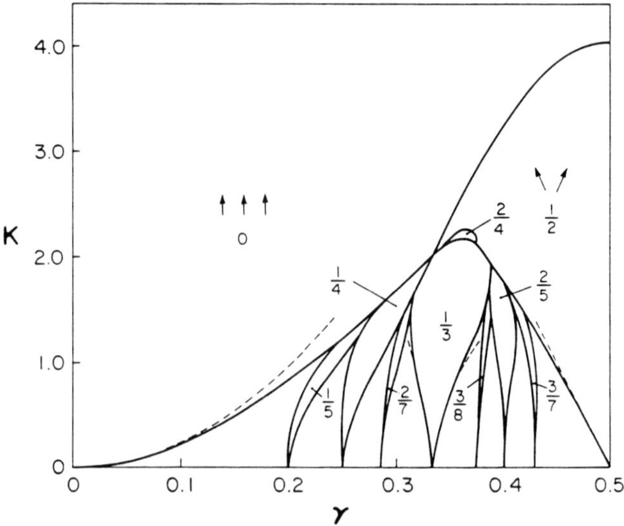

FIGURE 3.11
Phase diagram for the chiral
XY model in a magnetic field,
from Ref. 23. The dashed
lines should be ignored.

Obviously as K increases there must be a series of transitions where the infinite set of "tongues" characteristic of small K are somehow transformed into the much simpler structure at large K. The processes by which this happens are rather complicated, and we shall not discuss the details here. However, the following feature is worth pointing out. The "spring" potential W in (2.1.5) has both convex and non-convex parts, where its second derivative is or is not positive. One finds numerically that as long as all of the differences $u_{n+1} - u_n$ for the ground states fall in the convex part of W, the sort of phase diagram structure present in the standard model, the "devil's staircase" with a continuous ω, is present in the chiral XY model as well, but as soon as any of these differences $u_{n+1} - u_n$ fall in the non-convex part of W, qualitative changes occur in the phase transitions.

The third example[25] of Sec. (2.1), defined by (2.1.6) and (2.1.7), shows a similar behavior. For small values of K the ground states only "see" the convex parts of W, and it is possible to define a quantity ω which plays the role of the average spacing in the standard model. The phase diagram, Fig. 3.12, shows a series of "tongues" where ω takes on rational values. However, as K increases, a point comes where the ground state separations begin to sample the non-convex parts of W. The beginnings of these "non-convex" regions are indicated by the dashed lines in Fig. 3.12. These are not themselves lines of phase transitions, but where they meet the edges of their respective tongues, one finds a qualitative change in the character of the phase transitions.

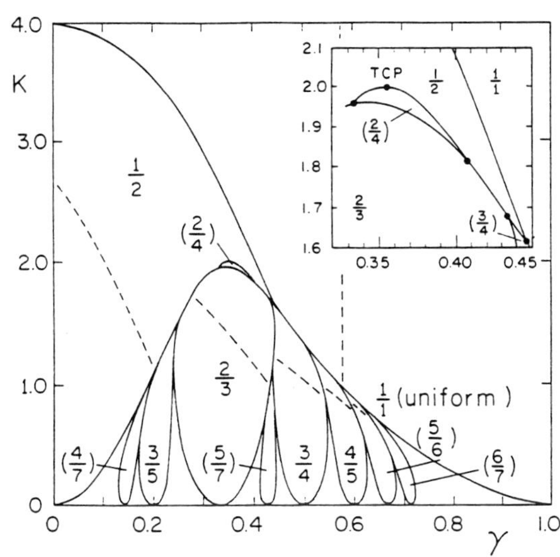

FIGURE 3.12
Phase diagram for the magnetoelastic model of Ref. 25, see (2.1.6) and (2.1.7); from Ref. 25(b). The dashed lines are not phase transitions, but indicate where the ground state begins to sample the non-convex parts of W.

4. DEFECTS AND PHASE TRANSITIONS

4.1. Defects and Their Properties

In this section we will consider certain types of defects in ground states of Frenkel-Kontorova models. These defects are of interest in themselves, but they also provide an intuitive understanding of the rather complicated sorts of phase transitions which occur in these models, as in Figs. 3.6 and 3.11.

To begin with, we consider configurations in which ω is very small, with σ slightly larger than the value at the right edge of the $\omega = 0$ plateau in Fig. 3.6. Such a configuration is sketched in Fig. 4.1 in two ways: (a) as atoms distributed among potential wells and (b) as a graph of u_n as a function of n.

As ω goes to zero, the number of atoms very near the bottom of each valley increases, whereas the number which are a significant fraction of the way up the side of one of the potential hills remains essentially unchanged. Or, in terms of Fig. 4.1(b), the separation δn of the successive regions where u_n is rising rapidly increases, whereas the width of one of these regions remains essentially fixed.

It is thus rather natural to think of the configuration in Fig. 4.1 as consisting of a series of defects, the regions where u_n increases rapidly as a function of n, separating portions of the configuration in which the atoms are, to a good approximation, in one of the $\omega = 0$ ground states at the bottom of one of the potential wells. Such defects are called "domain walls" or "kinks" or "solitons" by different authors. The term "discommensuration" has much to commend it, in terms of precision, even though it is a bit long.

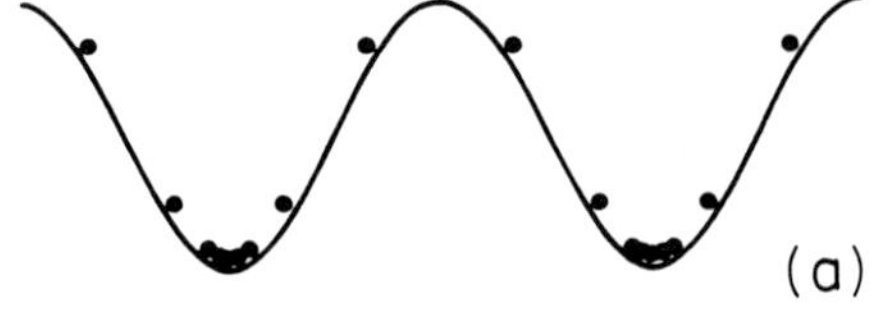

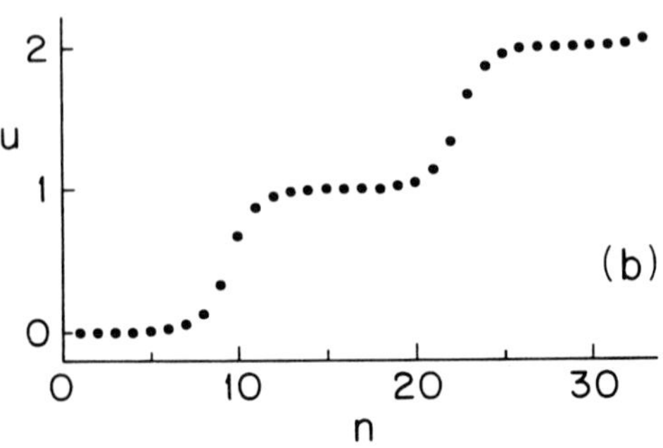

FIGURE 4.1
A configuration with ω slightly larger than zero. The positions of atoms in the potential wells are shown in (a); note that there are a large number of atoms at the bottom of every valley. In (b) the position is shown as a function of n.

A single, isolated defect of this type is a minimum energy configuration in which

$$u_n \to 0 \quad \text{as } n \to -\infty \, , \tag{4.1.1}$$

$$u_n \to 1 \quad \text{as } n \to +\infty \, , \tag{4.1.2}$$

where 0 and 1 can be replaced by any pair of successive integers. In Aubry's terminology this is an "advanced discommensuration, while a "retarded discommensuration" is the corresponding minimum energy configuration in which -1 replaces 1 in (4.1.2).

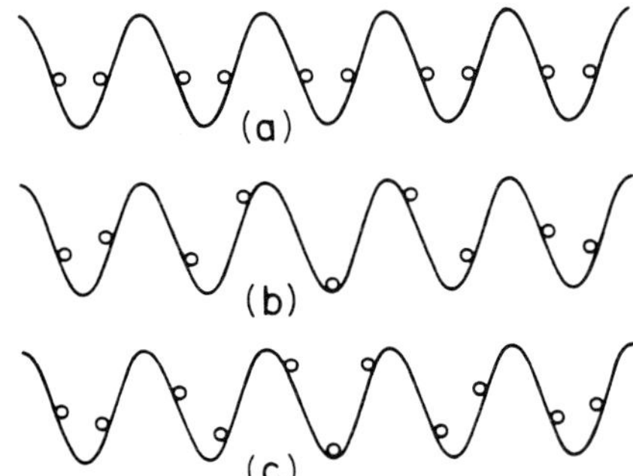

FIGURE 4.2
Positions of atoms (schematic) in: (a) an $\omega = 1/2$ ground state, (b) an advanced, and (c) a retarded discommensuration

Figure 4.2 shows advanced (b) and retarded (c) discommensurations in an $\omega = 1/2$ ground state (a). Similar defects can be defined for any commensurate ground state when W is convex. Note that while these discommensurations are minimum energy configurations, they are not recurrent, and hence they are not ground states in Aubry's terminology.

It is convenient at this point to introduce some properties of defects which will be needed in the later discussion. (For more details and greater precision, see the references in Sec. 5.4.)

A defect can be thought of as a "small" change in a <u>reference configuration</u>, which for present purposes will always be a periodic ground state with period Q. The <u>excess length</u> $\hat{u}$ of the defect is the average amount by which $u_N - u_{-N}$ in the defect configuration exceeds the corresponding quantity in the reference configuration, in the limit of large N. For example, $\hat{u}$ is +1/2 and -1/2 for the advanced and retarded discommensurations shown in Fig. 4.2(b) and (c). The <u>defect creation enthalpy</u> η_c is the difference between the enthalpy of the defect configuration and that of the reference configuration. (There are some subtleties, as one is dealing with infinite configurations, but we shall not give the details here.)

Let us suppose that the <u>center</u> of a defect, where there is a maximum deviation from the reference configuration, occurs at an atom n_O. (If the choice is not unambiguous, n_O can be fixed by convention.) Then if n is far from n_O, the positions of atoms in the defect will be close to those of a refence configuration, perhaps modulo 1. If this deviation Δu_n behaves as

$$\Delta u_n \sim \exp\ [-\ |n-n_O|/\xi] \qquad (4.1.3)$$

for large $|n-n_O|$, we shall call ξ the <u>decay</u> <u>range</u>.

Because the reference configuration is periodic, it is possible by shifting atoms around to transform a defect centered at n_O into an identical defect centered at $n_O + Q$. However, some energy (or enthalpy; the difference does not matter in this context) barrier in configuration space must be surmounted in this process, and the minimum barrier height is known as the <u>pinning</u> <u>energy</u>.

The enthalpy of a configuration containing two or more defects minus that of the reference configuration will consist of the sum of the creation enthalpies of the different defects along with an <u>interaction</u> <u>energy</u> (or enthalpy; for this purpose the two need not be distinguished) which can be decomposed into a contribution from pairs, triples of adjacent defects, etc.

4.2. Phase Transitions Mediated by Defects

When W is convex, one can think of the phase transitions which occur in a Frenkel-Kontorova model as σ is varied as produced by discommensurations, in the following way. Consider a commensurate ground state with average atomic separation given by the rational number ω_b. As long as σ remains between the two limits σ' and σ'' of the plateau corresponding to ω_b (see Fig. 3.6), the ground state configuration is independent of σ, assuming the other parameters remain fixed. However, the creation enthalpies for the advanced and retarded discommensurations do depend on σ, and go to zero at the right and left edges of the plateau, respectively. Beyond the edge of the plateau, the creation enthalpy of the corresponding defect is negative, making the ω_b ground state unstable. The new ground state corresponds to a finite density of defects, with this density determined by a competition between two effects. The negative creation enthalpy favors the formation of as many defects as possible. However, the interaction energy (or enthalpy) of the defects is positive, and increases as the defects come closer together.

Consequently, when σ is only slightly beyond the edge of the plateau, and thus the creation enthalpy is negative by only a small amount, the density of defects is low and they are well separated. With increasing distance from the edge, the density of defects rises. Note that the main change, as long as the defect density is not too high, is in the number of defects, and not in the structure of individual defects.

Let us try and put these qualitative ideas on a more quantitative basis. For a given reference state (ground state) configuration, the creation enthalpy of a defect varies with σ as

$$\eta_c = \text{constant} - \sigma\hat{u} , \qquad (4.2.1)$$

where $\hat{u}$ is the excess length (§4.1). For a ground state of period Q, one can show that

$$\hat{u} = \pm\, 1/Q \qquad (4.2.2)$$

for the advanced (+) and retarded (−) discommensurations. Thus the variation of the creation enthalpies for these two defects is as sketched in Fig. 4.3. Note that (4.2.1) and (4.2.2) together imply that the sum of the creation enthalpies for the advanced and retarded discommensurations is $(\sigma''-\sigma')/Q$, independent of σ.

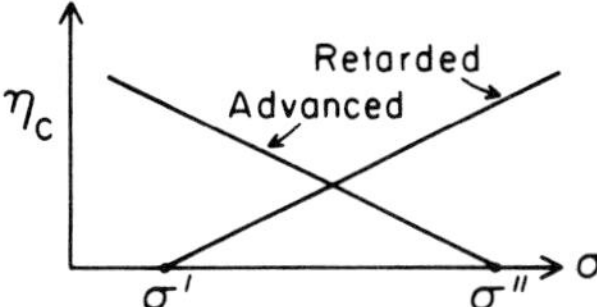

FIGURE 4.3
The creation enthalpy η_c as a function of σ for an advanced and a retarded discommensuration.

The interaction energy for two or more discommensurations of the same type (advanced or retarded) is positive (assuming W is convex), and vanishes as the defects become farther and farther apart. For widely separated discommensurations, the main interaction term is a pair interaction which decays exponentially with the separation δn between centers.

Suppose that σ is slightly larger than σ'',

$$\sigma = \sigma'' + \delta\sigma \qquad (4.2.3)$$

where σ'' is the right edge of the plateau corresponding to ω_0. If we think of the ground state configuration as consisting of a set of widely separated advanced discommensurations at a spacing δn, the enthalpy per particle relative to the ω_0 reference configuration is given by

$$\Delta\eta = [-\hat{u}\delta\sigma + Ce^{-2\delta n/\xi}]/\delta n \qquad (4.2.4)$$

where C is some constant and ξ is the decay range introduced in §4.1. By treating δn as a continuous variable (which should be a good approximation when it is large) and differentiating (4.2.4) we find that $\Delta\eta$ is a minimum provided

$$\hat{u}\delta\sigma = C(1+2\delta n/\xi)e^{-2\delta n/\xi} . \qquad (4.2.5)$$

In this configuration the average separation between neighboring atoms ω differs from its value in the ω_0 ground state by

$$\delta\omega = \omega - \omega_0 = \hat{u}/\delta n . \qquad (4.2.6)$$

Solving (4.2.5) for δn to leading order and combining this with (4.2.6) yields the result

$$\delta\omega \propto -1/\log(\delta\sigma) \qquad (4.2.7)$$

when $\delta\sigma$ is small. This accounts for the rapid variation of ω with σ at the edge of a plateau, as is evident in Fig. 3.6 for the plateaus with $\omega = 0$ and $\omega = 1/2$.

To be sure, $\delta\omega$ does not, in fact, vary smoothly with $\delta\sigma$, and so (4.2.7) only holds in some average sense if one ignores the additional plateaus which are present in this devil's staircase structure. Can our picture of interacting discommensurations also provide some sort of intuitive picture for the origin of this finer structure? Indeed it can. The defects are, in a first approximation, similar to particles moving in a continuum, which is the basis of the analysis following (4.2.4). But the "continuum" is really a periodic structure, the ω_0 ground state, and the discommensurations minimize their energy if they take up certain favored positions relative to this periodic structure, due to the positive pinning energy. Thus the pinning energy prevents δn from taking on a continuum of values, and it can thus stabilize certain commensurate configurations. To be sure, the situation is complicated, because there is no reason to expect that the separations between all adjacent defects will be the same, as we assumed when writing (4.2.4). Indeed, one expects these separations to vary, and this variation allows for more complicated commensurate and also for incommensurate phases. Additional details are beyond the scope of the present discussion. The main point is that defects provide a very useful tool for thinking about this type of ground state structure and the corresponding phase transitions.

Next let us consider the case in which W is <u>not</u> convex. As noted in Sec. 3.5, first-order (discontinuous) phase transitions are now possible in which ω changes discontinuously as a function of σ or some other parameter. Again, defects and defect interactions provide a useful way of thinking about at least some phase transitions of this kind.

The important point is that when W is not convex, the pair interaction between identical defects can be negative, at least at certain separations. To see how this can produce a first-order transition, consider the modification of Fig. 4.3 shown in Fig. 4.4, where for simplicity we

consider only one type of defect, not two. It may happen that as σ
increases towards σ'', where the creation enthalpy for this defect goes to
zero, it reaches a value σ^* at which the negative interaction energy produced
by introducing a periodic array of these defects at a particular spacing is
enough to compensate exactly for the positive enthalpy required to create
the defects. The result will be an abrupt change to a new ground state as σ
passes σ^*, and although $\eta(\sigma)$ is continuous, it has a change of slope at
$\sigma = \sigma^*$, as is typical for a first-order transition.

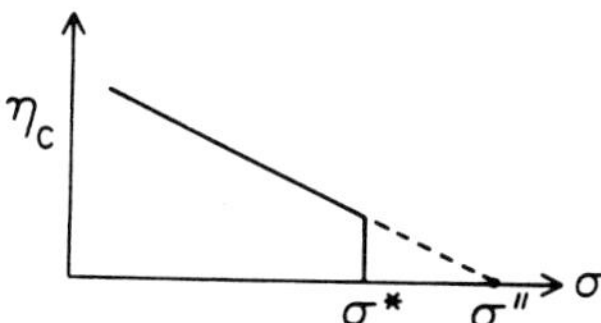

FIGURE 4.4
If W is not convex, a first-order
phase transition can occur at σ^*
when the defect creation enthalpy
η_c is still positive.

A little thought shows that not only is such a first-order transition
possible, but it is also certain to occur before σ reaches σ'' in those cases
in which the defects have negative pair interaction energies, or
interactions which oscillate in sign, at arbitrarily large separations –
provided (as seems plausible) the pair interactions are dominant over triple
and higher-order interactions at large separations. Consequently defects
provide a plausible explanation for the observation, based on numerical
studies (Sec. 3.5), that when interparticle separations begin to "sample"
the non-convex part of W, a devil's staircase structure with its infinite
number of phases and commensurate-incommensurate transitions tends to be
replaced with a finite number of phases separated by first-order transitions
(or sometimes second-order transitions of the more conventional type).

4.3. Defects and the Minimization Eigenvalue Equation

A solution to the minimization eigenvalue equation not only yields the
ground state of a Frenkel-Kontorova model, it also provides information
about certain low energy defects by means of the effective potential F,
(3.2.10).

Thus if W is convex and σ is inside but close to the edge of the plateau
corresponding to some commensurate ground state, F(u) will often have the
sort of behavior shown in Fig. 4.5, which was calculated for the standard
model with K = 1.5, σ = 0.435. Along with the absolute minima occurring at
positions (modulo 1) of atoms in the ground state, there are a set of local
minima, with F taking the same value at each one, which occur at the
positions (modulo 1) of some of the atoms in that discommensuration which
will mediate the phase transition occurring at the corresponding edge of the

plateau.

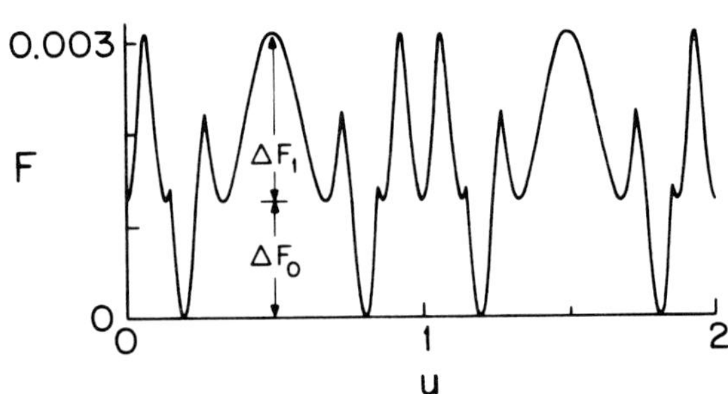

FIGURE 4.5
The effective potential F(u) for
the standard model with K = 1.5,
σ = .435. The absolute minima
correspond to the ω = 1/2 ground
state, and the secondary minima
to a retarded discommensuration
with creation enthalpy ΔF_0 and
pinning energy ΔF_1.

That this should occur becomes plausible when one considers the intuitive
significance of F as noted at the end of Sec. 3.3: it is the effective
potential seen by an atom at the center of a doubly infinite chain which is
constrained to be at position u while the other atoms are free to move so as
to minimize the enthalpy.

A discommensuration such as the one shown in Fig.4.2(b) is a minimum
energy configuration, and hence each atom in this configuratin is at a local
energy minimum in the sense that a small displacement will increase the total
energy. This minimum should also be a local minimum of F unless the enthalpy
can be further reduced by rearranging the atoms on one side or the other side
of at atom in question, by displacements which need not be small. Whether
such an enthalpy-lowering rearrangement exists depends on the atom in
question and also the value of σ. Nonetheless, it should not be surprising if
for certain atoms (typically those near the center of the defect) and certain
σ, no such rearrangement is possible, so that F also has a local minimum.

The difference ΔF_0, Fig. 4.5, between the value of F at one of the
secondary minima and its value on the ground state is the creation enthalpy
of this defect (for that value of σ used to compute F). In addition one can
sometimes identify a smooth maximum in F whose height ΔF_1 above the
secondary minima gives the pinning energy of the discommensuration. A
determination of ΔF_0 in this way can be used to locate the edge of the
plateau in the case of a continuous transition, that is one in which the
situation indicated in Fig. 4.3 holds, without actually carrying out a
computation with σ equal to the value at the edge. For high precision, it
is best to calculate the creation enthalpy from the precise positions of
the atoms in a discommensuration as determined by a numerical solution of
the equilibrium equations. The minimization eigenvalue approach gives ΔF_0
with lower precision because of the grid, though it does provide a first
approximation to the positions of the atoms in the defect, which can then be

further refined.

While the same procedure cannot be used to locate the edge of a plateau in a situation where a first-order transition occurs, as in Fig. 4.4, it is still useful to make plots of F. For example, a discontinuous change in the set of u values where F takes its minimum is a sign of such a transition.

5. LITERATURE REFERENCES

5.1. General References on Commensurate and Incommensurate Phases

The 1982 review article by Bak,[4] although out of date in certain respects, is still an excellent place to obtain an overview of the whole field of commensurate and incommensurate phases, and the transitions between them. The 1987 review by Janssen and Janner[5] can also be strongly recommended: it has brief discussions of several systems observed experimentally, and provides an introduction to the corresponding theories. Quasicrystals are included in the discussion. In addition Janssen has contributed a chapter to the first volume edited by Blinc and Levanyuk,[7] and is coauthor with Currat of a review article on excitations in incommensurate phases.[6]

R. Blinc and A. P. Levanyuk have edited two volumes[7] on incommensurate phases in dielectrics. The first is devoted to theory and experimental techniques; the second has chapters devoted to experimental results for a number of specific systems. H. Z. Cummins[8] has recently written a very extensive survey of experimental results on incommensurate structures, with a comprehensive bibliography.

The ANNNI model and chiral clock models are discussed in two recent reviews by Yeomans[9] and Selke.[10]

5.2. Frenkel-Kontorova Models

Research on this model has been dominated by the work of S. Aubry and his collaborators, of which Refs. 11 to 16 represent a small selection. His articles are often not easy to read - physicists, in particular, may find some of the mathematics unfamiliar - but studying them is a worthwhile effort for serious students of the subject. A good place to begin is his 1983 article (in English) in J. de Physique.[11] For those who can read French and afford the price of the book (or borrow a copy), his 1985 summer school lectures[14] are a source of much valuable material. In particular, one will find there the theorems cited in Sec. 2.3, and an extensive discussion of area preserving maps. Both topics are also treated in his 1983 articles in Physica;[15,16] the second coauthored with Le Daeron. Along with abstract theorems, Aubry has made a major contribution to the subject through an exactly soluble model in which $V(u)$ in (2.1.1) is a scalloped potential of repeating parabolas.[12,13]

Area preserving maps are themselves of mathematical interest as examples of nonlinear dynamical systems. As well as the work of Aubry cited earlier, the appropriate sections of articles by Lauwerier[17a], Henon[17b], and Berry[18] may be consulted for some of the lore in this fascinating field.

5.3. Minimization Eigenvalue Method

This method was introduced in the context of Frenkel-Kontorova models by Griffiths and Chou.[19] However, the type of eigenvalue problem found in (3.2.2) had been used previously in other contexts. A detailed discussion and references to the literature will be found in the book by Cuninghame-Green.[22] Reading this book is, unfortunately, not easy because of the unfamiliar notation. Chapters 23, 24, and 25 treat the minimization eigenvalue problem, (3.2.2) in my notation, in the case where u and u' are discrete variables taking on a finite number of values. This, of course, is the case which corresponds to numerical studies in which there is a finite grid. Proofs of the results needed in the case of continuous u and u' have not been published apart from the article by Chou and Duffin.[21]

Numerical procedures for solving the eigenvalue problem by iteration have been discussed by Chou and Griffiths[20] and by Hood,[30] while an exact method based on ideas of Karp[28] and von Golitschek[29] is described by Floria and Griffiths.[27]

5.4. Defects and Phase Transitions

The significance of discommensurations for phase transitions in Frenkel-Kontorova models has been emphasized by Aubry,[14-16] who has established various results concerning those discommensurations which are minimum energy configurations. Fisher and Szpilka[33] have shown how a formalism based on interacting defects ("domain walls") can be used to analyze the low temperature phase transitions in the ANNNI and chiral clock models. Yokoi et al.[23] have used defect ("soliton") interactions in order to analyze the phase diagram of the chiral XY chain in a magnetic field.

The analysis I present in Ch. 4 makes use of the article on defects by Tang and Griffiths,[31] and some unpublished results in Tang's thesis[32] related to the interaction of defects. The "excess length" $\hat{u}$ of a defect is related to the phase shift (q,s) of Tang and Griffiths[31] through the formula

$$\hat{u} = s - q\omega . \tag{5.4.1}$$

ACKNOWLEDGMENTS

It is a pleasure to acknowledge collaboration with the following individuals: Kevin Bassler, Weiren Chou, Richard Duffin, L. Mario Floria, Kazuo Sasaki, Leihan Tang, and Carlos Yokoi. In addition I have benefitted from discussions with Serge Aubry, Claude Godreche, Kevin Hood, and Mario Marchand. My research has been supported financially by the United States National Science Foundation under grants DMR 8108310 and DMR 8613218.

REFERENCES

1) J.J. Rhyne and T.R. McGuire, IEEE Trans. on Magnetics, MAG-8 (1972) 105; W.C. Koehler, Magnetic structures of rare earth metals and alloys, in: Magnetic Properties of Rare Earth Metals, ed. R.J. Elliott (Plenum Press, London, 1976) pp. 81-128.

2) A. Drillat et al., J. Magnetism Mag. Mat. 44 (1984) 232; D. Gibbs et al., Phys. Rev. B34 (1986) 8182; J. Bohr et al., Physica 140A (1986) 349.

3) L.D. Landau and E.M. Lifshitz, Statistical Physics, 2d ed. (Pergamon Press, London, 1969) Ch. XIV.

4) P. Bak, Rep. Prog. PHys. 45 (1982) 587.

5) T. Janssen and A. Janner, Adv. Phys. 36 (1987) 519.

6) R. Currat and T. Janssen, Excitations in incommensurate crystal phases, in: Solid State Physics, Vol. 41, eds. H. Ehrenreich and D. Turnbull (Academic Press, New York, 1988) pp. 201-302.

7) R. Blinc and A.P. Levanyuk, eds., Incommensurate Phases in Dielectrics 1. Fundamentals 2. Materials (North Holland, Amsterdam, 1986).

8) H.Z. Cummins, Physics Repts. (to appear).

9) J. Yeomans, The theory and application of axial Ising models, in: Solid State Physics, Vol. 41, eds. H. Ehrenreich and D. Turnbull (Academic Press, New York, 1988) pp. 151-200.

10) W. Selke, Phys. Repts. 170 (1988) 213.

11) S. Aubry, J. de Physique 44 (1983) 147.

12) S. Aubry, The new concept of transitions by breaking of analyticity in a crystallographic model, in: Solitons and Condensed Matter Physics, eds. A.R. Bishop and T. Schneider (Springer, Berlin, 1981) pp. 264-277.

13) S. Aubry, J. Phys. C16 (1983) 2497.

14) S. Aubry, Structures incommensurables et brisure de la symétrie de translation. I, in: Structures et Instabilités, ed. C. Godreche (Editions de Physique, Les Ulis, France, 1985) pp. 73-194.

15) S. Aubry, Physica 7D (1983) 240.

16) S. Aubry and P.Y. Le Daeron, Physica 8D (1983) 381.

17a) H.A. Lauwerier, Two-dimensional iterative maps, in: Chaos, ed. A.V. Holden (Princeton University Press, 1986) pp. 58-95.

17b) M. Henon, Numerical exploration of Hamiltonian systems, in: Chaotic Behaviour of Deterministic Systems, eds. G. Iooss, R.H.G. Helleman and R. Stora (North-Holland, Amsterdam, 1983) pp. 53-170.

18) M.V. Berry, Regular and irregular motion, in: Topics in Nonlinear Dynamics, ed. S. Jorna, A.I.P. Conference Proceedings 46 (American Institute of Physics, New York, 1978) pp 16-120.

19) R.B. Griffiths and W. Chou, Phys. Rev. Lett. 56 (1986) 1929.

20) W. Chou and R.B. Griffiths, Phys. Rev. B34 (1986) 6219.

21) W. Chou and R.J. Duffin, Adv. Appl. Math. 8 (1987) 486.

22) R. Cuninghame-Green, Minimax Algebra (Springer, Berlin, 1979).

23) C.S.O. Yokoi, L.H. Tang and W. Chou, Phys. Rev. B37 (1988) 2173.

24) K. Sasaki and L.M. Floria, J. Phys. Condensed Matter 1 (1989) 2179.

25) M. Marchand, K. Hood and A. Caillé, Phys. Rev. Lett. 58 (1987) 1660; and Phys. Rev. B37 (1988) 1898.

26) M. Marchand and A. Caillé, Phys. Rev. B38 (1988) 4845.

27) L.M. Floria and R.B. Griffiths, Numer. Math. 55 (1989) 565.

28) R.M. Karp, Discrete Math. 23 (1978) 309.

29) M. von Golitschek, Numer. Math. 39 (1982) 65.

30) K. Hood, J. Comp. Phys. (to appear).

31) L.H. Tang and R.B. Griffiths, J. Stat. Phys. 53 (1988) 853.

32) L.H. Tang, PhD Dissertation (unpublished), Carnegie Mellon University, 1987.

33) M.E. Fisher and A.M. Szpilka, Phys. Rev. B36 (1987) 644, 5343, 5363.

FUNDAMENTAL PROBLEMS IN STATISTICAL MECHANICS VII
H. van Beijeren, Editor
© *Elsevier Science Publishers B.V., 1990*

CONFORMAL INVARIANCE

H.J.F. KNOPS

Institute for Theoretical Physics, University of Nijmegen, Toernooiveld 1, 6525 ED Nijmegen,

The Netherlands

1. INTRODUCTION

Renormalization theory has been one of the major advances in our understanding of the theory of critical systems. (For reviews see e.g. references 1-3.) This theory, built on the scale-invariance of critical systems, has among its principle results:

(i) *Scaling laws* by which all possible critical exponents of a given system can be expressed in terms of a small number (two for an ordinary critical point) of fundamental eigenvalues.

(ii) The principle of *universality* according to which large classes of systems share the same fundamental eigenvalues and therefore the same critical behaviour.

Moreover, there are, within the renormalization theory, powerful techniques available (like the ε-expansion) for a perturbative calculation of the values of the fundamental eigenvalues. However, scale invariance by itself is not a constraint restrictive enough to fix the values of the fundamental eigenvalues a priori.

In 1984 Polyakov and co-workers (BPZ) published a seminal paper[4] in which the notion of *global* scale invariance of a critical system was extended to a *local* scale plus rotation invariance that is to say conformal invariance. For systems of dimensionality $d \geq 3$ the group of conformal transformations is not particularly large and only limited extra information can be gained. However, for two-dimensional critical systems the conformal group consists of all analytic mappings of the plane and is infinite dimensional. It turns out that the assumption of conformal invariance is so restrictive that it fixes the complete set of critical exponents for many of these two-dimensional systems.

In these lectures I intend to review the work of BPZ and a few directly related results. (An excellent and more extensive review by Cardy can be found in reference 5.) I will only have time to shortly mention recent developments like the employment of modular invariance.

In the first part of these lectures we shall see how the conformal invariance leads to a representation of the Virasoro algebra which is the (infinite dimensional) Lie-algebra of the conformal group (or rather its central extension). The commutation relations of the Virasoro algebra are characterized by a single real number: the central charge c. States in the representation space correspond to scaling fields of the critical theory and associated eigenvalues to critical exponents. The spectrum of these eigenvalues is determined by the commutation relations of the Virasoro algebra just like the familiar Lie-algebra of the rotation group fixes the spectrum of the angular momentum operator.

The second part of these lectures will be restricted to systems with a central charge $0 < c < 1$. It can be shown[6] that the natural demand of unitarity (i.e. positive definite transfer matrix) can only be appeased with the representation of the Virasoro algebra for $c = 1 - 6/p(p+1)$, $p = 3, 4, 5, \ldots$. Moreover, the number of allowed elementary states (scaling fields) is finite and all critical exponents are determined (once p is given). The most simple case, with $p = 3$ and only two elementary states, corresponds to the Ising model. Other values of p correspond to other universality classes like the tri-critical Ising-model or the Potts-model.

2. BASIC NOTIONS IN RENORMALIZATION THEORY

In this section some fundamentals of the renormalization theory will be reviewed as a quick reminder and to fix the notation. In the study of critical systems one is interested in the asymptotic behaviour of correlation functions. Asymptotic means at length scales large compared to the lattice constant a but small compared to the correlation length ξ (which is infinite for a truly critical system). It is then often convenient to consider the system as a (classical) field theory by taking the continuum limit $a \to 0$. A well-known example is the Gaussian model with Hamiltonian*

$$\mathcal{H} \equiv \frac{1}{2} \int (\vec{\nabla} S(\vec{r}))^2 \, d^2 r \qquad (2.1)$$

in which $S(\vec{r})$ is a fluctuating scalar field. A general field is a local expression in $S(\vec{r})$ e.g.

---

*In field theory in $1 + 1$ dimension one has the correspondence $x \to x$; $y \to it$. The quantity $(\vec{\nabla} S(r))^2 = (\frac{\partial S}{\partial x})^2 - (\frac{\partial S}{\partial t})^2$ then corresponds to the Lagrangian density and $\mathcal{H}$ to the action.

$\psi_1(r) = \vec{\nabla}S(\vec{r})$ or $\psi_2(r) = e^{i\alpha S(r)}$ etc. Correlation functions are defined by:

$$< \psi_1(\vec{r}_1) \dots \psi_n(\vec{r}_n) >= \frac{\int DS(\vec{r})e^{-\mathcal{H}}\psi_1(\vec{r}_1)\dots\psi_n(\vec{r}_n)}{\int DS(\vec{r})e^{-\mathcal{H}}} . \qquad (2.2)$$

A critical system has correlation length $\xi = \infty$ and since this is the only relevant length scale once the continuum limit is taken, the system is scale invariant. More precisely the Hamiltonian is invariant under a renormalization step[†] R_b that scales the system by a scale factor b:

$$\mathcal{H} \xrightarrow{R_b} \mathcal{H}. \qquad (2.3)$$

Under a renormalization step a field $\psi_j(r)$ transforms to a new field that can be expressed in terms of the old fields:

$$\psi_j(\vec{r}) \xrightarrow{R_b} \psi'_j(\vec{r}') = \sum_i M_{j,i}(b)\psi_i(\vec{r}'). \qquad (2.4)$$

The fields $\{\varphi_j(r)\}$ that form a basis for which the matrix $M_{j,i}(b)$ is diagonal are called *scaling fields*. Scaling fields transform according to:

$$\varphi_j(\vec{r}) \xrightarrow{R_b} \lambda_j(b)\varphi_j(\vec{r}'). \qquad (2.5)$$

The group property $R_{b_1}R_{b_2} = R_{b_1 b_2}$ then implies that

$$\lambda_j(b) = b^{-x_j} \qquad (2.6)$$

in which x_j is called the *anomalous dimension* of the scaling field φ_j.

Since rotations $U(\delta)$ commute with dilatations one can choose $\{\varphi_j\}$ to be also eigenvectors of $U(\delta)$:

$$\varphi_j(\vec{r}) \xrightarrow{U(\delta)} e^{i\ell_j\delta}\varphi_j(\vec{r}') \qquad (2.7)$$

in which ℓ_j is the *spin-index* of the field φ_j.

[†] Due to the fact that the continuum limit $a \to 0$ is a formal limit that usually only can be taken after all other calculations (like phase space integrals etc.) have been carried out a renormalization step must in fact be defined for a system with a *finite* cutoff. It consists of two steps:

(i) Integrate out all degrees of freedom that pertain to length scales between a and ba.

(ii) Scale the system by setting $r' = rb$.

Two point correlation functions of scaling fields are completely determined by the transformation properties of these fields under translations, rotations and dilatations. Consider as an example two scalar ($\ell = 0$) scaling fields, one has:

$$
\begin{aligned}
< \varphi_1(\vec{r}_1)\varphi_2(\vec{r}_2) > \ &= \ < \varphi_1(0)\varphi_2(\vec{r}_1 - \vec{r}_2) > \ \text{(translations)} \\
&= \ < \varphi_1(0)\varphi_2(|\vec{r}_1 - \vec{r}_2|) > \ \text{(scalar fields)} \\
&= \ b^{-x_1}b^{-x_2} < \varphi_1(0)\psi_2(|\vec{r}_1 - \vec{r}_2|/b) > \ \text{(dilatation)}.
\end{aligned}
$$

$$(2.8)$$

Taking $b = |\vec{r}_1 - \vec{r}_2|$ and defining $A = < \varphi_1(0)\varphi_2(1) >$ shows that the two-point correlator is a simple power law:

$$
< \varphi_1(\vec{r}_1)\varphi_2(\vec{r}_2) > = \frac{A}{|\vec{r}_1 - \vec{r}_2|^{x_1 + x_2}}
\tag{2.9}
$$

with a power determined by the anomalous dimensions of the fields φ_1 and φ_2. Critical exponents are connected with the divergence of (second order) derivatives of the free energy which in turn can be related to integrals over two-point correlators. A knowledge of the possible dimensions of the scaling fields of a theory implies therefore at the same time the values of all critical exponents.

Another important notion of the conventional renormalization theory that we shall need in the sequel is the 'operator algebra hypothesis' of Kadanoff and Wilson[7]. This hypothesis assumes that the product of two nearby scaling fields can be expanded in terms of all other scaling fields. Formally:

$$
\varphi_i(\vec{r})\varphi_j(0) = \sum_k C_{i,j}^k(\vec{r})\varphi_k(0).
\tag{2.10}
$$

More precisely this equation should be read inside a correlator as:

$$
< \varphi_i(\vec{r})\varphi_j(0)\varphi_1(\vec{r}_1)\ldots\varphi_n(\vec{r}_n) > = \sum_k C_{i,j}^k(\vec{r}) < \varphi_k(0)\varphi_1(\vec{r}_1)\ldots\varphi_n(\vec{r}_n) >
\tag{2.11}
$$

in which a $(n + 2)$-correlator is expanded in terms of $(n + 1)$-correlators. The expansion is convergent in a finite domain around $\vec{r} = 0$ determined by the location of $\vec{r}_1,\ldots,\vec{r}_n$ (this defines 'nearby'). It is a simple matter to show that the functions $C_{i,j}^k(\vec{r})$ are fixed by the scaling properties of the fields. For scalar fields one obtains:

$$
C_{i,j}^k(\vec{r}) = \frac{C_{i,j}^k}{|r|^{x_i + x_j - x_k}} \, .
\tag{2.12}
$$

The constants $C_{i,j}^k$ play the role of structure constants of the theory.

3. CONFORMAL TRANSFORMATIONS

A transformation which *locally* is the product of a scaling and a rotation is called a conformal transformation. For space dimensions $d \geq 3$ the group of these transformations has only a finite number of generators. Little extra information can be obtained from conformal invariance in that case. However, for $d = 2$ the conformal group is infinite dimensional. Consider a mapping $(x, y) \rightarrow (x', y')$ of the plane. Locally the mapping is given by the matrix:

$$J \equiv \begin{pmatrix} \frac{\partial x'}{\partial x} \;, & \frac{\partial x'}{\partial y} \\ \frac{\partial y'}{\partial x} \;, & \frac{\partial y'}{\partial y} \end{pmatrix} . \tag{3.1}$$

The demand that this corresponds to a product of a rotation and a dilatation amounts to:

$$J = \frac{1}{b(x, y)} \begin{pmatrix} cos\delta(x, y) \;, & sin\delta(x, y) \\ -sin\delta(x, y) \;, & cos\delta(x, y) \end{pmatrix} \tag{3.2}$$

which is equivalent to the Cauchy-Riemann relations $\frac{\partial x'}{\partial x} = \frac{\partial y'}{\partial y}$ and $\frac{\partial x'}{\partial y} = -\frac{\partial y'}{\partial x}$ for the mapping $(x, y) \rightarrow (x', y')$. If one identifies the Euclidean plane with the complex plane $\mathbf{C}$ by $z = x + iy$ and $\bar{z} = x - iy$, one concludes that conformal mappings correspond with analytic functions $z \rightarrow w(z)$. The local scale factor is given by

$$b^{-1} = \left| \frac{dw(z)}{dz} \right| \tag{3.3}$$

and the local rotation angle by

$$\delta = \arg \left[\frac{dw(z)}{dz} \right] . \tag{3.4}$$

An analytic function generates a conformal transformation that in general maps the complex plane to a different geometry (e.g. $z \rightarrow z^2$ extends $\mathbf{C}$ with a second Riemann sheet and $z \rightarrow \sqrt{z}$ is a mapping onto a half plane). However, there exists a subgroup of conformal mappings the *projective transformations* given by:

$$z \rightarrow \frac{Az + B}{Cz + D} \quad (AD - BC = 1) \tag{3.5}$$

that map the complex plane one to one on itself. It are precisely these transformations that can be extended to conformal mappings in higher dimensional spaces. The distinction between the geometry conserving subgroup of projective transformations and the other conformal mappings will be important in the sequel.

An infinitesimal conformal transformation can be written as:

$$z \rightarrow z + \varepsilon f(z) . \tag{3.6}$$

Notice that, since an analytic function can never be bounded everywhere, this transformation is only infinitesimal in a local sense. Expanding the function $f(z)$ in a Laurant series

$$z \to z + \sum_{n=-\infty}^{\infty} \varepsilon_n\, z^{n+1} \tag{3.7}$$

exhibits the infinitesimal generators of the conformal group as:

$$z \xrightarrow{\varepsilon_n} z + \varepsilon_n\, z^{n+1}, \quad n = \ldots, -2, -1, 0, 1, 2, \ldots \tag{3.8}$$

Notice that the projective subgroup is generated by the elements ε_n with $n = -1, 0, 1$ where $n = -1$ corresponds to translations, $n = 0$ to (uniform) rotations and dilatations and $n = 1$ to the inversion.

The conformal group has a canonical representation in the space of analytic functions $\{\varphi(z)\}$ via $\varphi(z) \xrightarrow{w} \varphi \circ w^{-1}(z)$. For the infinitesimal generators this yields:

$$\varphi(z) \xrightarrow{\varepsilon_n} \varphi(z - \varepsilon_n\, z^{n+1}) = \varphi(z) - \varepsilon_n\, z^{n+1} \frac{\partial}{\partial z} \varphi(z) . \tag{3.9}$$

Hence the Lie-algebra of the conformal group is represented by the elements $\ell_n = -z^{n+1} \frac{\partial}{\partial z}$ with commutation relations:

$$[\ell_n, \ell_m] = (n - m)\ell_{n+m} . \tag{3.10}$$

The Lie-algebra $\ell_{-1}, \ell_0, \ell_1$ corresponding to the projective group is a closed subalgebra.

4. CONFORMAL INVARIANCE OF CRITICAL SYSTEMS

A conformally invariant critical system is assumed to have the following properties:

(i) The Hamiltonian (action) is invariant $\mathcal{H} \to \mathcal{H}$ under the mapping $z \to w(z)$.

(ii) There exists a *subset* of the scaling fields called *primary fields* that transform under local scalings and rotations as they do under global ones by using the local values of the scale factor and rotation angle:

$$\varphi(z, \bar{z}) \to |w'(z)|^x\, e^{i\ell\, \arg[w'(z)]} \varphi(w, \bar{w}) \quad (w'(z) \equiv \frac{dw}{dz}) . \tag{4.1}$$

This can be conveniently rewritten as

$$\varphi(z, \bar{z}) \to w'(z)^\Delta\, \bar{w}'(\bar{z})^{\bar{\Delta}} \varphi(w, \bar{w}) \tag{4.2}$$

with

$$\Delta \equiv (x + \ell)/2 \ , \quad \bar{\Delta} = (x - \ell)/2 \ . \tag{4.3}$$

In what follows we shall characterize the primary scaling fields with the dimensions Δ and $\bar{\Delta}$ rather than with x and ℓ.

Under an infinitesimal conformal transformation $z \to z + \varepsilon(z)$ a primary field behaves as:

$$\begin{aligned}
\varphi(z, \bar{z}) \quad &\to \quad [1 + \varepsilon'(z)]^{\Delta}[1 + \bar{\varepsilon}'(\bar{z})]^{\bar{\Delta}} \varphi(z + \varepsilon(z), \bar{z} + \bar{\varepsilon}(\bar{z})) \\
&\simeq \quad \varphi(z, \bar{z}) + (\Delta \varepsilon'(z) + \varepsilon(z)\frac{\partial}{\partial z})\varphi(z, \bar{z}) \\
&+ \quad (\bar{\Delta}\bar{\varepsilon}'(\bar{z}) + \bar{\varepsilon}(\bar{z})\frac{\partial}{\partial \bar{z}})\varphi(z, \bar{z}) \ .
\end{aligned} \tag{4.4}$$

Hence one has:

$$\varphi(z, \bar{z}) \to \varphi(z, \bar{z}) + \partial_{\varepsilon}\varphi + \partial_{\bar{\varepsilon}}\varphi$$

with

$$\partial_{\varepsilon}\varphi \equiv \left\{ \Delta \varepsilon'(z) + \varepsilon(z)\frac{\partial}{\partial z} \right\} \varphi(z, \bar{z}) \tag{4.5}$$

and similarly for $\partial_{\bar{\varepsilon}}\varphi$.

4.1. Invariance under Projective Transformations

Let us consider the consequences of the conformal invariance for correlation functions involving primary fields. If one takes an infinitesimal mapping $z \to z + \varepsilon(z)$ one finds for a N-point correlator:

$$\begin{aligned}
< \varphi_1(z_1, \bar{z}_1) \ldots \varphi_N(z_N, \bar{z}_N) > \ &= \ < (1 + \partial_{\varepsilon} + \partial_{\bar{\varepsilon}})\varphi_1(z_1, \bar{z}_1) \ldots \\
&(1 + \partial_{\varepsilon} + \partial_{\bar{\varepsilon}})\varphi_N(z_N, \bar{z}_N) > \ .
\end{aligned} \tag{4.6}$$

The left and right hand side of this equation refer to an expectation value evaluated with respect to the same Hamiltonian due to conformal invariance. But in order that also the geometry of left and right hand side be the same we must restrict the transformations for the moment to the projective group. In the next section we shall see how this can be extended to general conformal transformations. Since it is precisely the projective group that can be generalized to higher dimensions, the conclusions of this subsection have also relevance for higher dimensional systems. Consider therefore in (4.6) $\varepsilon(z) = \varepsilon_n z^{n+1}$ with $n = -1, 0, 1$. Expanding this equation to order ε gives:

$$< \partial_{\varepsilon_n}\varphi_1 \cdot \varphi_2 \ldots \varphi_N > + \ldots < \varphi_1 \cdot \varphi_2 \ldots \partial_{\varepsilon_n}\varphi_N > = 0 \tag{4.7}$$

using (4.5) then yields:

$$\sum_{j=1}^{N}\left\{\Delta_j(n+1)z_j^n + z_j^{n+1}\frac{\partial}{\partial z_j}\right\} < \varphi_1(z_1,\bar{z}_1)\ldots\varphi_N(z_N,\bar{z}_N) >= 0 \tag{4.8}$$

and similarly for the $\bar{z}$ dependence with $\Delta_j \to \bar{\Delta}_j$. Let us first consider the consequences of these differential equations for the two-point functions. Employing the differential equations for $n=-1$ and $n=0$ (translations and dilatations) gives, as to be expected, the same result as known from renormalization theory:

$$< \varphi_1(z_1,\bar{z}_1)\varphi_2(z_2,\bar{z}_2) >= \frac{A_{1,2}}{(z_1-z_2)^{\Delta_1+\Delta_2}(\bar{z}_1-\bar{z}_2)^{\bar{\Delta}_1+\bar{\Delta}_2}}. \tag{4.9}$$

New information comes from the case $n=1$ (inversion), insertion of (4.9) in (4.8) with $n=1$ gives:

$$\frac{(\Delta_1-\Delta_2)A_{1,2}}{(z_1-z_2)^{\Delta_1+\Delta_2-1}} = 0 \tag{4.10}$$

which leads to the conclusion: $A_{1,2}=0$ unless $\Delta_1=\Delta_2$. By properly normalizing the fields one obtains:

$$< \varphi_i(z_1,\bar{z}_1)\varphi_j(z_j,\bar{z}_j) >= \frac{\delta_{i,j}}{(z_1-z_2)^{2\Delta}(\bar{z}_1-\bar{z}_2)^{2\bar{\Delta}}}. \tag{4.11}$$

Primary fields are in this sense orthogonal.

Also the three point correlators of primary fields are determined by (4.8) but now up to an amplitude:

$$< \varphi_1(z_1,\bar{z}_1)\varphi_2(z_2,\bar{z}_2)\varphi_3(z_3,\bar{z}_3) >=$$

$$\frac{A_{1,2,3}}{(z_1-z_2)^{\Delta_1+\Delta_2-\Delta_3}(z_1-z_3)^{\Delta_1+\Delta_3-\Delta_2}(z_2-z_3)^{\Delta_2+\Delta_3-\Delta_1}}$$
$$\bullet\{\bar{z}-\text{dependence}\}. \tag{4.12}$$

It will be important for the following that the amplitudes $A_{i,j,k}$ are directly related to the structure constants $C_{i,j}^k$ introduced above. This can be seen as follows. From the operator algebra hypothesis one has the relation

$$< \varphi_1(z)\varphi_2(0)\varphi_3(z_3) >= \sum_k \frac{C_{1,2}^k}{z^{\Delta_1+\Delta_2-\Delta_k}} < \varphi_k(0)\varphi_3(z_3) > \tag{4.13}$$

in which equation the $\bar{z}$ dependence is suppressed. If we split the sum in (4.13) into contributions of primary and non-primary fields and use the orthonormality of primary fields one obtains:

$$< \varphi_1(z)\varphi_2(0)\varphi_3(z_3) > = \frac{C_{1,2}^3}{z^{\Delta_1+\Delta_2-\Delta_3}} \cdot \frac{1}{z_3^{2\Delta_3}}$$
$$+ \quad \text{contributions non-primary fields}. \tag{4.14}$$

One can now compare this with the result (4.12) by setting $z_2 = 0$ and expanding with respect to $z_1 \equiv z$, this gives

$$< \varphi_1(z)\varphi_2(0)\varphi_3(z_3) > = \frac{A_{1,2,3}}{z^{\Delta_1 + \Delta_2 - \Delta_3} \cdot z_3^{2\Delta_3}} \left\{ 1 + a \frac{z_1}{z_3} + \dots \right\}. \qquad (4.15)$$

Comparing the leading[‡] terms in (4.14) and (4.15) then gives $A_{1,2,3} = C_{1,2}^3$ or in general

$$A_{i,j,k} = C_{i,j}^k . \qquad (4.16)$$

In particular this implies that the structure constants are symmetric in the indices i, j, k.

4.2. Invariance under General Conformal Transformations

A general conformal transformation, applied globally, changes the geometry in which correlation functions are evaluated. This can be very useful in relating properties of systems with boundaries to those of the infinite system[8,9]. However, presently we are interested in relations between correlation functions in the same geometry. For that purpose we consider a transformation of the plane that is only locally conformal. More specifically, in the study of the correlation function $< \varphi_1(z_1, \bar{z}_1) \dots \varphi_n(z_n, \bar{z}_n) >$ we take a contour K around the points $z_1, \dots z_n$. Consider now an infinitesimal transformation of the plane $z \to z + \varepsilon(z, \bar{z})$ with:

(i) $\varepsilon(z, \bar{z}) = \varepsilon(z)$ inside K. So inside K the mapping is analytic and thus conformal.

(ii) $\varepsilon(z, \bar{z})$ differentiable everywhere.

(iii) $\varepsilon(z, \bar{z}) \to 0$ as $|z| \to \infty$.

Notice that in view of (iii) ε can be taken sufficiently small that the mapping is uniquely invertible on $\mathbf{C}$. So there is no change of geometry, but we have now a mapping which is not conformal outside K. Hence we need the response of the Hamiltonian to a non-conformal transformation $z \to z + \varepsilon(z, \bar{z})$. A general result in field theory (compare also Electrodynamics or the theory of Elasticity) is that this response can be expressed in terms of a tensor field, called the stress-energy tensor. In our present complex notation it reads:

$$\mathcal{H} \to \mathcal{H} + \partial \mathcal{H} \qquad (4.17)$$

[‡]Strictly speaking this involves the assumption that there are no non-primary fields with the same dimension as the primary field φ_3. Later we shall see that this is in general the case as all non-primary fields have dimensions that differ an integer from that of a primary field.

with $\partial\mathcal{H}$ given by

$$\partial\mathcal{H} = \frac{1}{2\pi i} \int dz d\bar{z} \left\{ \frac{\partial\varepsilon}{\partial z} T^z{}_z + \frac{\partial\varepsilon}{\partial\bar{z}} T^{\bar{z}}{}_z + \frac{\partial\bar\varepsilon}{\partial z} T^z{}_{\bar{z}} + \frac{\partial\bar\varepsilon}{\partial\bar{z}} T^{\bar{z}}{}_{\bar{z}} \right\} . \qquad (4.18)$$

Here the stress-tensor $\vec{T}$ has the components:

$$\vec{T} = \begin{pmatrix} T^z{}_z & , & T^{\bar{z}}{}_z \\[6pt] T^z{}_{\bar{z}} & , & T^{\bar{z}}{}_{\bar{z}} \end{pmatrix} . \qquad (4.19)$$

When a theory is conformally invariant, as is presently the case, $\partial\mathcal{H}$ should vanish for $\varepsilon(z,\bar{z}) = \varepsilon(z)$. This implies $T^z{}_z = T^{\bar{z}}{}_{\bar{z}} = 0$. So only two components of the stress-tensor remain, they are usually abbreviated as:

$$T \equiv T^{\bar{z}}{}_z \quad \text{and} \quad \bar{T} \equiv T^z{}_{\bar{z}} . \qquad (4.20)$$

Hence the relation (4.18) reduces to $\partial\mathcal{H} = \partial_\varepsilon\mathcal{H} + \partial_{\bar\varepsilon}\mathcal{H}$ with:

$$\partial_\varepsilon\mathcal{H} = \frac{1}{2\pi i} \int dz d\bar{z} \, \frac{\partial\varepsilon}{\partial\bar{z}} T \qquad (4.21)$$

and similarly for $\partial_{\bar\varepsilon}\mathcal{H}$. It is left as an exercise to show that for the example of the Gaussian model (with Hamiltonian in complex notation given by $\mathcal{H} = i\int dz d\bar{z} \frac{\partial s}{\partial z} \frac{\partial s}{\partial\bar{z}}$) the elements of the stress tensor are given by $T = \left(\frac{\partial s}{\partial z}\right)^2$ and $\bar{T} = \left(\frac{\partial s}{\partial\bar{z}}\right)^2$.

Let us now apply the transformation $z \to z + \varepsilon(z,\bar{z})$ defined above to the n-point correlation function. It yields:

$$< \varphi_1 \cdot \varphi_2 \cdot \ldots >_{\mathcal{H}} = < (1 + \partial_\varepsilon + \partial_{\bar\varepsilon})\varphi_1 (1 + \partial_\varepsilon + \partial_{\bar\varepsilon})\varphi_2 \ldots >_{\mathcal{H}+\partial_\varepsilon\mathcal{H}+\partial_{\bar\varepsilon}\mathcal{H}} \qquad (4.22)$$

in which $\partial_\varepsilon\varphi$ stands for the response of the field φ to a transformation which is conformal (inside K). Expanding now in ε, and using that ε and $\bar\varepsilon$ can be made to vary independently gives:

$$< \partial_\varepsilon\varphi_1 \cdot \varphi_2 \cdot \ldots > + < \varphi_1 \cdot \partial_\varepsilon\varphi_2 \cdot \ldots > + \ldots$$

$$= \; < \partial_\varepsilon\mathcal{H} \cdot \varphi_1 \cdot \varphi_2 \cdot \ldots >$$

$$= \; \frac{1}{2\pi i} \int_{\{\text{outside } K\}} dz d\bar{z} \, \frac{\partial\varepsilon}{\partial\bar{z}} < T(z,\bar{z})\varphi_1(z_1,\bar{z}_1)\ldots > . \qquad (4.23)$$

The last equation follows from (4.21) since $\varepsilon = \varepsilon(z)$ inside K. By partial integration this equation can be rewritten as:

$$< \partial_\varepsilon\varphi_1 \cdot \varphi_2 \ldots > + \ldots - \frac{1}{2\pi i} \oint_K dz \, \varepsilon(z) < T(z,\bar{z})\varphi_1(z_1,\bar{z}_1)\ldots >$$

$$= -\frac{1}{2\pi i} \int_{\{\text{outside } K\}} dz \, d\bar{z} \, \varepsilon(z,\bar{z}) \frac{d}{d\bar{z}} < T(z,\bar{z})\varphi_1(z_1,\bar{z}_1)\ldots > .$$

$$(4.24)$$

Notice that for a given $\varepsilon(z)$ in the LHS the function $\varepsilon(z, \bar{z})$ in the RHS can be chosen freely. Hence one concludes:

$$\frac{d}{d\bar{z}} < T(z, \bar{z}) \varphi_1(z_1, \bar{z}_1) \ldots > = 0 \qquad (4.25)$$

for z outside K. Since K can be deformed to arbitrary small contours surrounding $z_1, \ldots, z_n$ this conclusion holds in fact for $z \neq z_1, z_2, \ldots, z_n$. The correlation function $< T(z) \varphi_1(z_1, \bar{z}_1) \ldots >$ is therefore, viewed as a function of z, analytic with possible poles at $z = z_1, z_2, \ldots, z_n$.

The precise form of this function can be obtained from (4.24). Since we know already that the RHS vanishes this equation reduces to:

$$< \partial_\varepsilon \varphi_1 \cdot \varphi_2 \ldots > \; + \; < \varphi_1 \cdot \partial_\varepsilon \varphi_2 \ldots > + \ldots$$
$$= \frac{1}{2\pi i} \oint_K \varepsilon(z) < T(z) \varphi_1(z_1, \bar{z}_1) \ldots > dz. \qquad (4.26)$$

(This relation is known as the conformal Ward identity.) Deformation of the contours to small circles surrounding $z_1, z_2, \ldots, z_n$ and using that $\varepsilon(z)$ can be chosen freely for each of these contours gives:

$$\partial_\varepsilon \varphi(z, \bar{z}_1) = \frac{1}{2\pi i} \oint dz \, \varepsilon(z) T(z) \varphi(z_1, \bar{z}_1) . \qquad (4.27)$$

A similar equation holds for $\partial_{\bar{\varepsilon}}$ with T replaced by $\overline{T}$. This is the first piece of information that can be obtained from the conformal Ward identity. It gives the response of an *arbitrary* field to a conformal transformation as a contour integral involving the field $T(z)$. In this sense the stress-tensor is the generator of conformal transformations. The second piece of information concerns correlation functions of the type $< T(z) \varphi_1(z_1, \bar{z}_1) \ldots \varphi_n(z_n, \bar{z}_n) >$ where the φ_i are *primary* fields. For a primary field the response to a conformal transformation is given by (4.5) as

$$\partial_\varepsilon \varphi(z_1, \bar{z}_1) = \left\{ \Delta \varepsilon'(z_1) + \varepsilon(z_1) \frac{\partial}{\partial z_1} \right\} \varphi(z_1, \bar{z}_1) . \qquad (4.28)$$

This can also be written as a contour integral:

$$\partial_\varepsilon \varphi(z_1, \bar{z}_1) = \frac{1}{2\pi i} \oint dz \, \varepsilon(z) \left\{ \frac{\Delta}{(z - z_1)^2} + \frac{1}{z - z_1} \frac{\partial}{\partial z_1} \right\} \varphi(z_1, \bar{z}_1) \qquad (4.29)$$

where the contour encircles the point z_1. Comparison with (4.27) then yields:

$$T(z) \varphi(z_1, \bar{z}_1) = \left\{ \frac{\Delta}{(z - z_1)^2} + \frac{1}{(z - z_1)} \frac{\partial}{\partial z_1} \right\} \varphi(z_1, \bar{z}_1) + \{\text{analytic part}\}. \qquad (4.30)$$

This relation as it stands is a formal relation which should in fact be read inside a correlation function. It expresses then the singular behaviour of that correlation function as $z \to z_1$.

Collecting similar behaviour at other poles and using the fact that correlation functions are regular at infinity leads to:

$$< T(z)\varphi_1(z_1, \bar{z}_1)\ldots\varphi_n(z_n, \bar{z}_n) >= \sum_{k=1}^{n}\left\{\frac{\Delta_k}{(z-z_k)^2} + \frac{1}{z-z_k}\frac{\partial}{\partial z_k}\right\} < \varphi_1(z_1, \bar{z}_1)\ldots\varphi_n(z_n, \bar{z}_n) >$$

$$(4.31)$$

This is the announced relation that exhibits all the poles of the function $< T(z)\varphi_1 \ldots >$. This relation, valid only for primary fields $\varphi_1, \varphi_2, \ldots$, is called the local Ward identity.

5. THE VIRASORO ALGEBRA

In the previous section we have seen that the conformal change of a general field is generated by the stress tensor $T(z)$. In order to obtain the structure of the Lie-algebra of the conformal group we need first to find the transformation properties of $T(z)$ itself.

5.1. Transformation Properties of $T(z)$

We start with the observation that from the conformal Ward identity (4.31) it is clear that $T(z)$ is a scaling field with dimensions $\Delta = 2$ and $\bar{\Delta} = 0$. This means $x = \Delta + \bar{\Delta} = 2$ and $\ell = \Delta - \bar{\Delta} = 2$ in accordance with the fact that $T(z)$ is a tensor field. Suppose for a moment that $T(z)$ is also a primary field. Then its two-point correlation function can be calculated as a simple application of (4.31):

$$< T(z)T(z') >= \left\{\frac{2}{(z-z')^2} + \frac{1}{(z-z')}\frac{\partial}{\partial z'}\right\} < T(z') >= 0. \qquad (5.1)$$

The result vanishes as the expectation value of any scaling field (not equal to the identity) vanishes. This result is contrary to the fact that for a critical theory this two-point correlation is given by:

$$< T(z)T(z') >= \frac{\frac{1}{2}c}{(z-z')^4} . \qquad (5.2)$$

It is left as an exercise to show that for the Gaussian model (with $T(z) = \left(\frac{\partial s}{\partial z}\right)^2$) this formula is obtained with $c = 1$. So we conclude that the field $T(z)$ must be non-primary. The transformation law (4.5) for primary fields can be generalized by including higher derivatives (which 'feel' the difference between a global and local rescaling). The response of the field $T(z)$ to a conformal transformation can then be written as:

$$\partial_\varepsilon T = \varepsilon(z)\frac{\partial T}{\partial z} + 2\varepsilon'(z)T(z) + \alpha\varepsilon''(z)V(z) + \beta\varepsilon'''(z)I + \ldots. \qquad (5.3)$$

Simple dimension counting shows that the first field $\frac{\partial T}{\partial z}$ in this formula has dimension $\Delta = 3$, the second one $T(z)$ has $\Delta = 2$. Continuing in this way we find that $V(z)$ has dimension

$\Delta = 1$ while I has dimension $\Delta = 0$ and corresponds therefore to the identity. The fields represented by the dots in (5.3) all come out to have negative dimensions which is not allowed in a physical theory and we are left with only four terms. A further plausible requirement is that the theory after inversion $z \to \frac{1}{z}$ is described by the same stress tensor. This means that the stress tensor $T(z)$ should behave like a primary field at least under the projective group (such fields are called quasi-primary). Since the inversion is generated by the infinitesimal transformation $z \to z + \varepsilon z^2$ this implies $\alpha = 0$. So we are left with:

$$\partial_\varepsilon T = \varepsilon(z)\frac{\partial T}{\partial z} + 2\varepsilon'(z)T(z) + \beta\varepsilon'''(z)I. \tag{5.4}$$

The extra term in this relation changes the conformal Ward identity precisely in such a way that one can now accommodate the result (5.2). Indeed rewriting (5.4) again as a contour integral

$$\partial_\varepsilon T = \frac{1}{2\pi i}\oint dz'\varepsilon(z')\left\{\frac{1}{z'-z}\frac{\partial T}{\partial z} + \frac{2}{(z'-z)^2}T(z) + \frac{6\beta}{(z'-z)^4}\right\} \tag{5.5}$$

gives on comparison with the general formula

$$\partial_\varepsilon T = \frac{1}{2\pi i}\oint dz'\,\varepsilon(z')T(z)T(z') \tag{5.6}$$

the result:

$$T(z)T(z') = \frac{1}{z-z'}\frac{\partial T}{\partial z} + \frac{2}{(z'-z)^2}T(z) + \frac{6\beta}{(z'-z)^4}. \tag{5.7}$$

On taking the expectation value this gives

$$< T(z)T(z') > = \frac{6\beta}{(z'-z)^4} \tag{5.8}$$

and one concludes $\beta = c/12$. So we are left with a transformation law for the stress-tensor that reads:

$$\partial_\varepsilon T(z) = \varepsilon(z)\frac{\partial T}{\partial z} + 2\varepsilon'(z)T(z) + \frac{c}{12}\varepsilon'''(z) . \tag{5.9}$$

The anomalous term appearing in this formula is called the 'Schwinger term' in field theory. The constant c that will turn out to be characteristic for a given universality class is called the *central charge*.

5.2. Operator Representation

In order to find a representation of the Lie-algebra of the conformal transformations it is useful to represent first the scaling fields $\{\varphi\}$ as operators in a Hilbert space. This is done as usual via the transfer matrix formalism. However, the *direction* of the transfer is chosen

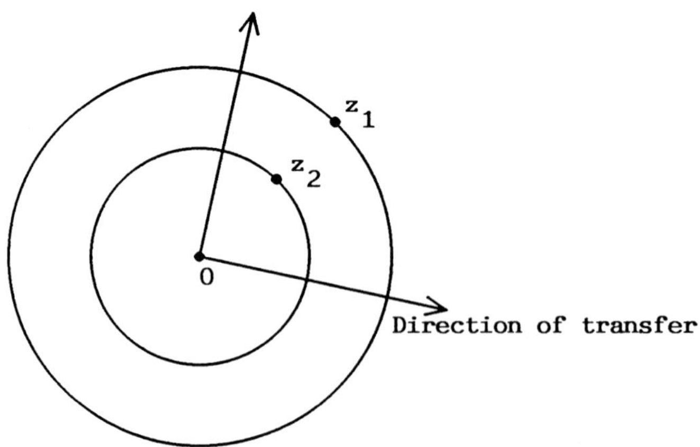

FIGURE 1

somewhat unusually here along the direction of the radius vector. In other words the transfer matrix connects configurations at concentric circles around the origin. In this transfer matrix formalism a correlation function can be written as:

$$< \varphi_1(z_1)\varphi_2(z_2)\ldots\varphi_n(z_n) >=< 0|R\{\hat{\varphi}_1(z_1)\hat{\varphi}_2(z_2)\ldots\hat{\varphi}_n(z_n)\}|0 > \qquad (5.10)$$

where each field $\varphi(z)$ is represented by a Hilbert space operator $\hat{\varphi}(z)$ (de $\bar{z}$ dependence has been suppressed) and R stands for *radial ordering* i.e. the operators in the RHS of (5.10) should be ordered according to decreasing $|z_i|$ value. The 'vacuum' state $|0>$ is the ground state of the transfer matrix.

It is consistent w17jith the inversion symmetry of (quasi-)primary fields that the Hermitian conjugate of an operator corresponding to such a field is given by:

$$\hat{\varphi}(z)^* = \bar{z}^{-2\Delta}\hat{\varphi}(\frac{1}{\bar{z}}) . \qquad (5.11)$$

Indeed, take $|z_1| > |z_2|\ldots > |z_n|$ then:

$$\begin{aligned}
< \varphi_1(z_1)\ldots\varphi_n(z_n) > &= < 0|\hat{\varphi}_1(z_1)\ldots\hat{\varphi}_n(z_n)|0 > \\
&= < 0|\hat{\varphi}_n(z_n)^*\ldots\hat{\varphi}_1(z_1)^*|0 >^{cc} .
\end{aligned} \qquad (5.12)$$

Assuming (5.11) one arrives at:

$$< 0|\hat{\varphi}_n(z_n)^* \ldots \varphi_1(z_1)^*|0 >^{cc} \; = \; < 0|\hat{\varphi}_n\left(\frac{1}{\bar{z}_n}\right) \ldots \hat{\varphi}_1\left(\frac{1}{\bar{z}_1}\right)|0 >^{cc} z_1^{-2\Delta_1} \ldots z_n^{-2\Delta_n}$$

$$= \; < \varphi_n\left(\frac{1}{z_n}\right) \ldots \varphi\left(\frac{1}{z_1}\right) > z_1^{-2\Delta_1} \ldots z_n^{-2\Delta_n}$$

$$= \; < \varphi_1(z_1) \ldots \varphi_n(z_n) > \tag{5.13}$$

where one uses in the second equation the fact that $\frac{1}{|z_n|} > \frac{1}{|z_{n-1}|} > \ldots \left|\frac{1}{z_1}\right|$. The last equation follows from the transformation law (4.2) applied to the inversion.

5.3. Definition of the Virasoro Algebra

One defines for every integer n an operator

$$\hat{L}_n \equiv \frac{1}{2\pi i} \oint \hat{T}(z) z^{n+1} dz \tag{5.14}$$

where the contour encircles the origin. In a similar way operators $\hat{\bar{L}}_n$ are associated with $\overline{T}(z)$. The $\{\hat{L}_n\}$ can be considered as the operators appearing in the Laurant expansion of the operator $\hat{T}(z)$ that represents the stress-tensor field $T(z)$. This is made explicit by the inverse formula:

$$\hat{T}(z) = \sum_{n=-\infty}^{\infty} \frac{\hat{L}_n}{z^{n+2}} \; . \tag{5.15}$$

From the relation $\hat{T}(z)^* = T\left(\frac{1}{\bar{z}}\right) \bar{z}^{-4}$ it can be directly derived that

$$\hat{L}_n^* = \hat{L}_{-n}. \tag{5.16}$$

The collection of operators $\{\hat{L}_n\}$ defines the Virasoro algebra. It is easy to see that the response of an operator $\hat{\varphi}(z)$ to an infinitesimal generator $z \to z + \varepsilon z^{n+1}$ of the conformal group can be expressed in terms of $\hat{L}_n$. For a field we have from the general relation (4.27) that:

$$\partial_{\varepsilon_n}\varphi(z) = \frac{\varepsilon}{2\pi i} \oint_c (z')^{n+1} T(z') \varphi(z) dz' \tag{5.17}$$

where the contour encircles the point z. In operator language this becomes:

$$\partial_{\varepsilon_n}\hat{\varphi}(z) = \frac{\varepsilon}{2\pi i} \oint_c (z')^{n+1} R\{\hat{T}(z')\hat{\varphi}(z)\} dz'. \tag{5.18}$$

Now one can deform the contour c into two contours c_1 and c_2 that are circles centered at the origin (see figure). One obtains in this way:

$$\partial_{\varepsilon_n}\hat{\varphi}(z) \; = \; \frac{\varepsilon}{2\pi i} \oint_{c_1} (z')^{n+1} \hat{T}(z')\hat{\varphi}(z) dz' - \frac{\varepsilon}{2\pi i} \oint_{c_2} (z')^{n+1} \hat{\varphi}(z)\hat{T}(z') dz'$$

$$= \; \varepsilon\{\hat{L}_n\hat{\varphi}(z) - \hat{\varphi}(z)\hat{L}_n\}. \tag{5.19}$$

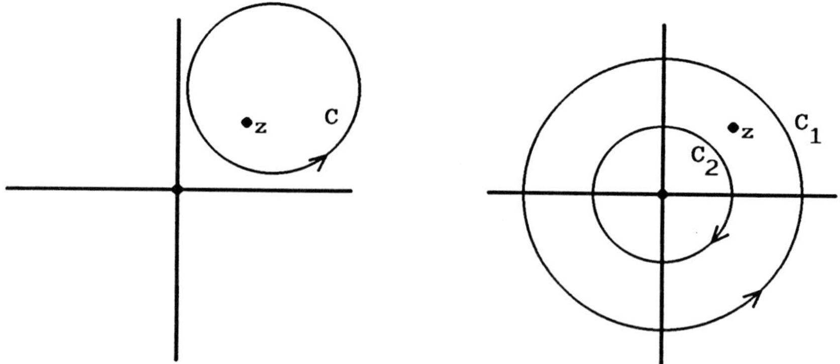

FIGURE 2

The order of the operators in this formula is dictated by the radial ordering operator R. Hence one arrives at the conclusion:

$$\partial_{\varepsilon_n}\hat{\varphi}(z) = \varepsilon \left[\hat{L}_n, \hat{\varphi}(z)\right]. \tag{5.20}$$

The next point is to investigate the structure of the Virasoro algebra $\{\hat{L}_n\}$. Commutation relations for this algebra follow by applying (5.20) for the conformal change of the operator $\hat{T}(z)$ itself:

$$\partial_{\varepsilon_n}\hat{T}(z) = \varepsilon \left[\hat{L}_n, \hat{T}(z)\right]. \tag{5.21}$$

After insertion of the Laurant expansion (5.15) this gives:

$$\partial_{\varepsilon_n}\hat{T}(z) = \varepsilon \sum_m \frac{[\hat{L}_n, \hat{L}_m]}{z^{m+2}}. \tag{5.22}$$

From the other hand we have for $\partial_\varepsilon T(z)$ the explicit relation (5.9) which for $z \to z + \varepsilon z^{n+1}$ yields:

$$
\begin{aligned}
\partial_{\varepsilon_n}\hat{T}(z) &= \varepsilon \left\{ z^{n+1}\frac{\partial \hat{T}}{\partial z} + 2(n+1)z^n \hat{T}(z) + \frac{c}{12}n(n^2-1)z^{n-2} \right\} \\
&= \varepsilon \left\{ \sum_m \frac{(n-m)\hat{L}_{n+m}}{z^{m+2}} + \frac{c}{12}\frac{n(n^2-1)}{z^{-n+2}} \right\}.
\end{aligned}
\tag{5.23}
$$

Comparing the expansions (5.22) and (5.23) gives:

$$[\hat{L}_n, \hat{L}_m] = (n-m)\hat{L}_{n+m} + \frac{c}{12}n(n^2-1)\delta_{n,-m}. \tag{5.24}$$

These are the commutation relations that constitute the Virasoro algebra. In the theory of Lie-algebra's this algebra is known as the central extension of the conformal algebra $\{\ell_n\}$ defined in (3.10).

5.4. Representation Virasoro Algebra

Representations of the Virasoro algebra can be constructed starting from a state $|\varphi >$ that can be associated to each primary operator $\hat{\varphi}(z)$ by:

$$|\varphi >\equiv \hat{\varphi}(0)|0 > . \qquad (5.25)$$

The state $|\varphi >$ acts as a vacuum for all $\hat{L}_n$ with $n > 0$, i.e.:

$$\hat{L}_n|\varphi >= 0 \ \text{ for } \ n > 0. \qquad (5.26)$$

This can be seen as follows. Consider primary fields at $z_2, z_3, \ldots z_m$ with $|z_m| > |z_{m-1}| > \ldots > |z_2|$. Then one has:

$$
\begin{aligned}
< 0|\hat{\varphi}_m(z_m)\ldots \hat{L}_n|\varphi > \ &= \ \frac{1}{2\pi i} \oint < 0|\hat{\varphi}_m(z_m)\ldots \hat{T}(z)\hat{\varphi}(0)|0 > z^{n+1}dz \\
&= \ \frac{1}{2\pi i} \oint < \varphi_m(z_m)\ldots T(z)\varphi(0) > z^{n+1}dz
\end{aligned}
$$

$$(5.27)$$

where the contour encircles the origin and $z_2, \ldots z_m$ are outside the contour. From the local Ward identity (4.31) one knows that the integrand of (5.27) has the structure:

$$< \varphi_m(z_m)\ldots T(z)\varphi(0) >= \left\{\frac{\Delta}{z^2} + \frac{1}{z}\frac{\partial}{\partial z_1}\right\} < \varphi_m(z_m)\ldots \varphi(z_1) >_{z_1=0} + \ \{\text{regular terms}\}.$$

$$(5.28)$$

As the contour integral (5.27) for $n > 0$ selects poles of order higher than two, one concludes that $\hat{L}_n|\varphi >= 0$. Notice further that the pole of order two has the residue $\Delta < \varphi_m(z_m)\ldots \varphi(0) >$ which implies:

$$\hat{L}_0|\varphi >= \Delta|\varphi > . \qquad (5.29)$$

So the dimension of the field φ is obtained as eigenvalue of the operator $\hat{L}_0$ with the eigenvector $|\varphi >$. The operators $\{\hat{L}_n\}$ with n *negative* built, when applied to the state $|\varphi >$ a representation space of the Virasoro algebra. In the following it will be important that matrix elements involving such states can be obtained from the action of a differential operator on a correlation function. This differential operator can be found by using the fact that the regular terms in (5.28) (which are selected in (5.27) once n is negative) come from the poles

$z = z_2, z_3, \ldots z_m$ and are explicitly given by (4.31).

Hence it follows that:

$$< 0|\hat{\varphi}_m(z_m)\ldots\hat{L}_n|\varphi >= \mathcal{L}_n < \varphi_m(z_m)\ldots\varphi(0) > , \quad n < 0$$

with the differential operator $\mathcal{L}_n$ given by:

$$\mathcal{L}_n = -\sum_{k\neq 1}\left\{\frac{\Delta_k(n+1)}{z_k^{-n}} + \frac{1}{z_k^{-n-1}}\frac{\partial}{\partial z_k}\right\} < \varphi_m(z_m)\ldots\varphi(0) > . \qquad (5.30)$$

The states in the representation space can be ordered according to their dimension (i.e. eigenvalue of $\hat{L}_0$). Consider for example the state $\hat{L}_{-n}|\varphi > (n > 0)$, it is a direct consequence of the commutation relations (5.24) that:

$$\hat{L}_0\hat{L}_{-n}|\varphi >= \hat{L}_{-n}\hat{L}_0|\varphi > +n\hat{L}_{-n}|\varphi >= (\Delta + n)\hat{L}_{-n}|\varphi > . \qquad (5.31)$$

Thus $\hat{L}_{-n}|\varphi >$ has dimension $\Delta + n$. More generally the states $\hat{L}_{-n_1}\hat{L}_{-n_2}\ldots\hat{L}_{-n_k}|\varphi >$ with $n_1 + n_2 + \ldots n_k = n$ also have dimension $\Delta + n$. The collection of all these states form a representation space of the Virasoro algebra which is known as the 'conformal tower' (or Verma-module) and may be represented in the following table:

TABLE 1

States	Dimension
$|\varphi >$	Δ
$\hat{L}_{-1}|\varphi >$	$\Delta + 1$
$\hat{L}_{-2}|\varphi >, \hat{L}^2_{-1}|\varphi >$	$\Delta + 2$
$\hat{L}_{-3}|\varphi >, \hat{L}_{-2}\hat{L}_{-1}|\varphi >, \hat{L}^3_{-1}|\varphi >$	$\Delta + 3$
$\ldots$	$\ldots$

Notice that the operators $\hat{L}_{-n}$ (with n positive) can be considered as creation operators that bring one to lower levels of the tower. The operators $\hat{L}_n = \hat{L}^*_{-n}$ are the corresponding annihilation operators while $|\varphi >$ serves as a new vacuum.

All states appearing in this tower (except the top one $|\varphi >$) correspond to non-primary fields. This can again be seen from the example of the state $\hat{L}_{-n}|\varphi >$. Calculate $\hat{L}_n\hat{L}_{-n}|\varphi >$, from the commutation relations one has:

$$\hat{L}_n\hat{L}_{-n}|\varphi> = \hat{L}_{-n}\hat{L}_n|\varphi> + 2n\hat{L}_0|\varphi> + \frac{c}{12}n(n^2-1)|\varphi> = \left\{2n\Delta + \frac{c}{12}n(n^2-1)\right\}|\varphi> \neq 0.$$

$$(5.32)$$

Since for a primary state one has $\hat{L}_n|\psi> = 0$ this shows that $\hat{L}_{-n}|\varphi> = |\psi>$ cannot be primary. Conversely it can be shown that a state that satisfies $\hat{L}_n|\psi> = 0$ $(n > 0)$ does arise from a primary field. This can be used to argue that every vector in the Hilbert space constructed for the transfer matrix formalism, must belong to one of the conformal towers. Consider a basis $|\psi>$ in which $\hat{L}_0$ is diagonal so that each $|\psi>$ has a given dimension Δ. Now acting on $|\psi>$ with operator $\hat{L}_n$ leads to new states with dimensions $\Delta - n$. Since in a physical theory negative dimensions are excluded, one must finally arrive at a state $|\psi_0>$ for which $\hat{L}_n|\psi_0> = 0$ for al positive n. This state is the primary state of which $|\psi>$ is a descendant.

6. MINIMAL AND UNITARY THEORIES

We have now formulated the general framework of conformally invariant theories. From here one can proceed from either of two assumptions. In the first one the interest is directed towards systems that are particularly simple in the sense that the structure constants $C_{i,j}^k$ that appear in the operator product of primary fields are such that there exists a *finite* closed subalgebra of these fields. These theories are called 'minimal'. In the second one the demand that the theory be unitary (positive definite transfer matrix) is used to restrict the possible number of primary fields. It will turn out that these unitary theories form a subclass of the minimal theories. For both a study of the degenerate representations of the Virasoro algebra will be essential.

6.1. Degenerate fields

Suppose $|\varphi>$ is the state corresponding to a primary field with dimension Δ. In general the descendants $\hat{L}_{-n_1}\hat{L}_{-n_2}\ldots|\varphi>$, $n_1 + n_2 + \ldots = n$ form a linearly independent basis at level n of the conformal tower of $|\varphi>$. However for special values of Δ and the central charge c these states turn out to be dependent. The field φ is then called to be degenerate at level n. Let us consider as an example the degeneracy at level $n = 2$. At this level there are only two states to consider: $\hat{L}_{-2}|\varphi>$ and $\hat{L}_{-1}^2|\varphi>$. The linear relationship between these two states is obtained from the matrix

$$M_2 = \begin{pmatrix} <\varphi|\hat{L}_{-1}^{*2}\hat{L}_{-1}^{2}|\varphi> & , & <\varphi|\hat{L}_{-2}^{*}\hat{L}_{-1}^{2}|\varphi> \\ <\varphi|L_{-1}^{*2}\hat{L}_{-2}|\varphi> & , & <\varphi|\hat{L}_{-2}^{*}\hat{L}_{-2}|\varphi> \end{pmatrix} . \tag{6.1}$$

The field φ is degenerate at level $n = 2$ when $\det M_2 = 0$. The elements of this matrix can be evaluated with the use of the commutation relations. For example:

$$\begin{aligned} <\varphi|\hat{L}_{-2}^{*}\hat{L}_{-2}|\varphi> &= <\varphi|\hat{L}_2\hat{L}_{-2}|\varphi> \\ &= <\varphi|4\hat{L}_0 + \frac{1}{2}c|\varphi> \\ &= (4\Delta + \frac{1}{2}c) . \end{aligned} \tag{6.2}$$

In this way one obtains

$$M_2 = \begin{pmatrix} 8\Delta^2 + 4\Delta & 6\Delta \\ 6\Delta & 4\Delta + \frac{1}{2}c \end{pmatrix} . \tag{6.3}$$

Hence $\det M_2 = 0$ leads to

$$8\Delta^2 + (c - 5)\Delta + \frac{1}{2}c = 0 \tag{6.4}$$

with solutions:

$$\Delta_{\pm} = \frac{5 - c}{16} \pm \frac{1}{16}\sqrt{(25 - c)(1 - c)} . \tag{6.5}$$

So, for a theory with a given value of the central charge, fields with dimension $\Delta = \Delta_+$ or $\Delta = \Delta_-$ are degenerate at level $n = 2$. For such a field there exists a linear combination of $\hat{L}_{-1}^2|\varphi>$ and $\hat{L}_{-2}|\varphi>$ that gives the null-state:

$$|\chi> \equiv \hat{L}_{-1}^2|\varphi> - \lambda\hat{L}_{-2}|\varphi> = 0 \quad \text{for} \quad \lambda = \frac{2}{3}(2\Delta_{\pm} + 1) . \tag{6.6}$$

This implies that correlation functions, involving the field φ, satisfy a new differential equation (in addition to the ones that follow from invariance under the projective transformations). Indeed one has:

$$\begin{aligned} 0 &= <0|\hat{\varphi}_m \ldots \hat{\varphi}_2|\chi> \\ &= <0|\hat{\varphi}_m \ldots \hat{\varphi}_2\hat{L}_{-1}^2|\varphi> - \lambda <0|\hat{\varphi}_m \ldots \hat{\varphi}_2\hat{L}_{-2}|\varphi> \\ &= \{\mathcal{L}_{-1}^2 - \lambda\mathcal{L}_{-2}\} <\varphi_m(z_m) \ldots \varphi_2(z_2)\varphi(0)> \end{aligned} \tag{6.7}$$

where $\mathcal{L}_n$ is the differential operator defined in (5.30). Hence the correlation function $<\varphi_m(z_m) \ldots \varphi(0)>$ satisfies a second order differential equation. More generally a correlation function with a field that is degenerate at level n obeys a nth-order differential equation. These differential equations can be used to determine four and higher point correlators[4, 10, 11], but as noted originally by BPZ they can also be used to obtain selection rules for the structure constants $C^i_{j,k}$ of degenerate, primary fields.

The result (6.5) is generalized to fields degenerate at level n by a result due to V. Kac[12]: Fields $\varphi_{k,m}$ with a dimension $\Delta_{k,m}$ given by

$$\Delta_{k,m} = \frac{(c-1)}{24} + (\frac{1}{2}\alpha_+(c)k + \frac{1}{2}\alpha_-(c)m)^2$$
$$\text{with } \alpha_\pm(c) = \frac{\sqrt{1-c} \pm \sqrt{25-c}}{\sqrt{24}} \tag{6.8}$$

are degenerate at level $n = k \cdot m$. In particular one has the correspondence $\Delta_{1,2} = \Delta_-$ and $\Delta_{2,1} = \Delta_+$. In the remainder of these lectures I shall restrict the discussion to theories with a central charge $0 < c < 1$ such that $\alpha_\pm$ are real. It is then customary to represent the degenerate fields in a so-called 'conformal grid' with lattice points (k, m), see figure 3. The line drawn has an angle θ with $tg\theta = -\alpha_-/\alpha_+$ with the m-axis. The dimension of the field $\varphi_{k,m}$ may then be found as:

$$\Delta_{k,m} = \frac{c-1}{24} + d^2_{k,m} \tag{6.9}$$

where $d_{k,m}$ is the distance of the point (k, m) to the line drawn.

6.2. Minimal Theories

The construction of minimal theories opens with a discussion of the selection rules for the structure constants that appear in the operator product expansion of degenerate fields. Consider the operator product expansion of two degenerate fields for example $\varphi_{1,2}$ and $\varphi_{k,m}$. It reads:

$$\varphi_{1,2}(0)\varphi_{k,m}(z) = \sum_\alpha \frac{C(1,2|k,m|\alpha)}{z^{\Delta_{1,2}+\Delta_{k,m}-\Delta_\alpha}} \varphi_\alpha^{(0)} \tag{6.10}$$

where the sum over α runs in principle over all primary fields both degenerate and non-degenerate. The structure constants C may, as we have seen in (4.16), be obtained from the three-point correlator:

$$< \varphi_{1,2}(z)\varphi_{k,m}(z_1)\varphi_\alpha(z_2) >=$$
$$\frac{C(1,2|k,m|\alpha)}{(z_1 - z)^{\Delta_{1,2}+\Delta_{k,m}-\Delta_\alpha}(z_2 - z)^{\Delta_{1,2}+\Delta_\alpha-\Delta_{k,m}}(z_1 - z_2)^{\Delta_{k,m}+\Delta_\alpha-\Delta_{1,2}}} \cdot \tag{6.11}$$

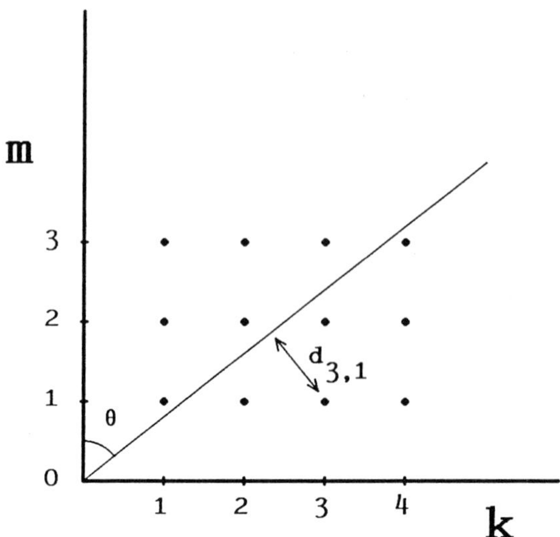

FIGURE 3

However, since $\varphi_{1,2}$ is degenerate at level 2 this correlator should obey the differential equation

$$\{\mathcal{L}^2_{-1} - \lambda \mathcal{L}_{-1}\} < \varphi_{1,2}(z)\ldots >= 0 \ . \tag{6.12}$$

This yields (for $C \neq 0$) a (second order) algebraic equation for the dimension Δ_α with the result:

$$\Delta_\alpha = \Delta_{k,m-1} \ \text{ or } \ \Delta_\alpha = \Delta_{k,m+2} \ . \tag{6.13}$$

So the conclusion is that only degenerate fields occur in the operator algebraic product of $\varphi_{1,2}$ and $\varphi_{k,m}$. The general result for the product of $\varphi_{p,q}$ and $\varphi_{k,m}$ (employing both the degeneracy of $\varphi_{p,q}$ and $\varphi_{k,m}$) is

$$\varphi_{p,q} \cdot \varphi_{k,m} = \{\varphi_{k',m'}\} \ \text{ with } \ k' \ = \ 1 + |k - p|, 1 + |k - p| + 2, \ldots, k + p - 1$$

$$m' \ = \ 1 + |m - q|, 1 + |m - q| + 2, \ldots, m + q - 1 \ . \tag{6.14}$$

Hence the degenerate fields form a (infinite) closed subalgebra.

BPZ now proceed by noting that for special values of c namely

$$c = 1 - 6 \frac{(p - q)^2}{qp} \qquad p, q = 1, 2, \ldots \tag{6.15}$$

this algebra contains *finite* closed subalgebra's. For these c-values one has

$$tg\theta = -\frac{\alpha_-}{\alpha_+} = p/q \tag{6.16}$$

and it follows from the Kac-formula (6.8) that $\Delta_{k,m} = \Delta_{p-k,q-m}$. Hence $\varphi_{k,m}$ is degenerate at level $k \cdot m$ *and* at level $(p-k) \cdot (q-m)$. It is then easy to show that the selection rules (6.14) imply that

$$\{\varphi_{k,m}\} \text{ with } 0 < k < p \text{ and } 0 < m < q \tag{6.17}$$

forms a finite closed subalgebra. It constitutes as such the most simple critical theory, it is called a minimal theory.

6.3. Unitary Theories

Suppose that the inner product of the Hilbert space is positive definite. Physically this corresponds to the assumption that all eigenvalues of the transfer matrix are positive. This implies in particular that all dimensions of the theory are positive. The reason is that dimensions occur as eigenvalues of the operator $\hat{L}_0 : \hat{L}_0|\varphi> = \Delta|\varphi>$. From the commutation relations one has $2\hat{L}_0 = [\hat{L}_1, \hat{L}_{-1}]$ and using that for a primary field $\hat{L}_1|\varphi> = 0$ this gives $\hat{L}_0|\varphi> = 2\hat{L}_1 \cdot \hat{L}_{-1}|\varphi> = 2\hat{L}_{-1}^*\hat{L}_{-1}|\varphi>$ and hence Δ is positive for a positive definite inner product. Now in the (infinite) algebra of degenerate fields one has (6.9) $\Delta_{k,m} = (c-1)/24 + d_{k,m}^2$. The first term is for $c < 1$ negative, while the second term (the distance between a lattice point and a line) can, by an appropriate choice of k, m be made arbitrary small. So negative dimensions are bound to occur in the algebra of degenerate fields. The only way to avoid this, is to take a value of c for which a closed *finite* algebra exists. It is therefore a necessary condition for unitarity of a theory containing degenerate fields to be minimal. Friedan, Qiu and Shenker[6] have sharpened this (by employing the change in sign for the eigenvalues of the matrices M_n implied by the Kac-formula) to the following result: A conformal invariant theory with $0 < c < 1$ can only be unitary when

$$\text{(i) } c = 1 - 6/p(p+1) \quad p = 3, 4, \ldots$$

$$\text{(i.e. } q = p+1 \text{ in the minimal theory)}$$

$$\text{(ii) } \varphi = \varphi_{k,m} \text{ with } 0 < k < p; \; 0 < m < p+1 \tag{6.18}$$

i.e. the only allowed fields are the degenerate fields of the minimal theory with dimensions given by the Kac-formula.

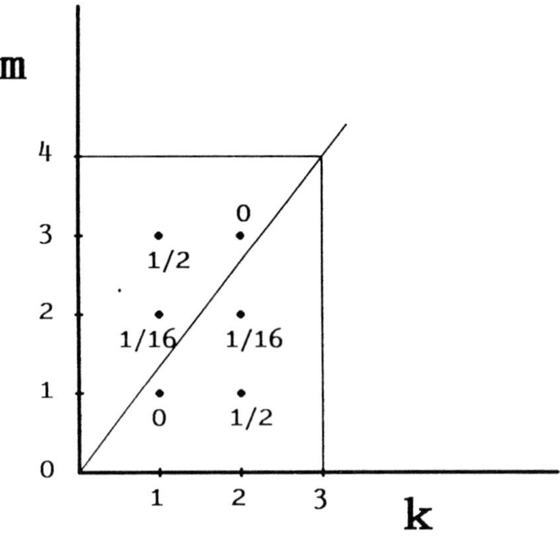

FIGURE 4

7. APPLICATION TO SPECIFIC MODELS

We have seen that conformal invariance together with unitarity severely restricts the possible scaling dimensions of a critical theory. In order to apply this to a specific model one must still find out:

(i) the value of p for the central charge of this model

(ii) investigate which of the fields, allowed by unitarity, do in fact occur.

Let us first consider the most simple case $p = 3$.

7.1. The Ising Model

The unitary theory with $p = 3$ $(c = 1/2)$ has only two non-trivial scaling fields as shown in the conformal grid for this case (figure 4). It follows from the Kac-formula (with $\alpha_+ = 2/3\sqrt{3}$ and $\alpha_- = -1/2\sqrt{3}$) that

$$
\begin{aligned}
\Delta_I &\equiv \Delta_{1,1} = \Delta_{2,3} = 0 \\
\Delta_H &\equiv \Delta_{2,2} = \Delta_{1,2} = 1/16 \\
\Delta_E &\equiv \Delta_{2,1} = \Delta_{1,3} = 1/2 \; .
\end{aligned}
\tag{7.1}
$$

The same values of course apply for the dimensions $\bar{\Delta}$ that describe the $\bar{z}$ dependence of the fields. Scalar fields with $\Delta = \bar{\Delta}$ have therefore in this theory the scaling dimensions

$$
\begin{aligned}
x_H &= 2\Delta_H = 1/8 \\
x_E &= 2\Delta_E = 1
\end{aligned}
\tag{7.2}
$$

these are precisely the scaling dimensions of the two relevant scaling fields in the Ising model. Also the multiplication table of fields in the Ising model matches the selection rules (6.14). For example:

$$
\varphi_H \cdot \varphi_H = \{\varphi_I, \varphi_E\} \ .
\tag{7.3}
$$

Finally, a direct calculation[13] of the value of c for the Ising model gives $c = 1/2$. So one concludes that the conformal invariant theory with $c = 1/2$ describes the Ising model.

7.2. Modular invariance

The operator content of a conformal invariant theory, that is the fields from the conformal grid that actually occur, can in principle be found from a study of the partition sum of the model. This can be seen as follows. If one maps a critical system by the conformal map $w = L/2\pi \, \ell n \, z$ to a strip of width L, it can be shown[5] that the transfer matrix of this strip can be written as $\hat{T} = e^{-\hat{H}}$ with

$$
\hat{H} = \frac{2\pi}{L}(\hat{L}_0 + \hat{\bar{L}}_0 - c/12) \ .
\tag{7.4}
$$

Thus the partition function on a torus (which is a finite $L \times M$ strip with periodic boundary conditions) is given by:

$$
\begin{aligned}
Z(q) \equiv Tr \, e^{-M\hat{H}} &= Tr \left\{ e^{-2\pi M/L(\hat{L}_0 - c/24)} \cdot e^{-2\pi M/L(\hat{\bar{L}}_0 - c/24)} \right\} \\
&= Tr \left\{ q^{\hat{L}_0 - c/24} \, q^{\hat{\bar{L}}_0 - c/24} \right\} \\
&\text{with } q \equiv e^{-2\pi M/L} \ .
\end{aligned}
\tag{7.5}
$$

For a unitary theory the Hilbert space over which this trace should be taken consists of conformal towers belonging to a finite set of (degenerate) primary fields. In fact to each primary field $\varphi_{\Delta,\bar{\Delta}}$ there belongs a tensor product of two conformal towers one (V_Δ) constructed from $\{\hat{L}_{-n}\}$ the other ($\bar{V}_{\bar{\Delta}}$) from $\{\hat{\bar{L}}_{-n}\}$. The trace over these towers can be evaluated using the structure that is known from the commutation relations and leads to the Rocha-Caridi[14] character formulae

$$\chi_\Delta(q)\chi_{\bar\Delta}(q) = Trace_{V_\Delta \otimes \bar V_{\bar\Delta}}\, q^{\hat L_0 - c/24} \cdot q^{\hat{\bar L}_0 - c/24} \tag{7.6}$$

where $\chi_\Delta(q)$ is known explicitly. For the partition sum one then obtains:

$$Z(q) = \sum N_{\Delta,\bar\Delta}\chi_\Delta(q)\chi_{\bar\Delta}(q) \tag{7.7}$$

where the sum runs over all primary fields in the conformal grid and $N_{\Delta,\bar\Delta}$ denotes the multi-plicities of these fields.

As noted by Cardy[15] the operator content of the theory (i.e. the possible values of $N_{\Delta,\bar\Delta}$) is severely restricted by the fact that the partition sum should not depend on the parametrization of the torus that one chooses. That is Z should be invariant under *modular* transformations. In particular $Z(q) = Z(q')$ with $q' = e^{-2\pi L/M}$. It turns out[16] that this invariance, together with the fact that $N_{\Delta,\bar\Delta}$ should be positive integers allows, for a unitary theory with $c = (p)$, in general only two solutions together with six exceptional ones (for $p = 11, 12, 18, 19, 29, 30$). This is in striking correspondence with the Cartan-Dynkin classification of simple Lie-algebra's of types A, D and E. In the solution that corresponds with the A-series each primary field $\varphi_{\Delta,\bar\Delta}$ from the conformal grid appears precisely once in a combination with no spin i.e. $\Delta = \bar\Delta = \Delta_{k,m}$. The first member of this series is the Ising model for $p = 3$. The next one is the tri-critical Ising model with $p = 4$. The higher members $(p = 5, \ldots)$ of this series describe general multicritical points and correspond as noted by Huse[17] to the exactly solved[18] RSOS models of Andrews, Baxter and Forrester.

The other type of solutions, that is the D-series and the exceptional E-cases can be associated[19] with critical models constructed on the basis of Dynkin diagrams. The operator content is more complicated: some fields from the conformal grid are absent, others occur twice and also combinations with integral spin are present. The lowest member of the D-series is found at $m = 5$, it is the three-states Potts-model. Notice that for a given value of $c(p)$ in principle two (or three) critical theories that are unitary, conformal and modular invariant, are possible. For example for $p = 5$ one has the four-critical Ising model from the A-series and the three-states Potts-model from the D-series.

ACKNOWLEDGEMENT

It is a pleasure to thank O. Foda, B. Nienhuis and J. Thijssen for many discussions on the subject of these lectures.

REFERENCES

1) K.G. Wilson and J. Kogut, Phys. Rep. 12C (1974) 75.

2) M.N. Barber, Phys. Rep. 29C (1977) 1.

3) S. Ma, Modern Theory of Critical Phenomena (Benjamin, New York, 1976).

4) A.A. Belavin, A.M. Polyakov and A.B. Zamolodchikov, Nucl. Phys. B241 (1984) 333.

5) J.L. Cardy in: Phase Transitions and Critical Phenomena, Vol. 11, eds. C. Domb and J. Lebowitz (Academic Press, London, 1986).

6) D. Friedan, Z. Qiu and S. Shenker, Phys. Rev. Lett. 52 (1984) 1575.

7) L.P. Kadanoff, Phys. Rev. Lett. 23 (1969) 1430.

8) J.L. Cardy, Nucl. Phys. B240 (1984) 514.

9) T.W. Burkhardt and E. Eisenriegler, J. Phys. A18 (1985) L83.

10) VI.S. Dotsenko, Nucl. Phys. B235 (1984) 54.

11) VI.S. Dotsenko and V.A. Fateev, Nucl. Phys. B240 (1984) 312.

12) V.G. Kac, Lecture Notes in Physics 94 (1979) 441.

13) H. Blöte, J. Cardy and M. Nightingale, Phys. Rev. Lett. 56 (1986) 742.

14) A. Rocha-Caridi, in: Vertex Operators in Mathematics and Physics (Springer Verlag, 1985).

15) J.L. Cardy, Nucl. Phys. B270 (1986) 186.

16) A. Cappelli, C. Itzykson and J.B. Zuber, Comm. Math. Phys. 113 (1987) 1.

17) D.A. Huse, Phys. Rev. B30 (1984) 3908.

18) G.E. Andrews, R.J. Baxter and P.J. Forrester, J. Stat. Phys. 35 (1984) 193.

19) V. Pasquier, Nucl. Phys. B285 (1987) 162.

FUNDAMENTAL PROBLEMS IN STATISTICAL MECHANICS VII
H. van Beijeren, Editor
© *Elsevier Science Publishers B.V., 1990*

SHAPE FLUCTUATIONS AND CRITICAL PHENOMENA

Reinhard Lipowsky

Sektion Physik der Universität München,
Theresienstr. 37, 8000 München 2, FRG

I. INTRODUCTION AND OUTLINE

Low–dimensional objects or manifolds such as interfaces, vortices or flux lines, membranes, and polymers are usually soft and flexible and, thus, undergo shape fluctuations which lead to a variety of critical phenomena. First of all, a fluctuating manifold represents a scale–invariant state which can be characterized by critical exponents. /1/ Depending on the strength of the fluctuations, this state may be *smooth, rough or crumpled.* In some systems, the manifold can undergo a transition between these different states; e.g., interfaces can undergo *roughening transitions* from smooth to rough states /2/ while (model) membranes can undergo *crumpling transitions* from rough to crumpled states /3/.

Many physical phenomena such as, e.g., wetting, adhesion and adsorption are governed by the mutual interaction of these low–dimensional manifolds. /4/ Quite generally, the shape fluctuations of these objects *renormalize* their direct interaction arising from intermolecular forces. This renormalization acts to increase the repulsive part of the interaction. In fact, sufficiently strong fluctuations overcome the attractive part of the direct interaction and lead to phase transitions from bound to unbound states of the manifolds. For interfaces, membranes, and polymers, these *unbinding transitions* represent wetting, adhesion, and adsorption transitions, respectively.

This paper is organized as follows. First, Sec. II contains a brief introduction to the physics of shape fluctuations. In Sec. III, the scale–invariance of rough and crumpled states, and the critical behavior at roughening and crumpling transitions is reviewed. In Sec. IV, a nontrivial example is described in some detail: the roughening of interfaces in quasiperiodic systems. Finally, the interplay of direct and fluctuation–induced interactions and the associated unbinding transitions are discussed in Sec.V.

II. SHAPE FLUCTUATIONS.
A. Fluctuations of interfaces and flux lines governed by tension.

An *interface* or domain wall represents the contact region between two bulk phases of matter. /5/ Its macroscopic shape is governed by the interfacial free energy or *tension*, Σ, which is the work (per unit area) required to create new interfacial area. If the two phases, say α and β, are both fluid, Σ is isotropic, and a macroscopic chunk of β phase surrounded by α phase has a spherical shape (in the absence of gravity). On the other hand, if one of the two phases is crystalline, the tension Σ is anisotropic and leads to the possibility of facets.

On mesoscopic scales, i.e., on scales which are larger than the molecular size but smaller than the size of the α and β domains, the interface does not have a fixed shape but undulates and thus undergoes shape fluctuations which lead to a certain interfacial *roughness*. These fluctuations do *not* conserve the total (intrinsic) area of the manifold and, thus, are governed by the tension Σ. For a liquid–vapor interface, these excitations represent *capillary waves*. For a crystal surface, i.e., for a crystal–vapor interface, the shape fluctuations are built up from microscopic *steps* or *ledges* which separate atomically flat terraces on the surface.

The interfaces described so far have been 2–dim surfaces separating 3–dim bulk domains. Condensed matter physics also provides a variety of 2–dim bulk systems containing 1–dim interfaces. /6/ Examples are domain boundaries (i) in monolayers of small molecules adsorbed onto a crystal surface, or (ii) in monolayers or bilayers of amphiphilic molecules adsorbed onto the air–water interface. Sometimes, these 1–dim domain boundaries form a superlattice such as the striped phase near a commensurate–incommensurate transition.

A vortex or *flux line* within a type–II superconductor represents a thin rod or thread of normal conducting material which contains one or several magnetic flux quanta. /7/ These flux lines start to penetrate the superconductor at a lower critical value of the external magnetic field. Close to this critical field, the lines are well separated and form a more or less ordered array. The undulations of these flux lines are controlled by their *line tension*.

Two different excitation mechanism for the fluctuations of interfaces and flux lines must be distinguished: (i) At finite temperatures, $T > 0$, these manifolds fluctuate in order to increase their configurational *entropy* ; and (ii) In the presence of quenched impurities (or frozen randomness), they adapt their shape to the randomness in order to minimize their *energy* . The latter mechanism acts even at $T = 0$.

B. Fluctuations of membranes and polymers governed by bending elasticity.

Membranes are ultrathin sheets or plates of molecules. Particularly important examples are monolayers and bilayers which form spontaneously in solutions of amphiphilic molecules. /8/ Monolayers arise, e.g., in microemulsions, i.e. in mixtures of water, oil, and surfactant where they separate water from oil domains. Bilayers are typically formed when lipid molecules are dissolved in water. /9/ These latter membranes represent the universal building block of all biomembranes and, thus, provide the spatial organization of biological systems.

Isolated bilayers in water form closed surfaces or vesicles which often have a linear size $\sim$ 10 μm. The shape of these vesicles has been studied in many experiments. These studies support the theoretical concept that the average shape of membranes is controlled by *bending rigidity.* /10/ Likewise, typical shape fluctuations of membranes represent thermally–excited *bending modes.*

The bending rigidity is mainly determined by the internal structure of the membrane. Lipid bilayers, e.g., typically exhibit a *crystalline* (L_β)–phase at low T and a *fluid* (L_α)–phase at high T. In a crystalline membrane, the lipid molecules form a fixed network while they diffuse freely within a fluid membrane. Therefore, crystalline membranes are more rigid than fluid ones. The rigidity can also be increased by polymerization of the lipid molecules which leads to a *tethered* membrane.

For sufficiently rigid membranes, the bending modes give rise to a certain *roughness* and, thus, play a role which is completely analogous to the capillary modes of interfaces. In principle, very soft membranes could become highly convoluted or *crumpled.* Indeed, such a behavior is expected from the standard model for fluid membranes /10/ which leads to a finite *persistence length,* ξ_p, for their normal vectors /11/. However, real membranes do not self–intersect and this acts against crumpling. In fact, recent computer simulations indicate that *tethered* membranes do not crumple at finite T because of their self–avoidance. /12/

Crumpled states are, of course, well established for *polymers,* i.e., for long rodlike molecules. /13/ Such macromolecules are formed by chemical reactions in which a huge number of monomers is linked together. Both artificial polymers and biopolymers often consist of $> 10^5$ monomers. On small scales, they should behave like elastic rods and, thus, should also exhibit thermally–excited bending modes. However, with the possible exception of polyelectrolytes, all polymers are expected to have a finite persistence length, ξ_p, for their tangent vectors. This length is often microscopic, and the polymer then forms a random coil, when viewed on the scale of μm's. In this crumpled state, the polymer behaves like a chain of mass points which are connected by harmonic springs with an *entropically generated* spring constant $\sim$ T.

III. SCALE INVARIANCE OF LOW – DIMENSIONAL MANIFOLDS
A. Rough manifolds.

Now, let us consider a *single* manifold such as an interface, flux line, membrane or polymer embedded in a d–dim system. The system is taken to be in thermal equilibrium at temperature T. Furthermore, let us first assume that the system does not contain any type of frozen randomness. Then, the manifold will attain a completely smooth (or flat) state at zero temperature, T = 0, in order to minimize its energy. At finite T > 0 , on the other hand, the manifold will usually develop bumps or wiggles and thus will acquire a certain roughness in order to increase its configurational entropy. Alternatively, one can introduce some frozen randomness into the system which roughens the manifold even at T = 0.

To proceed, let us choose a parametrization for the fluctuations of the manifold from its completely smooth state. First, the space dimensionality, d, is decomposed according to

$$d = d_{\parallel} + d_{\perp} \tag{3.1}$$

where $d_{\parallel}$ is the intrinsic dimensionality of the manifold. Its fluctuations are then described by a displacement field $z = (z_1,...,z_{d_\perp})$ with $z = z(x)$ and $x = (x_1,...,x_{d_\parallel})$. The completely smooth state at $T = 0$ corresponds to $z = $ const. Since all components of z are taken to be single–valued functions of x, manifold fluctuations with overhangs are not included. This means that *gradients* of the fluctuations, ∇z_α with $\nabla \equiv (\partial/\partial x_1, \; ... \; ,\partial/\partial x_{d_\parallel})$, are implicitly assumed to be *small*.

The shape fluctuations of the manifold can now be characterized by the behavior of the difference correlation function,

$$\Delta C_z(x) \equiv \tfrac{1}{2} < [z(x){-}z(0)]^2 > \tag{3.2}$$

In some systems, the function $\Delta C_z(x)$ remains bounded from above for arbitrarily large x. This behavior defines a *smooth state* of the manifold. On the other hand, the difference correlation function $\Delta C_z(x)$, may grow without bounds for large x corresponding to a *rough state* of the manifold. Such a situation is typically characterized by /14/

$$\Delta C_z \sim x^{2\zeta} \quad \text{for large x} \tag{3.3}$$

where ζ is the socalled *roughness exponent*. Such a behavior of ΔC_z implies that $\Delta C_z(x/b) \approx \Delta C_z(x)/b^{2\zeta}$ for $x >> a$ where b is an arbitrary rescaling factor and a the small–scale cutoff. Therefore, the shape fluctuations of a rough manifold are (asymptotically) scale–invariant under the rescaling transformation

$$x \to x/b \quad \text{and} \quad z \to z/b^\zeta \quad . \tag{3.4}$$

In order to visualize this scale invariance, choose a manifold segment of linear size, L_{in}, and intrinsic area, $L_{in}^{d_\parallel}$. Then, the linear size, $L_{\parallel}$, of the projected area, $L_{\parallel}^{d_\parallel}$, and the typical amplitude, $L_{\perp}$, of the manifold fluctuations, see Fig. 1(a) below, scale as

$$L_{\parallel} \sim L_{in} \quad \text{and} \quad L_{\perp} \sim L_{\parallel}^\zeta \sim L_{in}^\zeta \quad . \tag{3.5}$$

For *thermally–excited* fluctuations, a single manifold with roughness exponent ζ can be described by the effective Hamiltonian

$$\mathcal{H}\{z\} = \int d^{d_\parallel}x \, \tfrac{1}{2} K \sum_{\alpha=1}^{d_\perp} (\nabla^n z_\alpha)^2 = \int \frac{d^{d_\parallel}p}{(2\pi)^{d_\parallel}} \, \tfrac{1}{2} K \, p^{2n} \sum_{\alpha=1}^{d_\perp} |\tilde{z}_\alpha(p)|^2 \tag{3.6}$$

with

$$n = \zeta + d_{\parallel}/2 \tag{3.7}$$

where K represents an effective tension or bending rigidity.

For the model as given by (3.6), the difference correlation function ΔC_z defined in (3.2) is

$$\Delta C_z(x) = d_\perp (T/K) \int \frac{d^{d_\|}p}{(2\pi)^{d_\|}} [1 - \exp(i\mathbf{p}\cdot\mathbf{x})]/p^{2n}. \tag{3.8}$$

where a high–momentum cutoff $\sim 1/a$ is implicitly contained. For $d_\| > 2n$, the difference correlation function as given by (3.8) is bounded for large x and the manifold is *smooth*. For $d_\| = 2n$, one has a marginally rough manifold with $\Delta C_z(x) \sim ln(x/a)$ for large x. Finally, for $2n > d_\| > 2-2n$, (3.8) leads to a *rough* manifold characterized by

$$\Delta C_z(x) \approx d_\perp(T/K) c_\infty x^{2\zeta} \quad \text{with} \quad \zeta = n-d_\|/2 \tag{3.9}$$

for large x where the coefficient c_∞ depends on $d_\|$ but is independent of the cutoff, a.

For an interface with $d = d_\|+1$, the roughness exponent ζ depends on d (or $d_\|$) and on the nature of the two phases separated by the interface. /1/ If both phases are fluid, the interfacial fluctuations are governed by the effective Hamiltonian (3.6) with $n = 1$ (and $d_\perp = 1$) which leads to $\zeta = 0(\sqrt{\log})$ in $d = 3$ and $\zeta = (3- d)/2$ in $1 < d < 3$, see (3.9).

The same value of ζ applies to an interface which feels a *periodic* lattice potential provided T exceeds the roughening temperature, T_r, see Sec. III.C below. *Quasiperiodic* lattice potentials, on the other hand, can *lower* the value of ζ /15,16/ as will be explained in Sec. IV. If the two phases separated by the interface contain quenched impurities, the roughness exponent ζ is typically *increased* : a random field system is believed to exhibit /17,18/ $\zeta = (5-d)/3$ for $2 < d < 5$ while a random bond system (with short–range correlated randomness) is characterized by /19,20/ $\zeta = 2/3$ in $d = 1+1$.

A flux line in $d = 1+d_\perp$ subjected to thermally–excited fluctuations, is governed by the effective Hamiltonian (3.6) with $n = 1$ and $d_\| = 1$ which implies $\zeta = 1/2$ according to (3.9). /21/ In the presence of quenched impurities (with short–range correlations), the flux line in $d = 1+2$ should exhibit $\zeta \simeq 0.6$. /22/

The roughness of membranes in $d = 2+d_\perp$ depends on their internal structure. *Fluid* membranes with vanishing shear modulus are governed by bending elasticity alone which corresponds to an effective Hamiltonian as in (3.6) with $d_\| = 2$ and $n = 2$. /10/ It then follows from (3.9) that $\zeta = 1$. /23,24/ *Tethered* membranes, on the other hand, are more rigid and have $\zeta < 1$ /25,26/ : for zero and finite bending rigidity, numerical simulations in $d = 3$ gave the estimates /12/ $\zeta \simeq 0.8$ and /27/ $\zeta = 0.63 \pm 0.02$, respectively. Finally, polymers in $d = 1+d_\perp$, when treated as stiff rods governed by bending elasticity, are described by the effective Hamiltonian (3.6) with $d_\| = 1$ and $n = 2$ which implies $\zeta = 3/2$ according to (3.9).

As mentioned, the parametrization in terms of displacement fields, $\mathbf{z}(\mathbf{x})$, assumes that the gradients of $\mathbf{z}$ are small. The overall gradient of a hump as shown in Fig.1(a) is $\sim$

                    R. Lipowsky

$(L_\perp/L_\parallel) \sim L_{in}^{\zeta-1}$ where L_{in} measures the intrinsic size of the manifold segment. Thus, for $\zeta < 1$, the overall gradient *decreases* on large scales and the parametrization in terms of z should be adequate. On the other hand, for $\zeta \geq 1$, the overall gradient grows for large L_{in}. In the latter case, overhangs of the manifold fluctuations are essential and the parametrization $z(\mathbf{x})$ is no longer appropriate on large scales.

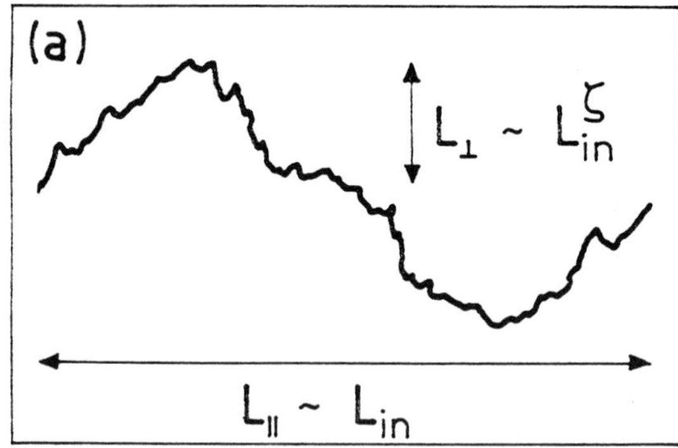

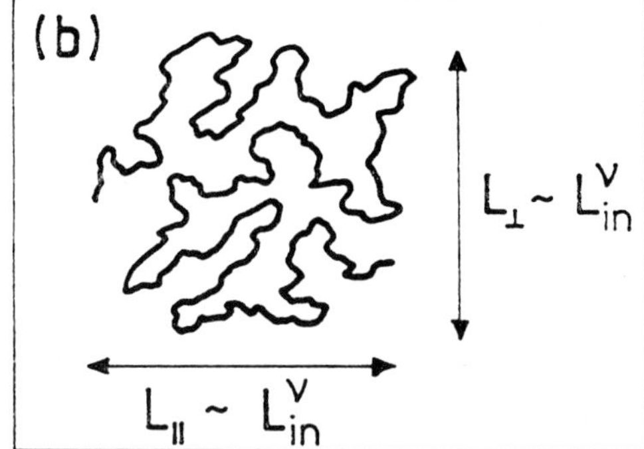

Fig.1 :   (a) Typical *hump* of a rough manifold, and (b) typical *blob* of a crumpled manifold with intrinsic size L_{in}. These states are characterized by the roughness exponent ζ and the crumple exponent ν, respectively, see (3.5) and (3.17).

B. Scale invariance of crumpled manifolds.

For roughness exponent $\zeta \geq 1$, the manifold develops many overhangs on sufficiently large scales and, thus, becomes highly convoluted or *crumpled*, see Fig.1(b). If the topology of the $d_\parallel$–dim manifold is not changed by the fluctuations, such a crumpled state can be parametrized by $\mathbf{r} = [r,(s),..,r_d(s)]$ with an internal coordinate $\mathbf{s} = (s_1,..,s_{d\parallel})$.

For a (linear) polymer chain with $d_\parallel = 1$, one has $\mathbf{r} = [r,(s),..,r_d(s)]$ where the coordinate s is taken to be the *arclength*. The thermally–excited bending modes can now be characterized by the more general difference correlation function

$$\Delta C_r(s) \equiv \tfrac{1}{2} < [\,\mathbf{r}(s)-\mathbf{r}(0)\,]^2 > \; = \; \tfrac{1}{2} \int_0^s ds_1 \int_0^s ds_2 <(d\hat{t}/ds_1)\cdot(d\hat{t}/ds_2)> \qquad (3.10)$$

where $\hat{t}(s) \equiv d\mathbf{r}/ds$ is the unit tangent vector. This function contains a characteristic *persistence length*, ξ_p, and is expected to exhibit the scaling form

$$\Delta C_r(s) \approx \xi_p^2 \, \mathscr{D}(s/\xi_p) \qquad . \qquad (3.11)$$

For large $s \gg \xi_p$, the polymer is crumpled and the shape function $\mathscr{D}(y)$ behaves as

$$\mathscr{D}(y) \approx c_\infty \, y^{2\nu} \quad \text{for large } y \gg 1 \tag{3.12}$$

which defines the *crumple exponent* (or Flory exponent or radius of gyration exponent) ν. If the persistence length ξ_p is large compared to microscopic scales, one has a crossover to a rough state of the polymer characterized by

$$\mathscr{D}(y) \approx y^2 - c_0 \, y^{2\zeta} \quad \text{for } y \ll 1 \quad \text{with } \zeta = 3/2 \; . \tag{3.13}$$

As a simple example consider a self–intersecting polymer in $d = 1+1$ parametrized by $\mathbf{r}(s) = [r_1(s), r_2(s)]$. The thermally–excited bending modes are then governed by

$$\mathscr{H}\{\hat{\mathbf{t}}\} = \int ds \, \tfrac{1}{2} \, K(d\hat{\mathbf{t}}/ds)^2 = \int ds \, \tfrac{1}{2} \, K(d^2\mathbf{r}/ds^2)^2 \tag{3.14}$$

where $d^2\mathbf{r}/ds^2$ is the local *curvature* of the polymer and K represents an effective bending rigidity. For this simple model, the difference correlation function as given by (3.10) can be calculated explicitly. One then finds the persistence length $\xi_p = 2K/T$ and the scaling form (3.11) with

$$\mathscr{D}(y) = y - 1 + \exp(-y) \quad . \tag{3.15}$$

For $y \gg 1$, this leads to $D(y) \approx y$ and, thus, to the crumple exponent $\nu = 1/2$. For $y \ll 1$, on the other hand, one recovers (3.13) with $\zeta = 3/2$.

The value $\nu = 1/2$ applies, in fact, to linear polymers in general d provided one allows for self–intersections. For *self–avoiding* polymers, ν depends on d and one has $\nu = 3/4$ and $\nu \simeq 3/5$ in $d = 1+1$ and $d = 1+2$, respectively, as obtained from the Flory argument. /13/

The behavior $\Delta C_r(s) \sim s^{2\nu}$ for large s implies that the shape fluctuations of a crumpled polymer are (asymptotically) scale–invariant under the rescaling transformation

$$s \to s/b \quad \text{and} \quad \mathbf{r} \to \mathbf{r}/b^\nu \quad . \tag{3.16}$$

Thus, a polymer segment of intrinsic length $s \simeq L_{in}$ has an extrinsic size $L_\perp \sim L_\parallel$ with

$$L_\perp \sim L_\parallel \sim L_{in}^\nu \tag{3.17}$$

see Fig. 1(b). The scale $L_\perp \sim L_\parallel$ is conveniently measured by the *radius of gyration*. The relation (3.17) is, in fact, quite general. It also applies to *branched* polymers provided L_{in}/a is taken to be the number of momomers within a connected piece of the branched structure.

Crumpled states are also expected for membranes. First, consider a *tethered* membrane, i.e., a fixed network of molecules connected by tethers or springs. /28,3/ The fluctuations of such a membrane are quasi–isometric, i.e., the intrinsic distance between two molecules in the membrane is roughly independent of the membrane configuration. For an ideal (or phantom) membrane with self–intersections, a segment with intrinsic area $\sim L_{in}^2$ has the size $L_\perp \sim L_\| \sim [ln(L_{in}/a)]^{1/2}$ in $d = 2+d_\perp$. /28/ However, self–avoidance represents a very strong constraint. This can be understood if one considers strictly isometric deformations of a plane, which are obtained by folding the membrane in one direction only. The corresponding bending energy is proportional to the bending energy of a self–avoiding polymer in $d = 1+1$ which implies that a membrane segment with intrinsic area $\sim L_{in}^2$ has $L_\perp \sim L_{in}^{3/4}$ and $L_\| \sim L_{in}$. This leads to the estimate $\nu \simeq 7/8$ since $(L_\perp L_\|)^{1/2} \sim L_{in}^{7/8}$. As mentioned, recent computer simulations indicate, in fact, that self–avoiding tethered membranes have $\nu = 1$ and are *not* crumpled. /12/

For a *fluid* membrane, the molecules can diffuse freely. The shape of these membranes is believed to be governed by curvature energies alone. The standard model of Helfrich /10/ leads to a finite persistence length $\xi_p \simeq a \exp(2\pi K/T)$ where K is the bending rigidity. /11/ The state of a fluid membrane for $L \gg \xi_p$ is not well understood. It is possible that the membrane develops many fingers in order to increase its entropy and becomes a branched polymer. The typical diameter of these fingers should be set by the persistence length, ξ_p. Then, a membrane segment of intrinsic area L_{in}^2 can form fingers with an overall length $L_0 \sim L_{in}^2/\xi_p$. A self–avoiding branched polymer with L_0/a monomers has a radius of gyration, $L_\perp \sim L_\| \sim L_0^\nu$, with /29/ $\nu \simeq 0.5$ in $d = 3$. This implies that the radius of gyration of a crumpled fluid membrane scales as $L_\perp \sim L_\| \sim L_{in}^\nu$ with $\nu \simeq 1$ if the membrane indeed behaves as a branched polymer on large scales.

C. Roughening and Crumpling Transitions.

In the last two sections, we encountered smooth, rough, and crumpled states of fluctuating manifolds. These different states can be distinguished by their long–range order (LRO) *within the d–dim* (embedding) *space.*

For a manifold with $d_\| = 1$, *orientational* LRO means that its tangent vectors have a preferred direction. Likewise, a manifold with $d_\perp = 1$ is orientationally ordered if its normal vectors point, on average, into a certain direction. One can then choose a straight line ($d_\| = 1$) or a planar hypersurface ($d_\perp = 1$), as a reference state which has the same orientation as the fluctuating manifold. *Translational* order or disorder is now defined in terms of the displacements of the manifold from this reference state. A manifold exhibits translational LRO if these displacements are bounded from above on arbitrarily large scales.

By definition, a *smooth* state exhibits both translational and orientational LRO within the d–dim system. A *rough* manifold, on the other hand, is translationally disordered since its displacements from the smooth reference state become arbitrarily large. For roughness exponent $\zeta < 1$, a rough manifold still exhibits orientational LRO. For $\zeta \geq 1$, on the other hand, the manifold becomes orientationally disordered as soon as the linear scales exceed the persistence length, ξ_p. In the latter case, the manifold is *crumpled*.

In some systems, the low–dim manifolds undergo a *roughening transition* from a *smooth* state at low T to a *rough* state at high T. At the roughening temperature, $T = T_r$, the manifold is rough with a roughness exponent ζ_r which, in general, can differ from ζ for $T > T_r$. As T_r is approached from below, the large–scale configurations of the manifold can be viewed as an ensemble of humps. The largest humps are statistically independent and characterized by a lateral extension, $\xi_{\parallel}$, and a transverse extension, $\xi_{\perp}$, see Fig. 2(a). For scales $L_{\parallel} \ll \xi_{\parallel}$, the manifold already exhibits the scale–invariance governed by ζ_r and $L_{\perp} \sim L_{\parallel}^{\zeta_r}$. Therefore, a large hump, which has an intrinsic size ξ_{in}, is characterized by $\xi_{\parallel} \sim \xi_{in}$ and $\xi_{\perp} \sim \xi_{in}^{\zeta_r}$, compare Fig.1(a) and 2(a). For a continuous roughening transition, this implies that the critical exponents ν_{in}, $\nu_{\perp}$ and $\nu_{\parallel}$ defined by

$$\xi_{in} \sim (T_r - T)^{-\nu_{in}}, \quad \xi_{\perp} \sim (T_r - T)^{-\nu_{\perp}} \quad \text{and} \quad \xi_{\parallel} \sim (T_r - T)^{-\nu_{\parallel}} \tag{3.18}$$

satisfy the scaling relation

$$\nu_{\perp} = \zeta_r \, \nu_{\parallel} = \zeta_r \, \nu_{in} . \tag{3.19}$$

If the manifold is only marginally rough with $\zeta_r = 0(\sqrt{\log})$ at $T = T_r$, one has

$$\xi_{\perp} \sim [ln(\xi_{\parallel}/a)]^{1/2} . \tag{3.20}$$

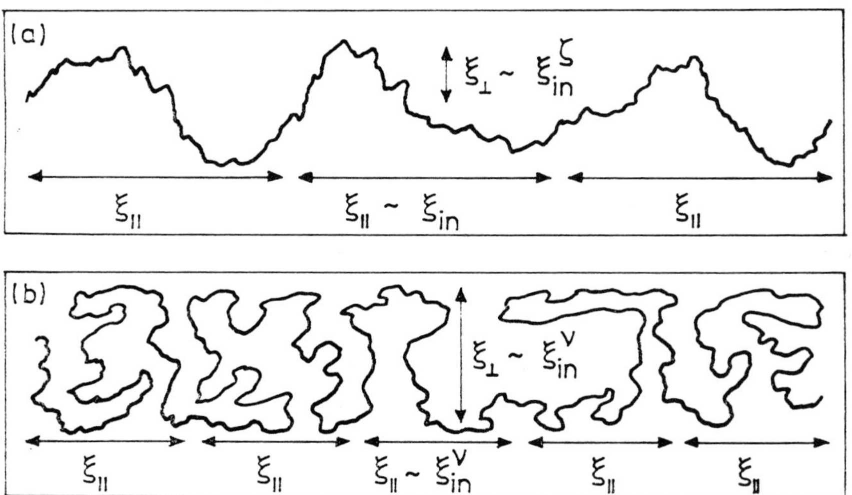

Fig.2 :   (a) Smooth manifold close to a *roughening* transition as an ensemble of humps, and (b) Rough manifold close to a *crumpling* transition as an ensemble of blobs. The length scales, ξ_{in}, $\xi_{\perp}$, and $\xi_{\parallel}$ diverge at the transitions, see (3.18) and (3.21)

A roughening transition occurs for an interface in d = 2+1 in a *periodic* lattice potential. /2/ In this case, the interface is marginally rough with $\zeta = \zeta_r = 0(\sqrt{\log})$ for $T \geq T_r$. However, $\xi_\parallel \sim \exp(c/\sqrt{T_r-T})$ which implies $\xi_\perp \sim (T_r-T)^{-\nu_\perp}$ with $\nu_\perp = 1/4$ via (3.20).

More recently, roughening transitions have been found for interfaces in d = 1+1 arising from *quasiperiodic* lattice potentials. /15,30,31/ In these systems, the roughness exponent ζ_r at $T = T_r$ satisfies $0 < \zeta_r < 1/2$ and, typically, differs from ζ for $T > T_r$. Thus, one has the power–law behavior given by (3.18) and (3.19) as will be explained in Sec. IV below.

Now, consider a manifold which undergoes a *crumpling transition* from a *rough* state at low T to a *crumpled* state at high T. At the crumpling temperature, $T = T_{cr}$, the manifold is crumpled with a crumple exponent ν_{cr} which, in general, will differ from the crumple exponent ν for $T > T_{cr}$. As T_{cr} is approached from below, the manifold configurations can be viewed as an ensemble of *blobs*. The largest blobs with intrinsic size, ξ_{in}, can be regarded as statistically independent with extrinsic size $\xi_\perp \sim \xi_\parallel \sim \xi_{in}^{\nu_{cr}}$ within the d–dim system, see Fig. 2(b). If the crumpling transition is continuous, one may define critical exponents ν_{in}, $\nu_\perp$ and $\nu_\parallel$ by

$$\xi_{in} \sim (T_{cr}-T)^{-\nu_{in}} \, , \, \xi_\perp \sim (T_{cr}-T)^{-\nu_\perp} \text{ and } \xi_\parallel \sim (T_{cr}-T)^{-\nu_\parallel} \tag{3.21}$$

in close analogy with (3.18) but now one has the scaling relation

$$\nu_\perp = \nu_\parallel = \nu_{cr} \, \nu_{in} \quad . \tag{3.22}$$

Then, a large manifold segment of intrinsic size $L_{in} \gg \xi_{in}$ has a projected size $L_\parallel \sim (\xi_\parallel/\xi_{in}) L_{in}$. As T_{cr} is approached from below, the prefactor $\xi_\parallel/\xi_{in}$ goes to zero as long as $\xi_\parallel \sim \xi_{in}^{\nu_{cr}}$ with $\nu_{cr} < 1$. Thus, one has

$$L_\parallel \sim (T_{cr} - T)^\beta \, L_{in} \tag{3.23}$$

with

$$\beta = \nu_{in} \, (1-\nu_{cr}) \quad . \tag{3.24}$$

Crumpling transitions have been theoretically found for *self–intersecting* manifolds with $d_\parallel > 2$ /32/, and for *self–intersecting* tethered membranes with $d_\parallel = 2$ and finite bending rigidity /25,33–35/. For the latter case, a mean–field theory gives /34/ $\beta = \nu_{cr} = 1/2$ and, therefore, $\nu_{in} = 1$ and $\nu_\parallel = 1/2$ according to (3.22) and (3.24), while an expansion around d = 2+$d_\perp$ = ∞ yields /35/ $\nu_{in} \approx d/2$, $\nu_{cr} \approx 1 - 1/d$ and, thus, $\nu_\parallel \approx d/2$.

E. Singular free energy arising from thermal fluctuations.

The fluctuations of a *rough* manifold give a contribution, F_{FL}, to its free energy. Consider a manifold segment of linear (intrinsic) size, L_{in}. Its largest humps then have a lateral and transverse extension $L_\parallel \sim L_{in}$ and $L_\perp \sim L_{in}^\zeta$, respectively, see (3.5) and Fig. 1(a). The thermal free energy, F_{FL}, of such a hump should be of order $\sim T$ as suggested by the equipartition theorem. Therefore, the thermal free energy per unit area is given by /1/

$$f_{FL} = F_{FL}/L_{in}^{d\parallel} \simeq T/L_{in}^{d\parallel} \simeq 1/L_\parallel^{d\parallel} \sim 1/L_\perp^\tau \qquad \text{with} \quad \tau = d_\parallel/\zeta \ . \tag{3.25}$$

Now, assume that the manifold is confined in such a way that its largest humps are characterized by an intrinsic scale, ξ_{in}, and two extrinsic scales, $\xi_\parallel \sim \xi_{in}$ and $\xi_\perp \sim \xi_{in}^\zeta$. Such a behavior occurs, e.g., close to a roughening transition (with $\zeta = \zeta_r$), see Fig. 2(a). Similar manifold states characterize the unbinding of manifolds as in wetting, adhesion, and adsorption phenomena as will be explained in Sec. V , see Fig.5(a) below. In these cases, the manifold can be viewed as an ensemble of essentially uncorrelated humps. Thus, a manifold of linear size, $L_{in} \sim L_\parallel$, consists of $(L_{in}/\xi_{in})^{d\parallel} \sim (L_\parallel/\xi_\parallel)^{d\parallel}$ independent humps. The thermal free energy of each hump should again be $\sim T$. This leads to $F_{FL} \simeq T(L_{in}/\xi_{in})^{d\parallel}$, and the thermal free energy per unit area arising from the fluctuations is now given by /4/

$$f_{FL} \simeq T/\xi_{in}^{d\parallel} \simeq T/\xi_\parallel^{d\parallel} \sim 1/\xi_\perp^\tau \qquad \text{with} \quad \tau = d_\parallel/\zeta \qquad \text{for } rough \text{ states} \tag{3.26}$$

as in (3.25). The first relation, $f_{FL} \sim 1/\xi_{in}^{d\parallel}$, is the socalled *hyperscaling* relation. For example, a power law behavior, $\xi_{in} \sim \xi_\parallel \sim (T_*-T)^{-\nu\parallel}$, as in (3.18) implies

$$f_{FL} \sim (T_*-T)^{2-\alpha} \quad \text{with} \quad 2-\alpha = d_\parallel \, \nu_\parallel = d_\parallel \, \nu_{in} \ . \tag{3.27}$$

For *crumpled* manifolds, the free energy arising from thermal fluctuations exhibits very similar behavior. Thus, a manifold segment of linear (intrinsic) size, L_{in}, forms large blobs of extrinsic size $L_\perp \sim L_\parallel \sim L_{in}^\nu$, see (3.17) and Fig. 1(b), which again give rise to a thermal free energy, $F_{FL} \simeq T$. This implies that the thermal free energy per unit (intrinsic) area is given by

$$f_{FL} = F_{FL}/L_{in}^{d\parallel} \simeq T/L_{in}^{d\parallel} \sim 1/L_\parallel^\tau \sim 1/L_\perp^\tau \qquad \text{with} \quad \tau = d_\parallel/\nu \ . \tag{3.28}$$

In the presence of some external constraint, the large–scale configurations of a crumpled manifold can be regarded as an ensemble of statistically independent blobs with $\xi_\perp \sim \xi_\parallel \sim \xi_{in}^\nu$, see Fig. 2(b) above and Fig. 5(b) below. Each blob has a free energy $\sim T$. Then, a manifold of linear size, L_{in}, has a free energy $F_{FL} \simeq T(L_{in}/\xi_{in})^{d\parallel}$ since $(L_{in}/\xi_{in})^{d\parallel}$ is the number of blobs. Therefore, the free energy per unit (intrinsic) area arising from the thermal

150   *R. Lipowsky*

fluctuations scales as /36,4/

$$f_{FL} \simeq T/\xi_{in}^{d\|} \sim 1/\xi_{\|}^{\tau} \sim 1/\xi_{\perp}^{\tau} \qquad \text{with} \quad \tau = d_{\|}/\nu \qquad \text{for } \textit{crumpled} \text{ states} \qquad (3.29)$$

as in (3.28). The first relation can again be regarded as hyperscaling. Thus, if $\xi_{in} \sim (T_*-T)^{-\nu_{in}}$, the free energy has a singular part $f_{FL} \sim (T_*-T)^{2-\alpha}$ with $2-\alpha = d_{\|}\,\nu_{in}$.

IV. INTERFACES IN (IDEAL) QUASIPERIODIC SYSTEMS.

A. Penrose tiling and quasiperiodic interface potentials of entropic origin.

A few years ago, a new class of crystalline materials has been discovered which exhibit sharp diffraction peaks and five–fold symmetry axes. /37/ Such a symmetry is, however, not compatible with a periodic structure which is built up from a single unit cell. It is now generally believed that these materials represent *quasicrystals* which are built up from two (or more) unit cells which are arranged in a quasiperiodic manner. The simplest models for such structures are provided by 2–dim Penrose tilings which are built up from two different tiles, namely a thin and thick rhombus, see Fig. 3 . /37/ These tilings exhibit orientational long–range order with a five–fold symmetry axis since all edges point in one out of five possible directions.

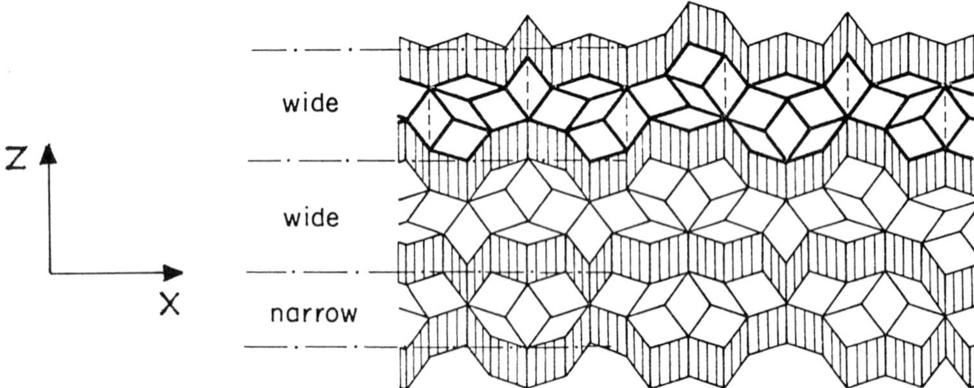

Fig.3 :   2–dim Penrose tiling : the shaded rows separate wide (W) and narrow (N) lanes.

Next, consider an Ising (or lattice gas) model on such a Penrose–tiling: on each tile, an Ising spin, $S_i = \pm 1$, is placed. Each spin interacts with four nearest neighbors on four adjacent tiles. The coupling constant, $J/2$, is taken to be ferromagnetic. Then, an interface separating an up–spin domain from a down–spin domain runs along the edges of the tiling. At $T = 0$, the interfacial configurations with the lowest energy have normal vectors which are parallel to one of the five possible edge orientations. /15,16/ In the following, I focus on one such orientation and choose the x– and the z–axis to be perpendicular and parallel to this orientation, see Fig. 3.

The tiling shown in Fig. 3 has been decomposed into (unshaded) lanes and shaded rows. /15/ The rows contain all tiles with two edges parallel to the z–axis while all other tiles are contained within the lanes. It turns out that all interfacial configurations, which stay within a given lane and which connect two vertices within this lane, have the same length and, thus, the same energy (provided there are no overhangs). Thus, the interfacial groundstate is degenerate at $T = 0$. For an interface consisting of M segments (or edges), this degeneracy behaves as $g_M \sim \exp(MS)$ for large M. There are essentially two types of lanes, narrow (N) and wide (W) ones, see Fig. 3, which have a ground state entropy per segment $S_N \simeq 0.45$ and $S_W \simeq 0.54$, respectively. /15,16/ These two types of lanes form a Fibonacci–sequence.

The Fibonacci–sequence of the two elements W and N can be expressed in terms of the function f(z) with $f(z) = W$ for $0 \leq z \leq 1$, $f(z) = N$ for $1 \leq z < \sigma = (1+\sqrt{5})/2$, and $f(z+\sigma) = f(z)$ for all real z. The Fibonacci sequence of W and N is then given by the sequence $f(z=n)$ for integer $n = 0,\pm 1,...$. Here, n is a discrete height variable which counts the number of lanes between the interface and the x–axis. Likewise, the ground state entropy (per segment) of the lane with height n is given by $S(z=n)$ with $S(z) = S_W$ for $0 \leq z < 1$, $S(z) = S_N$ for $1 \leq z < \sigma$ and $S(z+\sigma) = S(z)$.

At low $T > 0$, the interface makes many steps within a lane before it hops across a row into a neighboring lane. Furthermore, it prefers to stay in the wide lanes since it can then increase its configurational entropy. Therefore, the interface feels an effective quasiperiodic potential, $V_{QP}(z=n)$, with

$$
\begin{aligned}
V_{QP}(z) &= 0 && \text{for} \quad 0 \leq z < 1 \\
&= T\,(S_W - S_N) && \text{for} \quad 1 \leq z < \sigma \quad ,
\end{aligned}
\tag{4.1}
$$

and $V_{QP}(z+\sigma) = V_{QP}(z)$, which is *of entropic origin.* For higher temperatures, the precise form of $V_{QP}(z=n)$ within the Penrose tiling will be more complicated but it will still be quasiperiodic.

B. Interfacial behavoir in $d = 1+1$.

Inside the lanes, the interface feels a quasiperiodic potential as in (4.1). When it hops across a row, it has to increase its length by one edge and, thus, its energy by twice the nearest–neighbor coupling of the Ising model. Thus, the main effect of the Penrose tiling on the interface should be captured by the effective Hamiltonian

$$
\mathscr{H}\{z\}/T = \sum_{<ij>} (J/T)|z_i - z_j| + \sum V_{QP}(z_i)/T
\tag{4.2}
$$

for the discrete height variable $z_i \equiv z(x_i)$ where x_i labels the sites of a 1–dim lattice.

1. Transfer matrix approach.

Now, assume that the interface has the fixed height $z = 0$ at $x = 0$ and consider the total weight or probability, $P(x,z)$, to find the interface at $z(x_N) = z$. For the 1–dim model as given by (4.2) which contains only nearest–neighbor couplings, $P(x=x_N,z)$ depends only on $P(x=x_{N-1},z')$ for the previous step. The associated transfer matrix can be symmetrized by the change of variable, $W(x,z) \equiv \exp[\tfrac{1}{2} V_{QP}(z)] \, P(x,z)$. Furthermore, it is convenient to use the restriction $|z_i - z_j| = 0$ or 1. Then, the weight evolves according to /15/

$$W(x+1,z) - W(x,z) = - \mathscr{H} W(x,z) \qquad (4.3)$$

with the Schrödinger–type operator

$$\mathscr{H} W(x,z) \equiv - e^{-J/T} \left[t(z,z+1) \, W(x,z+1) + t(z,z-1) \, W(x,z-1) \right] + U(z) \, W(x,z) \qquad , \qquad (4.4)$$

the potential

$$U(z) \equiv 1 - exp[- V_{QP}(z)/T] \qquad , \qquad (4.5)$$

and the 'hopping' coefficients

$$t(z,z') \equiv exp[- \tfrac{1}{2} \{ V_{QP}(z) + V_{QP}(z') \}/T] = [1 - U(z)]^{1/2} \, [1 - U(z')]^{1/2} \qquad . \qquad (4.6)$$

Since the operator $\mathscr{H}$ is symmetric, its eigenvaules, E, are real. Then, the weight $W(x,z)$ may be expressed in terms of the eigenfunctions $\phi_E(z)$ according to

$$W(x,z) = \frac{1}{N} \int d\mu(E) \, e^{-(E-E_0)x} \, \phi_E(z) \qquad (4.7)$$

as follows from (4.3) in the continuum limit, $W(x+1,z) - W(x,z) \approx \partial W/\partial x$.

2. Interfacial roughness and coherence length ξ_c.

As discussed in Sec.III.A, the fluctuations of an interface can be characterized by the difference correlation function, $\Delta C_{z'}$, as defined in (3.2). In the present context, one has $L_\perp^2 = \Delta C_z(L_\parallel) = \tfrac{1}{2} \sum W(L_\parallel,z) \, z^2$. Thus. $L_\perp$ measures the width of the weight function W. For large $L_\parallel$, W is governed by the eigenfunctions $\phi_E(z)$ close to the groundstate. Indeed, inspection of (4.7) shows that, for large $x = L_\parallel$, only eigenfunctions with $E \lesssim E_1$ and

$$E_1 - E_0 \sim 1/L_\parallel \qquad (4.8)$$

contribute to the weight $W(L_\parallel,z)$ because of the exponential factor $\sim exp[-(E-E_0)L_\parallel]$.

If the eigenstates close to the groundstate, ϕ_{E_0}, are *localized*, the probability $W(L_\parallel,z)$ will attain a stationary distribution for large $L_\parallel$, and the interface is *smooth*. A *rough* interface, on the other hand, corresponds to the situation where the ground state and the eigenstates close to it are *extended*. In the latter case, the width, $L_\perp$, of $W(L_\parallel,x)$ can be estimated by /15/

$$L_\perp \sim \xi_c(E_1, E_0) \qquad (4.9)$$

where ξ_c is the *coherence length* of the eigenstates $\phi_{E_1}(z)$ and $\phi_{E_0}(z)$ with E_1 defined by (4.11). For $z \ll \xi_c$, the two states look indistinguishable (apart from an overall prefactor).

3. Coherence length and integrated density of states.

As E_1 approaches E_0, the coherence length ξ_c must diverge (provided the states are extended as assumed). It turns out that the singular behavior of ξ_c is intimately related to the scaling behavior of the integrated density of states. In order to understand this relation, consider the Schrödinger–type operator $\mathscr{O}$ as given by (4.4)–(4.6) but with constant hopping coefficients $t(z,z') = 1$. Then, one has to study the eigenvalue problem

$$-e^{-J/T}[\phi_E(z+1) + \phi_E(z-1)] + U(z|\sigma)\,\phi_E(z) = E\,\phi_E(z). \qquad (4.10)$$

The potential $U(z|\sigma)$ is quasiperiodic and satisfies $U(z+\sigma|\sigma) = U(z|\sigma)$ for real z where σ is an irrational number such as the golden mean.

The spectrum of a quasiperiodic potential as in (4.10) can be determined numerically from a sequence of *rational approximants*, $U^{(m)}(z) \equiv U(z|\sigma^{(m)})$ with $\sigma^{(m)} = p^{(m)}/q^{(m)} \approx \sigma$ for large m where $p^{(m)}$ and $q^{(m)}$ are integers, see e.g. Ref. /38/. The m–th approximant, $U^{(m)}(z)$, has period $p^{(m)}$. One then has a Bloch theorem and the spectrum of $U^{(m)}(z)$ consists of $p^{(m)}$ bands where each band has the same number of states,

$$\mathscr{D}^{(m)} \sim 1/p^{(m)} \qquad . \qquad (4.11)$$

Furthermore, numerical calculations have shown that the band widths, $\Delta E^{(m)}$, shrink according to /38/

$$\Delta E^{(m)} \sim [1/p^{(m)}]^{1/\alpha(E_<)} \qquad (4.12)$$

for large m where $E_<$ is the lower edge of the band of eigenvalues.

Now, consider a large but fixed value for the longitudinal extension, $L_\parallel$, of the interface, which implies a small energy gap $E_1 - E_0 \sim 1/L_\parallel$ as in (4.11). As long as this energy gap satisfies $E_1 - E_0 \ll \Delta E^{(m)}$, the eigenstates ϕ_{E_1} and ϕ_{E_0} are *coherent* over the period $p^{(m)}$ of the potential $U^{(m)}(z)$. For $E_1 - E_0 \gg \Delta E^{(m)}$, on the other hand, the two states are separated by many band edges at which the states undergo a phase shift, and thus are *incoherent* over the period $p^{(m)}$. Therefore, one can estimate the coherence length, ξ_c, by

$$\xi_c(E_1,E_0) \sim p^{(M)} \quad \text{with} \quad E_1{-}E_0 \sim \Delta E^{(M)} \tag{4.13}$$

where the lower band edge of $\Delta E^{(M)}$ is now the groundstate energy : $E_< = E_0$.

It then follows from (4.11) and (4.12) that the total number of states, $\mathscr{D}^{(M)}$, in the lowest band is $\mathscr{D}^{(M)} \sim 1/p^{(M)} \sim [\Delta E^{(M)}]^{\alpha(E_0)}$. This together with (4.13) implies that the integrated density of states has the asymptotic behavior $\mathscr{D}(E_1) - \mathscr{D}(E_0) \sim (E_1{-}E_0)^{\alpha(E_0)}$ in the limit of small $E_1{-}E_0$ as has been numerically found in many studies. In the same way, one concludes that the coherence length behaves as /15/

$$\xi_c(E_1,E_0) \sim (E_1{-}E_0)^{-\alpha(E_0)} \tag{4.14}$$

for small $E_1{-}E_0$. Thus, the divergence of ξ_c is governed by the same exponent as the decay of the integrated density of states.

4. Roughness exponent ζ.

The various scaling arguments described in the previous subsections can now be combined in order to obtain a relation between the interfacial roughness, $L_\perp$, and the (projected) length, $L_\parallel$, of the interface. Indeed, a combination of (4.9), (4.14), and (4.8) leads to

$$L_\perp \sim \xi_c(E_1,E_0) \sim (E_1 - E_0)^{-\alpha(E_0)} \sim L_\parallel^{\alpha(E_0)} \quad . \tag{4.15}$$

This means that the *roughness exponent*, ζ, as defined by (3.3)–(3.5), is given by /15/

$$\zeta = \alpha(E_0) \tag{4.16}$$

where $\alpha(E_0)$ governs the decay of the integrated density of states at the ground state.

As a first example, consider the interface Hamiltonian (4.2) with the restriction $|z_i{-}z_j| = 0$ or 1 as suggested by the Penrose tiling for low T. It then follows from (4.1) and (4.5) that the quasiperiodic potential $U(z|\sigma)$ has the form

$$
\begin{aligned}
U(z|\sigma) \quad &= 0 &&\text{for } 0 \leq z < 1 \\
&= U_N \equiv 1 - exp[S_N - S_W] &&\text{for } 1 \leq z < \sigma = (1{+}\sqrt{5})/2
\end{aligned} \tag{4.17}
$$

In the following, this case will be referred to as the *Fibonacci potential*. If one includes only the leading term, $t(z,z') \approx 1$, of the hopping coefficients (valid for small U), one has to calculate the E–spectrum of (4.10) with $U(z|\sigma)$ as given by (4.17). This can be done analytically using methods of nonlinear dynamics. As a result, one finds the exponent $\alpha(E)$ for the integrated density of states /39/ and, thus, the roughness exponent /15/

$$\zeta = \alpha(E_0) = ln[\,(1+\sqrt{5})/2\,]\,/\,ln\,\delta \tag{4.18}$$

with

$$\delta = \tfrac{1}{2}[K_I + (K_I^2 + 4)^{1/2}] \quad \text{and} \quad K_I = (25 + 4e^{2J/T}\,U_N^2)^{1/4} \quad. \tag{4.19}$$

This implies $\zeta < \tfrac{1}{2}$ for all parameter values and /15,16/

$$\zeta \approx 2\,ln[\,(1+\sqrt{5})/2\,]\,(T/J) \quad \text{for} \quad \text{small T/J}\,. \tag{4.20}$$

The nonlinear dynamics approach can be generalized to handle the case of general hopping coefficients $t(z,z+1) = t_{WW}$ or t_{WN} as in the model defined by (4.2) and (4.1). The resulting expression for $\zeta = \alpha(E_0)$ is slightly more complicated and now explicitly depends on E_0. /15/ However, the inequality $\zeta < \tfrac{1}{2}$ and the low–T behavior (4.20) are still valid.

Thus, for the Fibonacci–potential as given by (4.17), the roughness exponent ζ is found to be *nonuniversal*, i.e., to depend on the parameters of the system. This nonuniversality arises from the fact that the eigenstates close to the groundstate are *critical scattering states*. Such states are self–similar and form the singularly continuous part of the spectrum.

Quite generally, the character of the low–lying eigenstates will determine the value of the roughness exponent ζ as is evident from the preceding discussion. Three cases must be distinguished: (i) these states are *localized* corresponding to a smooth interface and, thus, $\zeta = 0$; (ii) these states are *normal scattering* states which implies a rough interface with the *universal* exponent $\zeta = 1/2$; and (iii) the states close to the groundstate are *critical scattering states* which leads to nonuniversal values for ζ.

5. Roughening transitions in d = 1+1.

For some quasiperiodic systems, the character of the low–lying states close to the ground state changes as a function of temperature (or any other parameter). Thus, these states may be localized for some values of the parameters but extended for others. Such a behavior was found for the 1–dim almost Mathieu (or Harpers) potential. For electrons moving in such a potential, this implies a metal–insulator transition. /40/ In the present context, such a behavior leads to a *roughening transition* from a smooth to rough state of the interface at some roughening temperature, $T = T_r$. /15/

The almost–Mathieu (or Harpers) potential has the form

$$U(z\,|\,\sigma) = -U\,cos[2\pi z/\sigma] \quad \text{with} \quad U > 0 \tag{4.21}$$

and $\sigma = (1+\sqrt{5})/2$. The choice $U > 0$ ensures that the global minimum of $U(z)$ is at $z = 0$.

The corresponding eigenvalue problem as given by (4.10) remains unchanged under a duality transformation with $\tilde{U} \equiv U \, exp(J/T) \rightarrow 4/\tilde{U}$. /40/ If there is a unique transition, it must then occur at the self–dual point

$$\tilde{U}_* = (e^{J/T}U)_* = 2 \quad . \tag{4.22}$$

Now, consider the interface model (4.2) with the restriction $|z_i - z_j| = 0$ or 1 and with $V_{QP}(z)/T = U(z|\sigma)$ as given by (4.21). One then has to study the Schrödinger–type equation (4.4)–(4.6) which, in the limit of small U, reduces to the form (4.10). It then follows from (4.22) that the interface undergoes a roughening transition at a roughening temperature $T = T_r$ which behaves as /15/

$$T_r \approx J \, / \, ln(2/U) \quad \text{for small } U \quad . \tag{4.23}$$

For constant hopping coefficients as in (4.10), the roughening exponent ζ can be obtained, via (4.16), from numerical studies of the density of states /38/. One then finds that /15/

$$\begin{aligned} \zeta \; &= \; \zeta_r \simeq 0.421 \quad \text{for} \quad T = T_r \quad (\text{or } \tilde{U} = 2) \\ &= \; 1/2 \qquad \quad \text{for} \quad T > T_r \quad (\text{or } \tilde{U} < 2) \; . \end{aligned} \tag{4.24}$$

For $\tilde{U} > 2$, the eigenstates are exponentially localized with a localization length $\sim 1/ln(\tilde{U}/2)$ which diverges as $\tilde{U}_* = 2$ is approached from above. /40/ This implies that the interfacial roughness, $\xi_\perp$, behaves as $\xi_\perp \sim (T_*-T)^{-\nu_\perp}$ with $\nu_\perp = 1$ as the roughening temperature, T_r, is approached from below. It then follows from the general scaling relation (3.19) (and $\zeta_r > 0$) that the parallel correlaction length, $\xi_\|$, behaves as $\xi_\| \sim (T_r-T)^{-\nu_\|}$ with $\nu_\| = \nu_\perp/\zeta_r \simeq 2.38$. Likewise, hyperscaling as described in Sec.III.E implies for a rough interface that the singular part of the interfacial free energy, $\Sigma_{FL} \equiv f_{FL} \sim 1/\xi_\|$. Then, the interfacial specific heat, C_I, has the singular part $\sim d^2\Sigma_{FL}/dT^2$ and behaves as $C_I \sim (T_r-T)^{-\alpha_r}$ with $\alpha_r = 2-1/\zeta_r \simeq -0.375$. Thus, the interfacial specific heat exhibits a weak cusp–like singularity at the roughening transition.

This scaling behavior shows that roughening transitions of 1–dim interfaces in quasiperiodic systems are *second–order* transitions and, thus, are very different from the well–known roughening transition in 3–dim periodic systems which are of the Kosterlitz–Thouless type. Furthermore, the critical behavior of the roughening transitions considered here will depend on the details of the quasiperiodic potentials (and hopping coefficients). Similar roughening transitions with somewhat different exponents have been found (i) for a quasiperiodic potential intermediate between the Fibonacci and the almost–Mathieu potential /30/, and (ii) for interface models in which the interface potential varies quasiperiodically both with z and with x /31/.

C. Interfacial behavior in d = 2+1.

1. Continuum model

In d = 2+1, quasiperiodic structures have been constructed by a variety of methods. One important example is the (ideal) icosahedral tiling as generated, e.g., by the dual–grid method. /37/ This tiling is built up from parallelepipeds (or rhombohedra) where each parallelepiped has three out of six possible edge orientations. Such a tiling can be decomposed, for each edge orientation, into a stack of sheets and slabs which are, on average, parallel. /41/ The sheets in d = 2+1 play the role of the rows in d = 1+1 (see Fig. 3) and contain all tiles with four faces parallel to the chosen edge orientation; the slabs play the role of the lanes (see Fig. 3) and are composed of all the remaining tiles.

At T = 0, an interface will be confined to one of the slabs and, thus, will exhibit a finite ground–state entropy. Furthermore, this entropy should vary quasiperiodically from slab to slab. Thus, at T > 0, the interface in d = 2+1 will again feel an effective quasiperiodic potential of entropic origin. In this way, one is led to study the 3–dim analogue of the effective Hamiltonian as given by (4.2) for which the integer height variable, $z(\mathbf{x_i})$, counts the number of slabs perpendicular to a certain edge orientation. This discrete height variable is now replaced by a continuous one which feels a generic quasiperiodic potential such as

$$V_{QP}(z)/T = A_0 \; cos(q_0 \; z) + A_1 \; cos(q_1 \; z) \tag{4.25}$$

with $q_0 \equiv 2\pi/\ell_0$, $q_1 \equiv q_0/\sigma$ and $\sigma = (1+\sqrt{5})/2$. In addition, the discrete lattice, $\mathbf{x_i}$, is replaced by a continuous coordinate $\mathbf{x} = (x_1, x_2)$. Then, the effective Hamiltonian for the interfacial shape has the generic form

$$\mathcal{H}\{z\}/T = \int d^2x \; \{ \; \tfrac{1}{2} \; (K/T) \; (\nabla z)^2 + V_{QP}(z)/T \; \} \tag{4.26}$$

with $V_{QP}(z)$ as given by (4.25). It will be shown below that such a continuum model leads to a smooth interface for all T < ∞ and, thus, to a roughening temperature, $T_r = \infty$. /41–43/ This property should also hold for the discrete models and for more realistic models which include a quasiperiodic variation of V_{QP} with $\mathbf{x}$.

The global minimum of the quasiperiodic potential $V_{QP}(z)$ as given by (4.25) depends on the signs of the amplitudes A_0 and A_1. In fact, one could also allow for a phase shift between the two cosine terms such that $cos(q_1 z)$ is replaced by $cos(q_1 z + \Delta)$. Then, the global minimum of $V_{QP}(z)$ will be at z = ∞ for most choices of the phase Δ. In such a situation, the interface can only be metastable at finite z, and the interfacial roughness in thermal equilibrium is not a well–defined quantity. Therefore, I chose $\Delta = 0$ in (4.25) and take $A_0 <$

0 and $A_1 < 0$. /41/ This choice ensures that the global minimum is at $z = 0$ for any finite z–interval. This together with the symmetry $V_{QP}(-z) = V_{QP}(z)$ implies $\langle z \rangle = 0$.

2. Functional renormalization and cumulant expansion

The model as defined by (4.26) and (4.25) will now be studied by a functional renormalization group (RG) method. Such a RG method consists of three basic steps. First, the fluctuating field, $z(\mathbf{x})$, is divided up into two parts : $z(\mathbf{x}) = z_<(\mathbf{x}) + z_>(\mathbf{x})$ where $z_<$ and $z_>$ represent the small–wavenumber (or large–scale) and the large–wavenumber (or small–scale) fluctuations. Then, the small–scale fluctuations, $z_>$, are integrated out which gives rise to

$$exp[-\mathcal{H}'\{z_<\}] \equiv \frac{1}{N} \int \mathcal{D}\{z_>\}\; exp[-\mathcal{H}\{z_<+z_>\}] \quad . \tag{4.27}$$

(The factor $1/T$ has been absorbed into $\mathcal{H}$ for notational convenience.) Finally, the spatial coordinate, $\mathbf{x}$, and the fluctuating field, z, are rescaled according to

$$\mathbf{x} \to \mathbf{x}/b \quad \text{and} \quad z \to z/b^{\zeta} \quad \text{with} \quad b > 1 \quad \text{and} \quad \zeta = 0 \quad . \tag{4.28}$$

Obviously, the most difficult step in the RG is the partial trace over the small–scale fluctuations as in (4.27). In order to perform this step in a perturbative way, let us divide $\mathcal{H}$ into $\mathcal{H}_0\{z\} \equiv \int d^2x\; \tfrac{1}{2}\, (K/T)\, (\nabla z)^2$ and $\mathcal{H}_1\{z\} \equiv \int d^2x\; V_{QP}(z)/T$. One then has

$$\mathcal{H}\{z_<+z_>\} = \mathcal{H}_0\{z_<\} + \mathcal{H}_0\{z_>\} + \mathcal{H}_1\{z_<+z_>\} \quad . \tag{4.29}$$

When this expression is inserted into (4.27), one obtains

$$exp[-\mathcal{H}'\{z_<\}] = exp[-\mathcal{H}_0\{z_<\}]\; \langle\, exp[-\mathcal{H}_1\{z_<+z_>\}]\, \rangle \tag{4.30}$$

where the expectation value, $\langle ... \rangle$, is calculated with the harmonic weight, $exp[-\mathcal{H}_0\{z_>\}]$.

Now, *assume* that the interface undergoes a roughening transition at a finite roughening temperature, $T = T_r < \infty$. Then, for $T > T_r$, the quasiperiodic potential, V_{QP}, must be *irrelevant* and must eventually become *small* unter the RG transformation. Thus, in order to study the rough phase, one may perform an expansion of (4.30) in powers of V_{QP} or $\mathcal{H}_1$. This leads to the cumulant expansion as given by

$$\langle\, exp[-\mathcal{H}_1]\, \rangle = exp[\sum \frac{1}{n!} \langle (-\mathcal{H}_1)^n \rangle_c] \approx exp[-\langle \mathcal{H}_1 \rangle + \tfrac{1}{2} \langle \mathcal{H}_1 \mathcal{H}_1 \rangle_c] \tag{4.31}$$

where the subscript c stands for cumulant. It then follows from (4.30) that, apart from the rescaling transformation (4.28), the renormalized Hamiltonian is given by

$$\mathcal{H}'\{z_<\} \approx \mathcal{H}_0\{z_<\} + <\mathcal{H}_1\{z_<+z_>\}> - \tfrac{1}{2} <\mathcal{H}_1\{z_<+z_>\}\,\mathcal{H}_1\{z_<+z_>\}>_c \quad (4.32)$$

3. Cumulant expansion up to second order in $\mathcal{H}_1$

The first–order term of the cumulant expansion for the quasiperiodic potential (4.25) is

$$<\mathcal{H}_1\{z_<+z_>\}> = \int d^2x\,[\,A_0 <cos[q_0(z_<+z_>)]> + A_1 <cos[q_1(z_<+z_>)]>\,] \quad (4.33)$$

The expectation values which involve the harmonic weight $exp[-\mathcal{H}_0\{z_>\}]$ are easily calculated. The result is $<cos[q_j(z_<+z_>)]> = cos[q_j z_<]\,exp[-\tfrac{1}{2}q_j^2 <z_>^2>\,]$. In addition, rescaling as in (4.28) yields a factor b^2. Therefore, the renormalized amplitudes are

$$A_j' = A_j\,b^2\,exp[-\tfrac{1}{2}\,q_j^2 <z_>^2>\,] \quad . \qquad (4.34)$$

Since $<z_>^2> = (T/K)c(b)$, one has $A_j' = A_j$ at $T = T_r(q_j)$ with

$$T_r(q_j)/K = 4\,ln[b]\,/\,c(b)\,q_j^2 \approx 8\pi\,/\,q_j^2 \quad \text{for} \quad b \to 1 + \Delta s\,. \qquad (4.35)$$

Thus, to leading order in the cumulant expansion, the potential term $A_j\,cos[q_j z]$ is relevant and irrelevant for $T < T_r(q_j)$ and $T > T_r(q_j)$, respectively.

Therefore, starting with two components of V_{QP} as in (4.25), one concludes that both components are irrelevant for $T > T_r(q_1)$. However, this does *not* represent the roughening temperature of the system since higher–order terms of the RG *generate* additional components of V_{QP} which have *smaller* wavenumbers q_j and, thus, *higher* temperatures $T_r(q_j)$ according to (4.35). Even though the second–order calculation is straight forward, the details are somewhat tedious and I will only indicate its main ingredients.

The additional relevant components of the potential arise from the cross term

$$V_2(q_0,q_1) \equiv <\int d^2x_0\,A_0\,cos[q_0 z(\mathbf{x}_0)]\int d^2x_1\,A_1\,cos[q_1 z(\mathbf{x}_1)] >_c \qquad (4.36)$$

The main contribution to the integrations comes from $\mathbf{x}_0 \simeq \mathbf{x}_1$. To leading order in a gradient expansion, one then obtains

$$V_2(q_0,q_1) \approx \int d^2x\,A_0 A_1\{\,g_+ cos[(q_0+q_1)z_<] + g_- cos[(q_0-q_1)z_<]\,\} \qquad (4.37)$$

with parameter–dependent coefficients $g_\pm$. The first term $\sim cos[(q_0+q_1)z]$ has a larger wavenumber and, thus, is irrelevant for $T \geq T_r(q_1)$, However, the second term in (4.37) has the wavenumber $q_2 \equiv q_0 - q_1 = q_0(1-1/\sigma) = q_0/\sigma^2$ with $\sigma = (1+\sqrt{5})/2$ which is smaller than both q_0 and $q_1 = q_0/\sigma$. It then follows from (4.35) that this term is relevant up to $T_r(q_2) > T_r(q_1)$.

Now, further iterations of the RG transformation generate components with smaller and smaller wavenumbers. Indeed, it is not difficult to see that the (j–1)th iteration step generates a term $A_j cos[q_j z]$ with $q_j = (q_{j-2} - q_{j-1}) = q_0/\sigma^j$ which is relevant up to $T = T_r(q_j) \sim \sigma^{2j}$ according to (4.35). Therefore, for any finite T, the interface feels an effective quasiperiodic potential on sufficiently large scales. This implies a *smooth* interface at $T < \infty$ and an *infinite roughening temperature*, $T_r = \infty$. /41–43/ This prediction of the RG has been confirmed by recent Monte Carlo simulations. /44/

V. INTERACTIONS OF MANIFOLDS AND UNBINDING TRANSITIONS

A. Wetting, Adhesion, and Adsorption Phenomena

A variety of physical phenomena is governed by the mutual interaction of interfaces, membranes, and polymers. /4/ The geometry of such phenomena is shown in Fig. 4. *Wetting* phenomena occur when an interface between two macroscopic bulk phases contains a thin film or layer of a third phase, see Fig. 4(a). /45/ The intermediate layer is bounded by two interfaces, and its thickness is determined by the mutual interactions of these interfaces. *Adhesion* of an oriented membrane onto an interface is shown in Fig. 4(b). This membrane could be, e.g., a small segment of a large lipid vesicle which adheres to the interface. *Adsorption* of a polymer and of a crumpled membrane is shown in Fig. 4(c) and 4(d).

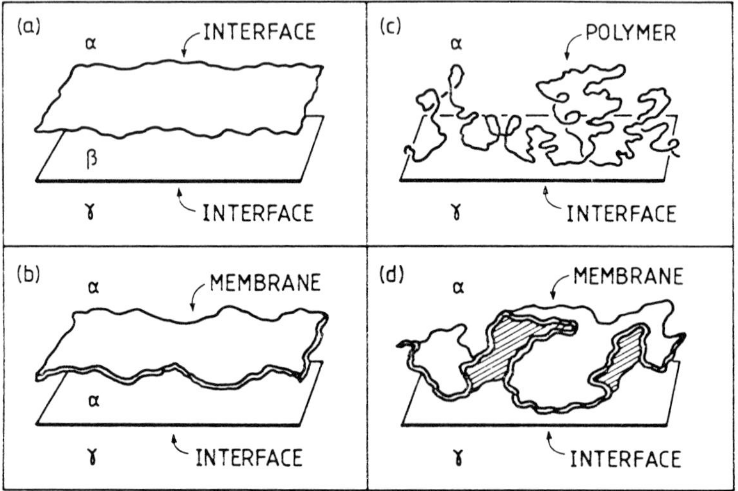

Fig.4 :   Bound states of low–dim manifolds : (a) Wetting, surface melting and related phenomena; (b) Adhesion of an oriented membrane; and (c),(d) Adsorption of a polymer and a crumpled membrane. (α, β, and γ denote distinct phases)

The manifolds shown in Fig. 4 form *bound states* in the sense that their mean separation has a finite value. These states require the presence of some *attractive* interaction between the manifolds. For a given shape of the interfaces, membranes, or polymers, their mutual interaction directly reflects microscopic intermolecular forces. This *direct* interaction consists of several short–ranged and long–ranged contributions such as electrostatic , van der Waals or structural interactions. There is in fact a huge literature on intermolecular forces and the resulting direct interactions. However, it has been realized only recently that these *interactions are strongly renormalized by shape fluctuations.*

Quite generally, the renormalization arising from shape fluctuations acts to decrease the attractive part of the direct interaction and to increase its repulsive part. Now, consider the case of thermally–excited fluctuations. At low T, these fluctuations are weak and the renormalized interaction closely resembles the direct interaction. However, as T is increased, the renormalization becomes more and more effective up to a characteristic unbinding temperature, $T = T_u$. For $T > T_u$, the renormalized interaction no longer has an attractive part, and the two manifolds are completely unbound.

As T_u is approached from below, the system undergoes an *unbinding transition* : the mean separation, ℓ, of the manifolds goes to infinity and typically behaves as $\ell \sim (T_u{-}T)^{-\psi}$. The critical exponent ψ depends, to some extent, on the form of the direct interaction, V_{DI}. In fact, one must distinguish several scaling regimes or universality classes for V_{DI}. This can be understood in a rather simple way if the manifolds are again described in terms of humps and blobs, compare Figs. 1 and 2. In the present context, these scaling pictures lead to effective fluctuation–induced interactions.

B. Fluctuation–induced Interactions

First, consider a *rough* manifold bound to a smooth one as in Fig. 4(a) and 4(b). Then, the manifold can again be viewed as an ensemble of essentially uncorrelated humps, see Fig. 5(a). Each hump with intrinsic size, ξ_{in}, has a lateral and transverse extension $\xi_\parallel \sim \xi_{in}$ and $\xi_\perp \sim \xi_{in}^\zeta$, respectively. Then, the scaling arguments of Sec. III.E.1 lead to a singular contribution of the free energy per unit area which is given by /14,46,1/

$$V_{FL} \equiv f_{FL} \sim 1/\xi_\perp^\tau \quad \text{with} \quad \tau = d_\parallel/\zeta \tag{5.1}$$

as in (3.27).

Likewise, a *crumpled* manifold bound to a smooth one can be regarded as an ensemble of essentially uncorrelated blobs, see Fig. 5(b). Each blob with intrinsic size, ξ_{in}, has an extrinsic size $\xi_\perp \sim \xi_\parallel \sim \xi_{in}^\nu$. Now, the scaling arguments of Sec. III.E.2 lead to a free energy per unit (intrinsic) area which behaves as /36,4/

$$V_{FL} \equiv f_{FL} \sim 1/\xi_\perp^\tau \quad \text{with} \quad \tau = d_\parallel/\nu , \tag{5.2}$$

compare with (3.31).

The free energies V_{FL} as given by (5.1) and (5.2) represent *excess* free energies of the *bound* manifolds. Therefore, they may be interpreted as effective, *fluctuation–induced* interactions. These interactions have two important properties: (i) they are *repulsive* and, thus, act to drive the manifolds apart; and (ii) they are *long–ranged* (for $\zeta > 0$ and $\nu > 0$) and, thus, may compete even with long–ranged components of the direct interactions.

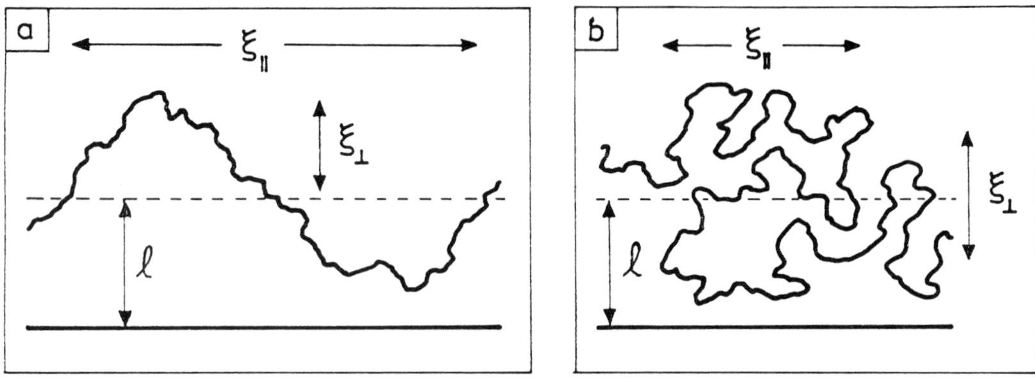

Fig.5 :   (a) A *rough* manifold with $\xi_\perp \sim \xi_\parallel^\zeta \sim \xi_{in}^\zeta$, and (b) a *crumpled* manifold with $\xi_\perp \sim \xi_\parallel \sim \xi_{in}^\nu$, which are bound to a smooth manifold at mean separation ℓ.

C. Different Scaling Regimes

Now, the fluctuation–induced interaction will be compared with the direct interaction arising from intermolecular forces. For the geometries as shown in Fig. 4, this direct interaction, V_{DI}, depends on the coordinate z which measures the distance of the fluctuating manifold from the smooth interface. Then, the interaction free energy (per unit area) of the bound manifold can be estimated by

$$< V_{DI}(z) > \sim V_{DI}(\ell) \quad \text{with} \quad \ell \equiv < z > . \tag{5.3}$$

Now, consider a certain pair of manifolds characterized by a certain value of τ and, thus, by the fluctuation–induced interaction $V_{FL} \sim 1/\xi_\perp^\tau$. Then, the space of all possible direct interactions, V_{DI}, consists of four different scaling regimes. These regimes can be identified by a simple superposition of V_{FL} and V_{DI}. One then finds the following four regimes /1,4/: (i) The *mean–field* (MF) regime characterized by

$$V_{DR}(\ell) \gg V_{FL}(\ell) \sim 1/\ell^\tau \quad \text{for large } \ell. \tag{5.4}$$

where V_{DR} represents the *repulsive* part of V_{DI}. In this situation, the mean separation, ℓ, of the manifolds is not affectd by the shape fluctuations; (ii) The *weak–fluctuation* (WFL) regime defined by

$$V_{DR}(\ell) \ll V_{FL}(\ell) \sim 1/\ell^T \ll V_{DA}(\ell) \quad \text{for large } \ell \tag{5.5}$$

where V_{DA} represents the *attractive* part of V_{DI}. The critical behavior within this regime can be obtained in a rather simple way by minimization of $V_{FL} + V_{DI}$; (iii) The *intermediate–fluctuation* (IFL) regime with

$$|V_{DI}(\ell)| \sim V_{FL}(\ell) \sim 1/\ell^T \quad \text{for large } \ell . \tag{5.6}$$

In this case, the superposition Ansatz indicates that the critical behavior depends both on the long–ranged tail $\sim 1/\ell^T$ and on the short–ranged part of $V_{DI}(\ell)$; and (iv) The *strong–fluctuation* (SFL) regime characterized by sufficiently short–ranged interactions with

$$|V_{DI}(\ell)| \ll V_{FL}(\ell) \sim 1/\ell^T \quad \text{for large } \ell . \tag{5.7}$$

In this regime, the superposition Ansatz predicts a first–order transition while the transition is, in fact, often continous and then governed by characteristic critical exponents.

The classification as given by eqs. (5.4) $-$ (5.7) which has been obtained from the superposition of direct and fluctuation– induced interactions is fully confirmed by more systematic methods. On the other hand, the superposition Ansatz fails for the critical behavior within the IFL and the SFL regimes. As explained below, functional renormalization group (RG) methods reveal that the critical behavior within these regimes is, in fact, rather unusual.

D.   Renormalized Interactions

1. Functional renormalization : nonperturbative methods

To proceed, let us again consider the geometry displayed in Figs. 4(a) and 4(b). As before, the shape of the fluctuating manifold will be parametrized by $z = z(\mathbf{x})$ where $\mathbf{x}$ is a $d_{\parallel}$–dim coordinate parallel to the smooth interface. The effective Hamiltonian then has the generic form

$$\mathcal{H}\{z\} = \int d^{d_{\parallel}}x \ \{ \tfrac{1}{2} K \, (\nabla^n z)^2 + V_{DI}(z) \} \tag{5.8}$$

with $n = \zeta + d_{\parallel}/2$ as in (3.7), and an implicit small–distance cutoff, a. The first term in (5.8) describes the elastic energy associated with the shape fluctuations, compare (3.6), while the second term represents the direct interactions of the manifolds. The statistical properties which follow from this model can be studied by a variety of theoretical methods. So far, the most useful approach has been functional renormalization of the direct interaction, V_{DI}.

In the context of roughening transitions, functional renormalization of the lattice potential, V_{QP}, can be based on a cumulant expansion in powers of V_{QP}, see Sec. IV.C .

For unbinding phenomena, such a perturbative approach is, however, *not* reliable in general since the direct interaction, $V_{DI}(z)$, contains a hard wall at $z = 0$. Indeed, the RG transformations for V_{DI}, which are discussed below, are *non*perturbative.

The first such method which has been used represents an extension /47,46/ of Wilson's approximate recursion relations /48/. For infinitesimal rescaling factor $b \to 1 + \Delta s$, this functional RG leads to the nonlinear flow equation /46/

$$\partial V / \partial s = d_{\parallel} V + \zeta\, z \partial V / \partial z + \tfrac{1}{2}\, v\, ln[\, 1 + (a_{\perp}^2 / v)\, \partial^2 V / \partial z^2\,] \tag{5.9}$$

for the renormalized interactions $V(z|s)$ with scale factors v and $a_{\perp}$. The 'initial' interaction at $s = 0$ is given by the direct interaction: $V(z|s{=}0) = V_{DI}(z)$.

As discussed further below, the RG transformation (5.9) leads to *whole lines of RG fixed points* which describe various types of unbinding transitions. /26,49/ More recently, we have also studied two other functional RG transformations: (i) an *exact* transformation for wetting transitions in $d = 1{+}1$ with $\tau = 2$ which acts in a somewhat enlarged function space /50/; and (ii) another approximate transformation /51,52/ which is based on a *smooth* cutoff procedure. Apart from the regimes where the unbinding transitions are first–order, the results of these different functional RG methods are in fair agreement.

2. Lines of renormalization group fixed points

Now, let us apply the RG transformation as given by (5.9) within the space of all interactions $V(z)$ which decay to zero for large z at least as $\sim 1/z^{\tau}$. This function space contains both the SFL and the IFL regimes as defined in (5.6) and (5.7). It is convenient to use the dimensionless variables $y \equiv \sqrt{2\zeta}\, z\, /\, a_{\perp}$ and $U(y) \equiv 2\zeta\, V(a_{\perp} y/\sqrt{2\zeta})\, /\, v$. Then, the flow equation (5.9) becomes /26/

$$\partial U / \partial s = \zeta[\, \tau\, U + y\, \partial U / \partial y + ln(\, 1 + \partial^2 U / \partial y^2\,)\,]\, . \tag{5.10}$$

The fixed points, $U^*(z)$, of this RG transformation satisfy

$$\tau\, U^* + y \partial U^* / \partial y + ln[\, 1 + \partial^2 U^* / \partial y^2\,] = 0\, . \tag{5.11}$$

Therefore, the rescaled fixed points depend only on *one* parameter, namely τ. This implies , e.g., that the adhesion of fluid membranes in $d = 3$ is characterized by the same RG fixed points as wetting in $d = 2$ since $\tau = 2$ in both cases. Likewise, (5.10) implies that both systems are characterized, apart from a factor ζ, by the same scaling indices and thus by the same critical exponents.

The fixed point equation (5.11) must be supplemented by appropriate boundary conditions. For unbinding transitions, one wants to include a hard wall at $y = 0$ and a tail which decays to zero for large y. One then finds a line of fixed points, $U^*(y|\sigma)$, parametrized by a parameter $\sigma > 0$ which behave as /26/

$$U^*(y) = \sigma/y^\tau + \frac{\tau+2}{\tau} \, ln(y) + \mathcal{O}(z^\tau) \qquad (\text{with} \quad \sigma > 0) \tag{5.12}$$

for small y and as

$$U^*(y) \approx \rho_L(\sigma)/y^\tau + \rho_S(\sigma) \, z^{\tau-1} \exp(-z^2/2) \tag{5.13}$$

for large y.

For sufficiently small τ, the function $\rho_L(\sigma)$ is found to be convex downwards with a unique minimum at $\sigma = \sigma_{E1}$. An example is shown in Fig. 6(a) for $\tau = 2$. /26,49/ In this case, the function $\rho_L = \rho_L(\sigma)$ has two zeros at $\sigma = \sigma_{S0}$ and $\sigma = \sigma_{S1}$ (with $\sigma_{S1} < \sigma_{E1} < \sigma_{S0}$) which correspond to short–ranged fixed points with a Gaussian tail, see (5.14). Such a convex downwards piece of $\rho_L(\sigma)$ persists to larger values of τ. However, as τ is increased, the function $\rho_L = \rho_L(\sigma)$ develops more structure. /52/ First, at a certain value $\tau = \tau_2$, it exhibits a point of inflection with zero slope at $\sigma = \sigma_2$ with $\sigma_2 < \sigma_{S1}$ which bifurcates, for $\tau > \tau_2$, into an additional minimum and maximum at $\sigma = \sigma_{\overline{E2}}$ and σ_{E2}. The new minimum at $\sigma = \sigma_{\overline{E2}}$ moves down with increasing τ and touches the σ–axis at $\tau = \tau_{S2} > \tau_2$. An example for the shape of $\rho_L(\sigma)$ with $\tau > \tau_{S2}$ is shown in Fig. 6(b). /52/

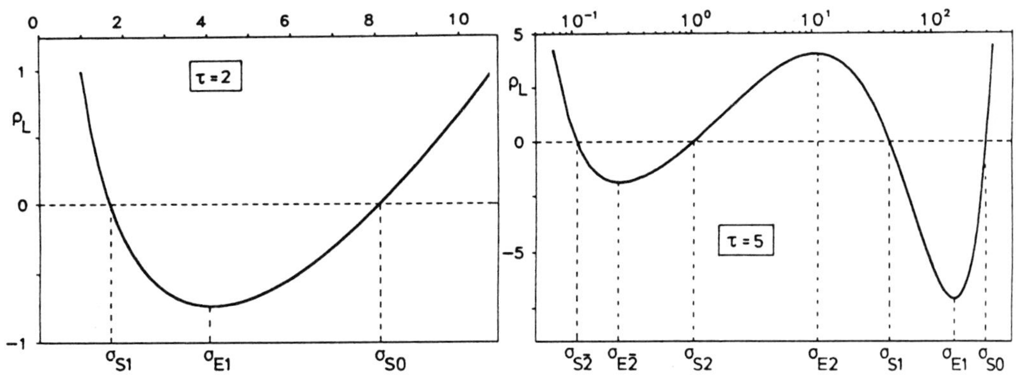

Fig.6 : The amplitude $\rho_L = \rho_L(\sigma)$ of the power law tail of the fixed points, see (5.13), for two values of $\tau = d_\parallel/\zeta$.

As τ is increased even further, another point of inflection with zero slope appears for $\tau = \tau_3$ which lies at $\sigma = \sigma_3$ with $\sigma_{\overline{E2}} < \sigma_3 < \sigma_{E2}$, i.e, between the minimum and the maximum of the previous bifurcation. /52/ This inflection point develops into another pair of extrema of $\rho_L(\sigma)$ which are located at $\sigma = \sigma_{\overline{E3}}$ and σ_{E3}. This type of bifurcation continues infinitely many times as τ is increased towards $\tau = \infty$: the point of inflection with zero slope for the jth bifurcation at $\tau = \tau_j$ occurs between the two extrema generated in the (j–1)th bifurcation but it only appears after one of these extrema has touched the σ–axis with $\rho_L = 0$. As a consequence, the function $\rho_L(\sigma)$ exhibits more and more zeroes located at $\sigma = \sigma_{S0}$, σ_{S1} , σ_{S2} and $\sigma_{S\bar{2}}$, σ_{S3} and $\sigma_{S\bar{3}}$ etc . /52,53/ These values of σ are ordered according to $\sigma_{S\bar{2}} < \sigma_{S\bar{3}} < ... < \sigma_{S3} < \sigma_{S2} < \sigma_{S1} < \sigma_{S0}$. They determine the overall shape of the fixed points, $U^*(y)$, as a function of y and , in particular, the number of zeroes of $U^*(y)$ at finite y : for $\sigma \geq \sigma_{S0}$, the fixed points have no extremum and no zero ; for $\sigma_{S1} \leq \sigma < \sigma_{S0}$, they have one minimum and one zero ; for $\sigma < \min(\sigma_{S\bar{3}}, \sigma_{S2})$ and $\sigma_{S2} \leq \sigma < \sigma_{S1}$, they have one minimum and one maximum and thus two zeros, etc.

3. Relevant scaling fields and scaling indices

The physical meaning of the various fixed points depends primarily on the number and on the character of their relevant perturbations or scaling fields. In the following, I will only consider those perturbations which do not decay more slowly than $1/y^\tau$ (or $1/z^\tau$) for large y (or z), and thus belong to the same subspace of interactions as the fixed points.

For sufficiently large σ, all fixed points are completely stable : all perturbations are irrelevant apart from the marginal one which corresponds to a shift of σ and, thus, to a translation along the line of fixed points. As σ is decreased, a relevant perturbation with eigenvalue $\lambda_1 > 0$ first appears at the location, $\sigma = \sigma_{E1}$, of the minimum of $\rho_L(\sigma)$. This perturbation is in fact short–ranged and has a Gaussian tail as can be shown from the RG transformation (5.10) when linearized around the fixed points. /49/ For $\tau < \tau_2$, no other relevant perturbation is present (within the IFL and SFL regimes) for any value of σ.

For $\tau_2 < \tau < \tau_3$, the function $\rho_L(\sigma)$ has an additional minimum at $\sigma = \sigma_{E\bar{2}}$, and an additional maximum at $\sigma = \sigma_{E2}$, see Fig. 6(b). In the intermediate σ–regime with $\sigma_{E\bar{2}} < \sigma < \sigma_{E2}$, a second relevant perturbation appears with eigenvalue $\lambda_2 > 0$ (which again has a Gaussian tail). As τ is increased beyond τ_3 , the function $\rho_L(\sigma)$ exhibits another maximum and minimum located at $\sigma = \sigma_{E\bar{3}}$ and $\sigma = \sigma_{E3}$, as mentioned. Then, one has three relevant scaling fields for the intermediate σ–range given by $\sigma_{E\bar{3}} < \sigma < \sigma_{E3}$, two relevant scaling fields for $\sigma_{E\bar{2}} < \sigma < \sigma_{E\bar{3}}$ and $\sigma_{E3} < \sigma < \sigma_{E2}$, and one relevant scaling field for $\sigma < \sigma_{E\bar{2}}$ and $\sigma_{E2} < \sigma < \sigma_{E1}$. Thus, each additional inflection point of $\rho_L(\sigma)$ and the associated bifurcation into a maximum and minimum corresponds to the appearance of an additional relevant scaling field (with a Gaussian tail). /52/

4. Critical behavior at unbinding transitions

Within the IFL and the SFL regimes, all fixed points with $\sigma > \sigma_{E1}$ have a domain of attraction with codimension zero, and describe completely *unbound states* of the manifolds. On the other hand, all fixed points with $\sigma < \sigma_{E1}$ have a domain of attraction with a nonzero codimension which is given by the number of their relevant scaling fields. These latter fixed points describe *unbinding transitions* within the IFL and the SFL regimes.

Now, consider a physical system at temperature T with reduced stiffness, K/T, and reduced direct interaction, $V_{DI}(z)/T$. A change of T now leads to a temperature–trajectory within the interaction space which intersects one of the attraction–domains at the unbinding temperature $T = T_u$. The corresponding fixed point and its relevant scaling fields then govern the critical behavior at T_u . If V_{DI} belongs to the SFL regime defined by (5.7), this fixed point has a Gaussian tail and $\sigma = \sigma_{Sj}$ with $j \geq 1$. If V_{DI} belongs to the IFL regime with $V_{DI}(z) \sim U_{DI}(y) \approx W/y^{\tau}$, the corresponding fixed point also has such a power law tail with amplitude $\rho_L = W$.

In general, $V_{DI}(z)$ may be a complicated function of z and thus may have several minima. Here, I will focus on the situation where $V_{DI}(z) \sim U_{DI}(y)$ achieves a *single minimum* at finite or infinite $z \sim y$. In this case, the critical behavior at the unbinding transitions is governed by the fixed points with $\sigma_{S1} \leq \sigma \leq \sigma_{E1}$.

As described above, the function $\rho_L(\sigma)$ has a unique minimum at $\sigma = \sigma_{E1}$ for $\sigma \geq \sigma_{S1}$. Furthermore, the amplitude $\rho_S(\sigma)$ of the Gaussian tail of the fixed points, see (5.13), satisfies $\rho_S(\sigma = \sigma_{S1}) < 0$ and $\rho_S(\sigma = \sigma_{S0}) > 0$. Therefore, for $\sigma_{S1} \leq \sigma \leq \sigma_{S0}$, $\rho_S(\sigma)$ should increase monotonically with σ , and the fixed points can then be parametrized by $\rho_L = \rho_L(\rho_S)$. /49/ This function again has a unique minimum at $\rho_S = \rho_S^*$, and behaves as $\rho_L(\rho_S) \approx \rho_L^* + \frac{1}{2} R_2 (\rho_S - \rho_S^*)^2$ for small $\rho_S - \rho_S^*$. Close to this minimum, the RG flow within the ∞–dim interaction space can be reduced to a flow within the 2–dim (ρ_L, ρ_S) – subspace, where it has a parabolic character. /49,4/ Integration of the RG trajectories then leads to two different subregimes, (A) and (B), for the unbinding transitions.

Subregime (A) is governed by the fixed point at (ρ_L^*, ρ_S^*) corresponding to $\sigma = \sigma_{E1}$. A temperature trajectory which intersects the attraction–domain of this fixed point exhibits an unbinding transition of *infinite order* : the mean separation, e.g., behaves as /49/

$$\ell \sim exp[\, 2\pi\zeta \,/\, \omega \,\{\rho_L^* - \rho_L\}^{1/2} \,] \sim exp[\, c \,/\, \sqrt{T_u - T} \,] \quad . \tag{5.14}$$

Subregime (B) is governed by the line of fixed points with $\sigma_{E2} < \sigma < \sigma_{E1}$. In this case, one obtains the power law behavior /49/

$$\ell \sim (T_u - T)^{-\psi} \quad \text{with} \quad \psi \approx \zeta/\omega\{\rho_L^* - \rho_L\}^{1/2} \tag{5.15}$$

as T_u is approached from below. Thus, this regime is characterized by *second–order* transitions but with non–universal critical exponents. It contains the case $\rho_L = 0$ or $\sigma = \sigma_{S1}$ which describes second–order transitions in the SFL regime. In the latter case, the RG considered here leads to a universal value for ψ which depends only on τ. /54/

VI. OUTLOOK

In summary, the shape fluctuations of low–dim manifolds lead to a variety of critical phenomena. As explained in Sec. III, a fluctuating manifold is a scale–invariant object which can be smooth, rough, or crumpled. A rough state consists of humps, a crumpled state of blobs, see Fig. 1. Furthermore, the manifolds may undergo *roughening* , *crumpling* , or *unbinding* transitions, as discussed in Sec. III.C and IV, Sec. III.C, and Sec. V, respectively.

Roughening and crumpling transitions can be characterized by the *delocalization* of humps and blobs, see Fig. 2. Such a delocalization also occurs at unbinding transitions where it acts to drive the manifolds apart, see Fig. 5. From a theoretical point of view, the basic scaling properties of these transitions are now understood. As far as experiments are concerned, both roughening /2/ and unbinding transitions /55/ have been observed but more work is highly desirable. In fact, there are many theoretical and experimental problems which deserve further study. Here are just a few of them : (i) *Roughening of quasicrystals* − The results described in Sec. IV imply that the surfaces of *ideal* quasicrystals in d = 3 are always smooth. However, real quasicrystals may contain a certain amount of disorder. It has been argued that interfaces (or domain walls) within *random* quasicrystals are rougher than in periodic structures. /41/ It remains to be shown that this applies also to the *surfaces* of random quasicrystals; (ii) *Crumpling of self−avoiding membranes* − As mentioned, tethered (or polymerized or crystalline) membranes with self−avoidance probably do not crumple. The effect of self−avoidance remains to be understood for more flexible membranes such as fluid or hexatic /25/ ones. One interesting possibility is that the latter membranes undergo a crumpling transition; and (iii) *Adhesion (or unbinding) of vesicles* − Real membranes of amphiphilic molecules usually form closed surfaces or vesicles which often adhere to interfaces or other membranes. We recently found theoretically that such vesicles can also undergo adhesion (or unbinding) transitions. /56/ These transitions should be accessible to various experimental techniques.

Acknowledgements

I thank all my collaborators, and especially Stefan Grotehans and Udo Seifert, for stimulating interactions, and the organizers of this summer school for their invitation. Partial support by the DFG through the SFB 266 is gratefully acknowledged.

References

/1/ A review is in R. Lipowsky, in *Random Fluctuations and Growth*, ed. by H.E. Stanley und N. Ostrowsky (Kluwer Academic Publishers, Dordrecht 1988)

/2/ A review is in H. van Beijeren und I. Nolden, in *Structure and Dynamics of Surfaces II*, ed. by W. Schommers und P. von Blanckenhagen (Springer−Verlag, 1987)

/3/ See reviews in *Statistical Mechanics of Membranes and Surfaces*, ed. by D. Nelson , T. Piran, and S. Weinberg (World Scientific, to be published)

/4/ A review is in R. Lipowsky, Physica Scripta T **29**, 259 (1989)

/5/ See, e.g., J.S. Rowlinson und B. Widom, *Molecular Theory of Capillarity*, Clarendon Press, Oxford 1982

/6/ A review is in M.E. Fisher, J. Chem. Soc. Faraday Trans. 2, **82**, 1569 (1986)

/7/   See any textbook on superconductivity.

/8/   See, e.g., *Physics of Amphiphilic Layers*, ed. by J. Meunier, D. Langevin und N. Boccara, Springer Proc. in Physics, Vol. 21 (Springer–Verlag, 1987)

/9/   A review is E. Sackmann, in *Biophysics*, ed. by W. Hoppe, W. Lohmann, H. Markl und H. Ziegler, (Springer–Verlag, 1982)

/10/   W. Helfrich, Z. Naturforsch. **28c**, 693 (1973).

/11/   P.–G. de Gennes und C. Taupin, J. Phys. Chem. **88**, 2294 (1982)

/12/   F.F. Abraham, W.E. Rudge, and M. Plischke, Phys. Rev. Lett. **62**, 1757 (1989), and J.–S. Ho and A. Baumgärtner (to be published)

/13/   P.G. de Gennes, *Scaling Concepts in Polymer Physics*, Cornell University Press, New York 1979

/14/   R. Lipowsky und M.E. Fisher, Phys. Rev. Lett. **56**, 472 (1986),

/15/   C.L. Henley und R. Lipowsky, Phys. Rev. Lett. **59**, 1679 (1987)

/16/   A. Garg und D. Levine, Phys. Rev. Lett. **59**, 1683 (1987)

/17/   G. Grinstein and S.–K. Ma, Phys. Rev. B **28**, 2588 (1983)

/18/   J. Villain, J. Physique Lett. **43**, L551 (1982)

/19/   D. A. Huse and C. L. Henley, Phys. Rev. Lett. **54**, 2708 (1985)

/20/   T. Nattermann and W. Renz, Phys. Rev. B **38**, 5184 (1988)

/21/   The unbinding of flux lines has been studied by D. R. Nelson, Phys. Rev. Lett. **60**, 1973 (1988), and K. Ziegler, Europhys. Lett. **9**, 277 (1989)

/22/   T. Nattermann and R. Lipowsky, Phys. Rev. Lett. **61**, 2508 (1988)

/23/   W. Helfrich, Z. Naturforsch. **33a**, 305 (1978)

/24/   R. Lipowsky und S. Leibler, Phys. Rev. Lett **56**, 2541 (1986), and **59**, 1983 (E) (1987)

/25/   D.R. Nelson and L. Peliti, J. Physique **48**, 1085 (1987)

/26/   R. Lipowsky, Europhys. Lett. **7**, 255 (1988)

/27/   S. Leibler and A. Maggs, Phys. Rev. Lett. **63**, 406 (1989)

/28/   Y. Kantor, M. Kardar, and D. R. Nelson, Phys. Rev. A **35**, 3056 (1987).

/29/   G. Parisi and N. Sourlas, Phys. Rev. Lett. **46**, 871 (1981)

/30/   A. Garg, Phys. Rev. B **37**, 10003 (1988)

/31/   B. Yang, W. F. Saam, and J. A. Jaszczak (to be published)

/32/   L. Peliti and S. Leibler, Phys. Rev. Lett. **54**, 690 (1985)

/33/   Y. Kantor and D. R. Nelson, Phys. Rev. Lett. **58**, 2774 (1987)

/34/   M. Paczuski, M. Kardar, and D. R. Nelson, Phys. Rev. Lett. **60**, 2638 (1988)

/35/   E. Guitter, F. David, S. Leibler, and L. Peliti, Phys. Rev. Lett. **61**, 2949 (1988)

/36/   R. Lipowsky and A. Baumgärtner, Phys. Rev. A **40**, 2078 (1989)

/37/   See the collection of reprints in P. J. Steinhardt and S. Ostlund, *The Physics of Quasicrystals* (World Scientific, 1987)

/38/   C. Tang and M. Kohmoto, Phys. Rev. B **34**, 2041 (1986)

/39/   M. Kohmoto, B. Sutherland, and C. Tang, Phys. Rev. B **35**, 1020 (1987)

/40/   S. Aubry and G. Andre, in *Group Theoretical Methods in Physics*, ed. by L. Horwitz and Y. Ne'eman, Annals of the Israel Phys. Soc., Vol. 3, and Ref. /37/

/41/   R. Lipowsky and C. L. Henley, Phys. Rev. Lett. **60**, 2394 (1988)

/42/   D. A. Huse, (unpublished)

/43/   Similar results have been obtained by L. V. Mikheev, Phys. Lett. A **132**, 137 (1988)

/44/   J. A. Jaszczak, W. F. Saam, and B. Yang, Phys. Rev. B **39** (in press)

/45/   See, e.g., R. Lipowsky, Phys. Rev. B **32**, 1731 (1985), and *Critical behavior of interfaces: wetting, surface melting and related phenomena*, Habilitations–Schrift, Universität München, 1987 (published as Juel–Spez–438 by KFA Jülich)

/46/   R. Lipowsky und M.E. Fisher, Phys. Rev. B**36**, 2126 (1987)

/47/   R. Lipowsky und M.E. Fisher, Phys. Rev. Lett. **57**, 2411 (1986)

/48/   K.G. Wilson, Phys. Rev. B**4**, 3184 (1971); K.G. Wilson and M.E. Fisher, Phys. Rev. Lett. **28**, 240 (1972)

/49/   R. Lipowsky, Phys. Rev. Lett. **62**, 704 (1989)

/50/   F. Jülicher, R. Lipowsky, and H. Müller–Krumbhaar, (to be published)

/51/   S. Scheidl, diploma thesis, Universität München, 1989

/52/   S. Grotehans and R. Lipowsky, (to be published)

/53/   The fixed points with $\rho_{\mathrm{L}} = 0$ have been independently studied by F. David and S. Leibler, (to be published)

/54/   This is consistent with the MC data on the adhesion transitions of fluid membranes as obtained by R. Lipowsky and B. Zielinska, Phys. Rev. Lett. **62**, 1572 (1989)

/55/   Adhesion transitions of fluid membranes have been experimentally observed by M. Mutz and W. Helfrich, Phys. Rev. Lett. **62**, 2881 (1989)

/56/   U. Seifert and R. Lipowsky (to be published)

FUNDAMENTAL PROBLEMS IN STATISTICAL MECHANICS VII
H. van Beijeren, Editor
© Elsevier Science Publishers B.V., 1990

RENORMALIZATION AND CONFORMAL INVARIANCE FOR POLYMERS

Bertrand DUPLANTIER

Service de Physique Théorique de Saclay*
F-91191 Gif-sur-Yvette Cedex, France

The essentials of the statistical mechanics of polymers (self-avoiding walks) are described. A direct method of renormalization is explained for the continuum model in space dimension $d = 4 - \varepsilon$. In two dimensions the critical properties are related to conformal invariance. Finally the statistical mechanics of polymers embedded in a fluctuating membrane is described via its link to conformal invariance in presence of "2D quantum gravity".

Contents

1. INTRODUCTION
2. GEOMETRICAL PROPERTIES
    2.1 First critical exponents
    2.2 Towards more complicated geometrical questions
    2.3 Multiple points of the Brownian motion
3. RENORMALIZATION OF THE EDWARDS MODEL
    3.1 Wiener measure
    3.2 Edwards measure
    3.3 Dimensional regularization
    3.4 Perturbation expansion
    3.5 Renormalization
4. POLYMERS IN 2D AND CONFORMAL INVARIANCE
    4.1 Introduction
    4.2 Conformal scaling dimensions of polymers
5. GEOMETRICAL EXPONENTS OF THE O(n) MODEL
    5.1 Watermelon configurations
    5.2 Coulomb gas
    5.3 Winding angle of a self-avoiding walk
6. POLYMERS OF HIGHER TOPOLOGY
    6.1 Polymer networks
    6.2 Contact exponents

---

* Laboratoire de l'Institut de Recherche Fondamentale du Commissariat à l'Energie Atomique

7. STATISTICAL MECHANICS ON A FLUCTUATING MEMBRANE

7.1 Brief history

7.2 O(n) model on a random lattice

7.3 Random matrices

7.4 Self-avoiding walks on a 2D-random lattice

7.5 Conformal invariance on a random surface

7.6 Conformal spectrum in 2D-quantum gravity

7.7 Higher topologies

REFERENCES

1. INTRODUCTION

A well-known model in statistical mechanics is that of a random walk (RW), for instance made of N independent unit steps in $\mathbb{Z}^d$ (or segments in $\mathbb{R}^d$). When the number of steps goes to infinity, while the lattice spacing goes to zero in $\mathbb{Z}^d$, or the size of the segments vanishes in $\mathbb{R}^d$, the process tends to the universal model of *Brownian motion* in $\mathbb{R}^d$. In biology, it is well known that it started with the observations by the Scottish naturalist R. Brown[1], who reported the endless movements of pollen particules in water. In physics we all know that Einstein[2] (and Schmoluchowski) gave the first statistical mechanics derivation of the Gaussian probability distribution of the Brownian motion. This law was earlier obtained in mathematics in 1-dimension by L. Bachelier in his thesis[3] "La théorie de la spéculation" in 1900. The rigorous construction is due to N. Wiener[4] in 1923. Random walks and Brownian motions are now ubiquitous in statistical physics, especially in diffusion processes, and we are not going to add information on that here.

We shall be concerned with Brownian motion as a first approximation or, better, as a building brick of the different and more difficult physics of polymers. The latter are made of the linear repetition $(-A-)_N$ (N times) of a peculiar chemical structure, the "monomer" A (like $-CH_2 = \underset{\underset{C_6H_5}{|}}{CH} -$ in the case of polystyrene). A first approximation of the configurations of a polymer embedded in a solvent is then given by the trajectory of a Brownian motion (Fig.1). This is easily understood. The angular correlations between successive monomers along a macromolecule are locally quite strong, but after, roughly, ten monomers they vanish, and at this larger scale the molecule appears as made of independent segments, which are freely rotating in space. If the polymer is quite long, and a number of monomers $N \simeq 10^5$ is currently feasible, the stastistical aspect will be that of a quasi-continuum Brownian motion. A very long polymer floating in 3-space probably appears as in Fig.1, i.e. as a very erratic curve. A Brownian motion is, in a crude approximation, a configuration of a polymer chain, in which all details at a small scale have been washed out. We observe here, in an approximate way, the phenomenon of scale invariance (Fig.1), which is central to the Brownian motion theory, but more generally to the physics of critical phenomena. The latter appear in second order phase transitions, when the correlation length of the fluctuations *diverge*. Then this scale is mathematically pushed to infinity and disappears. If the microstructure is also forgotten, then the physics has no scale and becomes scale invariant. The infinite Brownian motion is perhaps the simplest (but trivial) example of such a geometrical critical system.

Fig.1 : A 2D quasi-Brownian motion. A 3D polymer probably looks like that, when projected by vision. Illustration of scale invariance: a part is statistically indistinguishable from the whole set.

Truly, polymers in a good solvent are subject to a new and important physical effect, that of *excluded volume*, which completely modifies the statistics. This was recognized as early as 1934 by chemical physicists W. Kuhn, and E. Guth and H. Mark independently: a polymer chain *avoids itself*. The explanation is found in *thermodynamics*, and not in a purely quantum mechanical "steric" constraint forbidding the atoms to occupy the same place. When the solvent is good, the thermodynamics favorizes the contact between the

monomers and the solvent, at the expense of that between monomers. The net result is
an effective repulsion between monomers, which acts at short distance in the embedding
Euclidean space, but at long distance between monomers along the chain. The usual steric
effect of course also exists, but would be too weak for usual polymers to explain excluded
volume effects: one has to resort to this solvatation effect.

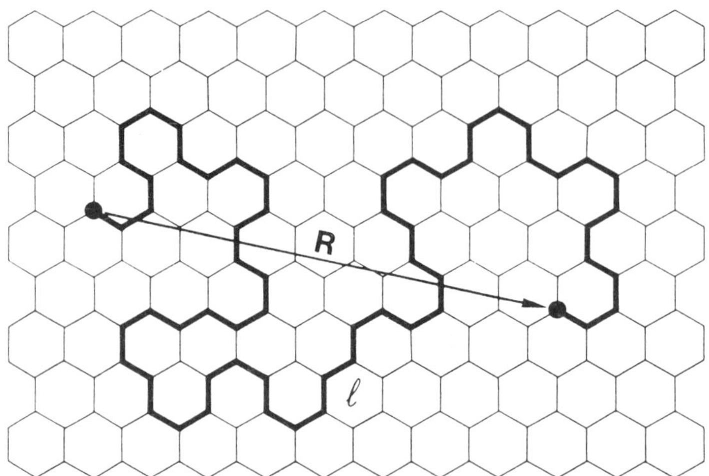

Fig.2 : Self-avoiding walk on the hexagonal lattice.

Before describing the theory, we would like to make two remarks. As in the case of
Brownian motion, which is the unique, thus universal, limit of random walks, it is often
convenient to start with a lattice realization of the polymer. One represents it by a "self-
avoiding walk" (SAW), made of a simple chain with no double points nor double bonds
(Fig.2). In the continuum limit when the number of steps diverges, and the lattice spacing
shrinks to zero, one expects to find a *universal* random process describing the self-avoiding
polymer. In two dimensions for instance, one easily sees that the statistics is totally modified
by self-avoidance (Fig.3). The fractal aspect of the chain is not Brownian anymore but is
more like a fractal coast-line[5].

An essential idea and assumption is that the limit continuum process should be universal and described by unique probability distributions, in the same way as Brownian paths
are described by Gaussian laws. To the best of my knowledge, it does not exist yet any
rigorous mathematical proof of this universality, but all theoretical, numerical and even experimental evidence speaks for it. It is also one of the foundations of the theory of critical
phenomena[6,7,8]. A second point is worth noticing. The expression "self-avoiding walk" is
now widely used to represent polymers. Strictly speaking, it should mean that the walk,
which is a kinetic concept, avoids at time t the previous steps, but not the forecoming ones.
This is not the case of a polymer, which is rather a chain existing as a whole, and whose

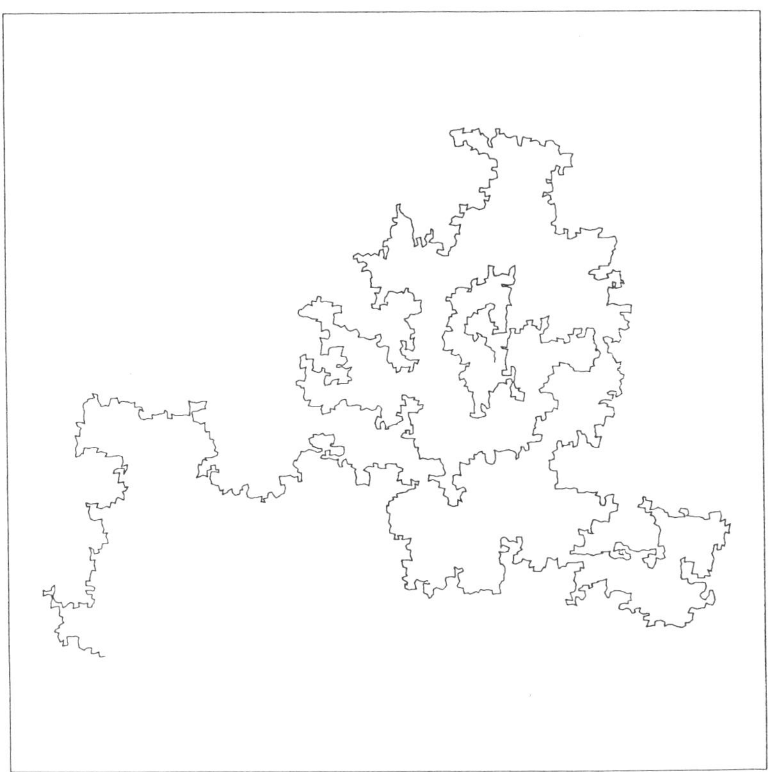

Fig.3 : A much longer SAW (2000 steps) in two dimensions. The chains already appears as a very irregular geometrical object, a fractal, which should be statistically universal.

configurations are entirely self-avoiding. The kinetic model is known as "kinetic growth walk" (KGW) and a priori could belong to a different universality class. Indeed the statistical weights of the configurations are different for SAW and KGW. In particular, cage effects exist in 2D KGW, when the head of the walk get trapped by spiralling inward. For a long time the numerical simulations concluded in this direction, up to the latest ones which, in the far asymptotic regime, observed a convergence of the two processes. So it is now *believed* that the KGW and SAW should belong to the same universality class, in the asymptotic limit.

2. GEOMETRICAL PROPERTIES

2.1 First critical exponents

In this kind of geometrical statistical physics, one always considers first the same simple quantities. The mean square end-to-end distance of a SAW of length ℓ scales as (Fig.3)

$$R^2 \underset{\ell \longrightarrow \infty}{\sim} \ell^{2\nu} \qquad (2.1)$$

where ν is the correlation length exponent. In 2D its exact value[9] is 3/4 (see Nienhuis' lecture in the same volume, and below). We shall describe below some techniques to compute ν. A second quantity of interest is the partition function $\mathcal{Z}$ of a SAW of length ℓ on a given lattice, i.e. the total number of self-avoiding configurations, with the origin fixed. First relax the constraint of self-avoidance, i.e. take a random walk. Then $\mathcal{Z} = 3^\ell$ on the hexagonal lattice, or $\mathcal{Z} = (2d)^\ell$ on the hypercubic $d-$dimensional lattice. Now for a SAW, it has been shown by Hammersley[10] in 1961 that

$$\lim_{\ell \longrightarrow \infty} \frac{1}{\ell} \ell n\, \mathcal{Z} = \mu < \infty$$

exists, and is called the effective connectivity constant. For a RW it is just the number of nearest neighbours. Notice that μ cannot be universal and depends strongly on the lattice, as already shown by the example of RW on various lattices. Now, for a SAW one expects an asymptotic behavior of $\mathcal{Z}$ of the type

$$\mathcal{Z} \underset{\ell \longrightarrow \infty}{\sim} \mu^\ell \ell^{\gamma-1} \tag{2.2}$$

where γ is a new configuration exponent, which should be universal and depending, as ν, only on the space dimension. For instance in 1D, one believes $\gamma = 1$, while in 2D[9] $\gamma = \frac{43}{32}$. As an illustration, on the square lattice one has

$$\mathcal{Z} \underset{\ell \longrightarrow \infty}{\simeq} Z_\square \,(2.63815 \cdots)^\ell\, \ell^{11/32}$$

while on the hexagonal 2D lattice[9]

$$\mathcal{Z} \underset{\ell \longrightarrow \infty}{\simeq} Z_H \left[\left(2+\sqrt{2}\right)^{1/2} \right]^\ell \ell^{11/32}$$

The coefficients Z above are depending on the geometry of the lattices alone. As indicated, the connectivity constant μ for the square lattice is not exactly known*, in contradistinction to the hexagonal case[9].

2.2 Towards more complicated geometrical questions

The two universal exponents ν and γ are actually the only first two of an infinite set. Take for instance a single polymer and two arbitrary points on it (Fig.4), either the two extremities ($a = 0$) or one extremity and an interior point ($a = 1$), or finally two interior points ($a = 2$). What is then the probability $P_a(r)$ that two such points are at a distance r? The full laws are not completely known but we have at our disposal scaling or asymptotic informations.

---

* During the Altenberg Summer School, a group of young dutch students tried hard to derive $\mu_\square$ from recently solved 2D $O(n)$ models, but apparently did not yet succeed!

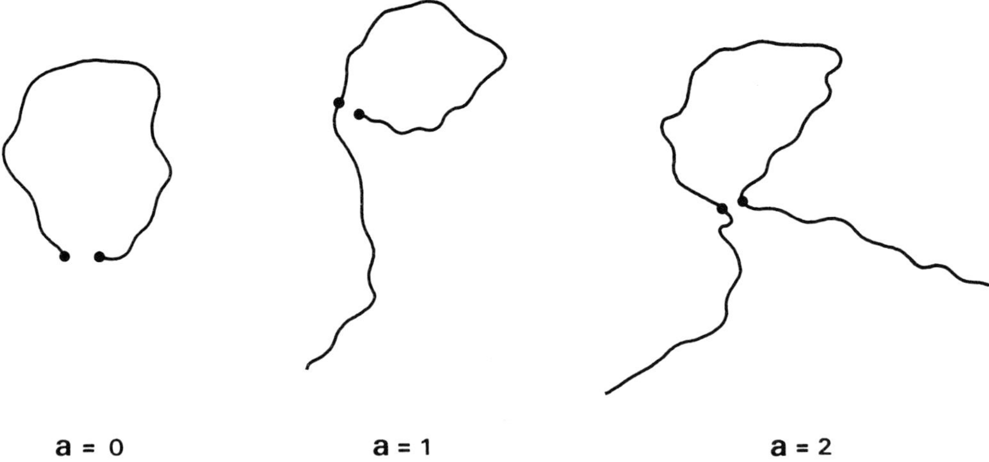

Fig.4 : The three types of possible contacts inside a linear chain.

For instance, the end-to-end distribution P_0 must read

$$P_0(r) = \frac{1}{(2\pi R^2/d)^{d/2}} F\left(\frac{r}{R}\right) \tag{2.3}$$

where R is given by (2.1) and where F is a totally universal function somewhere in the mathematical space. For Brownian motion F is Gaussian:

$$F(x) = \exp\left(-\frac{d}{2}x^2\right) \tag{2.4}$$

At short distance, $P_a(0) = 0$, for $a = 0, 1, 2$ since the polymer is strictly self-avoiding. Then one expects[11] the vanishing for $r \longrightarrow 0$

$$P_a(r) \sim r^{\theta_a} \tag{2.5}$$
$$r \longrightarrow 0$$

where θ_a are three universal exponents. One has in fact[11] $\theta_0 = \frac{\gamma-1}{\nu}$, while θ_1 and θ_2 are new. In 2D dimensions for instance[11] $\theta_0 = \frac{11}{24}$, $\theta_1 = \frac{5}{6}$, $\theta_2 = \frac{19}{12}$. θ_1 and θ_2 cannot be expressed in terms of ν and γ alone and are new.

The set of exponents is not exhausted. One can consider now star polymers made of L arms of equal lengths ℓ, which are self- and mutually avoiding (Fig.5). They exist in nature and one commonly produced[12]. Now, what is the number of configurations of an $L-$arm star? It scales as

$$\mathcal{Z}_L \sim \mu^{L\ell} \ell^{\gamma_L - 1} \tag{2.6}$$
$$\ell \longrightarrow \infty$$

Fig.5 : A star polymer with $L = 5$ branches (in 3D).

The total length is $L\ell$ and the dominant term $\mu^{L\ell}$ corresponds to an effective number of choices μ at each local step. The correction exponent γ_L depends on the *topology* and crucially on the fact that L branches are fused together at a vertex. It is a universal function $\gamma_L(d)$ of L and d only.

We shall see below that γ_L is exactly calculable in 2 dimensions[13]

$$\gamma_L(d=2) = \frac{68 + 9L(3-L)}{64} \tag{2.7}$$

Notice that a 1-star is a single polymer, while a $L = 2$ star is a chain divided into two parts by a point insertion, hence still a single polymer chain. Hence $\gamma_1 = \gamma_2 = \gamma$ which is readily verified on (2.7). In $d = 4 - \varepsilon$ dimensions, $\gamma_L(d)$ will be also calculable to $O\left(\varepsilon^2\right)$.

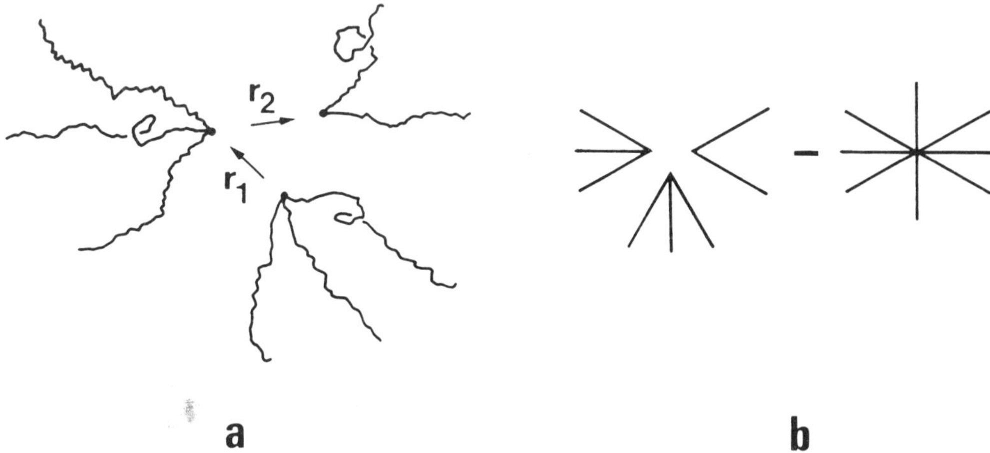

Fig.6 : a) Contacts between cores of star polymers, b) fusion rule (2.9).

The contact problem we have encountered above for the contacts inside a single linear chain can be generalized to the star polymers. Consider, for instance, a set $i = 1, ..., n$ of n

star polymers, each star i having L_i branches (Fig.6). Define the probability density $P\{\mathbf{r}_i\}$ $i = 1, ..., n-1$ for the set of relative positions $\mathbf{r}_i$ between the cores of stars i and $i+1$. All these stars are interacting together by excluded volume. Then there is a strong repulsion at short distance between the cores because of the high local monomer density. It manifests itself through the scaling law

$$P\{\lambda \mathbf{r}_i\} = \lambda^\theta P\{\mathbf{r}_i\} \qquad (2.8)$$
$$\lambda \longrightarrow 0$$

where $\theta = \theta\{L_i\}$ is a universal number depending only on the numbers of arm L_i, $i = 1, ..., n$ and on the dimension of space.

Actually, these contact exponents can be entirely calculated in terms of the configuration exponents γ_L of the corresponding stars. One has by scaling[11]

$$\nu\theta = \sum_{i=1}^{n}(\gamma_{L_i} - 1) - \left[\gamma\left(\sum_{i=1}^{n} L_i\right) - 1\right] \qquad (2.9)$$

i.e. one sums the configuration exponents of the stars before the contact, from which one has to subtract that of the new star made by the fusion after the contact (Fig.6).

As an application, we find for the contacts inside the linear chain (Fig.4) (forgetting the loops which do not play any role[11])

$$\begin{aligned}
\nu\theta_0 &= \gamma - 1 \\
\nu\theta_1 &= \gamma_2 - 1 + \gamma_1 - 1 - (\gamma_3 - 1) \\
&= 2\gamma - \gamma_3 - 1
\end{aligned} \qquad (2.10)$$

since $\gamma_2 \equiv \gamma_1 = \gamma$, and

$$\nu\theta_2 = 2(\gamma_2 - 1) - (\gamma_4 - 1) = 2\gamma - \gamma_4 - 1 \qquad (2.11)$$

From these relations and the exact values in 2D, $\nu = 3/4$, and γ_L (2.7), hence $\gamma = 43/32$, $\gamma_3 = 17/16$, $\gamma_4 = 1/2$ one finds the values given above for θ_0, θ_1 and θ_2 in 2D.

2.3 Multiple points of the Brownian motion

We have not discussed yet the role of the dimension in SAW problems. Since the deep analogy to critical phenomena already perspired, the reader probably anticipated that the standard upper critical dimension $d^* = 4$ will play a central role. Above it, the constraint of self-avoidance will be irrelevant, and SAW will be essentially Brownian paths. We show now how this comes from rigorous results on Brownian motion in space dimension d.

The multiple points of order n are the points visited n times by a Brownian motion (Fig.7). The double points were studied in mathematics as early as in 1950 by Dvoretsky, Erdös and Kakutani[14].

Fig.7 : Multiple points in a Brownian path.

A Brownian motion does not have double points (a fortiori multiple points) in dimension $d \geq 4$. The Hausdorff (or fractal[5]) dimension of the set of double points is given by the formula due to Taylor[15]

$$D_H(2) = \max\{0, 4 - d\} \tag{2.12}$$

More generally that of the set of order n is

$$D_H(n) = \max\{0, 2n - (n-1)d\} \tag{2.13}$$

One can give actually a heuristic derivation of these Hausdorff dimensions[16]. First for $n = 1$ $D_H(1)$ is formally the fractal dimension of the simple points, i.e. that of the Brownian path itself. According to (2.13) $D_H(1) = 2$. Hence the Brownian curve, even if parametrized by a 1-dimensional set (time) is embedded in space like a two-dimensional sheet.

Now consider the intersection of n Brownian paths $\mathbf{B}_i$, $i = 1, ..., n$ (or n parts of the same Brownian path) in Euclidean space $\mathbb{R}^d$ of dimension d. The "fractal codimensions" $\bar{D}_i = d - D_i$ of the $\mathbf{B}_i$ are $\bar{D}_i = d - 2$, and for standard Euclidean subspaces they would represent the number of equations needed to describe them in $\mathbb{R}^d$. Their intersection $\bigcap_{i=1}^n \mathbf{B}_i$ is described by the reunion of the equations in number $n \times (d - 2)$, which is thus also its codimension $\bar{D}\left(\bigcap_i \mathbf{B}_i\right) = n(d - 2)$. Hence from $\bar{D} \equiv d - D$ we find the dimension $D\left(\bigcap_i \mathbf{B}_i\right) = d - n(d - 2)$, which is nothing but (2.13), when, of course, D is positive.

So Eq.(2.13) can be seen as a *fractal generalization of a simple codimension rule for Euclidean subspaces.*

As a first consequence, we find that for $d \geq 4$ all these dimensions are identically zero. The space dimension being too big, the Brownian path has no chance to meet itself. So if one requires the self-avoiding constraint, the Brownian path for $d \geq 4$ already fullfills it, since the constraint is relevant for a set of configurations of zero probability measure. Hence for $d \geq 4$ all the exponents described above take their mean field Brownian values:

$$\nu = 1/2, \quad \gamma = 1, \quad \gamma_L = 1, \quad \forall L \geq 1$$
$$\theta_{\{L_i\}} = 0 \tag{2.14}$$

A second consequence of the set of Hausdorff dimensions (2.13) is the location of upper multicritical dimensions. Suppose that multiple points of order $n(> 2)$ are forbidden to the random walk, but not the crossings at levels $2 \leq n' < n$. This condition would be relevant only if $D_H(n) \geq 0$, hence

$$d \leq \frac{2n}{n-1} \equiv d^*(n) \tag{2.15}$$

which is the definiton of the upper multicritcal dimension of order n. (One can actually show easily in the field theoretic description of critical phenomena, that the multiple points of order n in a Brownian motion correspond to an interaction term of the form $\left(\varphi^2\right)^n$ in the Lagrangian of a Landau-Ginzburg-Wilson field theory: $\mathcal{L} = \int \mathrm{d}^d x \left[(\nabla \varphi)^2 + g \left(\varphi^2\right)^n\right]$, which is indeed the effective Lagrangian for a $n-$multicritical point).

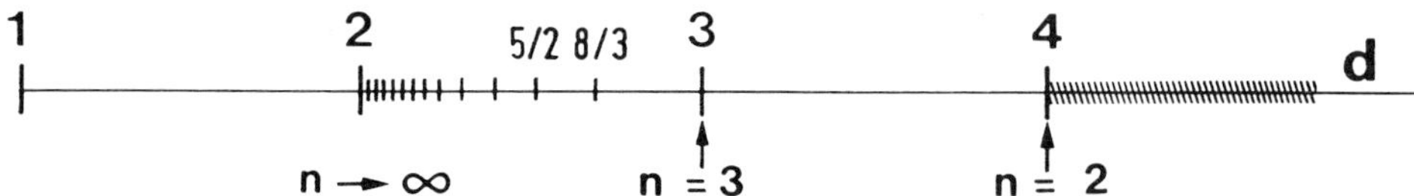

Fig.8 : The set of upper multicritical dimensions of order n.

Finally we find the set of upper critical dimensions as described in Fig.8. It is quite noticeable in (2.13) (2.15) that $d = 2$ is very special. There $D_H(n) = 2$ for any n. So there are multiple points of all orders with fractal dimension 2. As a consequence, all interactions are relevant in 2D and $d = 2$ is the lower accumulation point of the set $d^*(n)$ (2.15).

3. RENORMALIZATION OF THE EDWARDS MODEL

3.1 Wiener measure (à la physicienne)

Consider a Brownian path $\mathbf{B}_t$ in $\mathbb{R}^d$, starting at $\mathbf{0}$ at time 0: $\mathbf{B}_0 = \mathbf{0}$. Then the probability that it is located at times $(t_1, t_2, ..., t_n)$ at points $\mathbf{x}_1, ..., \mathbf{x}_n$ is given exactly by the Gaussian distribution law

$$\begin{aligned}
\mathcal{P}_0\left(\mathbf{x}_1, ..., \mathbf{x}_n\right) &= (2\pi)^{-nd/2} t_1^{-d/2} \left(t_2 - t_1\right)^{-d/2} \cdots \cdot \left(t_n - t_{n-1}\right)^{-d/2} \\
&\times \exp\left(-\frac{\mathbf{x}_1^2}{2t_1} - \frac{\left(\mathbf{x}_2 - \mathbf{x}_1\right)^2}{2\left(t_2 - t_1\right)} \cdots - \frac{\left(\mathbf{x}_n - \mathbf{x}_{n-1}\right)^2}{2\left(t_n - t_{n-1}\right)}\right)
\end{aligned} \tag{3.1}$$

This probability distribution is normalized with the Lebesgue measure on $\mathbb{R}^{d \times n}$

$$\mathrm{d}\mathcal{P}_0 \equiv \prod_{k=1}^{n} \mathrm{d}^d x_k \; \mathcal{P}_0\left(\mathbf{x}_1, ..., \mathbf{x}_n\right) \tag{3.2}$$

such that $\int_{\mathbb{R}^{d \times n}} \mathrm{d}\mathcal{P}_0 = 1$.

Now, take the formal continuum limit $n \longrightarrow \infty$, t fixed, such that

$$
\begin{aligned}
t_k - t_{k-1} &= \frac{t}{n}, \qquad k = 1, ..., n \\
t_k &= k\frac{t}{n} \longrightarrow s \in [0, t] \\
\mathbf{x}_k &= \mathbf{X}_s
\end{aligned}
\tag{3.4}
$$

Then the measure $d\mathcal{P}_0$ (3.2) formally tends toward the *Wiener measure*

$$
d\mathcal{P}_0 = \mathcal{D}\mathbf{X}_s \exp\left(-\frac{1}{2}\int_0^t \left(\dot{\mathbf{X}}_s\right)^2 ds\right)
\tag{3.5}
$$

where $\dot{\mathbf{X}}_s \equiv \frac{d\mathbf{X}_s}{ds}$, and where the functional integrand $\mathcal{D}\mathbf{X}_s$ can be thought of as the formal (diverging) limit

$$
\mathcal{D}\mathbf{X}_s = \text{``}\lim_{n \longrightarrow \infty}\text{''} \prod_{k=1}^n \frac{d^d x_k}{\left(2\pi\dfrac{t}{n}\right)^{d/2}}
\tag{3.6}
$$

Physicists like this expression very much, mathematicians avoid it with slight contempt. Actually it amounts to simulate Brownian trajectories $\mathbf{B}_s$, $s \in [0, t]$ by discrete sets of points $\mathbf{x}_{t_1}, ..., \mathbf{x}_{t_n}$ for any regular sequence $(t_1, ..., t_n)$ of $[0, t]$ and refining the intervals ad infinitum. It is also interesting to notice that the probability $\exp\left(-\frac{1}{2\eta}\int_0^t \left(\dot{\mathbf{X}}_s\right)^2 ds\right)$ is just that which appears in probability theory, under the name of *large deviations*[17], when considering the chances that a Brownian path $\mathbf{B}_s$ approachs a *smooth* curve $\mathbf{X}_s$ at a distance constantly less than η. The only ingredient one needs to calculate functional integrals with the continuum weight (3.5) is the following average

$$
\begin{aligned}
\int d\mathcal{P}_0 e^{i\mathbf{q}\cdot(\mathbf{X}_{s_1} - \mathbf{X}_{s_2})} &\equiv \int \mathcal{D}\mathbf{X}_s \exp\left[-\frac{1}{2}\int_0^t \left(\dot{\mathbf{X}}_s\right)^2 ds + i\mathbf{q}\cdot(\mathbf{X}_{s_1} - \mathbf{X}_{s_2})\right] \\
&= \exp\left(-\frac{1}{2}q^2 |s_2 - s_1|\right)
\end{aligned}
\tag{3.7}
$$

The last formula is easily obtained by replacing the continuum functional integral by its discrete origin (3.1) (3.2) and performing standard Gaussian integrals. With these tools, we are now in position to perform analytical calculations on self-avoiding random paths.

3.2 Edwards measure

In 1965, Sam Edwards[18] introduced a continuum representation of a self-avoiding path. He proposed to modify the continuum weight (3.5) by an interaction term measuring self-contacts of the path

$$
d\mathcal{P} \equiv d\mathcal{P}_0 \exp\left(-\frac{b}{2}I\right)
\tag{3.8}
$$

where

$$
I = \int_0^t ds \int_0^t ds' \, \delta^d\left(\mathbf{X}_s - \mathbf{X}_{s'}\right)
\tag{3.9}
$$

Hence the full Edwards measure is

$$d\mathcal{P} = \int \mathcal{D}\mathbf{X}_s \, \exp\left(-\frac{1}{2}\int_0^t \left(\dot{\mathbf{X}}_s\right)^2 ds - \frac{b}{2}\int_0^t ds \int_0^t ds' \, \delta^d\left(\mathbf{X}_s - \mathbf{X}_{s'}\right)\right) \qquad (3.10)$$

The coefficient b is positive, and I is called in mathematics the intersection local time. For $b \longrightarrow \infty$, clearly, only paths without multiple points should survive with non zero measure. As we shall see, the same is true when for b fixed, the "time" t, i.e. the length of the polymer goes to infinity, $t \longrightarrow \infty$. The Edwards model was not estimated for a while at its right value. It required functional integration, and furthermore a renormalization technique. Furthermore, the first studies in renormalization theory for polymers were based on the equivalence to an n-component field theory[7], in the limit $n \longrightarrow 0$. (We shall study this aspect later in two dimensions.) However, we may say now that the Edwards model is really the *minimal* continuum model for polymers, which embodies all the necessary mathematics. It is also directly generalizable to the study of self-avoiding membranes[19].

The dimensional analysis of (3.10) is instructive. Define

$$s = tx, \quad x \in [0,1]$$
$$\mathbf{X}_s = t^{1/2}\boldsymbol{\rho}_x \qquad (3.11)$$

Notice that for a Brownian motion as defined by (3.5) (3.1), one has

$$\left\langle (\mathbf{B}_t - \mathbf{B}_0)^2 \right\rangle_0 = d\,t, \qquad (3.12)$$

hence (3.11) is just a rescaling in terms of the natural Brownian diffusion scale. Then dimensional analysis of the Brownian action (or "kinetic term") (3.5) leaves it *invariant*

$$\mathcal{A}_0 = \frac{1}{2}\int_0^1 (\dot{\boldsymbol{\rho}}_x)^2 \, dx \qquad (3.13)$$

On the contrary the interaction term or intersection local time is rewritten as

$$I = t^{2-d/2}\int_0^1 dx \int_0^1 dx' \, \delta^d\left(\boldsymbol{\rho}_x - \boldsymbol{\rho}_{x'}\right) \qquad (3.14)$$

The natural extension of this reduced Brownian motion $\boldsymbol{\rho}_x$ is given by the analog of (3.12) $\left\langle (\boldsymbol{\rho}_x - \boldsymbol{\rho}_0)^2 \right\rangle_0 = d$, and stays finite when the "length" or "time" t diverges.

One expects the same for $\mathcal{A}_0$ and the integral in I (3.14). On the contrary, the value of the self-interaction I of the Brownian path with itself scales as $t^{2-d/2}$. Hence for $d > 4$, it vanishes for $t \longrightarrow \infty$, while for $d < 4$, any perturbation of the Wiener measure (3.5) (3.13) by an Edwards interaction term (3.9) (3.14), whatever small b is, will have a *dominant* effect in the limit of long polymer chains. For $d > 4$, the effect of the self-intersection term is negligible asymptotically, and the chains remain Brownian. We thus recover the result, of §2.3.

3.3 Dimensional regularization

A few mathematical comments: Varadhan[20] showed in 1969 the existence of the Edwards model in 2 dimensions, in an Appendix to the representation by Symanzik[20] of field theory by intersections of Brownian paths. In 1980, Westwater[21] established its existence in the more difficult three-dimensional case. The necessity of these proofs can be understood by noticing that the Edwards model involves a distribution δ^d which must be regularized, since it can be very singular when applied to Brownian paths. One must introduce a short-distance cut-off in order to properly define the model, for instance by forbidding two interacting points along the chain to be at less than s_0, and defining[22]

$$I_{s_0} = \int_0^t ds \int_0^t ds' \, \theta\left(|s - s'| - s_0\right) \delta^d \left(\mathbf{X}_s - \mathbf{X}_{s'}\right) \tag{3.15}$$

where s_0 is a small contour length. Now, one understands immediately the necessity of this cut-off by calculating the Brownian average $\langle I_{s_0} \rangle$. For this, one writes

$$\delta^d \left(\mathbf{X}_s - \mathbf{X}_{s'}\right) = \int_{\mathbb{R}^d} \frac{d^d q}{(2\pi)^d} e^{i\mathbf{q}\cdot(\mathbf{X}_s - \mathbf{X}_{s'})} \tag{3.16}$$

and uses the fundamental Brownian average (3.7) to get

$$\langle I_{s_0}\rangle_0 \equiv \int d\mathcal{P}_0 I_{s_0} = \int_0^t ds \int_0^t ds' \, \theta\left(|s - s'| - s_0\right) \int \frac{d^d q}{(2\pi)^d} e^{-q^2 |s - s'|/2}$$

Performing the **q**-integral gives

$$\langle I_{s_0}\rangle_0 = \int_0^t ds \int_0^t ds' \, \theta\left(|s - s'| - s_0\right) \left(2\pi |s - s'|\right)^{-d/2} \tag{3.17}$$

which can be computed by elementary means as

$$\langle I_{s_0}\rangle = 2(2\pi)^{-d/2} t^{2-d/2} \left[\frac{-x_0^{1-d/2}}{1 - d/2} + \frac{1}{(1 - d/2)(2 - d/2)} + \frac{x_0^{2-d/2}}{2 - d/2} \right] \tag{3.19}$$

where $x_0 \equiv s_0/t$ is a very small dimensionless parameter. In this simple formula we find all the information we need. For $d < 2$, the limit for $s_0 \longrightarrow 0$, $x_0 \longrightarrow 0$ exists

$$\begin{array}{c} \langle I_{s_0}\rangle_0 \\ s_0 \longrightarrow 0 \\ d < 2 \end{array} = 2(2\pi)^{-d/2} t^{2-d/2} \frac{1}{(1 - d/2)(2 - d/2)} \tag{3.20a}$$

while for $d > 2$, $\langle I_{s_0}\rangle_0$ diverges when $s_0 \longrightarrow 0$, as $\langle I_{s_0}\rangle \sim t \, s_0^{1-d/2}$. So the self-intersection time of a Brownian path diverges in dimension greater than 2, and the proper definition of the Edwards model requires a regularization for $2 \leq d(\leq 4)$, which is the domain of physical interest. Notice also that (3.19) exhibits a regular subdominant term which just scales as $t^{2-d/2}$, as in our heuristic dimensional argument (3.14). This is the conclusion of this simple Brownian calculation: For $d < 2$, the model is free of short-distance (ultraviolet) singularities. For $d \geq 2$, one has to regularize it. The interesting long range (infrared) effects due to self-avoidance are found in the *regular* part (like (3.20)).

The most drastic (and best) way to regularize the Edwards model is the *dimensional regularization*[22,23]. One computes analytical expressions, e.g. of the partition functions, for $d < 2$ without any cut-off, and continue analytically the formulae as meromorphic functions of d, for $d > 2$. In other words, one simply forgets ultraviolet divergences as well as the cut-off s_0, and take e.g. for $\langle I \rangle_0$ dim.reg. the finite result (3.20)

$$\langle I \rangle_0 \text{ dim.reg.} = 2(2\pi)^{-d/2} t^{2-d/2} \frac{1}{(1 - d/2)(2 - d/2)}, \quad \forall d \qquad (3.20b)$$

We shall follow this approach from now on, (and advise the reader to always use it in similar problems). Of course, there exists a connection between the partition functions of the two Edwards theories, one with s_0, the other dimensionally regularized, which has been established in detail[22,23], but we won't need it here.

Let us finally mention the following recent interesting theorem[24]. The space of configurations of the paths $\mathbf{X}_s$ on which the Edwards measure is non vanishing, has no intersection with the space of Brownian configurations $\mathbf{B}_s$ on which the Wiener measure does not vanish. This is the rigorous continuum version of the physical idea that self-avoiding walks and random walks have entirely different universality classes (in the asymptotic critical limit).

3.4 Perturbation expansion

Owing to (3.10) (3.14), the natural parameter which measures the strength of the interaction is

$$z = (2\pi)^{-d/2} b \, t^{2-d/2} \qquad (3.21)$$

(the factor $(2\pi)^{-d/2}$ being there only for convenience).

This parameter was introduced[25] in polymer physics as early as in 1953, much before the advent of the Edwards model. One does not know how to compute directly averages with the full Edwards measure (3.10). But one can expands *formally* in powers of b the weight (3.8)

$$\exp\left(-\frac{b}{2} I\right) = \sum_{p \in \mathbf{N}} \left(-\frac{b}{2}\right)^p \frac{I^p}{p!}$$

and average with the Wiener measure $d\mathcal{P}_0$ (3.5). Each integral I is represented by a Fourier transform like in (3.16) and then averaged in a functional integral weighted by $d\mathcal{P}_0$ with the help of Gaussian formula (3.7). One can make use of Feynman diagrams, where each term of order b^p is represented by a Brownian motion with p double points[22]. We illustrate the procedure in two particular cases.

The average end-to-end distance in the case of the Edwards model is modified with respect to its Brownian value as

$$R^2 \equiv \left\langle (\mathbf{X}_t - \mathbf{X}_0)^2 \right\rangle_{\text{Edwards}} = d\, t\, \mathcal{X}_0 \tag{3.22}$$

where the "swelling factor" $\mathcal{X}_0$ is a function of z and d, with the formal series expansion in z

$$\mathcal{X}_0(z, d) = 1 + \sum_{n \geq 1} a_n(d) z^n \tag{3.23}$$

(Notice that $\mathcal{X}_0$ is dimensionless, and the only dimensionless parameter is now z, whence (3.23)). The $a_n(d)$ are, in principle, calculable. To first order, one finds with the Brownian calculus sketched above

$$a_1(d) = \frac{1}{(3 - d/2)(2 - d/2)} \tag{3.24}$$

In the same way, the (dimensionally regularized) Edwards partition function $\mathcal{Z}$

$$\mathcal{Z} \equiv \int \mathrm{d}\mathcal{P}_0 \, \exp\left(-\frac{b}{2}I\right)$$

can be expanded as a formal series

$$\mathcal{Z} = 1 + \sum_{n \geq 1} a'_n(d) z^n \tag{3.25}$$

To first order

$$\mathcal{Z} = 1 - \frac{b}{2} \int \mathrm{d}\mathcal{P}_0 \, I + O\left(b^2\right),$$

hence we get from the (dimensionally regularized) formula (3.17) (setting formally $s_0 = 0$)

$$\mathcal{Z} = 1 - \frac{1}{2} z \int_0^1 \mathrm{d}x \int_0^1 \mathrm{d}x' |x - x'|^{-d/2} + O\left(z^2\right) \tag{3.26}$$

In these perturbation expansions, the space dimension d is now considered as a real parameter (or even complex). The double integral above converges for $\mathrm{Re}(d) < 2$, and a straightforward calculation gives (3.20)

$$\mathcal{Z} = 1 - \frac{z}{(2 - d/2)(1 - d/2)} + O\left(z^2\right) \tag{3.27}$$

For $2 < \mathrm{Re}(d) < 4$, we continue analytically (3.26) in variable d, remembering that this continuation is also the *regular* part of $\mathcal{Z}$ appearing in (3.19), which is obtained when one excludes for $\mathrm{Re}(d) > 2$, a small band $|x - x'| \geq x_0$ about the bissectrix $x = x'$ where ultraviolet divergences occur.

We also remark the most important fact that in $\mathcal{X}_0$ and $\mathcal{Z}$, coefficients $a_1(d)$ and $a_1'(d)$ possess simple poles at $d = 4$. This is recurrent: $a_n(d)$ and $a_n'(d)$ have poles of order $(4 - d)^{-n}$ and, of course, subleading ones $(4 - d)^{-p}$, $1 \le p \le n - 1$. One sets

$$\varepsilon = 4 - d$$

and performs a double *Taylor* expansion in powers of z and *Laurent* expansion in powers of ε. Let us illustrate this with the second order expansions[22]

$$\begin{aligned}
\mathcal{X}_0(z,d) &= 1 + z\left(\frac{2}{\varepsilon} - 1 + \cdots\right) + z^2\left(-\frac{6}{\varepsilon^2} + \frac{11}{2\varepsilon} + \cdots\right) + \cdots \\
\mathcal{Z}(z,d) &= 1 + z\left(\frac{2}{\varepsilon} + 1 + \cdots\right) + z^2\left(-\frac{6}{\varepsilon^2} - \frac{7}{\varepsilon} + \cdots\right) + \cdots
\end{aligned} \tag{3.28}$$

This double expansion is not at all random but contains all the information for obtaining the exponents. Indeed, let us recall that we expect for large t the behaviors $R^2 \sim t^{2\nu}$ and $\mathcal{Z} \sim t^{\gamma-1}$, which imply for $z \sim t^{\varepsilon/2}$ large

$$\mathcal{X}_0 \underset{z \to \infty}{\sim} z^{(2\nu-1)2/\varepsilon}, \qquad \mathcal{Z} \underset{z \to \infty}{\sim} z^{(\gamma-1)2/\varepsilon} \tag{3.29}$$

We show now where these exponents are hidden in the double expansions (3.28).

3.5 Renormalization

3.5a Fast one-loop renormalization

Take the leading divergences ε^{-p} of (3.28) to each order in z^p :

$$\begin{aligned}
\mathcal{X}_0(z,d) &= 1 + z\frac{2}{\varepsilon} - \frac{6}{\varepsilon^2}z^2 + \cdots \\
\mathcal{Z}(z,d) &= 1 + z\frac{2}{\varepsilon} - \frac{6}{\varepsilon^2}z^2 + \cdots
\end{aligned}$$

and compare it to the expansion of

$$\mathcal{X} = \left(1 + \frac{8z}{\varepsilon}\right)^{1/4} = 1 + \sum_{p \ge 1}\left(\frac{8z}{\varepsilon}\right)^p \frac{1}{p!}\frac{1}{4}\left(\frac{1}{4} - 1\right)\cdots\left[\frac{1}{4} - (p-1)\right] \tag{3.30}$$

They coincide to first three orders, and actually, to all orders. Hence (3.30) is the resummed form of all leading divergences $(z/\varepsilon)^p$. From it one gets the asymptotic limit

$$\mathcal{X} \underset{z \to \infty}{\simeq} \left(\frac{8z}{\varepsilon}\right)^{1/4} \sim t^{\varepsilon/8}$$

hence from (3.29) the universal critical exponents to first order in ε

$$\begin{aligned}
\gamma - 1 &= \varepsilon/8 + \cdots \\
\nu &= \frac{1}{2}(1 + \varepsilon/8 + \cdots)
\end{aligned}$$

We see that one can extract information about the asymptotic behaviour of the scaling functions from the first terms of the double Taylor-Laurent expansion, provided one knows the resummation properties, i.e. has information over the recurrent structure of the theory. This process is systematized under the name of *renormalization*.

3.5b Systematics

We define the effective exponents

$$
\begin{aligned}
\sigma_0(z,\varepsilon) &= \frac{\varepsilon}{2} z \frac{\partial}{\partial z} \ell n \, \mathcal{X}_0 \\
\sigma_1(z,\varepsilon) &= \frac{\varepsilon}{2} z \frac{\partial}{\partial z} \ell n \, \mathcal{Z} \ ,
\end{aligned}
\tag{3.31}
$$

(notice that $\frac{\varepsilon}{2} z \frac{\partial}{\partial z} \equiv t \frac{\partial}{\partial t}$). We expect for $z \longrightarrow \infty$, with ε fixed, the universal critical exponents

$$
\sigma_0(z \longrightarrow \infty, \varepsilon) = 2\nu - 1, \quad \sigma_1(z \longrightarrow \infty, \varepsilon) = \gamma - 1
$$

which are function of space dimension d only. From (3.28) one gets the second order expansions

$$
\begin{aligned}
\sigma_0(z,\varepsilon) &= z \left(1 - \frac{\varepsilon}{2}\right) + z^2 \left(-\frac{8}{\varepsilon} + \frac{15}{2}\right) \\
\sigma_1(z,\varepsilon) &= z \left(1 + \frac{\varepsilon}{2}\right) + z^2 \left(-\frac{8}{\varepsilon} - 9\right)
\end{aligned}
\tag{3.32}
$$

We are still facing the difficulty that the expansions diverge for $z \longrightarrow \infty$ and $\varepsilon \longrightarrow 0$. Once resummed they should reach finite limits $2\nu - 1$, $\gamma - 1$, which moreover should vanish at the upper critical dimension $d = 4$, $\varepsilon = 0$.

Now renormalization works order by order in perturbation theory. One shows there exists a unique Taylor-Laurent expansion

$$
g = z + \sum_{n \geq 2} \left(\frac{\alpha_{n,n-1}}{\varepsilon^{n-1}} + \cdots + \frac{\alpha_{n,1}}{\varepsilon}\right) z^n
$$

inverted to

$$
z = g + \sum_{n \geq 2} \left(\frac{\beta_{n,n-1}}{\varepsilon^{n-1}} + \cdots + \frac{\beta_{n,1}}{\varepsilon}\right) g^n
$$

such that substituting g to z in effective exponents like (3.31), leads to a double Taylor series in g and ε, without poles in $1/\varepsilon$, and this, to all orders in g. This is reordering the terms z^p of the expansion in powers of z, which lead to a singular series in ε, into an expansion in g, whose coefficients are regular at $\varepsilon = 0$. Expansions (3.32) and actually a third quantity, the second virial coefficient, which has to be calculated separately[22], give the universal g to third order

$$
\begin{aligned}
g &= z - \frac{8}{\varepsilon} z^2 + \left(\frac{64}{\varepsilon^2} + \frac{17}{\varepsilon}\right) z^3 + \cdots \\
z &= g + \frac{8}{\varepsilon} g^2 + \left(\frac{64}{\varepsilon^2} - \frac{17}{\varepsilon}\right) g^3 + \cdots
\end{aligned}
\tag{3.33}
$$

Substituting g to z in (3.32) gives regular expansions

$$\begin{aligned} \sigma_0[g,\varepsilon] &= g - \frac{1}{2}g^2 + O\left(g^3\right) \\ \sigma_1[g,\varepsilon] &= g - g^2 + O\left(g^3\right) \end{aligned}$$

(3.34)

This substitution, valid order by order, is called minimal subtraction in field theory.

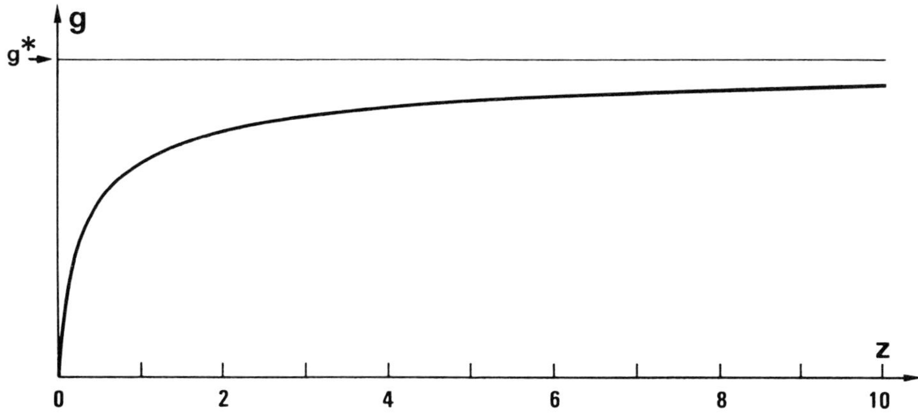

Fig.9 : Schematic behavior of $g(z)$ for $d < 4$.

Finally, while $z \sim t^{2-d/2}$ diverges for $t \longrightarrow \infty$, one expects g to reach a universal fixed point limit $g(z \longrightarrow \infty, \varepsilon) = g^*$ (Fig.9). To get g^*, one defines the Wilson function

$$W = \frac{\varepsilon}{2} z \frac{\partial}{\partial z} g$$

(3.35)

which is a formal double *Taylor-Laurent* series expansion in z and ε, which becomes a double *Taylor* expansion in g and ε. A simple algebra on Eqs.(3.33) gives indeed

$$W[g,\varepsilon] = \frac{1}{2}\varepsilon g - 4g^2 + 17g^3 + O\left(g^4\right)$$

(3.36)

Since $g(z \longrightarrow \infty) \longrightarrow g^* < +\infty$, one expects its derivative to vanish at infinity in variable z, hence $W(z \longrightarrow \infty, \varepsilon) = W[g^*,\varepsilon] = 0$ in variable g.

This *fixed point* equation is solved easily for (3.36)

$$g^* = \frac{\varepsilon}{8} + \frac{17}{4}\left(\frac{\varepsilon}{8}\right)^2 + O\left(\varepsilon^3\right)$$

(3.37)

Substituting then into (3.34) for getting the critical exponents $\sigma_0[g^*] = 2\nu - 1$ and $\sigma_1[g^*] = \gamma - 1$, one finds the celebrated Wilson-Fisher ε-expansion, here for polymers

$$\begin{aligned} 2\nu - 1 &= \frac{\varepsilon}{8} + \frac{15}{4}\left(\frac{\varepsilon}{8}\right)^2 + O\left(\varepsilon^2\right) \\ \gamma - 1 &= \frac{\varepsilon}{8} + \frac{13}{4}\left(\frac{\varepsilon}{8}\right)^2 + O\left(\varepsilon^3\right) \end{aligned}$$

(3.38)

Remarks:

i) For $d = 4$, $\varepsilon = 0$, one recovers the Brownian values.

ii) The renormalization process, here described to second order, which is the first non trivial one, is iterated to all orders in z, g and ε. The proof for polymers is derived[26] from the equivalence to a $n = 0$ component field theory, for which the dimensional renormalization has been established. A direct proof for polymers, starting from the Edwards model, is still lacking but should be feasible.

iii) We can now return to our *one-loop renormalization* to leading order (section 3.5a), and establish the scaling form (3.30), using the renormalization theorem. Indeed, suppose a partition function $\mathcal{X}$ and coupling constant g given by

$$
\begin{cases}
\mathcal{X} = 1 + \dfrac{a'}{\varepsilon} z + \cdots \\[2mm]
g = z - \dfrac{a}{\varepsilon} z^2 + \cdots, \quad z = g + \dfrac{a}{\varepsilon} g^2 + \cdots
\end{cases}
\tag{3.39}
$$

Then

$$
\begin{aligned}
\varepsilon\, z \frac{\partial}{\partial z} \ell n\, \mathcal{X} &= a' z + \cdots = a' g + O\left(g^2\right) \\[2mm]
\varepsilon\, z \frac{\partial}{\partial z} g &= \varepsilon g - a g^2 + O\left(g^3\right)
\end{aligned}
\tag{3.40}
$$

Recall that the latter are regular when $\varepsilon \longrightarrow 0$, to all orders in g. Integrating the above truncated Wilson equations leads first to the exact solution

$$
g = \frac{z}{1 + \dfrac{a}{\varepsilon} z} \, ,
\tag{3.41}
$$

hence

$$
\frac{\partial}{\partial z} \ell n\, \mathcal{X} = \frac{a'}{\varepsilon} \frac{1}{1 + \dfrac{a}{\varepsilon} z}
$$

is integrated into

$$
\ell n\, \mathcal{X} = \frac{a'}{a} \ell n \left(1 + \frac{a}{\varepsilon} z\right) + \text{const.}
$$

From the boundary condition $\mathcal{X}(0) = 1$, we get finally

$$
\mathcal{X} = \left(1 + \frac{a}{\varepsilon} z\right)^{a'/a}
\tag{3.42}
$$

Hence, keeping only the first diverging terms in (3.39) and assuming the validity of the renormalization process (3.40), leads us easily to the resummed one-loop formulae (3.41) (3.42). Notice that from (3.41) one gets the fixed point

$$
g(z \longrightarrow \infty) = g^* = \frac{\varepsilon}{a} + O\left(\varepsilon^2\right)
\tag{3.43}
$$

For polymers, we had (Eqs.(3.28) (3.33)) $a' = 2$, $a = 8$ and we just recover (3.30), as well as g^* (3.37) to first order, QED. Notice that this direct method of resumming leading

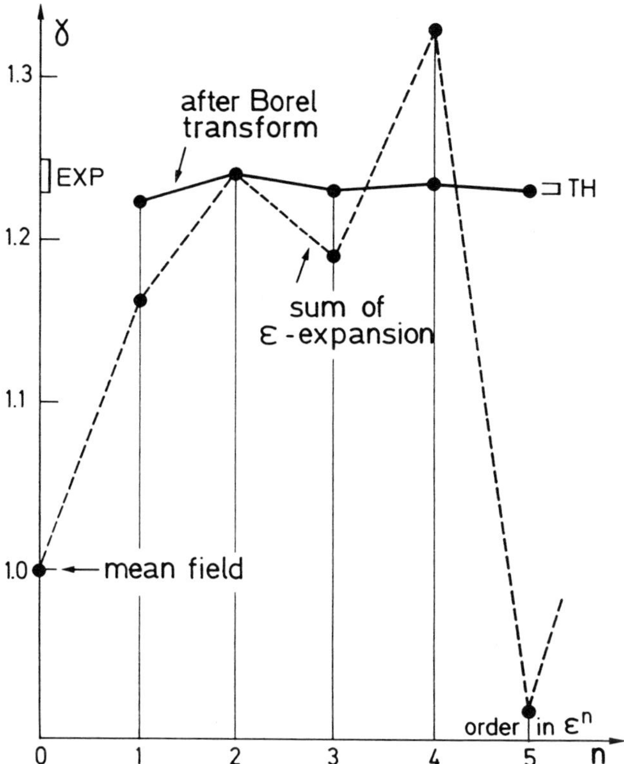

Fig.10 : Comparison of the (asymptotic) bare ε-expansion results to those after Borel transform and use of large orders. The illustrated case is the γ exponent of the 3D Ising model (courtesy of J.C. Le Guillou), but the polymer case would be entirely similar.

divergences, is directly applicable when one does not know if a model is renormalizable, like in the case of self-avoiding membranes, which are more alike string theories that field theories[27].

The ε-expansions of critical exponents like (3.38) are actually divergent series, which are only asymptotic (Fig.10). One believes these series to be Borel-summable, which means that if the series $\sum_n a_n \varepsilon^n$ diverges, the new $\sum_n a_n \varepsilon^n /n!$ converges. From this, one derives important numerical information on the actual functions $\nu(d)$, $\gamma(d)$. A real achievement of the Soviet school was the computation of ν and γ up to order ε^5(!) One finds for instance for ν to $O\left(\varepsilon^4\right)$

$$\frac{1}{\nu} = 2 - \frac{\varepsilon}{4} - \frac{11}{128}\varepsilon^2 - \left[\frac{83}{2048} - \frac{33}{256}\zeta(3)\right]\varepsilon^3$$
$$- \left[\frac{1395}{131072} - \frac{609}{8192}\zeta(3) - \frac{99}{1024}\zeta(4) + \frac{465}{1024}\zeta(5)\right]\varepsilon^4 + O\left(\varepsilon^5\right)$$

Quite remarkably, the expansion involves only rational numbers, and the values of the

Riemann ζ-function $\zeta(s) = \sum_{n \geq 1} n^{-s}$ for integers. The order $O\left(\varepsilon^5\right)$ contains similarly $\zeta(5)$ and $\zeta(7)$. It seems likely that this remains true to all orders.

Notice that all coefficients a_n in $\nu = \sum_n a_n \varepsilon^n$ above are obtained from configurations of a Brownian chain with n double points. So they are *characteristics of the Brownian motion* itself and the relation to the recurrent ζ-function is quite mysterious. It suggests also that perhaps the model would be exactly solvable at each dimension.

One knows also the *large order* behaviour of the perturbation series in z or ε. Then, thanks to the Borel-summability and to the exact knowledge of the $O\left(\varepsilon^5\right)$ terms, one can extract numerically very good approximate values of the exponents for ε finite[28]. The best results at $d = 3$ are for polymers

$$\nu = 0.5880 \pm 0.0015$$
$$(d = 3)$$
$$\gamma = 1.157 \pm 0.003$$

to be compared to the best experimental value (neutron diffraction) (Fig.11)

$$\nu_{\text{exp.}} = 0.586 \pm 0.004 \qquad (d = 3)$$

No experimental determination of γ exists, but experiments are under way (1989), which should measure the contact exponent $\theta_0 = \frac{\gamma - 1}{\nu}$ of the two ends of a polymer.

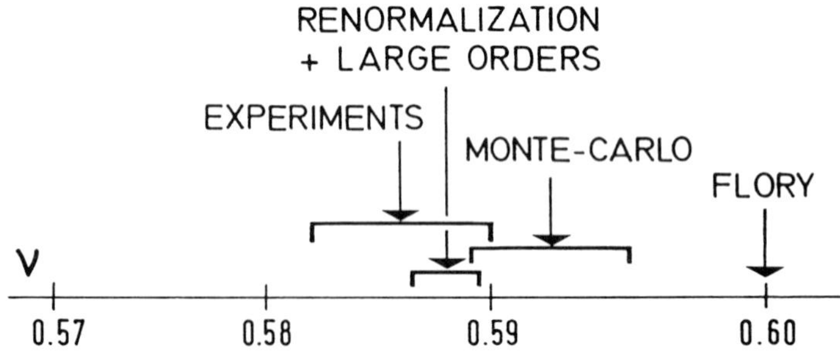

Fig.11 : Results for polymer exponent ν $(d = 3)$ (courtesy of J.C. Le Guillou).

The numerical treatment of the divergent series can be pushed up to $d = 2$, $\varepsilon = 2$ giving[28]

$$\nu = 0.76 \pm 0.03$$
$$(d = 2)$$
$$\gamma = 1.39 \pm 0.04$$

which compare well with the exact values $\nu = 3/4 = 0.75$ and $\gamma = 43/32 = 1.34375$. The agreement with renormalization group results can be considered as quite striking since $d = 2$ is quite far from $d = 4$! The results for ν can be summarized in a curve $\nu(d)$ for $1 \leq d \leq 4$ (Fig.12). One would like to know the geometrical meaning of non integer dimensions!

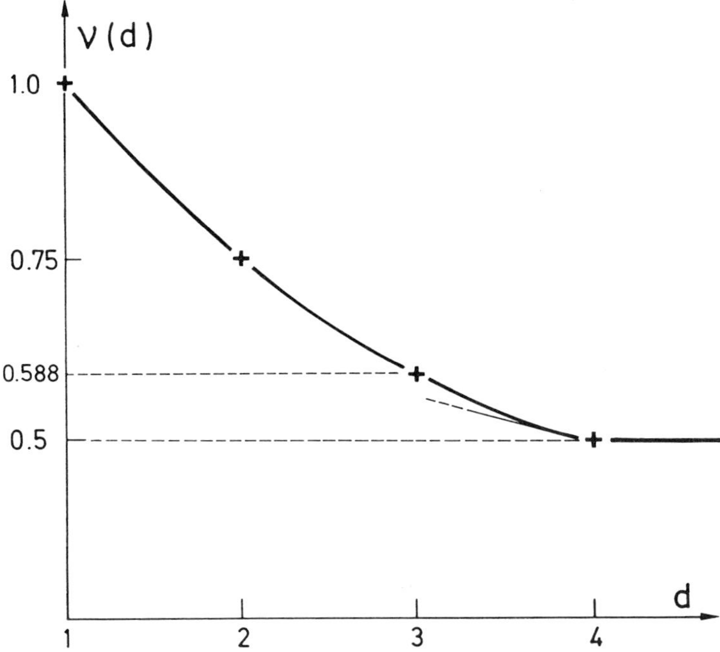

Fig.12 : The curve $\nu(d)$ for SAW.

Finally, let us mention that the star exponents (2.6) $\gamma_L(d = 4 - \varepsilon)$ have been calculated to second order[29,30] :

$$\gamma_L = 1 + \frac{\varepsilon}{8}\frac{L}{2}(3 - L) + \left(\frac{\varepsilon}{8}\right)^2 \frac{L}{8}\left(8L^2 - 37L + 55\right) + O\left(\varepsilon^3\right) \tag{3.44}$$

For $L = 1, 2$ one recovers $\gamma_1 = \gamma_2 = \gamma$ (3.38), as it must.

4. POLYMERS IN 2D AND CONFORMAL INVARIANCE

4.1 Introduction

This section is of course related to the lectures on conformal invariance by H. Knops in the same volume[31], and also by B. Nienhuis on polymers in two dimensions[32]. It will provide an illustration of the principles and applications of conformal invariance to 2D statistical systems at a critical point, in the specific case of polymers. The presentation will be more heuristic here than in Knops' lectures, in particular because they are still some gaps to be filled in the conformal theory of geometrical systems like polymers, percolation etc..., which are *non unitary* conformal field theories.

Let us start in a simple way and recall that conformal transforms in the plane are such that the shape of infinitesimal objets is conserved, e.g. a small patch of sides $ABCD$ is transformed into a new one $A'B'C'D'$ such that (Fig.13)

$$\frac{A'B'}{AB} = \frac{B'C'}{BC} = \frac{C'D'}{CD} = \frac{D'A'}{DA}$$

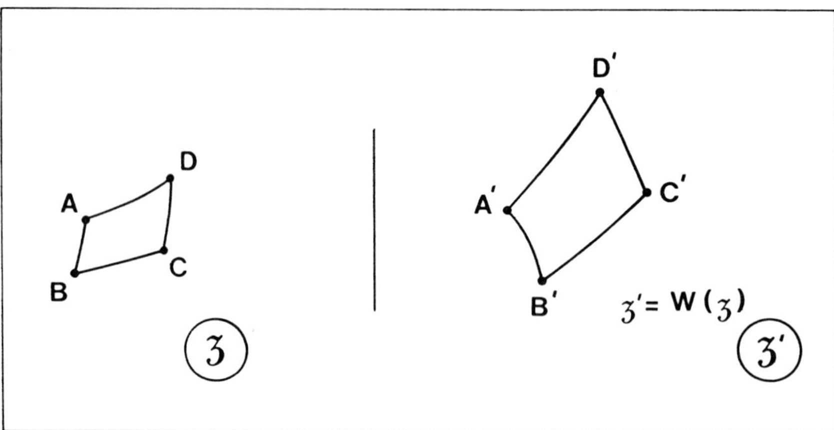

Fig.13 : A conformal mapping in the complex plane.

The mappings $z \longrightarrow z' = w(z)$ of the complex plane $\mathbf{C}$ onto itself which realize that property are all analytical functions $w(z)$ of the complex variable. For some problems where boundary conditions are important, one can be led to consider functions which are analytical only in certain domains. In electrostatics in 2D it is well known that many problems are solved by conformal mappings. The reason is that electrostatics is the theory of harmonic and analytical functions, solutions of the Laplace equation. Harmonic functions have a representation in terms of diffusion processes and Brownian motions. Hence the (infinite) Brownian motion itself should be a conformally invariant process, a property which had been remarked by P. Lévy[33] a long time ago. From Brownian motion to polymers the path is not too long, and we anticipate that SAW should be, in the infinite and continuum limit, conformally invariant.

This can be intuitively understood by noting that a Brownian path (BP) $\mathbf{B}_t$, $t \in [0, \infty[$, as well as a continuum infinite polymer $\mathbf{X}_s$, $s \in [0, \infty[$ have no length scale. So any conformal mapping which is locally a product of dilation and rotation, should transform the conformations of a BP or a SAW into new equivalent ones of the same system. This is the geometrical formulation of the breakthrough statement of Belavin, Polyakov and Zamolodchikov[34] (BPZ) in 1984 concerning the field theory of statistical systems at a critical point. Global scale invariance, augmented by global translational and rotational invariances, and the requirement of short range interactions in space, should lead to *conformal invariance*. In field theory, it implies[34,35,31] the conformal covariance of Green functions. If for instance $\{\phi_i\}$ is a set of conformal operators of scaling dimensions x_i of the field theory, than one expects the *conformal covariance* equation

$$\left\langle \prod_i \phi_i(z_i) \right\rangle = \prod_i |w'(z_i)|^{x_i} \left\langle \prod_i \phi_i(w(z_i)) \right\rangle \tag{4.1}$$

where $w(z)$ is a conformal mapping of the complex $z \in \mathbf{C}$. This relation is the obvious generalization to non uniform transformations $w(z)$ of the standard scaling. Indeed suppose $w(z) = \lambda z$. Then (4.1) reduces to the well-known form

$$\left\langle \prod_i \phi_i(z_i) \right\rangle = |\lambda|^{\sum_i x_i} \left\langle \prod_i \phi_i(\lambda z_i) \right\rangle \tag{4.2}$$

Recall that the scaling dimensions x_i above are also defined by the two point critical correlation functions

$$\langle \phi_i(z)\phi_i(z') \rangle = |z - z'|^{-2x_i} \tag{4.3}$$

for the same operator ϕ_i. Actually the requirement of conformal invariance puts much more constraints on the theory. BPZ[34] and Friedan, Qiu and Shenker[36] showed that the critical exponents and the critical correlation functions are associated with (finite-dimensional) representations of the conformal group. As a consequence the critical exponents, more precisely the scaling dimensions x of "primary" operators ϕ (see Knops' lectures) are given by the Kac table[37] $h_{p,q}$ where p and q are two labels. These $h_{p,q}$ are zeros of a determinant which, when vanishing, insures the degeneracy of the representation[31,34−37]. The quantized form of the scaling dimensions x is obtained[36] for unitary models (i.e. with a unitary transfer matrix) and a central charge $c < 1$ (see the definition in Refs.[31,35]) as $x = 2h$, with

$$h_{p,q} = \frac{[(m+1)p - mq]^2 - 1}{4m(m+1)}, \quad m \in \mathbb{N}, \quad p,q \in \mathbb{N} \tag{4.4}$$

where m parametrizes the universality class and the central charge

$$c = 1 - \frac{6}{m(m+1)}, \quad m \in \mathbb{N} \tag{4.5}$$

For unitary theories, the integer m satisfies $m \geq 3$ and the values of p and q are restricted to the "minimal block"

$$1 \leq p \leq m - 1, \quad 1 \leq q \leq p \tag{4.6}$$

The standard exponents η, ν, γ, α etc... are given in terms of the h's. For instance $m = 3$ and 4 correspond[36] to the critical and tricritical Ising models, and higher values of m to some RSOS models[38]. As we shall see, polymers correspond to a non unitary theory and are just outside this classification, filling in the two missing values $m = 1, 2$ corresponding to $c = -2, 0$.

One may understand intuitively the conformal quantization of the critical exponents, by analogy to the quantization of angular momentum in quantum mechanics (QM), as follows

Rotational invariance	Conformal invariance
Angular momentum	Conformal dimensions
J_z, m	h
Hilbert space,	H hermitian,
non negative norm	Unitary transfer matrix

In QM, there are operator J_z, and raising and lowering operators J_+, J_-. Then one defines states of minimal m such that $J_z |m_{\min}\rangle = m_{\min} |m_{\min}\rangle$ and $J_- |m_{\min}\rangle = 0$. One gets new eigenstates $(J_+)^n |m_{\min}\rangle$ as $n + m_{\min}$ eigenvectors of J_z. Then the nonexistence of states of negative norm (unitarity) implies that at some level $J_+ |m_{\max}\rangle = 0$. Hence the values of m are all quantized. A similar, but more complicated mechanism exists here in the case of the infinite group of conformal symmetries.

4.2 Conformal scaling dimensions of polymers

Let us now return to SAW. If we are looking for conformal operators, the most natural thing is to search for geometrical objects which are conserved in conformal transforms. So a guess is to consider "operators" which create various polymer vertices (Fig.14). With an L-leg vertex at point $\mathbf{x}$ should be associated an conformal field operator $\phi_L(\mathbf{x})$, with scaling dimension x_L. $L = 1$ corresponds to a free extremity, $L = 2$ to any point inside an infinite SAW, and $L = 4$ to the core of a $L = 4$ arm star. The essential idea is that a new critical singularity should be associated with each value of L.

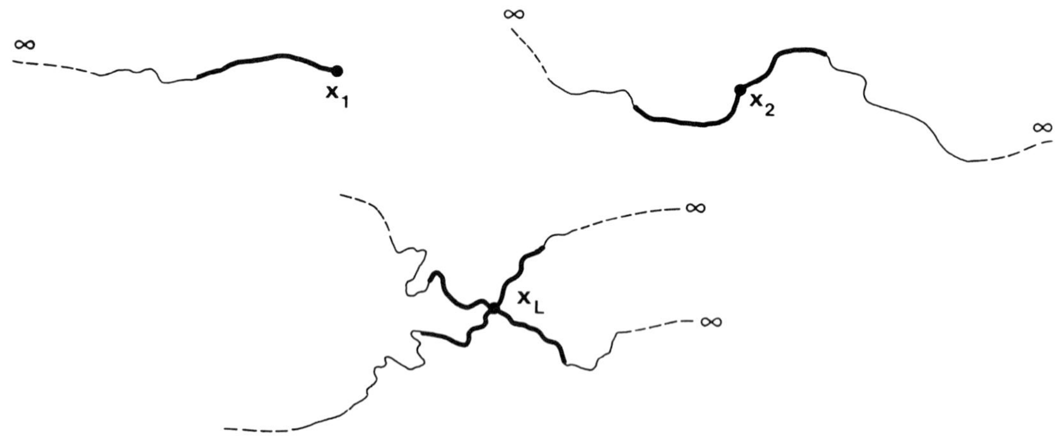

Fig.14 : One-, two- and four- polymer vertices and their scaling dimensions x_1 $(= 5/48)$, x_2 $(= 2/3)$ and x_4 $(= \frac{35}{12})$.

A more precise way to define the scaling dimensions is to define critical correlation functions of the watermelon polymers[13,39,40,41]. We consider e.g. on the honeycomb lattice $\mathcal{H}$ (Fig.15), L self- and mutually avoiding walks tied together at their extremities $\mathbf{x}$ and $\mathbf{y}$ in the plane, and with lengths $\ell_1, ..., \ell_L$. The partition function is simply

$$\mathcal{Z}_L\left(\mathbf{x} - \mathbf{y}, \ell_1, ..., \ell_L\right) = \sum_{\Gamma^L_{\mathbf{x},\mathbf{y}}} 1 \tag{4.7}$$

where $\Gamma^L_{\mathbf{x},\mathbf{y}}$ is the set of all L nonintersecting walks of lengths $(\ell_1, ..., \ell_L)$ joining $\mathbf{x}$ to $\mathbf{y}$. (On lattice $\mathcal{H}$, if $L > 3$ one has to start the walks at nearby points, which is irrelevant in the long distance limit).

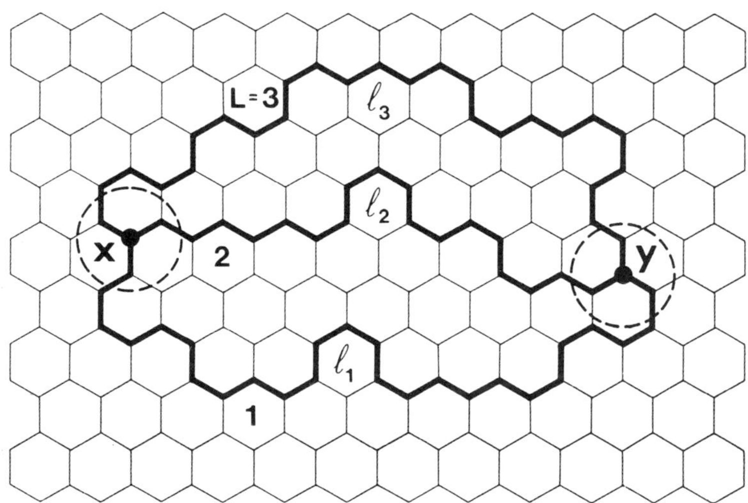

Fig.15 : Watermelon multipolymer, made of $L = 3$ linear chains tied together, of fluctuating lengths ℓ_i, $i = 1, 2, 3$.

Now the asymptotic behavior of $\mathcal{Z}$ for $\{\ell_i, \ i = 1, ..., L\}$ large is

$$\mathcal{Z}_L \sim \mu^{\sum_{i=1}^L \ell_i} \tag{4.8}$$

where μ is the same local effective connectivity constant of the lattice as for a single SAW. It remains in $\mathcal{Z}_L$ a non trivial dependance on $\mathbf{x} - \mathbf{y}$, which is best studied by Laplace transforming into the correlation function

$$G_L(\mathbf{x} - \mathbf{y}, K) = \sum_{\{\ell_i\}} K^{\sum_i \ell_i} \mathcal{Z}_L\left(\mathbf{x} - \mathbf{y}, \{\ell_i\}\right) \tag{4.9}$$

where K is the monomer fugacity. The sum converges for $K \le \mu^{-1}$ (because of (4.8)) and at $K_c = \mu^{-1}$ it has a critical power law behavior

$$G_L\left(\mathbf{x} - \mathbf{y}, K_c\right) = \langle \phi_L(\mathbf{x})\phi_L(\mathbf{y})\rangle = |\mathbf{x} - \mathbf{y}|^{-2x_L} \tag{4.10}$$

Owing to the discussion above we have also written it as a correlation function of two L-vertex operators ϕ_L, with scaling dimension x_L.

$\phi_L(\mathbf{x})$, of which we assume the existence, creates L infinite critical polymers at point $\mathbf{x}$. The factor 2 in $2x_L$ corresponds to the presence of two operators (see (4.3)). Notice that the form (4.10) is not characteristic of two dimensions and generalizes to d dimensions[28]. The values of the x_L should belong to the Kac table (4.4). For polymers, the value of the central charge is $c = 0$, easily obtained from the equivalence of SAW to an $n = 0$ component $O(n)$ spin model (see Ref.7 and Ref.32 in this volume). The $O(n)$ partition function in any geometry is for $n = 0 : Z_{O(0)} = 1$, hence the free energy $F_0 = -\ell \text{n } Z = 0$. The central charge can be seen a finite size universal coefficient[43] of the free energy in a periodic strip of width w, $w \longrightarrow \infty : F_{\text{strip}} = f_0 w - \frac{\pi c}{6w}$ where f_0 is the bulk free energy per unit area. For SAW, $f_0 = 0$, $F_{\text{strip}} = 0$, hence $c = 0$, QED. Hence we find the parameter $m = 2$ for polymers in (4.5). Then the values of the x_L can be found by numerical simulations[39], Coulomb gas methods[39-41] (see below), or more recently a Bethe Ansatz solution[42]

$$x_L = 2h_{L/2,0}^{c=0} = \frac{9L^2 - 4}{48}, \qquad m = 2 \qquad (4.11a)$$

The factor 2 in $x = 2h$ comes from the fact that we are in 2D and the operators ϕ_L are scalar. The derivation is heuristic and the operators $\phi_{L/2,0}$ are not really constructed here. The meaning of half-integer indices $(p, q) = (L/2, 0)$ is still unclear. We also note that the (p, q) are taken outside the minimal block (4.6).

When one considers in (4.9) $K > K_c$ and puts the polymer system in a box, one reaches the *dense* polymer phase. There the single SAW watermelon network fills a finite fraction of the box, even in the thermodynamic limit. This corresponds to a polymer melt[40,41,44,45], in a new universality class, independent of $K > K_c$. This universality class includes for $K \longrightarrow \infty$ the limit of Hamiltonian walks[41] which are SAW totally filling the lattice. The dense phase is critical, with a new central charge[44,45] $c = -2$, $m = 1$, and new watermelon scaling dimensions

$$x_L^D = 2h_{0,L/2}^{c=-2} = \frac{L^2 - 4}{16}, \qquad m = 1 \qquad (4.11b)$$

In the next section we shall give a sketch of the Coulomb gas derivation of these exponents within the $O(n)$ model, and later apply them to correlation functions of polymer networks of arbitrary topologies.

5. GEOMETRICAL EXPONENTS IN THE O(n) MODEL

5.1 Watermelon configurations

For definiteness we first consider the $O(n)$ model on the honeycomb two-dimensional lattice. Its partition function is defined as[46,9,32]

$$Z_{O(n)} = \int \prod_i d\vec{S}_i \prod_{\langle i,j \rangle} \left(1 + K \, \vec{S}_i \cdot \vec{S}_j \right) \qquad (5.1)$$

where the spin variables $\vec{S}_i$ are $n-$vectors with $S_i^2 = n$, and where $\langle i, j \rangle$ are nearest neighbour sites. One can show easily that on the honeycomb lattice $\mathcal{H}$ Eq.(5.1) reads

$$Z_{O(n)} = \sum_{\text{nonintersecting loops}} K^{\mathcal{N}_B} n^{\mathcal{N}_L} \tag{5.2}$$

where the sum runs on all configurations of closed and nonintersecting loops drawn on $\mathcal{H}$, with total numbers of bonds $\mathcal{N}_B$, and of loops $\mathcal{N}_L$ (Fig.16).

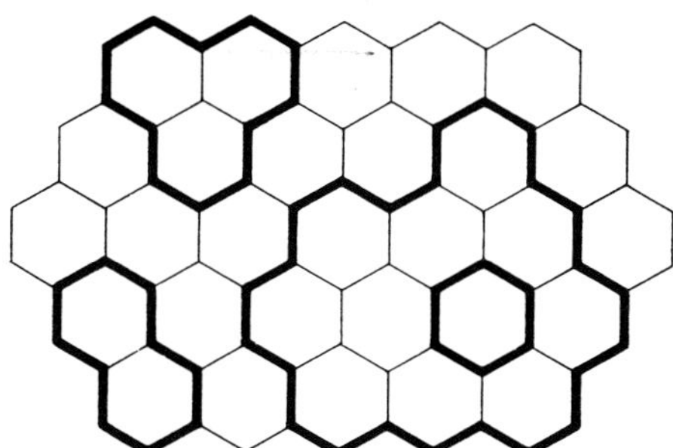

Fig.16 : Loop representation of the $O(n)$ model on the honeycomb lattice (Eq.(5.2)).

In this way, the high-temperature (i.e. small K) expansion of the $O(n)$ model gives a *geometrical loop model*. For a fixed value of $n \in [-2, 2]$ there is a critical point $K_c = \left[2 + (2 - n)^{1/2}\right]^{-1/2}$ (on the hexagonal lattice[9]) where the mean length $\langle \mathcal{N}_B \rangle \sim (K_c - K)^{-1}$ diverges. Geometrical correlation functions can be defined by considering *watermelon* configurations on the lattice (Fig.17). One specifies to configurations $\mathcal{C}_L(\mathbf{x}, \mathbf{y})$ such that L nonintersecting lines have a source point at $\mathbf{x}$ and a sink at $\mathbf{y}$ where they all meet together, in a presence of vacuum loop of the $O(n)$ model. The associated correlator is defined as

$$G_L(\mathbf{x} - \mathbf{y}, K) = \sum_{\mathcal{C}_L(\mathbf{x}, \mathbf{y})} W(\mathcal{C}_L) / Z_{O(n)} \tag{5.3}$$

where the statistical weight of the configuration $\mathcal{C}_L$ (watermelon + vacuum loops) is

$$W(\mathcal{C}_L) = K^{\mathcal{N}_B} n^{\mathcal{N}_L}$$

where now $\mathcal{N}_B$ is the total length of the watermelon and loop lines. At K_c one expects a critical algebraic decay

$$\begin{aligned} G_L(\mathbf{x} - \mathbf{y}, K_c) &\equiv \langle \phi_L(\mathbf{x}) \phi_L(\mathbf{y}) \rangle \\ &= |\mathbf{x} - \mathbf{y}|^{-2x_L} \end{aligned} \tag{5.4}$$

where x_L is the scaling dimension of the (conformal) operator $\phi_L(\mathbf{x})$ assumed to represent a source of L nonintersecting *infinite* lines at a point $\mathbf{x}$. As we shall see these x_L are not linear in L (these is no gap exponent) and are independent of each other. Hence a new geometrical critical exponent x_L appears for each star-like critical object. It is also important to **notice** that in the $O(n)$ model, a second critical phase appears, the *dense* phase, for any value $K > K_c$. Geometrically, this corresponds to taking infinite loops or lines filling the lattice with a finite (non vanishing) density, even in the thermodynamic limit. The physical properties thus describe those of a loop melt in two-dimensions. In this dense phase are associated other scaling dimensions x_L^D with each source operator ϕ_L.

5.2. Coulomb gas technique

It is known[39−41] that one can determine the geometrical exponents x_L by a Coulomb gas technique first devised in 1982[9] for obtaining the usual spin and energy exponents η and ν of the $O(n)$ model (see also Nienhuis, this volume and den Nijs[40]). As we shall see, the x_L also belong to the conformal Kac table of the (unitary) minimal theories with central charge $c \leq 1$. Lastly, let us mention that a derivation of the same dimensions is also possible through a Bethe Ansatz on the hexagonal lattice[43]. In the standard Coulomb gas technique, the $O(n)$ model can be transformed into a SOS model by orienting the loops and the watermelon lines and in the continuum critical limit, it renormalizes onto a Gaussian field with action $A = (g/4\pi) \int (\partial\varphi)^2 \mathrm{d}^2 x$, where g is the Coulomb gas coupling constant such that

$$n = -2 \cos(\pi g) \tag{5.5}$$

and $g \in [1, 2]$ at K_c, while $g \in [0, 1]$ in the *dense* phase[9].

Now, the watermelon correlation function generalized to the $O(n)$ model, can be associated in the SOS model with the existence of a dislocation. Indeed the L lines, once oriented, are interpreted as domain walls in a height model and a circuit around $\mathbf{x}$ or $\mathbf{y}$ gives a height discontinuity depending on L (Fig.17).

As a result, the correlation function G_L (5.4) is simply a Coulomb gas correlator of two electromagnetic operators

$$G_L(\mathbf{x} - \mathbf{y}) = \langle \mathcal{O}_{e_0, m'}(\mathbf{x}) \mathcal{O}_{e_0, -m'}(\mathbf{y}) \rangle = |\mathbf{x} - \mathbf{y}|^{-gm'^2 + e_0^2/g} \tag{5.6}$$

where $e_0 = 1 - g$ is the standard "floating electric charge"[47] and m' the magnetic contribution coming from the L defect lines: $m' = L/2$. The name "Coulomb gas" is justified when one remarks that the correlator (5.6) reads also

$$e^{-\left(gm'^2 - e_0^2/g\right)\ell n|\mathbf{x} - \mathbf{y}|}$$

i.e. is the *Gibbs weight* of pairs of magnetic and electric charges (m', e_0) and $(-m', e_0)$ located at $\mathbf{x}$ and $\mathbf{y}$ respectively, and interacting via the Coulomb potential in 2D $(\ell n|\mathbf{x} - \mathbf{y}|.)$ We refer the reader to Ref.[9,39-41] for precise geometric derivations of (5.6). Let us simply mention that the seemingly mysterious presence of the electric charge $e_0 = 1 - g$ is necessary

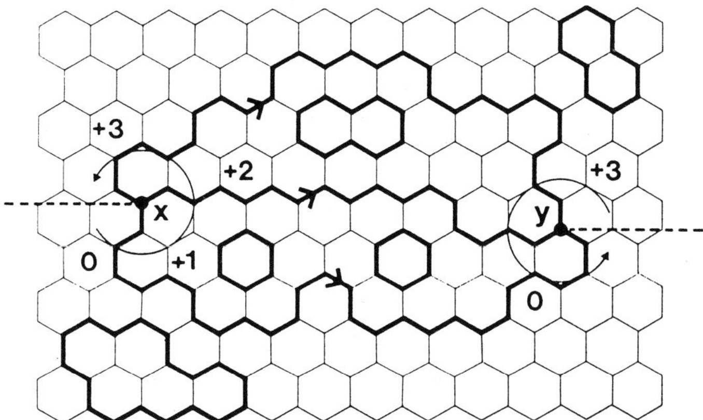

Fig.17 : Watermelon configuration of $L = 3$ nonintersecting lines, in presence of vacuum loops of the $O(n)$ model. The integers indicate the heights φ on the equivalent SOS model, once the SAW are oriented and represent domain walls. The dotted lines represent dislocations associated with the watermelon.

for correcting curvature effects along the watermelon lines, which appear in the SOS model but not in the original $O(n)$ model, when the watermelon *winds* about one of its origins (Fig.18).

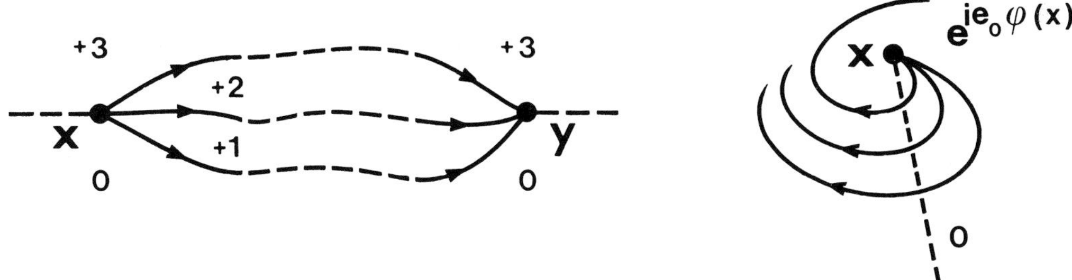

Fig.18 : Magnetic and electric operators. A set of 3 lines creates a discontinuity (dotted lines) +3 or -3, when winding about $\mathbf{x}$ or $\mathbf{y}$, hence (in proper units) magnetic charges $m'_{\mathbf{x}} = -m'_{\mathbf{y}} = 3/2$. The winding curvature in the SOS model is corrected by a phase factor (electric operator) $e^{ie_0\varphi(x)}$, where φ is the height counted in units of π along the dotted line, in and $e_0 = 1 - g$, $n = -2 \cos g$. In Nienhuis' notations for the SOS model (same volume), $e_0 \equiv -\frac{6\theta}{\pi}$.

Hence the watermelon scaling dimension is finally

$$x_L = \frac{1}{2}\left(gm'^2 - \frac{e_0^2}{g}\right) = g\,L^2/8 - (1-g)^2/2g \tag{5.7}$$

for the $O(n)$ model where n is parametrized by (5.5).

Conformal invariance. These scaling dimensions can be rewritten as a Kac formula[37]

$$
\begin{aligned}
h_{p,q} &= \frac{[(m+1)p - mq]^2 - 1}{4m(m+1)} \\
c &= 1 - 6/m(m+1) = 1 - 6(1-g)^2/g \\
x_L &= 2h_{L/2,0} \quad \text{(dilute phase, } g \in [1,2]) \\
x_L^D &= 2h_{0,L/2} \quad \text{(dense phase, } g \in [0,1])
\end{aligned}
\tag{5.8}
$$

The values of the conformal parameter m parametrizing the universality class are $m = \frac{1}{g-1} \in [1,\infty[$ for the dilute phase and $m = \frac{g}{1-g} \in [0,\infty[$ for the dense one. The $O(n=2)$ model corresponds to the XY model at the Kosterlitz-Thouless transition point, for which $g = 1$, $m = \infty$, $c = 1$. The dilute and dense phases coincide there. Only this model and the $n = 1$ standard Ising model, for which $m = 3$, are unitary. The other $O(n)$ models (n continuous) are analytic extensions of the spin model (5.1) and have a purely geometrical interpretation (5.2) in terms of systems of nonintersecting loops.

Polymers. When $n = 0$ in (5.2), all loops give a zero weight and only the empty graph contributes, leading to a trivial partition function $Z_{O(n=0)} = 1$. However, in the watermelon correlator (5.3), the L lines joining $\mathbf{x}$ to $\mathbf{y}$ survive when $n \longrightarrow 0$, and describe in the continuum limit the correlation function of L self- and mutually-avoiding polymer lines, tied at their extremities $\mathbf{x}$ and $\mathbf{y}$, and with fluctuating lengths $\ell_1, ..., \ell_L$

$$
G_L(\mathbf{x} - \mathbf{y}, K_c) = \int_0^\infty d\ell_1...d\ell_L \, K_c^{(\ell_1 + ... + \ell_L)} \mathcal{Z}_L(\mathbf{x} - \mathbf{y}, \ell_1, ..., \ell_L)
\tag{5.9}
$$

where $\mathcal{Z}_L$ is the partition function of the watermelon, and K_c^{-1} is the usual effective connectivity constant. The expression (5.9) is written in the continuum limit where the sums over bond numbers can be replaced by integrals over lengths.

The two possible phases for polymers ($n = 0$) correspond to $g = 3/2$ (dilute) or $g = 1/2$ (dense). Hence (5.7) (5.8) give the set of scaling dimensions (4.11):

$$
\begin{aligned}
x_L &= \left(9L^2 - 4\right)/48, & c &= 0, & g &= 3/2, & m &= 2 \ \text{(dilute)} \\
x_L^D &= \left(L^2 - 4\right)/16, & c &= -2, & g &= 1/2, & m &= 1 \ \text{(dense)}
\end{aligned}
\tag{5.10}
$$

It is interesting to remark that the usual conformal classification of "unitary" statistical systems started with integer values $m \geq 3$ of the parameter m appearing in $h_{p,q}$ (5.8), $m = 3$ corresponding to the Ising model, and models with higher values of m being identified later. The polymer case is special, since there is no symmetric transfer matrix, or equivalently it is an $n = 0$ vector model, and it is "non unitary". It fills then precisely the two missing values $m = 1, 2$ of Friedan, Qiu and Shenker classification[36]. Notice also that the interpretation of half-integer indices in $h_{p,q}$ is not yet clear. Formulas like (5.8) for the $O(n)$ model are thus to be interpreted as useful analytic continuations. The standard critical exponents ν and γ for SAW are obtained by scaling in terms of the first two x_L, namely

$$
\nu^{-1} = 2 - x_2, \quad \eta = 2x_1, \quad \gamma = (2 - \eta)\nu
\tag{5.11}
$$

which, owing to (5.10), give Nienhuis values $\nu = 3/4$, $\gamma = 43/32$. Notice that the thermal exponent x_2 (5.10) was conjectured before (from the $O(n)$ model) by Cardy and Hamber[48]. For the dense phase of polymers, we find instead from (5.10)

$$\nu^D = \frac{1}{2}, \quad \gamma^D = \frac{19}{16}$$

The interpretation of $\nu^D = \frac{1}{2}$ is immediate: the fractal dimension is $D_F = 1/\nu^D = 2$, as expected for a dense walk filling the plane. The interpretation of the γ exponent is more subtle, and requires a careful analysis[44,45] of boundary conditions for dense SAW, since these conditions are able to add nonuniversal terms to γ^D.

Up to now we used only the first exponents x_1, x_2. The higher order ones x_L, $L \geq 3$ will play a fundamental role, described below (section 6), in the making of branched polymer networks.

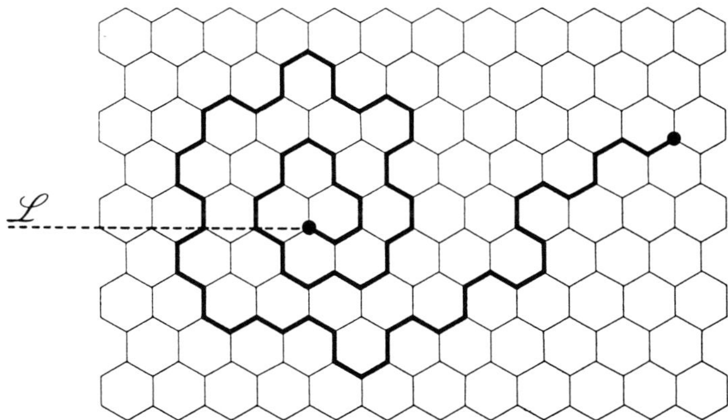

Fig.19 : Winding of a self-avoiding walk about its origin. Asymptotically the winding angle is $\theta = 2\pi n_1$ where n_1 is the algebraic number of crossings with an infinite half-line $\mathcal{L}$.

5.3 Winding angle of a self-avoiding walk

Another interesting geometrical application of Coulomb gas methods in 2D is found in the problem of windings of a SAW about its extremities[49] (Fig.19). On e.g. the hexagonal 2D lattice the winding angle θ_1 of a self-avoiding walk around its extremity $\mathbf{x}$ is defined as $\theta_1 = 2\pi n_1$, where n_1 is the algebraic number of times the walk crosses a line joining $\mathbf{x}$ to infinity. A similar definition holds for the winding angle θ_2 around the other extremity $\mathbf{y}$. One considers then the windings on presence of $O(n)$ loops, and shows by a geometrical construction, involving the passage to the SOS model, that in Fourier space and at the critical point

$$\langle \exp\left[ie\pi\left(n_1 - n_2\right)\right]\rangle_{O(n)} = |\mathbf{x} - \mathbf{y}|^{-e^2/g}$$
$$= \exp\left(-\frac{e^2}{g}\ell n|\mathbf{x} - \mathbf{y}|\right)$$

where, as before g is the CG coupling constant such that $n = -2 \cos \pi g$, and where the Fourier variable e is arbitrary. Hence the angular correlator appears as a Coulomb gas correlation function (5.6) of two electric charges e, $-e$ located at $\mathbf{x}$ and $\mathbf{y}$ respectively. Since it is purely *Gaussian* in variable e, its inverse Fourier transform, i.e. the probability density of a relative angle $\hat{\theta} = 2\pi \left(n_1 - n_2 \right)$ is also a Gaussian. Fourier inverting, going to finite chains of length ℓ by scaling $|\mathbf{x} - \mathbf{y}| \sim \ell^\nu$, and deconvoluting the single extremity winding distribution, one finds[49] for each angle $\theta = \theta_1$ or θ_2 *separately*

$$\underset{\ell \,\longrightarrow\, \infty}{P(\theta)} = \frac{1}{\sqrt{8\pi\nu g^{-1}\ell n\, \ell}}\exp\left(-\frac{g\theta^2}{8\nu\ell n\, \ell}\right)$$

For polymers $n = 0$, $g = 3/2$, $\nu = 3/4$ and finally the winding probability distribution is the *universal* law in reduced units

$$\underset{\ell \,\longrightarrow\, \infty}{P_{\text{SAW}}}\left(x = (4\ell n\, \ell)^{-1/2}\theta\right) = \frac{1}{\sqrt{\pi}}e^{-x^2}$$

where ℓ is the length of the SAW. The universal $\langle\theta^2\rangle = 2\ell n\, \ell$ agrees well with earlier numerical results[50]. It is also interesting to compare $P_{\text{SAW}}(x)$ to the Cauchy law for a planar Brownian path, derived by Spitzer[51] in 1958

$$P_{\text{Brown}}(x = 2\theta/\ell n\, \ell) = \frac{1}{\pi}\frac{1}{1 + x^2}\ ,$$

and to the result for 2D random walks, recently derived by Bélisle[52]

$$P_{\text{RW}}(x = 2\theta/\ell n\, \ell) = [2\,\text{ch}\,(\pi x/2)]^{-1},$$

which coincides with that for "big windings" of the planar Brownian motion[53].

6. POLYMERS OF HIGHER TOPOLOGY

6.1 Polymer networks

Consider a polymer graph $\mathcal{G}$ of arbitrary but fixed topology[13] (Fig.20), made of lines connected at some vertices i, $i = 1, ..., \mathcal{V}$, with L_i lines going out of vertex i.

This graph represents physically a polymer micronetwork, floating in a solvent. If the solvent is good, the lines are self- and mutually avoiding because of the standard excluded volume effect. From a mathematical point of view, one can put the network on a 2D lattice and ask about the asymptotic number of configurations when the length of the lines diverge. One can distinsguish "monodisperse" networks where the lengths of the chains are all identical and equal to ℓ, or scaling in the same way, from polydisperse networks where the lengths fluctuate with a fixed total length ℓ. The first case is of course the experimental situation,

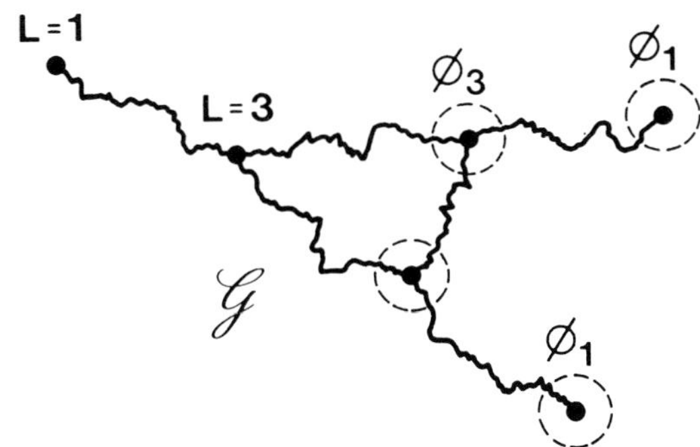

Fig.20 : A polymer graph $\mathcal{G}$ or micronetwork, made of three $L = 3$ vertices and three $L = 1$ vertices (dangling ends). The total number of chains is $\mathcal{N} = 6$. The γ−configuration exponent of $\mathcal{G}$ is here $\gamma_{\mathcal{G}} = -\frac{43}{32}$ (Eq.(6.18)).

while the second is mathematically suited for conformal field theory. Both numbers of configurations are trivially related by

$$Z_{\mathcal{G}}^{\text{poly}} \sim \ell^{\mathcal{N}-1} Z_{\mathcal{G}}^{\text{mono}} \tag{6.1}$$

where $\mathcal{N}$ is the total number of branches of $\mathcal{G}$, and where $\ell^{\mathcal{N}-1}$ is the "phase space" factor of the distribution of a total length ℓ among $\mathcal{N}$ chains

$$\ell^{\mathcal{N}-1} \sim \int \delta\left(\ell - \sum_{i=1}^{\mathcal{N}} \ell_i\right) d\ell_1 ... d\ell_{\mathcal{N}}$$

For a monodisperse network $\mathcal{G}$, one expects the total number of configurations to scale as

$$Z_{\mathcal{G}}^{\text{mono}} \underset{\ell \longrightarrow \infty}{\sim} \mu^{\mathcal{N}\ell} \ell^{\gamma_{\mathcal{G}}-1} \tag{3.2}$$

where $\mathcal{N}\ell$ is the total length of $\mathcal{G}$, μ is the entropy per monomer (an ultraviolet local and nonuniversal constant) and $\gamma_{\mathcal{G}}$ a universal configuration exponent. For a purely "Brownian" or "*phantom*" network on a $d-$dimensional hypercubic lattice $\mathbf{Z}^d$ without self-avoidance one has for instance

$$Z_{\mathcal{G}}^{\text{Brownian}} \sim (2d)^{\mathcal{N}\ell} \ell^{-\frac{d}{2}\mathcal{L}} \tag{6.3}$$

where $\mathcal{L}$ is the number of independent loops of the micronetwork $\mathcal{G}$, and $2d$ is the actual number of choices of a random walk at each step in $\mathbf{Z}^d$. When self-avoidance is present $\gamma_{\mathcal{G}}$ becomes a non trivial critical exponent depending on the topology of $\mathcal{G}$. We show now briefly how it can be expressed in terms of the watermelon exponents $\{x_L\}$, by decomposing the network $\mathcal{G}$ into its vertices[13].

One considers the multiple correlation function of the conformal operators ϕ_{L_i} associated with the vertices of $\mathcal{G}$,

$$G_{\mathcal{G}}\{z_i\} = \left\langle \prod_{i=1}^{\mathcal{V}} \phi_{L_i}(z_i) \right\rangle \tag{6.4}$$

where the $\{z_i\}$ denote the vertex positions in the plane. Dangling ends of $\mathcal{G}$ are to be included in the set of vertices I and are $L = 1$ one-leg vertices. The above multiple correlator represents the Laplace transform of the partition function of the polydisperse network with respect to the lengths ℓ_a, $a = 1, ..., \mathcal{N}$ of the branches

$$G_{\mathcal{G}}\{z_i, K\} = \left\langle \prod_{i=1}^{\mathcal{V}} \phi_{L_i}(z_i) \right\rangle_K = \int_0^{\infty} d\ell_1...d\ell_{\mathcal{N}} \, K^{\sum_{a=1}^{\mathcal{N}} \ell_a} \mathcal{Z}_{\mathcal{G}}\{\ell_a, z_i\} \tag{6.5}$$

and become critical at $K_c = \mu^{-1}$ (see (5.9)). Now the scaling behavior (6.2) is easily derived by simple power counting. The polydisperse partition function is obtained by Laplace inverting (6.5) and integrating it over the positions of $\mathcal{V} - 1$ vertices, keeping the last one fixed for eliminating the overall translational invariance. Hence

$$\mathcal{Z}_{\mathcal{G}}^{\text{poly.}}(\ell) = \int \prod_{i=1}^{\mathcal{V}-1} d^2 z_i \oint \frac{d\left[\ell n\left(K_c/K\right)\right]}{2\pi i} K^{-\ell} \left\langle \prod_{i=1}^{\mathcal{V}} \phi_{L_i}(z_i) \right\rangle_K \tag{6.6}$$

where the contour of integration lies at the ℓ.h.s. of singularities in the $K-$complex plane. Power counting implies that each conformal operator ϕ_{L_i} brings in a scaling power $[z]^{-x_{L_i}}$ in length unit (see Eq.(5.4)). So in the power counting or dimensional analysis sense

$$\mathcal{Z}_{\mathcal{G}}^{\text{poly.}}(\ell) \sim [z]^{2(\mathcal{V}-1)-\sum_{i=1}^{\mathcal{V}} x_{L_i}} \left[\ell n\left(K_c/K\right)\right] K_c^{-\ell} \tag{6.7}$$

where we have set K equal to its critical value $K_c = \mu^{-1}$ whenever possible. We just have to convert $[z]$ into length units $[\ell]$. This is done with the help of the well-known correlation length exponent ν such that

$$z \sim \ell^{\nu} \tag{6.8}$$

where z is the typical size of the network, or of any fraction of it. On the other hand the distance to the critical point K_c governs the actual length of the network in monomeric units, whence

$$\ell \sim \left[\ell n\left(K_c/K\right)\right]^{-1} \sim \left(K_c - K\right)^{-1}, \tag{6.9}$$

an infinite network corresponding to the critical point K_c. The usual basic equivalence is thus

$$z \sim \ell^{\nu} \sim \left(K_c - K\right)^{-\nu}, \tag{6.10}$$

which implies in (6.7)

$$\mathcal{Z}_{\mathcal{G}}^{\text{poly.}} \sim K_c^{-\ell}\left(K_c - K\right)^{1-\gamma_{\mathcal{G}}^{\text{poly.}}} \tag{6.11a}$$

$$\sim \mu^{\ell} \, \ell^{\gamma_{\mathcal{G}}^{\text{poly.}}-1} \tag{6.11b}$$

with

$$\gamma_{\mathcal{G}}^{\text{poly.}} \equiv \nu \left(2(\mathcal{V} - 1) - \sum_{i=1}^{\mathcal{V}} x_{L_i} \right) \tag{6.12}$$

Recall that in (6.11b) ℓ is the total length of the polydisperse network with fluctuating individual lengths of the $\mathcal{N}$ branches. The monodisperse case (6.2) is obtained from relation (6.1) yielding

$$\gamma_{\mathcal{G}} - 1 = \gamma_{\mathcal{G}}^{\text{poly.}} - \mathcal{N} \tag{6.13}$$

It is also convenient to use the numbers n_L of $L-$leg vertices in a given network $\mathcal{G}$, and the topological relations $\sum_{L \geq 1} n_L = \mathcal{V}$, $\sum_{L \geq 1} n_L \, L/2 = \mathcal{N}$, to write finally

$$\gamma_{\mathcal{G}} - 1 = -2\nu + \sum_{L \geq 1} n_L \left\{ \nu \left(2 - x_L \right) - L/2 \right\} \tag{6.14}$$

which illustrates the basic principle of the factorization[13] of ("infrared") divergences of the partition function (6.2) of a given graph $\mathcal{G}$ over the vertices of $\mathcal{G}$.

It is important to mention several points. All the above analysis is fairly general and applies to *any* $O(n)$ model, where the problem is generalized to the configuration number of a self-avoiding graph in presence of vacuum loops. In this case the correlation length exponent ν in (6.14) is replaced by the exponent ν_2

$$\nu_2^{-1} \equiv 2 - x_2 \tag{6.15}$$

associated with the fractal dimension D_F of the self-avoiding lines in the $O(n)$ model

$$D_F = \nu_2^{-1} = 2 - x_2 = 1 + \frac{1}{2g} \tag{6.16}$$

where $n = -2 \cos(\pi g)$, $g \in [1,2]$ (dilute), $g \in [0,1]$ (dense) (Eq.(5.5)). The exponent ν_2 differs from the standard thermal exponent ν associated with the energy operator, except for the polymer case $n = 0$. Notice that this choice (6.15) gives a curly bracket contribution for $L = 2$ in (6.14), $\nu_2 \left(2 - x_2 \right) - 1 \equiv 0$, as it must, since the number n_2 of two-leg insertions in a graph is arbitrary and corresponds to any occupied bond in the high-temperature expansion of the $O(n)$ model.

For standard dilute SAW, one plugs the values (5.10) of the x_L exponents into (6.14) using also (see (6.15)) the value $\nu = 3/4$ to get

$$\gamma_{\mathcal{G}} = -\frac{1}{2} + \frac{1}{64} \sum_{L \geq 1} n_L(2 - L)(9L + 50) \tag{6.17}$$

This formula is very well checked from numerical simulation results. For instance for network having the topology of a star, this gives $\gamma_L = \frac{68 + 9L(3 - L)}{64}$, hence $\gamma_1 = \gamma_2 \equiv \gamma = \frac{43}{32}$ (usual susceptibility exponent of a linear chain), $\gamma_3 = \frac{17}{16} = 1.0625$, $\gamma_4 = \frac{1}{2}$, $\gamma_5 = -\frac{11}{32} =$

-0.34374 and $\gamma_6 = -\frac{47}{32} = -1.46875$. The numerical results are[54] $\gamma_3 = 1.07 \pm 0.02$, $\gamma_4 = 0.52 \pm 0.04$, $\gamma_5 = -0.29 \pm 0.04$ and $\gamma_6 = -1.33 \pm 0.05$. The γ_H exponent of a $H-$comb is $\gamma_H = \frac{25}{32} = 0.78125$, and numerically[54] $\gamma_H = 0.79 \pm 0.02$.

Of course, the vertex decomposition formula generalizes[29] to any space dimension. A general proof in the realm of standard perturbative renormalization theory is still lacking however. It would require a careful selection of the irreducible vertex operator associated with the branching of L lines in the $O(n)$ model, among all operators mixing with φ^L in a φ^4 theory. Let us finally mention that we shall encounter in the case of the $O(n)$ model embedded in a fluctuating surface, a decomposition formula entirely similar to (6.14), which is established there by a direct solution.

6.2 Contact exponents

Let us now return briefly to the contact exponents θ introduced above (section **2.2**, Figs.4 and 5). We claimed that by scaling[11] one gets the formula (2.9) relating $\theta_{\{L_i\}}$ of a set of star cores i with L_i branches, to their configuration exponents γ_{L_i}. Using now the general formula (6.14) for the star exponents γ_{L_i} we find easily the very simple formula[41]

$$\theta_{\{L_i\}} = x_{\sum_i L_i} - \sum_i x_{L_i} \tag{6.18}$$

Hence a contact exponent is simply given by the scaling dimension of the *fused* vertex, from which one has to subtract the dimensions of the stars before contact. This formula is, of course, a formula of operator product expansion in field theory.

7. STATISTICAL MECHANICS ON A FLUCTUATING MEMBRANE

7.1 Brief history

The consideration of statistical mechanics on a fluctuating membrane modelized by a random lattice originated from studies of such two-dimensional lattices as models of random surfaces. The first example of critical phenomena on a surface with fluctuating geometry ($2d$ quantum gravity) is given by the bosonic string in Polyakov's formulation[55]. The matter field is represented by d free bosons (Gaussian fields) living on the surface, where d is the dimension of the equivalent embedding space. The discretization of the model is obtained by replacing the integration over the metrics representing the quantum gravitational field, by a sum over arbitrary (planar) discrete graphs[56−58]. Notice that one has to fix the topology of the world-surface in order to prevent divergences in the sum over topologies, otherwise unavoidable; and the planar (spherical) topology is the simplest one. It became also clear that the graphs generated by any set of local rules (like φ^3, φ^4 etc... vertices) would have the same critical behaviour[56,59,60].

Pure gravity (an abstract random surface with a fluctuating intrinsic metric and not embedded, i.e. in zero dimension) is thus described by the ensemble of abstract planar graphs. Exact results were obtained in $d = 0$ or $d = -2$ dimensions for the string suscepti-bility exponent γ_{string} of the free energy near criticality[57,61]. Then the idea of putting more general statistical models on a random surface occured in the first example of the Ising model treated by V.A. Kazakov[62], who obtained critical exponents different from those of Onsager in two-dimensions. Some exact results were also obtained for the $Q-$state Potts model[63].

Recently, another continuum approach to 2D-critical phenomena in presence of gravity has been developped by Polyakov[64], and Knizhnik, Polyakov and Zamolodchikov[65] (KPZ). It is the generalization of the standard conformal invariance in the plane associated with the Virasoro algebra, to conformal invariance in presence of an (annealed) fluctuating metric. A spectacular prediction by KPZ is that the conformal dimensions of the usual Kac spectrum (4.4)

$$h_{p,q} \equiv \Delta^{(0)}_{p,q} = \frac{(\kappa p - q)^2 - (\kappa - 1)^2}{4\kappa} \ , \quad \kappa \equiv 1 + \frac{1}{m} \tag{7.1}$$

which are associated with a central charge $c \leq 1$

$$c = 1 - 6\frac{(\kappa - 1)^2}{\kappa}, \quad \kappa \geq 1 \tag{7.2}$$

are *linearized in presence of (critical) quantum gravity* into the new ones

$$\Delta_{p,q} = \frac{1}{2}(|\kappa p - q| - (\kappa - 1)), \tag{7.3}$$

satisfying the KPZ quadratic relation

$$\Delta - \Delta^{(0)} = \Delta(1 - \Delta)/\kappa \tag{7.4}$$

We have also used the Russian notation Δ, and have distinguished here the dimensions $\Delta^{(0)} \equiv h$ in the plane from those Δ dressed by gravity. Also, we use the notation

$$\kappa \equiv \frac{m + 1}{m} \geq 1 \tag{7.5}$$

in place of the usual conformal parameter m in (4.4). The quite striking idea that critical phenomena in the 2D plane and on a critical random surface are actually organically related led then immediately to a burst of exact solutions for other statistical models on a random lattice: self-avoiding walks[66], $O(n)$ model[66,67], $Q = 1$ Potts model and bond percolation[68].

All these solvable models, including the former Ising case[62], produced conformal di-mensions in presence of gravity which fully satisfy KPZ relation (7.4). Of particular interest is the polymer, or more generally the $O(n)$ model case, where the complete spectrum of the watermelon dimensions Δ_L can be obtained, and compared then to the dimension $\Delta^{(0)}_L = \frac{1}{2}x_L$ in the regular plane (5.7) (5.8) obtained by Coulomb gas techniques. We shall describe here more precisely how the geometrical $O(n)$ loop model or SAW model can be resolved in a fluctuating 2D-lattice, and the watermelon exponents determined. Actually, all sorts of *geometrical* questions, like the embedding of a topological network $\mathcal{G}$ in a random surface, are very well suited to the method[67]. We shall in particular describe the analog on a fluctuating surface of the scaling properties of polymer networks described in section 6. One can there prove basic relationships like (6.12).

7.2 O(n) model on a random lattice

As an illustration, and to parallel the study of section **5** in the regular plane, we shall start from the $O(n)$ model on a fluctuating lattice. Instead of considering the partition function (5.1) in the infinite regular lattice, one first has to define it on an arbitrary 3-coordinate planar graph G (i.e. with the topology of the sphere)

$$Z_{O(n)}(G) = \int \prod_{i \in G} \mathrm{d}\vec{S}_i \prod_{\langle j,k \rangle} \left(1 + K\, \vec{S}_j \cdot \vec{S}_k \right) \tag{7.6}$$

where i, j, k are arbitrary sites of lattice G and $\langle j, k \rangle$ denotes nearest neighbours as before. Now we have to take into account the fluctuations of the lattice by summing over all planar graphs G made of e.g. φ^3−type vertices, possibly with a symmetry factor $1/S(G)$. The total *annealed* partition function is therefore

$$Z_{O(n)}(\beta, K) = \sum_{G} \frac{1}{S(G)} e^{-\beta |G|} Z_{O(n)}(G) \tag{7.7}$$

where β is the "cosmological constant" which is conjugated to the "area" or "volume" or "mass" $|G|$ of the fluctuating lattice

$$|G| = \# \text{ vertices of } G \tag{7.8}$$

The system is thus defined in terms of two fugacities: K (the inverse temperature of the $O(n)$ model) and $e^{-\beta}$ (that of the lattice). The K−expansion of the $O(n)$ model is of course still given by a self-avoiding loop representation on each graph G

$$Z_{O(n)}(G) = \sum_{\substack{\text{nonintersecting} \\ \text{loops on } G}} n^{\mathcal{N}_L} K^{\mathcal{N}_B}, \tag{7.9}$$

exactly as in Eq.(5.2) (Fig.21). Notice that we consider a φ^3−lattice for definiteness, but it can be shown by explicit calculations (like on a φ^n−lattice) that the critical behaviour described below is universal. Notice also that here the only difference from the standard honeycomb case (Fig.16) is that the local constraints of closing little hexagons after six e.g. right turning bonds are relaxed on the fluctuating lattice (Fig.21). This suppression of constraints makes the problem solvable.

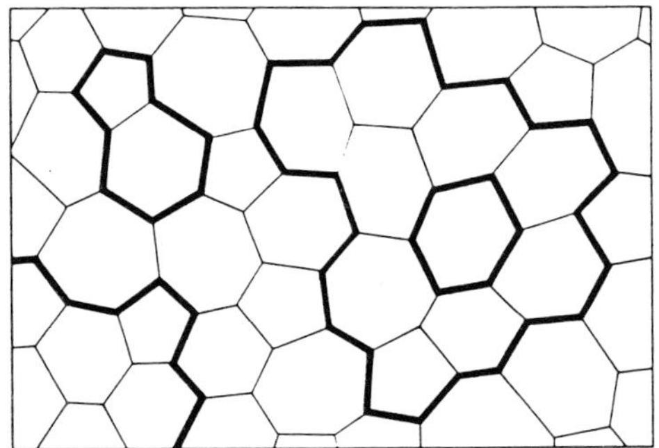

Fig.21 : Loop expansion (7.9) of the $O(n)$ model on the random φ^3—lattice.

When one considers only the empty loop configuration ($K = 0$), one gets the partition function of the random surface only (Fig.22)

$$Z(\beta) = \sum_G \frac{1}{S(G)} e^{-\beta|G|} \tag{7.10}$$

It exists a critical potential β_c such that the characteristic fluctuations of $|G|$ diverge, and where the partition function Z (5.10) develops a non analyticity

$$Z(\beta) \sim (\beta - \beta_c)^{2-\gamma_{\text{str}}} \quad , \beta > \beta_c, \tag{7.11}$$

γ_{str} being the string susceptibility exponent, and the random surface model being defined only for $\beta \geq \beta_c$. This string susceptibility γ_{str} depends on the topology of the random surface (see below).

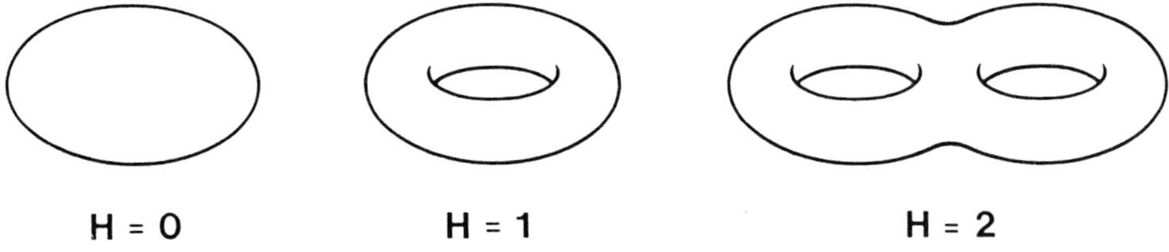

Fig.22 : Pure random surfaces of genus H (number of handles).

For the $O(n)$ model, the loop and graph expansion (7.7) (7.9) reads explicitly

$$Z_{O(n)}(\beta, K) = \sum_G \sum_{\text{loops on } G} n^{N_L} \left(e^{-\beta}\right)^{N_E} \left(e^{-\beta} K\right)^{N_B} \tag{7.12}$$

where $\mathcal{N}_L$ is the number of loops, $\mathcal{N}_B$ that of the occupied bonds on G, $\mathcal{N}_E$ that of the empty vertices of G. The two fugacities $g \equiv e^{-\beta}$, $z \equiv e^{-\beta}K$ thus control the complete phase behaviour of the system.

If, for a fixed and low enough value of z, g increases, one will enconter a critical point $g_c(z,n)$ where the fluctuations of the random surface diverge. This is the usual infinite random surface point, with the same universality class, and only the nonuniversal position of the critical point $g_c(z,n)$ depends on the presence of the $O(n)$ model at temperature K^{-1}. If one fixes on the contrary g (and small enough) then one shall meet by increasing K or z another critical phase $z_c(g)$ where the mean length of the $O(n)$ loops diverges but not the nonfilled part of the random lattice, still governed by a non critical g. This naturally corresponds to the dense phase of the $O(n)$ model (see part **5**) where the self-avoiding loops fill the (here fluctuating) lattice with a non vanishing density. Finally only when one approaches the intersection point of the two critical lines $g_c(z^*) = g^*$, $z_c(g^*) = z^*$, is a simultaneous double critical behaviour possible, when the lattice and loop volumes diverge independently. This will correspond to the standard dilute critical point of the $O(n)$ model, here on an infinite surface. The shape of the lines $g_c(z,n)$, $z_c(g,n)$ is in general complicated for arbitrary n. Notice that one expects a second order phase transition in the $O(n)$ model only for $-2 \leq n \leq 2$, even on the random surface[66]. We shall later concentrate on the polymer i.e. $n = 0$ model, in which case the phase diagram is simpler[67]. Indeed since for $n = 0$, there are no vacuum loops in the partition function (7.12) the critical surface line $g_c(z, n = 0)$ is a constant $g_c(z, n = 0) = g_c \ (= (12\sqrt{3})^{-1/2}$ in the case of a φ^3 lattice[59]), corresponding to the empty lattice critical point. The polymer critical line $z_c(g, n = 0)$ is determined below from an explicit solution.

7.3 Random matrices

Most statistical models on a fluctuating lattice can be represented in terms of models of random $N \times N$ matrices in the large N limit[59−63,66,68,69], the action of the matrix model depending on the statistical model considered, and on the type of connectivity of the lattice. We shall not develop this aspect here, which will not be used explicitly in the specific case of self-avoiding walks. For completeness in the general introduction of the $O(n)$ model (7.6) (7.7) (7.12), let us simply add that after a representation in terms of a large N matrix integral, evaluated at its saddle point, the partition function reads explicitly[66]

$$Z_{0(n)}(\beta, K) = \int_a^b d\lambda\, \rho(\lambda) \left(-\frac{1}{2}\lambda^2 + \frac{g}{3}\lambda^3 \right)$$

$$+ \int_a^b d\lambda\, \rho(\lambda) \int_a^b d\mu\, \rho(\mu) \left(\ell n|\lambda - \mu| - \frac{n}{2}\ell n\left(\frac{1}{z} - \lambda - \mu \right) \right) \quad (7.13)$$

where $\rho(\lambda)$ is the normalized eigenvalue density of the random matrix at the saddle point

$$\lambda - g\,\lambda^2 = 2\int_a^b \mathrm{d}\mu\,\rho(\mu)\left(\frac{P}{\lambda - \mu} + \frac{n}{2}\frac{1}{\frac{1}{z} - \lambda - \mu}\right) \tag{7.14}$$

$$\int_a^b \rho(\lambda)\mathrm{d}\lambda = 1$$

(P means principal value). Notice that solving ρ in Eq.(7.14) requires also the determination of the support $[a, b]$ outside of which ρ vanishes. The solution of (7.14) for general n is not an easy task[69] and is known only along the critical line $\frac{1}{z} = 2b$. From (7.13) (7.14) it is indeed possible to see the expected characteristics: the critical surface line $g_c(z, n)$ corresponds to the singularity of the end point b in terms of g : $\left.\frac{\partial g}{\partial b}\right|_z = 0$. This end-point b plays an important role since for $\lambda = \mu = b$ the nearest singularity of the integrand in (7.13) (7.14) is at $\frac{1}{z} = 2b(g, z)$, yielding $z_c(g, n)$. From (7.13) (7.14) one can show also that the equations make sense only for $n \in [-2, 2]$ as expected[66,69].

7.4 Self-avoiding walks on a 2D-random lattice

a) *Watermelon partition function.* We now consider specifically the $n = 0$ case and calculate the scaling dimensions Δ_L of the operator source if L self-avoiding lines on the random lattice. However the method we follow can be generalized identically to the $O(n)$ model. For this, we consider as usual the watermelon network, but this time on the fluctuating lattice (Fig.23).

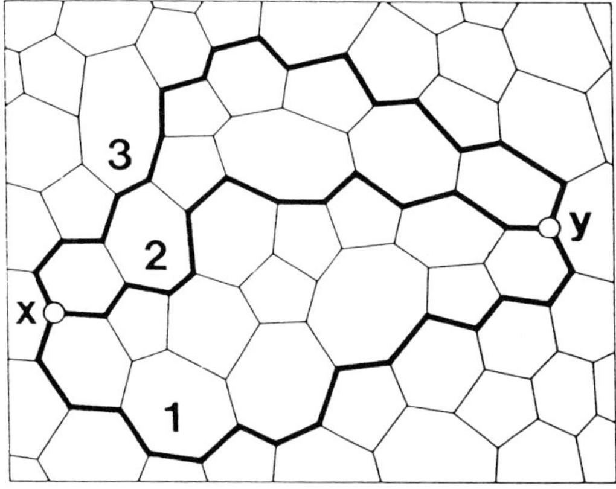

Fig.23 : Watermelon network on the random φ^3-lattice.

The total partition function of the lattice and of the L SAW is defined as

$$Z_L(\beta, K) = \sum_G \frac{1}{S(G)}\mathrm{e}^{-\beta|G|}\sum_{\mathbf{x},\mathbf{y}}\sum_{\Gamma_{\mathbf{x},\mathbf{y}}^{(\ell)}} K^{|\Gamma|} \tag{7.15}$$

$$\ell = 1, ..., L$$

where the sum extends on all graphs G with the topology of the sphere, made of φ^3 vertices, and over all polymer sets of lines $\Gamma_{\mathbf{x},\mathbf{y}}^{(\ell)}$, $\ell = 1, ..., L$, joining any two points $\mathbf{x}, \mathbf{y}$ on G. In the absence of SAW ($K = 0$) the random surface itself has a critical point β_c where the typical size $|G|$ diverges, with a partition function (7.10) behaving as in (7.11), where the string susceptibility exponent γ_{st} has been calculated exactly[56,57] for embedding space dimensions $d = -2$, $\gamma_{\mathrm{st}} = -1$, and $d = 0$, $\gamma_{\mathrm{st}} = -1/2$. Note that here the embedding space dimension is formally $d = 0$, but will be replaced by the central charges $c = -2$ or $c = 0$.

The key point is that the partition function Z_L can be calculated exactly. For this, one fixes first the SAW lengths and sums over the L lattice connected pieces between successive polymer lines ℓ and $\ell + 1$ (Fig.24).

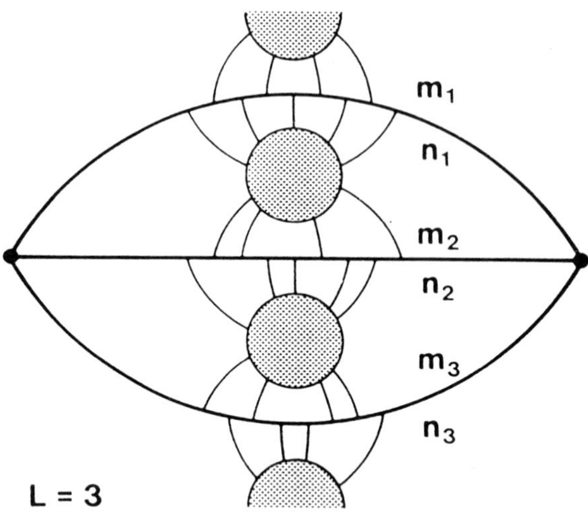

Fig.24 : Illustration of the convolutive sum (7.17) on the spherical lattice. Each bubble $\ell = 1, 2, 3$ corresponds to a connected piece of lattice with a Green function $G_{n_\ell + m_{\ell+1}}$ (7.16). The bold lines are the SAW.

Each of these planar pieces contributes a factor $G_{n_\ell + m_{\ell+1}}$ to the partition function, where G_n is the Green function of random graphs with n external legs

$$G_n(\beta) = \sum_{n\text{-leg planar } G} e^{-\beta|G|} \tag{7.16}$$

and where n_ℓ and $m_{\ell+1}$ are the number of legs attached to the ℓ and $\ell+1$ SAW respectively. As a result, Z_L is the convolutive sum

$$Z_L(\beta, K) = \sum_{m_\ell, n_\ell} (e^{-\beta} K)^{\sum_\ell m_\ell + n_\ell} \prod_{\ell=1}^{L} \binom{m_\ell + n_\ell}{m_\ell} G_{n_\ell + m_{\ell+1}}(\beta) \tag{7.17}$$

with the cyclic convention $L+1 \equiv 1$. (The combinatorial factors are the numbers of relative positions of m_ℓ and n_ℓ lattice legs branched on each side of the ℓth SAW). Now the second key point is that the planar random lattice Green function, G_n, can be calculated from the thermodynamic limit of matrix integrals. It is simply the nth moment of the eigenvalue density (7.14)[59]

$$G_n = \int_a^b d\lambda \, \lambda^n \rho(\lambda) \tag{7.18}$$

Notice that all the calculations so far are valid for any loop model when the connected pieces of lattice between the lines of the watermelon bear vacuum loops. The density ρ in (7.18) is then simply the solution of (7.14). Now a very simple algebra allows to resum (7.17) once (7.18) is inserted. One gets the integral representation of Z_L

$$Z_L(\beta, K) = \int_a^b \cdots \int_a^b \prod_{\ell=1}^{L} \rho(\lambda_\ell) \, d\lambda_\ell \prod_{\ell=1}^{L} [1 - z(\lambda_\ell + \lambda_{\ell+1})]^{-1} \tag{7.19}$$

where $z \equiv e^{-\beta} K$ is the effective polymer fugacity. The integrals are defined for $z < (2b)^{-1}$, $z_c(\beta) \equiv e^{-\beta} K_c(\beta) = [2b(\beta)]^{-1}$ giving the critical point where the SAW lengths diverge. Hence only the vicinity of the upper boundary $\lambda = b$ will matter for the critical behaviour of (7.19). In the case of polymers ($n = 0$) the solution of (7.14) has been calculated ten years ago for counting planar diagrams[59]

$$\rho(\lambda) = \frac{1}{2\pi} \left[1 - e^{-\beta} \left(\frac{a+b}{2} + \lambda \right) \right] \sqrt{(\lambda - a)(b - \lambda)} \tag{7.20}$$

and the endpoints a, b of the support of ρ are functions of β through

$$\begin{aligned}
2 \, e^{-2\beta} &= \sigma(1 - \sigma)(1 - 2\sigma) \\
a + b &= 2\sigma \, e^{\beta} \\
b - a &= 4(1 - 2\sigma)^{-1/2}
\end{aligned} \tag{7.20a}$$

The critical point of the empty surface corresponds to the first singularity of the parametric representation where $\frac{\partial g}{\partial b} = 0$, which is found to be $e^{-\beta_c} = \left(12\sqrt{3}\right)^{-1/2}$. So from (7.19) one already observes the two kinds of possible critical behaviours: one (that of the surface) is hidden in the singularities of $\rho(\lambda)$ (7.20) near β_c; the second one, associated with the SAW, occurs when z approaches its critical upper value $z_c(\beta) = 1/2b(\beta)$, the dominant contribution to the multiple λ−integral in (7.19) coming from all $\lambda_\ell \longrightarrow b^-$, $\ell = 1, ..., L$. Hence the only features we will need are the two distinct critical behaviors of ρ near the upper end b of interval $[a, b]$ for $\beta > \beta_c$ or $\beta = \beta_c$

$$\rho(\lambda) \sim (\beta - \beta_c)^{1/2} (b - \lambda)^{1/2} + \text{const.}(b - \lambda)^{3/2} \tag{7.21}$$

Notice the difference in the two critical exponents $1/2$ or $3/2$.

b) *Critical behaviours.* We shall not give here all details of the analysis of (7.19) (7.21), but it is easy to understand that there will be two different critical behaviours corresponding to (7.21). First, if $\beta > \beta_c$ the part of the surface unoccupied by the SAW remains finite, while one can reach the critical point $K_c(\beta)$ where the SAW become *infinite*. This corresponds to a *dense* polymer phase. Second, if one puts $\beta = \beta_c$ and $K = K_c(\beta_c)$, both the surface not occupied by the polymers and the SAW are infinite. Using the edge behaviour (7.21) in (7.19) and a standard power counting argument one finds the critical behaviour

$$Z_L(\beta, K) \sim \left[(K_c - K)^{1/2}(\beta - \beta_c)^{1/2} + \mathrm{const}\,(K_c - K)^{3/2}\right]^L \tag{7.22}$$

For $\beta > \beta_c$ the first term dominates the *dense* SAW phase. Hence

$$Z_L(\beta, K) \sim (K_c - K)^{L/2}, \quad K \longrightarrow K_c(\beta) \tag{7.23}$$

When the surface is first infinite, $\beta = \beta_c$ and in this dilute polymer phase

$$Z_L(\beta_c, K) \sim |K_c - K|^{3L/2}, \quad K \longrightarrow K_c(\beta_c) \tag{7.24}$$

Note that the two different exponents 1/2, 3/2 in (7.23) and (7.24) correspond directly to those governing the eigenvalue density $\rho(\lambda)$ in Eq.(7.21). Another way to reach the dilute polymer phase is to let the non-filled surface and polymers become infinite simultaneously in the *finite size scaling regime* $\beta - \beta_c \sim (K_c - K)^2$, such that both terms of (7.22) are equivalent. Then the characteristic surface and SAW sizes $|G| \sim (\beta - \beta_c)^{-1}$ and $|\Gamma| \sim |K_c - K|^{-1}$ respectively, are related by

$$\beta - \beta_c \sim |K_c - K|^{\nu D}, \quad \nu D = 2 \tag{7.25}$$

where ν is the polymer size exponent and D the fractal dimension of the surface. This shows that the respective surface size $|G|$ and polymer size $|\Gamma|$ scale like $|\Gamma| \sim |G|^{1/2}$, thus leading to a dilute polymer phase with vanishing density $|\Gamma|/|G| \sim |G|^{-1/2} \longrightarrow 0$. Notice that for the other dense phase one has $|G| \sim |\Gamma|$, hence

$$\nu D = 1 \tag{7.26}$$

7.5 Conformal invariance on a random surface

It remains to identify the conformal dimensions underlying the two exact behaviours (7.23) (7.24) of Z_L in the dense and dilute critical phases. The watermelon partition function Z_L can be viewed as that of two conformal operators $\phi_L(\mathbf{x})$, $\phi_L(\mathbf{y})$ associated with the $L-$leg polymer vertex, as in the plane, integrated over the points $\mathbf{x}$ and $\mathbf{y}$ of the random lattice. Hence one expects it to scale with the cosmological constant β like

$$Z_L(\beta, K) \sim (\beta - \beta_c)^{2\Delta_L - \gamma_{\mathrm{st}}} \tag{7.27}$$

where Δ_L is the conformal weight of ϕ_L on the random surface, the insertion of the two points on the random surface giving a further factor $Z''(\beta) \sim (\beta - \beta_c)^{-\gamma_{st}}$ in (7.27). Eq.(7.27) holds in the FSS regimes (7.25) or (7.26). Hence one can also rewrite it as

$$Z_L \sim |K_c - K|^{\nu D(2\Delta_L - \gamma_{st})} \tag{7.28}$$

where $\nu D = 1\,(\beta > \beta_c)$ or $\nu D = 2\,(\beta = \beta_c)$. To identifying the values of Δ_L requires the further information[66] $\nu D = (1 - \Delta_2)^{-1}$ which is nothing but hyperscaling. One thus finds for the dimensions and string susceptibilities

$$\Delta_L^D = (L - 2)/4, \quad \gamma_{st} = -1, \quad c = -2 \text{(dense phase)}$$
$$\Delta_L = (3L - 2)/8, \quad \gamma_{st} = -1/2, \quad c = 0 \quad \text{(dilute)} \tag{7.29}$$

Notice that γ_{st} itself is deterined here from the exact solution. We have recalled the associated central charges and the values of γ_{st} are also those of free random surfaces embedded in $d = -2$ or $d = 0$ dimensions since the latter have the same central charge $c = d = \#$ components of a Gaussian field.

Let us now compare these results to Knizhnik, Polyakov and Zamolodchikov's relationship (7.4) between a critical system in the plane and its counterpart on a random $2D$ lattice (i.e dressed by quantum gravity). The conformal dimensions $\Delta^{(0)}$ in the plane and Δ on the random surface of a conformal operator satisfy the simple quadratic relation (7.4) $\Delta - \Delta^{(0)} = \Delta(1 - \Delta)/\kappa$, where κ parametrizes the central charge (7.2) $c = 13 - 6\kappa - 6/\kappa$. Note that $\Delta^{(0)}$ here is KPZ notation for the Kac spectrum $h = x/2$ as in (5.8). Let us collect the conformal dimensions found for an L−leg vertex in the *plane* from Coulomb gas technique (Eq.(5.10)) $\Delta_L^{(0)D} = x_L^D/2 = \left(L^2 - 4\right)/32$ (dense), and $\Delta_L^{(0)} = x_L/2 = \left(9L^2 - 4\right)/96$ (dilute). It is easy to see that in the dense phase, for all L, $\Delta_L^{(0)D}$ and Δ_L^D satisfy KPZ relation for $c = -2$, $\kappa = 2$, while in the dilute phase $\Delta_L^{(0)}$ and Δ_L satisfy it for $c = 0$, $\kappa = 3/2$. In terms of the *linear* KPZ conformal dimensions (7.3), one finds explicitly for (7.29)

$$\Delta_L^D \equiv \Delta_{0,\frac{L}{2}}(\kappa = 2)$$
$$\Delta_L \equiv \Delta_{\frac{L}{2},0}(\kappa = \frac{3}{2}) \tag{7.30}$$

i.e. the same indices as in the plane (Eqs.(5.8) (5.10)), QED.

Notice that this provides a full series $(L \geq 1)$ of cross checks of the validity of Coulomb gas methods in the plane, of KPZ conformal theory on fluctuating continuum surfaces, and also of the equivalence of critical random lattices to the continuum Polyakov's summation over metrics in 2D quantum gravity. Note also that relation (7.4) can be derived in the conformal gauge of the Liouville theory of random surfaces[70,71].

5.6 Conformal spectrum in 2D quantum gravity

For the $O(n)$ model in the Euclidean plane, we know the values of the watermelon scaling dimensions (5.8)

$$x_L = g\,L^2/8 - (1 - g)^2/2g$$

$$n = -2 \cos \pi g$$

$$x_L = 2\Delta^{(0)}_{L/2,0} \equiv 2h_{L/2,0}, \quad g = \frac{m+1}{m}, \quad g \in [1,2] \ \ (\text{dilute})$$

$$= 2\Delta^{(0)}_{0,L/2} \equiv 2h_{0,L/2}, \quad g = \frac{m}{m+1}, \quad g \in [0,1] \ \ (\text{dense})$$

Comparing this to Eq.(7.5), we find that in the dilute and dense phases respectively

$$\begin{array}{ccc} \kappa = g & \text{for} & g \in [1,2] \\ \kappa = 1/g & \text{for} & g \in [0,1] \end{array}$$

and we expect the conformal dimensions (7.3) dressed by gravity predicted by KPZ theory

$$\begin{aligned} \Delta_L \equiv \Delta_{L/2,0} &= g\,L/4 - (g-1)/2 \\ \Delta_L^D \equiv \Delta_{0,L/2} &= \frac{L}{4} - \left(\frac{1}{g}-1\right)\frac{1}{2} \end{aligned} \tag{7.31}$$

These values can be obtained directly from the solution of the $O(n)$ model (7.12) (7.13) on the random 2D-lattice[66,69], by a technique entirely similar to that presented in section (**7.4**) for polymers, providing a check of both KPZ and Coulomb gas approaches in the case of all $O(n)$ models for $n \in [-2,2]$.

As an application, we can calculate the fractal dimension D_F of the loops in the high-temperature expansion of the $O(n)$ model. The critical exponent $\nu_2 D$ (where, as above, D is the fractal dimension of the random surface and where $\nu_2 = 1/D'_F$ is the inverse fractal dimension of the loops) is given by the two-leg conformal operator[67]

$$\nu_2 D = \frac{1}{1-\Delta_2} \tag{7.32}$$

From Eq.(7.31) we find in the dilute and dense phase respectively

$$\begin{aligned} \nu_2 D &= 2 & (\text{dilute } O(n)) \\ \nu_2 D &= 2g & (\text{dense } O(n)) \quad g \in \left[\frac{1}{2},1\right], \quad n \geq 0 \end{aligned} \tag{7.33}$$

in agreement with the values found for the polymer phases (7.25) (7.26). Hence we find for the fractal dimension $D'_F = 1/\nu_2$ in presence of a fluctuating metric

$$\frac{D'_F}{D} = \frac{1}{2} \ \ (\text{dilute}) \ , \quad \frac{D'_F}{D} = 1/2g \ \ (\text{dense}) \tag{7.34}$$

This is to be contrasted to the result in the Euclidean plane $(D=2)$ $\frac{1}{2}D_F = 1 - \frac{1}{2}x_2 = \frac{1}{2} + \frac{1}{4g}$, valid in both phases.

5.7 Higher topologies

a) *Polymer networks*. We consider a general polymer network $\mathcal{G}$ of arbitrary but fixed topology, embedded in the random surface (of e.g. genus zero) (Fig.25).

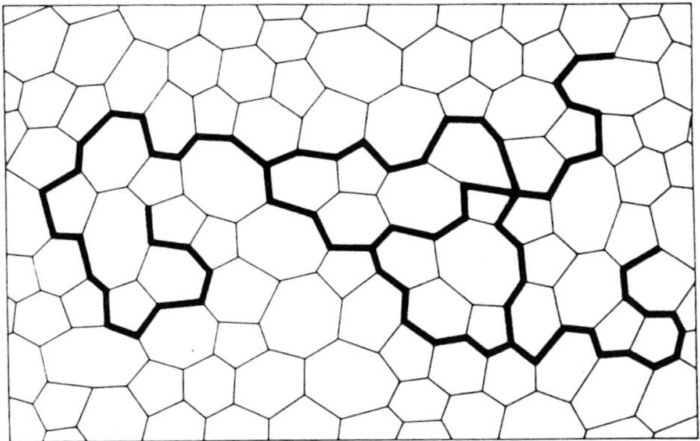

Fig.25 : A polymer network embedded in the planar random φ^3–lattice.

In section **6**, we have shown how general SAW networks can be handled in the plane, and in particular their configuration exponents $\gamma_{\mathcal{G}}$ calculated as functions of the set of scaling dimensions $\{x_L\}$ (Eqs.(6.12) (6.14)). On a random critical surface, a similar decomposition in terms of vertices exists. It can be shown by an exact solution[67] that on a random surface

$$\mathcal{Z}_{\mathcal{G}}^{\text{poly}} \sim |K_c - K|^{-\gamma_{\mathcal{G}} - \nu D} \tag{7.35}$$

with a critical exponent

$$\gamma_{\mathcal{G}}^{\text{random}} = \nu D \left[\mathcal{V} - 1 - (2 - \gamma_{\text{st}}) - \sum_{L \geq 1} n_L \Delta_L \right] \tag{7.36}$$

where, as in part **6**, $\mathcal{V}$ is the total number of vertices of $\mathcal{G}$, and n_L that of type L. Formula (7.38) for a network on a random surface is then very similar to that obtained in the Euclidean plane. Eqs.(7.38) and (6.12) are indeed identical, provided one replaces $2\nu(\equiv d\nu)$ by νD, $x_L \equiv 2\Delta_L^{(0)}$ by $D \, \Delta_L$, and adds a contribution $2 - \gamma_{\text{st}}$ coming from the free energy of the random surface itself (Eqs.(7.11)). This is of particular interest since the exact solution on a random surface provides an (indirect) check of the scaling theory (6.14) in the plane.

b) *Higher surface genus.* Formula (7.38) is also valid for any topology $\mathcal{T}$ of the fluctuating membrane itself. $\gamma_{\mathcal{G}}^{\text{random}}$ depends on $\mathcal{T}$ only through γ_{st}, since the conformal dimensions Δ_L are local and independent of the global topology. Eq.(7.38) is then interesting since it allows the determination of $\gamma_{\text{st}}(\mathcal{T})$, by a direct calculation of $\gamma_{\mathcal{G}}$ and then identification with (7.38). For example, if one considers a double torus (Fig.26), one can draw a "lasso" polymer network on it, whose only vertex conformal operator is ϕ_8 with known dimension Δ_8. The exponent $\gamma_{\mathcal{G}}$ can be computed by the generalization of technique (7.19). One finds

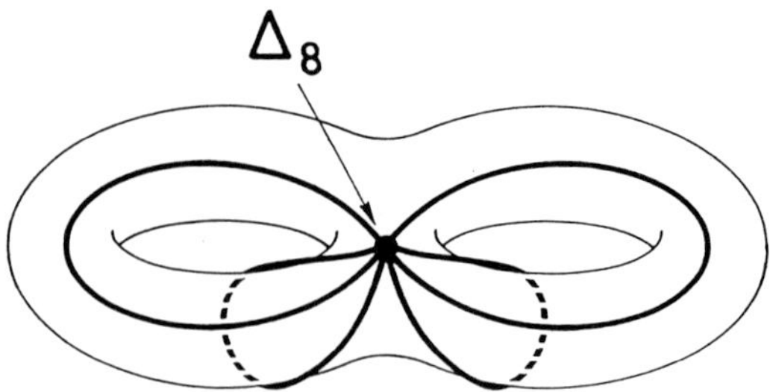

Fig.26 : A minimal polymer "lassoo" on the double torus. The extraction of the conformal dimension Δ_8 yields the value of γ_{st} for the double torus.

then for any topological genus of Euler characteristics $\chi = 2(1 - H)$ (H handle number or genus, Fig.22) the string susceptibility exponent

$$2 - \gamma_{\text{st}} = \frac{3}{2}\chi, \qquad d = c = -2$$

$$2 - \gamma_{\text{st}} = \frac{5}{4}\chi, \qquad d = c = 0$$

in agreement with a former exact solution for $d = 0, -2$[61]. All of this can be generalized to e.g. the $O(n)$ model with central charge $c = 1 - 6(1 - g)^2/g = 13 - 6\kappa - 6/\kappa$ and one finds[67]

$$2 - \gamma_{\text{st}} = \frac{1}{2}\chi(\kappa + 1) \tag{7.37}$$

where $\kappa = 1/g$, $g \in [0, 1]$ in the dense phase, and $\kappa = g$, $g \in [1, 2]$ at the critical point. Eq.(7.39) can be rewritten as

$$2 - \gamma_{\text{st}} = \chi \frac{1}{24}\left[25 - c + \sqrt{(1 - c)(25 - c)}\right] \tag{7.38}$$

For the sphere topology ($\chi = 2$) one recovers the result conjectured by KPZ[65]

$$\gamma_{\text{st}}(\text{sphere}) = \frac{1}{12}\left[c - 1 - \sqrt{(1 - c)(25 - c)}\right]$$

while the general result (7.40) for arbitrary genus was obtained in the conformal gauge in the continuum Liouville theory of random surfaces[70] (see also Ref.[71]). Notice that the geometrical methods presented here allow for a direct calculation of (7.38) (7.39) for any genus[67].

Acknowledgements

It is a pleasure to thank F. David, J. des Cloizeaux, I. Kostov and H. Saleur with whom different aspects presented here were developed.

REFERENCES

1. For a nice historical discussion, see e.g. E. Nelson "Dynamical theories of Brownian motion" (Princeton University Press, Princeton, 1972).

2. A. Einstein, "Investigations on the theory of the Brownian motion", Dover, 1956; groups five articles published between 1905 and 1908. See also A. Pais, "Subtle is the Lord..." (Oxford, New York, 1982).

3. L. Bachelier, "Théorie de la spéculation" Ann. Sci. Ecole Normale Sup. Sér. 3, $\underline{17}$, (1900) 21.

4. N. Wiener, Journal Math. and Physics Mass. Inst. Technology $\underline{2}$ (1923) 131.

5. B.B. Mandelbrot, "The fractal geometry of nature" (Freeman, New York, 1983).

6. H.E. Stanley, "Introduction to phase transitions and critical phenomena" (Oxford, London, 1971).

7. P.G. de Gennes, "Scaling concepts in polymer physics" (Cornell University Press, Ithaca, London, 1979); J. des Cloizeaux and G. Jannink, "Les polymères en solution, leur modélisation et leur structure" (Les Editions de Physique, Les Ulis, 1987). For original studies, see P.G. de Gennes, Phys. Lett. $\underline{38A}$ (1972) 339; J. des Cloizeaux, J. Physique $\underline{36}$ (1975) 281.

8. "Phase transitions and critical phenomena", Vol.6, eds. C. Domb and M.S. Green (Academic, London, 1976).

9. B. Nienhuis, Phys. Rev. Lett. $\underline{49}$ (1982) 1062; J. Stat. Phys. $\underline{34}$ (1984) 731.

10. J.M. Hammersley, Quat. J. Math. Oxford $\underline{12}$ (1961) 250.

11. For the introduction of θ exponents, see J. des Cloizeaux, J. Physique $\underline{41}$ (1980) 223, for the 2D case, see B. Duplantier, Phys. Rev. $\underline{B35}$ (1987) 5290.

12. M. Fetters, Exxon center for research, Annandale, USA.

13. B. Duplantier, Phys. Rev. Lett. $\underline{57}$ (1986) 941.

14. A. Dvoretzky, P. Erdös and S. Kakutani, Acta Sci. Math. (Szeged) $\underline{12}$ (1950) 64.

15. S.J. Taylor, Z. Wahr. verw. Gebiete $\underline{5}$ (1966) 247.

16. B. Duplantier, Phys. Rev. $\underline{A38}$ (1988) 3647 (see the appendix).

17. See for instance S.R.S. Varadhan, "Large deviations and applications" Society for Industrial and Applied Mathematics (1984).

18. S.F. Edwards, Proc. Phys. Soc. $\underline{85}$ (1965) 613.

19. B. Duplantier in "Statistical Mechanics of Membranes and Surfaces", Proceedings of the 5th Jerusalem Winter School, eds. D.R. Nelson, T. Piran and S. Weinberg (World Scientific, Singapore, 1989); for original works see M. Kardar and D.R. Nelson, Phys. Rev. $\underline{A38}$ (1988) 966; J. Aronowitz and T.C. Lubensky, Europhys. Lett. $\underline{4}$ (1987) 395.

20. S.R.S. Varadhan, appendix to "Euclidean Quantum Field Theory" by K.Z. Symanzik in "Local Quantum Theory", ed. R. Jost (Academic, New York, 1969).

21. J. Westwater, Comm. Math. Phys. $\underline{72}$ (1980) 131; $\underline{79}$ (1981) 53; $\underline{84}$ (1982) 459.

 B. Duplantier

22. J. des Cloizeaux, J. Physique $\underline{42}$ (1981) 635.

23. B. Duplantier, J. Chem. Phys. $\underline{86}$ (1987) 4233.

24. J.F. Le Gall, Comm. Math. Phys. $\underline{104}$ (1986) 471; 505.

25. B.H. Zimm, W.H. Stockmayer and M. Fixman, J. Chem. Phys. $\underline{21}$ (1953) 1716.

26. M. Benhamou and G. Mahoux, J. Physique $\underline{47}$ (1986) 559; B. Duplantie,r ibid 569.

27. B. Duplantier, Phys. Rev. Lett. $\underline{62}$ (1989) 2337.
    B. Duplantier, T. Hwa and M. Kardar, to be published.

28. J.C. Le Guillou and J. Zinn-Justin, J. Physique Lett. $\underline{46}$ (1985) L137.

29. B. Duplantier, J. Stat. Phys. $\underline{54}$ (1989) 581.

30. See also K. Ohno and K. Binder, J. Physique $\underline{49}$ (1988) 1329.

31. H.J.F. Knops, "Conformal Invariance", this volume.

32. B. Nienhuis, "Statistical Mechanics of Polymers" this volume.

33. P. Lévy, "Processus stochastiques et mouvement brownien" (Gauthier-Villars, Paris, 1965); see also B. Davies, Ann. Probab. $\underline{7}$ (1979) 913.

34. A.A. Belavin, A.M. Polyakov and A.B. Zamolodchikov, Nucl. Phys. $\underline{B241}$ (1984) 333.

35. See J.L. Cardy in "Phase Transitions and Critical Phenomena", Vol.11, eds. C. Domb and J. Lebowitz (Academic, London, 1987).

36. D. Friedan, Z. Qiu and S. Shenker, Phys. Rev. Lett. $\underline{52}$ (1984) 1575.

37. V.G. Kac, Lecture Notes in Physics $\underline{94}$ (Springer, Berlin, 1979) 441.

38. G.E. Andrews, R.J. Baxter and P.J. Forrester, J. Stat. Phys. $\underline{35}$ (1984) 193; D.A. Huse, Phys. Rev. $\underline{B30}$ (1984) 3908.

39. H. Saleur, J. Phys. $\underline{A20}$ (1987) 455, $\underline{A19}$ (1986) L807.

40. B. Nienhuis in "Phase Transitions and Critical Phenomena", Vol.11, eds. C. Domb and J.L. Lebowitz (Academic, London, 1987). For earlier works and a clear explanation of Coulomb gas technique, see also M. den Nijs, Phys. Rev. $\underline{B27}$ (1983) 1674, and references therein.

41. B. Duplantier, J. Stat. Phys. $\underline{49}$ (1987) 411.

42. M.T. Batchelor and H.W.J. Blöte, Phys. Rev. Lett. $\underline{61}$ (1988) 138.

43. H.W.J. Blöte, J. Cardy and M.P. Nightingale, Phys. Rev. Lett. $\underline{56}$ (1986) 742; I. Affleck, ibid. 746.

44. B. Duplantier and H. Saleur, Nucl. Phys. $\underline{B290}$ [FS20] (1987) 291.

45. B. Duplantier and F. David, J. Stat. Phys. $\underline{51}$ (1988) 327.

46. E. Domany, D. Mukamel, B. Nienhuis and A. Schwimmer, Nucl. Phys. $\underline{B190}$ [FS3] (1981) 279.

47. Vl.S. Dotsenko and V.A. Fateev, Nucl. Phys. $\underline{B240}$ [FS12] (1984) 312.

48. J. Cardy and H.W. Hamber, Phys. Rev. Lett. $\underline{45}$ (1980) 499.

49. B. Duplantier and H. Saleur, Phys. Rev. Lett. $\underline{60}$ (1988) 2343.

50. M.E. Fisher, V. Privman, S. Redner, J. Phys. $\underline{A17}$ (1984) L569.

51. F. Spitzer, Trans. Am. Math. Soc. $\underline{87}$ (1958) 187.

52. C. Bélisle, University of Michigan, preprint (1988).

53. J.W. Pitman and M. Yor, Ann. Probab. $\underline{14}$ (1986) 733.

54. M.K. Wilkinson, D.S. Gaunt, J.E.G. Lipson and S.G. Whittington, J. Phys. $\underline{A19}$ (1986) 789; D.S. Gaunt, J.E.G. Lipson, S.G. Whittington and M.K. Wilkinson, J. Phys. $\underline{A19}$ (1986) L811.

55. A.M. Polyakov, Phys. Lett. $\underline{103B}$ (1981) 207.

56. F. David, Nucl. Phys. $\underline{B257}$ (1985) 45; ibidem 543.

57. V.A. Kazakov, I.K. Kostov and A.A. Migdal, Phys. Lett. $\underline{157B}$ (1985) 295; D. Boulatov, V. Kazakov, I. Kostov and A. Migdal, Nucl. Phys. $\underline{B275}$ (1986) 641; Phys. Lett. $\underline{B174}$ (1986) 87.

58. J. Fröhlich, in Lecture Notes in Physics, Vol.216, L. Garrido Ed. (Springer, Berlin, 1985); J. Ambjorn, B. Durhuus and J. Fröhlich, Nucl. Phys. $\underline{B257}$ (1985) 433.

59. E. Brézin, C. Itzykson, G. Parisi, J.B. Zuber, Comm. Math. Phys. $\underline{59}$ (1978) 35.

60. I. Kostov, Nucl. Phys. B (Proc. Suppl.) $\underline{4}$ (1988) 98.

61. F. David, Phys. Lett. $\underline{B159}$ (1985) 303; I. Kostov and M. Mehta, Phys. Lett. $\underline{189B}$ (1987) 118.

62. V.A. Kazakov, Phys. Lett. $\underline{119A}$ (1986) 140; D.V. Boulatov and V.A. Kazakov, Phys. Lett. $\underline{186B}$ (1987) 379.

63. V.A. Kazakov, Nucl. Phys. B (Proc. Suppl.) $\underline{4}$ (1988) 93.

64. A.M. Polyakov, Mod. Phys. Lett. $\underline{A2}$ (1987) 899.

65. V.G. Knizhnik, A.M. Polyakov and A.B. Zamolodchikov, Mod. Phys. Lett. $\underline{A3}$ (1988) 819.

66. B. Duplantier and I. Kostov, Phys. Rev. Lett. $\underline{61}$ (1988) 1433; I. Kostov, Modern Phys. Lett. $\underline{A4}$ (1989) 217.

67. B. Duplantier and I. Kostov, Nucl. Phys. B[FS] (to appear).

68. V.A. Kazakov, Niels Bohr Institute preprint (1988).

69. M. Gaudin and I. Kostov, Phys. Lett. $\underline{B220}$ (1989) 200.

70. F. David, C.R. Acad. Sci. (Paris) $\underline{307}$ (1988) 1051, Mod. Phys. Lett. $\underline{A3}$ (1988) 1651.

71. J. Distler and H. Kawai, Nucl. Phys. $\underline{B321}$ (1989) 509.

FUNDAMENTAL PROBLEMS IN STATISTICAL MECHANICS VII
H. van Beijeren, Editor
© *Elsevier Science Publishers B.V., 1990*

HYDRODYNAMICS OF SUSPENSIONS

B. Ubbo FELDERHOF

Institut für Theoretische Physik A, R.W.T.H. Aachen, Templergraben 55, 5100 Aachen, F.R. Germany

1. INTRODUCTION

In these lectures we deal with the application of the methods of statistical mechanics to the problem of hydrodynamic interactions in colloidal suspensions. In the early days of statistical mechanics, at the time of Gibbs and Ehrenfest, one considered many-particle systems with dynamics described by Hamilton's equations of motion. With the advent of quantummechanics the theory was extended to account for dynamics described by Schrödinger's equation. In colloidal suspensions one deals with many-particle systems which in principle are described by microscopic equations on the molecular scale. However, in practice such a description is far too detailed, and involves information which is not relevant on the macroscopic length and timescale where measurements are done. A colloidal suspension, such as milk or blood, consists of large particles, composed of many molecules, immersed in a fluid of small molecules, usually water. The large particles may be fat globules, biological cells, liquid droplets, small polystyrene spheres, etc. The collection of large particles is the system of interest. For a description of its dynamics it suffices to regard the background fluid of small molecules as a continuum. Thus we are faced with a problem typical of mesoscopic systems. We must study a many-particle system in interaction with a continuum described by phenomenological equations, in our case the Navier-Stokes equations for the background fluid. Our task is to relate this mesoscopic description to measurements on the macroscopic length and timescale. In dealing with this task we use the methods of statistical physics.

The problem of hydrodynamic interactions in suspensions is of particular interest. Not only does it occur in many systems of practical importance, it also gives rise to peculiar theoretical problems due to both the long range and the many-body character of the hydrodynamic interactions. It was first noted by Smoluchowski[1] in 1912 that the long range causes divergence difficulties in the cal-

culation of the sedimentation velocity of a suspension influenced
by gravity. These difficulites have been resolved only recently [2-6]
The problem is akin to that arising in Maxwell's theory of dielectrics and must be resolved in two steps. First one must derive average equations valid on the macroscopic scale. Secondly one must
formulate constitutive equations relating the average fields and
calculate the transport coefficients occurring in these equations.
The average equations take care of the sample shape dependence on
the macroscopic scale. The transport coefficients depend only on
the local properties of the suspension and may be calculated from
absolutely convergent integrals derived on the mesoscopic scale.
The transport coefficients we shall encounter are the effective viscosity of the suspension, the collective mobility, and the self-
mobility.

The scheme sketched above and elaborated below is of wider theoretical interest. Similar problems occur in other linear transport
problems in disordered many-particle systems [7,8] and may be handled by analogous methods.

2. MESOSCOPIC EQUATIONS

We consider a collection of N identical spherical particles of
radius a immersed in an incompressible fluid of shear viscosity η.
The fluid is imagined to fill all space. A typical particle configuration is specified by the positions of the centers $(\vec{R}_1,...,\vec{R}_N)$.
We shall call this detailed level of description mesoscopic, in
accordance with the discussion in the introduction. In most cases
of practical interest it suffices to employ the equations of cree-
ping flow for the velocity $\vec{v}(\vec{r})$ and pressure $p(\vec{r})$ of the background
fluid. By definition these are the linear, time-independent Navier-
Stokes equations for an incompressible fluid [9]

$$\eta\nabla^2\vec{v} - \nabla p = 0 \quad , \qquad\qquad \nabla \cdot \vec{v} = 0 \quad . \qquad\qquad (2.1)$$

The pressure is a subsidiary quantity determined up to a constant
by the condition of incompressibility. We shall assume that the
fluid velocity satisfies stick boundary conditions at the surface
of the particles. This assumption may be relaxed, but is valid in
many cases of practical interest.

We consider the situation where external forces $(\vec{E}_1,....,\vec{E}_N)$ and
external torques $(\vec{N}_1,...,\vec{N}_N)$ are applied to the particle centers.

For each instantaneous configuration $(\vec{R}_1,\ldots,\vec{R}_N)$ the particles then have corresponding translational velocities $(\vec{U}_1,\ldots,\vec{U}_N)$ and rotational velocities $(\vec{\Omega}_1,\ldots,\vec{\Omega}_N)$, and they generate a velocity field $\vec{v}(\vec{r})$ and a pressure field $p(\vec{r})$ in the surrounding fluid. The velocity of particle j is given by the rigid body motion

$$\vec{w}_j(\vec{r}) = \vec{U}_j + \vec{\Omega}_j \times (\vec{r}-\vec{R}_j) \quad , \qquad |\vec{r}-\vec{R}_j| < a. \qquad (2.2)$$

The stick boundary conditions imply $\vec{v}(\vec{r}) = \vec{w}_j(\vec{r})$ at $|\vec{r}-\vec{R}_j| = a$ for all j.

Consider first a single particle centered at the origin and subjected to the applied force $\vec{E}$. It acquires a translational velocity $\vec{U}^{St}$ given by the Stokes relation

$$\vec{E} = 6\pi\eta a \; \vec{U}^{St} \quad , \qquad (2.3)$$

corresponding to the translational friction coefficient $\zeta_t = 6\pi\eta a$. The fluid velocity and pressure are given by

$$\vec{v}(\vec{r}) = \left[\frac{\vec{1}+\hat{r}\hat{r}}{r} + \frac{1}{3} a^2 \frac{\vec{1}-3\hat{r}\hat{r}}{r^3} \right] \cdot \frac{\vec{E}}{8\pi\eta} \;,$$

$$p(\vec{r}) = \frac{1}{4\pi} \frac{\hat{r}}{r^2} \cdot \vec{E} \quad , \qquad\qquad r > a \quad . \qquad (2.4)$$

It is easily seen by use of (2.3) that the stick boundary conditions at r=a are satisfied. Note that the fluid disturbance is of very long range.

The external force is transmitted to the fluid. The first equation in (2.1) may be rewritten

$$\nabla \cdot \vec{\vec{\sigma}} = 0 \;, \qquad (2.5)$$

where the stress tensor $\vec{\vec{\sigma}}$ has cartesian components

$$\sigma_{\alpha\beta} = \eta \left[\partial_\alpha v_\beta + \partial_\beta v_\alpha \right] - p\delta_{\alpha\beta} \quad . \qquad (2.6)$$

The force exerted on the fluid is given by

$$\vec{\mathcal{F}} = - \int \vec{\vec{\sigma}} \cdot \hat{r} \; dS \quad , \qquad (2.7)$$

where the integral is over an arbitrary surface surrounding the particle. Choosing a spherical surface and using (2.4) one finds $\vec{\mathcal{F}} = \vec{E}$.

From (2.1) and (2.4) one may deduce the Green's function defined as the solution of the equations

$$\eta\nabla^2\vec{v} - \nabla p = -\vec{\mathcal{F}}\delta(\vec{r}) \quad , \qquad\qquad \nabla \cdot \vec{v} = 0 \quad . \qquad (2.8)$$

The solution is

$$\vec{v}(\vec{r}) = \vec{\vec{T}}(\vec{r}) \cdot \vec{\mathcal{F}} \quad , \qquad p(\vec{r}) = \vec{Q}(\vec{r}) \cdot \vec{\mathcal{F}} \qquad (2.9)$$

with

$$\vec{\vec{T}}(\vec{r}) = \frac{1}{8\pi\eta} \frac{\vec{\vec{1}}+\hat{r}\hat{r}}{r} \quad , \qquad \vec{Q}(\vec{r}) = \frac{1}{4\pi} \frac{\hat{r}}{r^2} \quad . \qquad (2.10)$$

The Green's function $\vec{\vec{T}}(\vec{r})$ is called the Oseen tensor. We note that the part of the velocity field in (2.4) which decays as $1/r^3$ does not give rise to a pressure field, since it has zero divergence.

Next we consider a single particle centered at the origin and subjected to an applied torque $\vec{N}$. The particle acquires an angular velocity $\vec{\Omega}^{St}$ given by

$$\vec{N} = 8\pi\eta a^3 \, \vec{\Omega}^{St} \qquad (2.11)$$

corresponding to the rotational friction coefficient $\zeta_r = 8\pi\eta a^3$. The fluid velocity and pressure are given by

$$\vec{v}(\vec{r}) = \frac{1}{8\pi\eta} \frac{\vec{N}\times\vec{r}}{r^3} \quad , \qquad p(\vec{r}) = 0 \quad , \qquad r > a \quad . \qquad (2.12)$$

Again the external torque is transmitted to the fluid. The torque exerted by the particle on the fluid is given by

$$\vec{\mathcal{J}} = - \int \vec{r} \times (\vec{\vec{\sigma}}\cdot\hat{r})dS \quad . \qquad (2.13)$$

Calculating the integral for a spherical surface surrounding the particle one finds $\vec{\mathcal{J}} = \vec{N}$.

For the rigid body motion (2.2) we may obtain the velocity and pressure by linear superposition of the fields given in (2.4) and (2.12). We denote the Stokes flow pattern generated by particle j centered at $\vec{R}_j$ as $(\vec{v}_j^{St}(\vec{r}), p_j^{St}(\vec{r}))$. Thus we have

$$\vec{v}_j^{St}(\vec{r}) = \left[\frac{\vec{1}+\hat{r}_j\hat{r}_j}{r_j} + \frac{1}{3}\,a^2\,\frac{\vec{1}-3\hat{r}_j\hat{r}_j}{r_j^3}\right]\cdot\frac{\vec{E}_j}{8\pi\eta} + \frac{\vec{N}_j\times\hat{r}_j}{8\pi\eta r_j^2} \quad ,$$

$$p_j^{St}(\vec{r}) = \frac{1}{4\pi}\,\frac{\hat{r}_j}{r_j^2}\cdot\vec{E}_j \quad , \qquad\qquad r_j > a \quad , \qquad\qquad (2.14)$$

where $\vec{r}_j = \vec{r}-\vec{R}_j$ and $\hat{r}_j = \vec{r}_j/r_j$.

If there are N bodies, then a superposition of these flow patterns yields

$$\vec{v}^{St}(\vec{r}) = \sum_{j=1}^{N}\vec{v}_j^{St}(\vec{r}) \quad , \qquad p^{St}(\vec{r}) = \sum_{j=1}^{N}p_j^{St}(\vec{r}) \quad . \qquad (2.15)$$

However, this cannot be the total flow, since the velocity field does not satisfy the stick boundary conditions at the surface of the particles. We write the actual flow pattern as

$$\vec{v}(\vec{r}) = \vec{v}^{St}(\vec{r}) + \hat{\vec{v}}(\vec{r}) \quad , \qquad\qquad p(\vec{r}) = p^{St}(\vec{r}) + \hat{p}(\vec{r}) \quad , \qquad (2.16)$$

where $(\hat{\vec{v}}(\vec{r}), \hat{p}(\vec{r}))$ is a solution of the creeping flow equations chosen such that the boundary conditions are satisfied. Similarly we write the particle velocities as

$$\vec{U}_j = \vec{U}_j^{St} + \hat{\vec{U}}_j \quad , \qquad \vec{\Omega}_j = \vec{\Omega}_j^{St} + \hat{\vec{\Omega}}_j \quad . \qquad (2.17)$$

The velocities $\hat{\vec{U}}_j$ and $\hat{\vec{\Omega}}_j$ must be chosen such that the force and torque exerted by particle j on the fluid remains equal to $\vec{E}_j$ and $\vec{N}_j$, respectively. Thus we must choose $\hat{\vec{U}}_j$ and $\hat{\vec{\Omega}}_j$ such that from the additional stress tensor $\hat{\vec{\sigma}}$, calculated from $(\hat{\vec{v}},\hat{p})$ according to (2.6), we find

$$\int_j \hat{\vec{\sigma}}\cdot\hat{r}_j dS = 0 \quad , \qquad \int_j \hat{r}_j\times(\hat{\vec{\sigma}}\cdot\hat{r}_j)dS = 0 \quad , \quad j=1,\ldots,N. \qquad (2.18)$$

For each configuration $(\vec{R}_1,\ldots,\vec{R}_N)$ and for given applied forces $(\vec{E}_1,\ldots,\vec{E}_N)$ and applied torques $(\vec{N}_1,\ldots,\vec{N}_N)$ the particle velocities $(\vec{U}_1,\ldots,\vec{U}_N)$, $(\vec{\Omega}_1,\ldots,\vec{\Omega}_N)$ and the fluid flow pattern are uniquely determined, provided the latter is assumed to vanish at infinity. The particle velocities and the fluid flow pattern are linear in the applied forces and torques. The particle velocities may therefore be written

$$\vec{U}_j = \sum_k \overset{\rightarrow}{\tilde{\mu}}{}^{tt}_{jk}(\vec{R}_1,\ldots,\vec{R}_N)\cdot\vec{E}_k + \sum_k \overset{\rightarrow}{\tilde{\mu}}{}^{tr}_{jk}(\vec{R}_1,\ldots,\vec{R}_N)\cdot\vec{N}_k ,$$

$$\vec{\Omega}_j = \sum_k \overset{\rightarrow}{\tilde{\mu}}{}^{rt}_{jk}(\vec{R}_1,\ldots,\vec{R}_N)\cdot\vec{E}_k + \sum_k \overset{\rightarrow}{\tilde{\mu}}{}^{rr}_{jk}(\vec{R}_1,\ldots,\vec{R}_N)\cdot\vec{N}_K , \qquad (2.19)$$

with four 3Nx3N mobility matrices, which are complicated functions of the particle configuration. On general grounds, related to the Lorentz reciprocity theorem [9], one can derive the symmetry relations

$$\overset{\rightarrow}{\mu}{}^{tt}_{jk} = \overset{\approx}{\mu}{}^{tt}_{kj} , \qquad \overset{\rightarrow}{\mu}{}^{tr}_{jk} = \overset{\approx}{\mu}{}^{rt}_{kj} , \qquad \overset{\rightarrow}{\mu}{}^{rr}_{jk} = \overset{\approx}{\mu}{}^{rr}_{kj} . \qquad (2.20)$$

The mobility matrices have many-body character and we must use the methods of statistical mechanics to extract the desired information from the mesoscopic equations (2.19).

3. AVERAGE EQUATIONS

As we shall see, we can find a complete formal solution of the mesoscopic equations [10,11]. From this solution we must extract what is relevant on the macroscopic level In statistical mechanics we employ methods which allow one to derive and calculate the macroscopic properties of many-body systems. As a matter of course, these methods involve statistical averaging.

In our case we consider instead of a precise configuration $(\vec{R}_1,\ldots,\vec{R}_N)$ a probability distribution $P(\vec{R}_1,\ldots,\vec{R}_N)$. A complication is that the configurations change in the course of time. Hence we must also consider the time-evolution of the probability distribution. However, if the forces and torques are applied suddenly, then for short times the distribution function has no time to change. In the last section we return to the question of time evolution. Presently we assume that the distribution function is known and is given, for example, by an equilibrium distribution of hard spheres.

For simplicity we assume that the applied torques vanish. Also we assume that the applied forces may be derived from a vector field $\vec{E}(\vec{r})$ according to

$$\vec{E}_j = \vec{E}(\vec{R}_j) = \int \vec{E}(\vec{r})\delta(\vec{r}-\vec{R}_j)d\vec{r} . \qquad (3.1)$$

The field $\vec{E}(\vec{r})$ is assumed to be independent of the particle configuration. The problem of sedimentation in the field of gravity

could be formulated this way. A quantity of interest then is the
local current density, defined by

$$\vec{J}(\vec{r}) = \sum_{j=1}^{N} \vec{U}_j \; \delta(\vec{r}-\vec{R}_j) \quad . \tag{3.2}$$

From (2.19) we find the linear relation

$$\vec{J}(\vec{r}) = \int \sum_{jk} \delta(\vec{r}-\vec{R}_j)\vec{\mu}_{jk}^{tt}(\vec{R}_1,\ldots,\vec{R}_N)\delta(\vec{r}\,'-\vec{R}_k) \cdot \vec{E}(\vec{r}\,')d\vec{r}\,' \quad . \tag{3.3}$$

Averaging this equation over the probability distribution
$P(\vec{R}_1,\ldots,\vec{R}_N)$ we find

$$< \vec{J}(\vec{r}) > = \int \vec{\vec{Y}}(\vec{r},\vec{r}\,') \cdot \vec{E}(\vec{r}\,')d\vec{r}\,' \tag{3.4}$$

with the average response kernel

$$\vec{\vec{Y}}(\vec{r},\vec{r}\,') = < \sum_{jk} \delta(\vec{r}-\vec{R}_j)\vec{\mu}_{jk}^{tt}(\vec{R}_1,\ldots,\vec{R}_N)\delta(\vec{r}\,'-\vec{R}_k) > \quad . \tag{3.5}$$

The problem is that this kernel has a very long range. If we consider just the direct pair contribution to the mobility tensor $\vec{\mu}_{jk}^{tt}$,
then we see from (2.10) that at large distances the latter behaves
as

$$\vec{\mu}_{jk}^{tt} \approx \frac{\vec{1}+\hat{R}_{jk}\hat{R}_{jk}}{8\pi\eta R_{jk}} \qquad \text{for } j{\neq}k \text{ and } R_{jk} \gg a \quad , \tag{3.6}$$

where $\vec{R}_{jk} = \vec{R}_k-\vec{R}_j$. In dense suspensions the prefactor will be modified by an effective viscosity, but the distance dependence will
remain the same. The kernel $\vec{\vec{Y}}(\vec{r},\vec{r}\,')$ in (3.5) has a corresponding
long range. Hence it follows that the kernel does not have a proper
thermodynamic limit. We imagine that the N spherical particles are
confined to a volume Ω of simple shape and are distributed approximately uniformly with number density n = N/Ω. From the long range
behavior it follows that for large N and Ω the kernel $\underset{\approx}{Y}$ depends on
the shape of Ω. Hence $\underset{\approx}{Y}$ is not a proper transport kernel.

The problem arises because we have not yet achieved a complete
macroscopic description. Besides the average current density $<\vec{J}>$
we must also consider the average flow velocity $< \vec{v} >$. Before deriving the complete set of macroscopic equations we discuss a tech-
nical point. We recall that in the space occupied by the fluid the
velocity field and pressure satisfy the linear Navier-Stokes equa-

tions (2.1). These equations are supplemented with stick boundary
conditions to be satisfied at the surface of the particles. In order
to derive equations for $< \vec{v} >$ and $< p >$ it is convenient to choose
a slightly different formulation. We replace (2.1) by equations
valid in all space

$$\eta \nabla^2 \vec{v} - \nabla p = -\vec{F}(\vec{r}) \quad , \qquad \nabla \cdot \vec{v} = 0 \quad , \tag{3.7}$$

with the conditions

$$\vec{v}(\vec{r}) = \vec{w}_j(\vec{r}) \quad ,$$
$$\qquad\qquad \text{for } r_j < a \, , \quad j = 1,\ldots,N$$
$$p(\vec{r}) = 0 \quad . \tag{3.8}$$

The vector field $\vec{F}(\vec{r})$ is called the induced force density [12]. It
is localized on the surface of the particles and can be written as

$$\vec{F}(\vec{r}) = \sum_{j=1}^{N} \vec{F}_j(\vec{r}) \quad , \qquad \vec{F}_j(\vec{r}) = \vec{f}_j(\hat{r}_j)\delta(r_j - a) \tag{3.9}$$

with a surface force density

$$f_{j\alpha} = -\sigma_{n\alpha}(r_j = a+) \quad , \tag{3.10}$$

where $\sigma_{n\alpha}$ is defined from the stress tensor (2.6) as $\hat{r}_j \cdot \vec{\vec{\sigma}} \cdot \vec{e}_\alpha$.
The induced force density may therefore be calculated from the
solution $(\vec{v}, p)$ in the space occupied by the fluid.

It may be helpful to consider also a slightly different model
in which the particles are permeable to the flow, but offer a re-
sistance proportional to the local velocity difference with the
fluid. For such a model the basic equations are

$$\eta \nabla^2 \vec{v} - \nabla p - \lambda \sum_{j=1}^{N} [\vec{v}(\vec{r}) - \vec{w}_j(\vec{r})]\theta(a - r_j) = 0, \quad \nabla \cdot \vec{v} = 0 \, , \tag{3.11}$$

where λ^{-1} is the permeability. In the limit of zero permeability
$\vec{v}$ and $\vec{w}_j$ must coincide inside particle j and the model becomes
identical to that for hard spheres with stick boundary conditions.
The equations (3.11) are identical to (3.7) with the induced force
density

$$\vec{F}(\vec{r}) = -\lambda \sum_{j=1}^{N} [\vec{v}(\vec{r}) - \vec{w}_j(\vec{r})]\theta(a - r_j) \quad . \tag{3.12}$$

In the limit $\lambda \to \infty$ the force density $\vec{F}_j$ is localized on the particle surface and the flow velocity $\vec{v}(\vec{r})$ tends to $\vec{w}_j(\vec{r})$ for $r_j < a$.

The advantage of the formulation with the induced force density is that we can now average (3.7) over the distribution $P(\vec{R}_1 \ldots, \vec{R}_N)$ to obtain equations for $< \vec{v} >$ and $< p >$ valid in all space. The averaged equations are simply

$$\eta \nabla^2 < \vec{v} > - \nabla < p > = -< \vec{F} >, \qquad \nabla \cdot < \vec{v} > = 0. \qquad (3.13)$$

It should be noted that because of (3.8) we cannot identify $< \vec{v} >$ as the average fluid velocity, since it contains a contribution from the particles. We must interpret $< \vec{v} >$ as the average suspension velocity field.

We note that the general solution of (3.7) may be written with the aid of the Green's function

$$\vec{v}(\vec{r}) = \vec{v}_0(\vec{r}) + \int \vec{\vec{T}}(\vec{r}-\vec{r}') \cdot \vec{F}(\vec{r}')d\vec{r}' \quad ,$$

$$p(\vec{r}) = p_0(\vec{r}) + \int \vec{Q}(\vec{r}-\vec{r}') \cdot \vec{F}(\vec{r}')d\vec{r}' \quad , \qquad (3.14)$$

where $(\vec{v}_0, p_0)$ is a solution of the homogeneous equations and $\vec{\vec{T}}(\vec{r})$ and $\vec{Q}(\vec{r})$ are given by (2.10). Of course, the solution of the average equations (3.13) reads correspondingly

$$< \vec{v}(\vec{r}) > = \vec{v}_0(\vec{r}) + \int \vec{\vec{T}}(\vec{r}-\vec{r}') \cdot < \vec{F}(\vec{r}') > d\vec{r}' \quad ,$$

$$< p(\vec{r}) > = p_0(\vec{r}) + \int \vec{Q}(\vec{r}-\vec{r}') \cdot < \vec{F}(\vec{r}') > d\vec{r}' \quad . \qquad (3.15)$$

The averaged equations (3.13) must yet be supplemented with a constitutive equation relating the average force density $< \vec{F} >$ to the averages $< \vec{v} >$ and $< \vec{J} >$, and to the applied force field $\vec{E}(\vec{r})$. The situation is similar to that for dielectrics. For a medium consisting of polarizable molecules it is easy to derive Maxwell's equations for the average electric and magnetic fields from the microscopic Lorentz equations [13]. The difficult part is to derive a constitutive equation for the average polarization in terms of the average field.

A natural form for the constitutive equations is [6]

$$< \vec{F}(\vec{r}) > = \int \vec{\vec{X}}_{MM}(\vec{r},\vec{r}') \cdot < \vec{v}(\vec{r}') > d\vec{r}' + \int \vec{\vec{X}}_{ME}(\vec{r},\vec{r}') \cdot \vec{E}(\vec{r}')d\vec{r}'$$

$$< \vec{J}(\vec{r}) > = \int \vec{\vec{X}}_{JM}(\vec{r},\vec{r}') \cdot < \vec{v}(\vec{r}') > d\vec{r}' + \int \vec{\vec{X}}_{JE}(\vec{r},\vec{r}') \cdot \vec{E}(\vec{r}')d\vec{r}' \quad , \tag{3.16}$$

or for brevity

$$< \vec{F} > = \underset{\approx}{X}_{MM} < \vec{v} > + \underset{\approx}{X}_{ME} \vec{E}$$

$$< \vec{J} > = \underset{\approx}{X}_{JM} < \vec{v} > + \underset{\approx}{X}_{JE} \vec{E} \quad . \tag{3.17}$$

The four kernels $\underset{\approx}{X}_{MM}$, $\underset{\approx}{X}_{ME}$, $\underset{\approx}{X}_{JM}$ and $\underset{\approx}{X}_{JE}$ can be shown to have short range with properties determined by the local geometrical structure of the suspension. For a spatially uniform system they have a well-defined thermodynamic limit. In this limit the kernels become translationally invariant and dependent only on the difference $\vec{r}-\vec{r}'$. Moreover one can show the symmetry properties

$$X_{MM,\alpha\beta}(\vec{r},\vec{r}') = X_{MM,\beta\alpha}(\vec{r}',\vec{r}) \quad ,$$

$$X_{ME,\alpha\beta}(\vec{r},\vec{r}') = X_{JM,\beta\alpha}(\vec{r}',\vec{r}) \quad ,$$

$$X_{JE,\alpha\beta}(\vec{r},\vec{r}') = X_{JE,\beta\alpha}(\vec{r}',\vec{r}) \quad . \tag{3.18}$$

If we compare (3.4) and (3.16) we see that the long range dependence in the response of $< \vec{J} >$ has been absorbed in $< \vec{v} >$. The average flow velocity $< \vec{v} >$ itself has a long range dependence on $< \vec{F} >$, as shown in (3.15).

We can rearrange the equations to read

$$\eta\nabla^2 < \vec{v} > + \underset{\approx}{X}_{MM} < \vec{v} > - \nabla < p > = -\underset{\approx}{X}_{ME} \vec{E} \quad , \qquad \nabla \cdot < \vec{v} > = 0,$$

$$< \vec{J} > - \underset{\approx}{X}_{JM} < \vec{v} > = \underset{\approx}{X}_{JE} \vec{E} \quad . \tag{3.19}$$

In the thermodynamic limit and for long wavelengths, i.e. for slowly varying fields, the equations become

$$\eta_{eff} \, \nabla^2 < \vec{v} > - \nabla < p > = -n\vec{E} \quad , \qquad \nabla \cdot < \vec{v} > = 0 \quad ,$$

$$< \vec{J} > - n < \vec{v} > = n\mu_C \vec{E} \quad . \tag{3.20}$$

In this limit the kernels $\underset{\approx}{X}_{ME}$ and $\underset{\approx}{X}_{JM}$ become rather trivial and the dynamics of the suspension is described by just two transport coefficients, the effective viscosity η_{eff} and the collective mobility μ_C.

We are faced with two questions:
a) how does one derive the constitutive equations (3.16)?
b) how does one calculate the transport kernels, or more modestly, the transport coefficients η_{eff} and μ_C?

To answer these questions we must turn to the complete formal solution of the mesoscopic equations in terms of a multiple scattering formalism.

4. MULTIPLE SCATTERING

For any given configuration $(\vec{R}_1,\ldots,\vec{R}_N)$ of spheres and applied field $\vec{E}(\vec{r})$ with corresponding forces $\vec{E}_j = \vec{E}(\vec{R}_j)$ ($j = 1,\ldots,N$) the solution of the flow equations and the particle velocities are uniquely determined if we impose the condition that the flow velocity vanishes at infinity. The induced force density $\vec{F}$ and the current density $\vec{J}$ are linear in the applied field, so that we may write

$$\vec{F}(\vec{r};1,\ldots,N) = \int \vec{\vec{T}}_{FE}(\vec{r},\vec{r}';1,\ldots,N) \cdot \vec{E}(\vec{r}')d\vec{r}' \quad ,$$

$$\vec{J}(\vec{r};1,\ldots,N) = \int \vec{\vec{T}}_{JE}(\vec{r},\vec{r}';1,\ldots,N) \cdot \vec{E}(\vec{r}')d\vec{r}' \quad . \tag{4.1}$$

We shall analyze the linear operators $\underset{\approx}{T}_{FE}$ and $\underset{\approx}{T}_{JE}$ by the methods of multiple scattering theory.

Suppose only particle j is subjected to a force $\vec{E}_j$. In the absence of the other particles this would create a Stokes flow pattern centered at $\vec{R}_j$, given by (2.4). This flow pattern acts on the other particles and causes them to move in such a way that no force or torque is exerted on any of these particles. The incident flow pattern induces a force density on each of the particles which in turn generates a new flow pattern. In this way we get a sequence of flow patterns building up to the complete flow. The whole process may be analyzed as a sum of elementary scattering events. Each event involves the response of a single spherical particle to an incident flow.

Thus we consider a single spherical particle centered at $\vec{R}_j$, subject to an incident flow pattern $(\vec{v}_o(\vec{r}), p_o(\vec{r}))$, which is a solution of (2.1). The particle may have a rigid body motion $\vec{w}_j(\vec{r})$ given by (2.2). The induced force density $\vec{F}_j(\vec{r})$ will be linear in the difference $\vec{v}_o(\vec{r})-\vec{w}_j(\vec{r})$, and may be written as

$$\vec{F}_j(\vec{r}) = - \int_{r'_j<a} \overset{\leftrightarrow}{Z}(j;\vec{r},\vec{r}') \cdot [\vec{v}_o(\vec{r}')-\vec{U}_j-\vec{\Omega}_j\times(\vec{r}'-\vec{R}_j)]\,d\vec{r}' \tag{4.2}$$

with a linear integral kernel $\underset{\approx}{Z}(j)$. This kernel may be found from the solution of the one-body problem for an arbitrary incident flow $(\vec{v}_o(\vec{r}), p_o(\vec{r}))$. For a spherical particle with stick boundary conditions, and for a number of other cases, the kernel is known in full detail [14].

It is convenient to split off the force density corresponding to the Stokes solution generated by the applied force $\vec{E}_j$ with corresponding Stokes velocity $\vec{U}_j^{St} = (6\pi\eta a)^{-1}\vec{E}_j$ and $\vec{\Omega}_j^{St} = 0$. Thus we write

$$\vec{F}_j = \vec{F}_j^{St} + \hat{\vec{F}}_j \quad , \tag{4.3}$$

where

$$\vec{F}_j^{St}(r) = \int_{r'_j<a} \overset{\leftrightarrow}{Z}(j;\vec{r},\vec{r}') \cdot \vec{U}_j^{St}\,d\vec{r}' \quad . \tag{4.4}$$

The velocities $\vec{U}_j$ and $\vec{\Omega}_j$ follow from two theorems by Faxén which read [9]

$$\mathcal{F}_j = 6\pi\eta a\left[\vec{U}_j - \vec{v}_o(\vec{R}_j) - \frac{1}{6}\,a^2\nabla^2\vec{v}_o|_{\vec{R}_j}\right],$$

$$\mathcal{T}_j = 8\pi\eta a^3\left[\vec{\Omega}_j - \frac{1}{2}\nabla\times\vec{v}_o|_{\vec{R}_j}\right] \quad . \tag{4.5}$$

In our case $\mathcal{F}_j = \vec{E}_j$ and $\mathcal{T}_j = 0$, so that we find

$$\vec{U}_j = \vec{U}_j^{St} + \hat{\vec{U}}_j \quad , \qquad \vec{\Omega}_j = \hat{\vec{\Omega}}_j \quad , \tag{4.6}$$

with

$$\hat{\vec{U}}_j = \vec{v}_o(\vec{R}_j) + \frac{1}{6}\,a^2\nabla^2\vec{v}_o|_{\vec{R}_j} \qquad \hat{\vec{\Omega}}_j = \frac{1}{2}\nabla\times\vec{v}_o|_{\vec{R}_j} \quad . \tag{4.7}$$

Substituting in (4.2) we may therefore write

$$\hat{\vec{F}}_j(\vec{r}) = - \int_{r_j < a} \hat{\vec{Z}}(j;\vec{r},\vec{r}') \cdot \vec{v}_0(\vec{r}')d\vec{r}' \tag{4.8}$$

with a modified friction kernel, which is called the convective friction kernel [15]. This kernel has the properties

$$\hat{\vec{F}}_j = \int \hat{\vec{F}}_j d\vec{r} = 0 \quad , \quad \hat{\vec{T}}_j = \int (\vec{r}-\vec{R}_j) \times \hat{\vec{F}}_j d\vec{r} = 0 \quad , \tag{4.9}$$

so that $\hat{\vec{F}}_j(\vec{r})$ comprises only higher order force moments.

The first nonvanishing moment of $\hat{\vec{F}}_j$ is the symmetric force dipole moment

$$\hat{\vec{F}}_j^{(2s)} = \frac{1}{2} \int \left[(\vec{r}-\vec{R}_j)\hat{\vec{F}}_j + \hat{\vec{F}}_j(\vec{r}-\vec{R}_j) \right]d\vec{r} \quad . \tag{4.10}$$

Its value may again be found from a theorem of Faxén type [16]

$$\hat{F}_{j,\alpha\beta}^{(2s)} = - \frac{10}{3} \pi\eta(\partial_\alpha v_{0\beta}+\partial_\beta v_{0\alpha})\big|_{\vec{R}_j} - \frac{1}{3} \pi\eta a^2 (\partial_\alpha \nabla^2 v_{0\beta}+\partial_\beta \nabla^2 v_{0\alpha})\big|_{\vec{R}_j} \quad . \tag{4.11}$$

The dipole moment generates a flow

$$\hat{v}_{j\alpha}^{(2s)}(\vec{r}) = \frac{3}{8\pi\eta} \frac{\hat{r}_{j\alpha}\hat{r}_{j\beta}\hat{r}_{j\gamma}}{r_j^2} \hat{F}_{j,\beta\gamma}^{(2s)} \quad , \qquad r_j > a \quad . \tag{4.12}$$

It is clear that this is again a velocity disturbance of long range.

Even in a situation of zero forces and torques, which occurs for example in a suspension subjected to a uniform shear flow, the fluid disturbances are of long range. Since each flow pattern of the form (4.12) induces at least a symmetric force dipole moment on each of the other spheres, we deal with interactions which have intrinsic many-body character. The situation is analogous to that in dielectrics with electrically polarizable particles.

We are now in a position to construct the formal multiple scattering solution of the many-body problem [10,11]. Consider again the N spheres subject to applied forces $(\vec{E}_1,...,\vec{E}_N)$ and zero applied torques. We write the force density induced on sphere j again as a sum of two terms, as in (4.3). The part $\hat{\vec{F}}_j(\vec{r})$ is the force density induced by the flow velocity acting on sphere j. Thus we may write

$$\hat{\vec{F}}_j(\vec{r}) = - \int \hat{\vec{Z}}(j;\vec{r},\vec{r}') \cdot \vec{v}_j^a(\vec{r}')d\vec{r}' \quad , \qquad (4.13)$$

where the flow velocity acting on sphere j is given by

$$\vec{v}_j^a(\vec{r}) = \sum_{k \neq j} \vec{v}_k^{St}(\vec{r}) + \sum_{k \neq j} \hat{\vec{v}}_k(\vec{r}) \quad . \qquad (4.14)$$

Here $\vec{v}_k^{St}(\vec{r})$ is the Stokes flow pattern due to the force $\vec{E}_k$ acting on sphere k, and $\hat{\vec{v}}_k(\vec{r})$ is given by

$$\hat{\vec{v}}_k(\vec{r}) = \int \vec{T}(\vec{r}-\vec{r}') \cdot \hat{\vec{F}}_k(\vec{r}')d\vec{r}' \quad . \qquad (4.15)$$

The equations may be solved by iteration. The forces $\{\vec{E}_j\}$ generate the Stokes flow patterns $\{\vec{v}_j^{St}(\vec{r})\}$, these induce new force densities, which generate flows, etc.

We may cast the solution in the form of a multiple scattering expansion. For generality we write

$$\underset{\approx}{M}(j) = -\underset{\approx}{\hat{Z}}(j) \quad , \qquad \underset{\approx}{G}_0 = \underset{\approx}{T} \quad . \qquad (4.16)$$

The initial step is expressed by an operator $\underset{\approx}{M}_E(j)$ such that

$$\vec{F}_j^{St}(\vec{r}) = \int \vec{M}_E(j;\vec{r},\vec{r}') \cdot \vec{E}(\vec{r}')d\vec{r}' \quad . \qquad (4.17)$$

Explicitly

$$\vec{M}_E(j;\vec{r},\vec{r}') = \int \vec{Z}(j;\vec{r},\vec{r}'')d\vec{r}''(6\pi\eta a)^{-1}\delta(\vec{r}'-\vec{R}_j) \quad . \qquad (4.18)$$

Subsequent scattering processes are expressed by means of the Green's function $\underset{\approx}{G}_0$ and the one-body scattering kernels $\{\underset{\approx}{M}(j)\}$. Finally we wish to evaluate the force density $\vec{F}(\vec{r})$ and the current $\vec{J}(\vec{r})$.

We write the current density as a sum of two terms

$$\vec{J} = \vec{J}^{St} + \hat{\vec{J}} \qquad (4.19)$$

with

$$\vec{J}^{St}(\vec{r}) = \sum_{j=1}^{N} \vec{U}_j^{St} \delta(\vec{r}-\vec{R}_j) \quad , \qquad \hat{\vec{J}}(\vec{r}) = \sum_{j=1}^{N} \hat{\vec{U}}_j \delta(\vec{r}-\vec{R}_j) \quad .(4.20)$$

The first term may be expressed by

$$\vec{J}^{St} = \sum_{j=1}^{N} \underset{\approx}{M}_{JE}(j) \cdot \vec{E} \tag{4.21}$$

with the linear operator

$$\overset{\rightarrow}{\underset{\approx}{M}}_{JE}(j;\vec{r},\vec{r}\,') = (6\pi\eta a)^{-1}\,\vec{\mathbf{1}}\,\delta(\vec{r}-\vec{R}_j)\delta(\vec{r}\,'-\vec{R}_j) \quad . \tag{4.22}$$

The second term is given by

$$\hat{\vec{J}} = \sum_{j=1}^{N} \underset{\approx}{M}_J(j) \cdot \vec{v}_j^a \quad , \tag{4.23}$$

with

$$\left[\underset{\approx}{M}_J(j) \cdot \vec{v}_j^a\right](\vec{r}) = \delta(\vec{r}-\vec{R}_j)\left[\vec{v}_j^a(\vec{R}_j) + \frac{1}{6}\,a^2 \nabla^2 \vec{v}_j^a\,|_{\vec{R}_j}\right] \quad . \tag{4.24}$$

The linear operator $\underset{\approx}{T}_{JE}(1,\ldots,N)$ defined in (4.1) may now be expressed in terms of the multiple scattering expansion

$$\underset{\approx}{T}_{JE}(1,\ldots,N) = \sum_{\substack{j=1}}^{N} \underset{\approx}{M}_{JE}(j) + \sum_{\substack{j=1 \\ k \neq j}}^{N} \underset{\approx}{M}_J(j)\underset{\approx}{G}_0\,\underset{\approx}{M}_E(k)$$

$$+ \sum_{\ell=2}^{\infty} \underset{[j]}{\Sigma'}\,\underset{\approx}{M}_J(j_1)\left[\prod_{i=2}^{\ell}\underset{\approx}{G}_0\,\underset{\approx}{M}(j_i)\right]\underset{\approx}{G}_0\,\underset{\approx}{M}_E(j_{\ell+1}) \quad , \tag{4.25}$$

where the last sum is over all sequences [j] of $\ell+1$ labels with
the condition that no label be repeated in succession.

A somewhat more transparent description is obtained with the
aid of a matrix formalism. Let us define the NxN operator matrices

$$\underset{\approx}{\mathfrak{m}} = \begin{pmatrix} \underset{\approx}{M}(1) & & & \\ & \underset{\approx}{M}(2) & & 0 \\ & & \ddots & \\ 0 & & & \underset{\approx}{M}(N) \end{pmatrix}, \underset{\approx}{\mathfrak{G}} = \begin{pmatrix} 0 & \underset{\approx}{G}_0 & \underset{\approx}{G}_0 & \cdots & \underset{\approx}{G}_0 \\ \underset{\approx}{G}_0 & 0 & & & \vdots \\ \vdots & & \ddots & & \vdots \\ \vdots & & & 0 & \underset{\approx}{G}_0 \\ \underset{\approx}{G}_0 & \cdots & & \underset{\approx}{G}_0 & 0 \end{pmatrix} \quad . \tag{4.26}$$

Then we may write

$$\underset{\approx}{T}_{JE}(1,\ldots,N) = \sum_{j,k}\underset{\approx}{T}_{JE,jk}(1,\ldots,N) \tag{4.27}$$

with

$$\underset{\approx}{T}_{JE,jk}(1,\ldots,N) = \underset{\approx}{M}_{JE}(j)\delta_{jk} + \underset{\approx}{M}_J(j)\underset{\approx}{V}_{jk}\,\underset{\approx}{M}_E(k) \ , \qquad (4.28)$$

where the connector operator matrix $\underset{\approx}{V}$ is defined by

$$\underset{\approx}{V} = \underset{\approx}{G}\left[\underset{\approx}{I} - \underset{\approx}{m}\,\underset{\approx}{G}\right]^{-1} \ . \qquad (4.29)$$

Similarly the operator $\underset{\approx}{T}_{FE}(1,\ldots,N)$ defined in (4.1) is given by

$$\underset{\approx}{T}_{FE}(1,\ldots,N) = \sum_{j,k} \underset{\approx}{T}_{FE,jk}(1,\ldots,N) \qquad (4.30)$$

with

$$\underset{\approx}{T}_{FE,jk}(1,\ldots,N) = \underset{\approx}{M}_E(j)\delta_{jk} + \underset{\approx}{M}(j)\underset{\approx}{V}_{jk}\,\underset{\approx}{M}_E(k) \ . \qquad (4.31)$$

The above formulation is very similar to that arising in the theory of multiple scattering of waves. Consider for example the wave equation

$$\nabla^2\psi + k^2\psi - \sum_{j=1}^{N} V(j)\psi = 0 \qquad (4.32)$$

for a scalar wavefunction scattering from a potential V which is composed of a sum of identical non-overlapping one-body potentials centered at arbitrarily distributed scattering-centers $(\vec{R}_1,\ldots,\vec{R}_N)$. The formal solution of (4.32) is given by

$$\psi(1,\ldots,N) = \psi_0 + \underset{\approx}{G}_0\,\underset{\approx}{T}(1,\ldots,N)\psi_0 \ , \qquad (4.33)$$

where ψ_0 is the incident wave, which satisfies the wave equation for V = 0. The free space Green's function $\underset{\approx}{G}_0$ is here given by

$$G_0(\vec{r}-\vec{r}\,') = \frac{-1}{4\pi}\,\frac{e^{ik|\vec{r}-\vec{r}\,'|}}{|\vec{r}-\vec{r}\,'|} \qquad (4.34)$$

and the T-operator is defined by

$$\underset{\approx}{T}(1,\ldots,N)\psi_0 = \sum_{j=1}^{N} V(j)\psi(1,\ldots,N) \ . \qquad (4.35)$$

Denoting the one-body T-operator corresponding to center j by $\underset{\approx}{M}(j)$ and introducing operator matrices as in (4.26) we find by iteration the formal solution

$$\underset{\approx}{T}(1,\ldots,N) = \sum_{j,k} \underset{\approx}{T}_{jk}(1,\ldots,N) \tag{4.36}$$

with

$$\underset{\approx}{T}_{jk}(1,\ldots,N) = \left[\underset{\approx}{\mathcal{M}}(\underset{\approx}{I} - \underset{\approx}{\mathcal{G}}\,\underset{\approx}{\mathcal{M}})^{-1}\right]_{jk}$$

$$= \underset{\approx}{M}(j)\delta_{jk} + \underset{\approx}{M}(j)\underset{\approx}{\mathcal{V}}_{jk}\,\underset{\approx}{M}(k) \quad . \tag{4.37}$$

In the case of hydrodynamics there are slight modifications of the operators for the first and last scatterer, but otherwise the situation is very similar. Our next task is to consider the averaging of the multiple scattering expansion over a probability distribution $P(\vec{R}_1,\ldots,\vec{R}_N)$, which is assumed to be known.

5. ORNSTEIN-ZERNIKE TYPE EQUATIONS

In this section we show that after averaging the multiple scattering expansion we can formulate integral equations of Ornstein-Zernike type, which lead to expressions for the transport kernels introduced in (3.16). The analysis extends a formalism developed by Wertheim for systems of polarizable point dipoles [17].

For simplicity we consider first the equations (4.32-37) corresponding to the scalar wave equation. Afterward we can easily make the modifications necessary for the hydrodynamics of suspensions. If we average (4.33) and (4.35) we obtain the pair of equations

$$< \psi > = \psi_0 + \underset{\approx}{G}_0 < \underset{\approx}{T} > \psi_0 \quad ,$$

$$< \underset{\approx}{T} > \psi_0 = < \underset{\approx}{V}\psi > \quad , \tag{5.1}$$

where we have used that ψ_0 and $\underset{\approx}{G}_0$ are independent of the configuration $(\vec{R}_1,\ldots,\vec{R}_N)$. The average T-operator is a kernel $< T >(\vec{r},\vec{r}')$ with contributions from widely different points $\vec{r}$ and $\vec{r}'$. The reason is that the points $\vec{r}$ and $\vec{r}'$ may be located in particles which are uncorrelated in position and connected only by a single Green's function $\underset{\approx}{G}_0$. Such loose connections are eliminated if we consider the operator $\underset{\approx}{X}$ defined by

$$< \underset{\approx}{V}\psi > = \underset{\approx}{X} < \psi > \tag{5.2}$$

with the operator expression

$$\underset{\approx}{X} = <\underset{\approx}{T}> \left[\underset{\approx}{I} + \underset{\approx}{G}_0 <\underset{\approx}{T}>\right]^{-1} \quad . \tag{5.3}$$

In scattering theory $\underset{\approx}{X}$ is called the self-energy operator. That the simple operation (5.3) eliminates single bonds may be seen by the following analysis [18].

   We write (5.3) as

$$<\underset{\approx}{V}\psi> = <\underset{\approx}{T}> \psi_0 = [\ \int d1\ n(1)\underset{\approx}{B}(1) + \int d1d2\ n(1)n(2)\underset{\approx}{A}(1,2)]\psi_0\ , \tag{5.4}$$

where $n(1)$ is the one-particle density. Here $\underset{\approx}{B}(1)$ is the so-called bridge operator which describes all scattering processes beginning and ending at the same scatterer, labeled 1, averaged over the positions of all the other scatterers. The operator $\underset{\approx}{A}(1,2)$ describes all scattering processes beginning at a particle, labeled 2, and ending at another particle, again with an average over all intermediate scatterers. The latter operator may be decomposed as

$$\underset{\approx}{A}(1,2) = \underset{\approx}{B}(1)\underset{\approx}{H}(1,2)\underset{\approx}{B}(2) \quad , \tag{5.5}$$

which defines the pair connector $\underset{\approx}{H}(1,2)$. We may write this as a sum of chains of operators by formulating an equation of Ornstein-Zernike type

$$\underset{\approx}{H}(1,2) = \underset{\approx}{C}(1,2) + \int d3\ n(3)\underset{\approx}{C}(1,3)\underset{\approx}{B}(3)\underset{\approx}{H}(3,2) \quad , \tag{5.6}$$

where the operator $\underset{\approx}{C}(1,2)$ is the so-called direct connector. One of the contributions to the direct connector is provided by direct propagation between the two scatterers 1 and 2, as given by the Green's function $\underset{\approx}{G}_0$. We subtract this contribution and define the short-range connector $\underset{\approx}{S}(1,2)$ by

$$\underset{\approx}{C}(1,2) = \underset{\approx}{G}_0 + \underset{\approx}{S}(1,2) \quad . \tag{5.7}$$

Finally we define the pair connector $\underset{\approx}{F}(1,2)$ from the Ornstein-Zernike type equation

$$\underset{\approx}{F}(1,2) = \underset{\approx}{S}(1,2) + \int d3\ n(3)\underset{\approx}{S}(1,3)\underset{\approx}{B}(3)\underset{\approx}{F}(3,2) \quad . \tag{5.8}$$

Substituting the above chain of operations in (5.4) we find that we may write

$$< \underset{\approx}{V}\psi > = < \underset{\approx}{T} > \psi_0 = \underset{\approx}{X}(\underset{\approx}{I} - \underset{\approx 0}{G}\,\underset{\approx}{X})^{-1}\psi_0 \qquad (5.9)$$

with the operator $\underset{\approx}{X}$ given by

$$\underset{\approx}{X} = \int d1\ n(1)\underset{\approx}{B}(1) + \int d1\ d2\ n(1)n(2)\underset{\approx}{B}(1)\underset{\approx}{F}(1,2)\underset{\approx}{B}(2) \quad . \qquad (5.10)$$

It is easily seen that (5.9) is equivalent to (5.3). It is evident from the construction that in the operator $\underset{\approx}{X}$ no single bonds between uncorrelated particles occur. We may expect the kernel $\underset{\approx}{X}$ to be short-ranged and to possess a well-defined thermodynamic limit independent of sample shape.

Further analysis shows that the bridge operator $\underset{\approx}{B}(1)$ may be expressed as

$$\underset{\approx}{B}(1) = \underset{\approx}{M}(1)\left[\underset{\approx}{I} - \underset{\approx}{S}(1)\underset{\approx}{M}(1)\right]^{-1} \quad , \qquad (5.11)$$

where the so-called reaction field operator $\underset{\approx}{S}(1)$ may be expressed in terms of scattering processes averaged over correlation functions.

The above expressions are exact, but we cannot hope to be able to perform an exact calculation of the operator $\underset{\approx}{X}$. For low density systems we can make an expansion in powers of the density and limit ourselves to the first few terms. At high density we must be bold and make judicious approximations.

Before turning to that question we show which modifications of the formalism are needed in the application to the hydrodynamics of suspensions. In hydrodynamics the single-body scattering operator $\underset{\approx}{M}(j)$ is given by the convective friction kernel $-\underset{\approx}{\hat{Z}}(j)$, as shown in (4.16), and the Green's function $\underset{\approx 0}{G}$ is given by the Oseen kernel $\vec{T}(\vec{r}-\vec{r}')$, as defined in (2.10). The wavefunction ψ is replaced by the velocity field $\vec{v}$, and the product $V\psi$ is replaced by the induced force density $\vec{F}$. Hence the kernel $\underset{\approx MM}{X}$, appearing in (3.16), is given precisely by the expression (5.10).

The kernels $\underset{\approx ME}{X}$, $\underset{\approx JM}{X}$ and $\underset{\approx JE}{X}$ in (3.16) are given by slight modifications of this expression. The kernel $\underset{\approx ME}{X}$ is given by

$$\underset{\approx ME}{X} = \int d1\ n(1)\underset{\approx E}{B}(1) + \int d1\ d2\ n(1)n(2)\underset{\approx}{B}(1)\underset{\approx}{F}(1,2)\underset{\approx E}{B}(2) \quad , \qquad (5.12)$$

where the operator $\underset{\approx}{B}_E(1)$ is defined by

$$\underset{\approx}{B}_E(1) = \left[\underset{\approx}{I} - \underset{\approx}{M}(1)\underset{\approx}{S}(1) \right]^{-1} \underset{\approx}{M}_E(1) \quad , \tag{5.13}$$

with $\underset{\approx}{M}_E(1)$ as defined in (4.18). Similarly the operator $\underset{\approx}{X}_{JM}$ is given by

$$\underset{\approx}{X}_{JM} = \int d1\ n(1)\underset{\approx}{B}_J(1) + \int d1d2\ n(1)n(2)\underset{\approx}{B}_J(1)\underset{\approx}{F}(1,2)\underset{\approx}{B}(2) \quad , \tag{5.14}$$

where the operator $\underset{\approx}{B}_J(1)$ is defined by

$$\underset{\approx}{B}_J(1) = \underset{\approx}{M}_J(1) \left[\underset{\approx}{I} - \underset{\approx}{S}(1)\underset{\approx}{M}(1) \right]^{-1} \tag{5.15}$$

with $\underset{\approx}{M}_J(1)$ as defined in (4.24). Finally the operator $\underset{\approx}{X}_{JE}$ is given by

$$\underset{\approx}{X}_{JE} = \int d1\ n(1)\underset{\approx}{B}_{JE}(1) + \int d1\ d2\ n(1)n(2)\underset{\approx}{B}_J(1)\underset{\approx}{F}(1,2)\underset{\approx}{B}_E(2) \quad , \tag{5.16}$$

where the operator $\underset{\approx}{B}_{JE}(1)$ is defined by

$$\underset{\approx}{B}_{JE}(1) = \underset{\approx}{M}_{JE}(1) + \underset{\approx}{B}_J(1)S(1)\underset{\approx}{M}_E(1) = \underset{\approx}{M}_{JE}(1) + \underset{\approx}{M}_J(1)\underset{\approx}{S}(1)\underset{\approx}{B}_E(1) \quad , \tag{5.17}$$

with $\underset{\approx}{M}_{JE}(1)$ as defined in (4.22).

Above we have derived formal expressions for the four transport kernels appearing in the macroscopic equations (3.19). There remains the task of explicit evaluation of these kernels. As pointed out already, we can hope to aim only at an approximate calculation. It is slightly easier to calculate the transport coefficients appearing in (3.20). In the next section we derive expressions for the transport coefficients.

6. TRANSPORT COEFFICIENTS

In order to obtain well-defined transport coefficients we must consider the thermodynamic limit. Thus we consider a probability distribution $P(1,\ldots,N)$ for which the centers $(\vec{R}_1,\ldots,\vec{R}_N)$ are restricted to lie within a volume Ω of simple shape and for which the average number density $n(1)$ is approximately uniform inside Ω. We then consider the thermodynamic limit $N \to \infty$, $\Omega \to \infty$ at constant $n = N/\Omega$. We expect that the four transport kernels $\underset{\approx}{X}_{MM}$, $\underset{\approx}{X}_{ME}$, $\underset{\approx}{X}_{JM}$

and $\underset{\approx}{X}_{JE}$ possess a well-defined limit independent of the shape of Ω. In that case their expressions are said to be absolutely convergent. We also expect that in the limit the four kernels become translationally invariant and dependent only on the coordinate difference $\vec{r}-\vec{r}'$. We can then define wavevector-dependent transport coefficients in the following manner.

For the kernel $\underset{\approx}{X}_{MM}$ we consider the matrix element

$$(\vec{q}|\underset{\approx}{X}_{MM}|\vec{q}') = \int \left[\exp(-i\vec{q}\cdot\vec{r})\right]\vec{\tilde{X}}_{MM}(\vec{r},\vec{r}')\left[\exp(i\vec{q}'\cdot\vec{r}')\right]d\vec{r}\ d\vec{r}' \quad . \quad (6.1)$$

This is the wavevector representation of the operator. On account of translational invariance the matrix becomes diagonal in the thermodynamic limit, so that

$$\lim_{\substack{N\to\infty \\ \Omega\to\infty}} (\vec{q}|\underset{\approx}{X}_{MM}|\vec{q}') = 8\pi^3 \vec{\overset{\infty}{\chi}}(\vec{q})\delta(\vec{q}-\vec{q}') \quad , \tag{6.2}$$

which defines the tensor $\vec{\overset{\infty}{\chi}}(\vec{q})$. From isotropy it follows that the tensor has the form

$$\vec{\overset{\infty}{\chi}}(\vec{q}) = \chi_L(q)\hat{q}\hat{q} + \chi_T(q)(\vec{1}-\hat{q}\hat{q}) \quad . \tag{6.3}$$

The Fourier transform of the average equations (3.19) becomes in the thermodynamic limit

$$-\eta q^2 < \vec{v}_{\vec{q}} > + \vec{\overset{\infty}{\chi}}(\vec{q}) \cdot < \vec{v}_{\vec{q}} > - i\vec{q} < p_{\vec{q}} > = -\vec{\overset{\infty}{\chi}}_{ME}(\vec{q}) \cdot \vec{E}_{\vec{q}} \quad ,$$

$$\vec{q} \cdot < \vec{v}_q > = 0 \tag{6.4}$$

where $\vec{v}_{\vec{q}}$ is defined by

$$\vec{v}_{\vec{q}} = \frac{1}{8\pi^3} \int \vec{v}(\vec{r}) \exp(-i\vec{q}\cdot\vec{r})d\vec{r} \tag{6.5}$$

and the tensor $\vec{\overset{\infty}{\chi}}_{ME}(\vec{q})$ is defined as in (6.2). Due to the condition of incompressibility the term with $\chi_L(q)$ in (6.3) has no effect in (6.4). The effective viscosity appearing in (3.20) is given by

$$\eta_{eff} = \eta - \lim_{q\to 0} \chi_T(q)/q^2 \quad . \tag{6.6}$$

To lowest order in the density

$$\overset{\rightarrow}{\chi}{}^{\infty}(\vec{q}) \approx n(\vec{q}|\underset{\approx}{M}|\vec{q}) \quad , \tag{6.7}$$

where $\underset{\approx}{M}$ is the single-body convective friction kernel for a sphere taken to be centered at the origin. For a hard sphere with stick boundary conditions (6.7) leads to Einstein's expression for the effective viscosity [19]

$$\eta_{eff} = \eta[1 + \frac{5}{2}\phi + 0(\phi^2)] \quad , \tag{6.8}$$

where $\phi = (4\pi/3)na^3$ is the volume fraction occupied by spheres.

The expression (6.7) is the analogue of the expression derived by Foldy [20] and Lax [21] for the self-energy of a quantummechanical particle in a dilute random array of scatterers. The Ornstein-Zernike analysis presented in the preceding section allows an exact generalization of the expression [22], namely

$$\overset{\rightarrow}{\chi}{}^{\infty}(\vec{q}) = n(\vec{q}|\underset{\approx}{M} \ [\underset{\approx}{I} - n \ \underset{\approx}{R}(\vec{q})\underset{\approx}{M}]^{-1}|\vec{q}) \quad , \tag{6.9}$$

where the recurrence operator $\underset{\approx}{R}(\vec{q})$ may be expressed in terms of the operators $\underset{\approx}{S}(1)$ and $\underset{\approx}{S}(1,2)$ defined in (5.7) and (5.11). Explicitly

$$n \ \vec{R}(\vec{q};\vec{s},\vec{s}') = \vec{S}{}^{(1)}(\vec{s},\vec{s}') + n \ \vec{S}{}^{(2)}(\vec{q};\vec{s},\vec{s}') \quad , \tag{6.10}$$

where $\underset{\approx}{S}{}^{(1)}$ is defined by

$$\underset{\approx}{S}{}^{(1)} = \lim_{\substack{N\to\infty \\ \Omega\to\infty}} \underset{\approx}{S}(1)|_{\vec{R}_1=0} \tag{6.11}$$

and $\underset{\approx}{S}{}^{(2)}(\vec{q})$ by

$$\vec{S}{}^{(2)}(\vec{q};\vec{s},\vec{s}') = \lim_{\substack{N\to\infty \\ \Omega\to\infty}} \int d\vec{R} \ e^{i\vec{q}\cdot\vec{R}} \ \vec{S}(1,2;\vec{R}_1 + \vec{s},\vec{R}_2 + \vec{s}')|_{\vec{R}_1 = 0} \quad , \tag{6.12}$$

where $\vec{R} = \vec{R}_2 - \vec{R}_1$.

For the tensors $\overset{\rightarrow}{\chi}{}^{\infty}_{ME}(\vec{q})$, $\overset{\rightarrow}{\chi}{}^{\infty}_{JM}(\vec{q})$ and $\overset{\rightarrow}{\chi}{}^{\infty}_{JE}(\vec{q})$ one finds expressions analogous to (6.9). For example

$$\overset{\rightarrow}{\chi}{}^{\infty}_{JE}(\vec{q}) = n(\vec{q}|\underset{\approx}{M}_{JE}|\vec{q}) + n^2(\vec{q}|\underset{\approx}{M}_J \ \underset{\approx}{R}(\vec{q})[\underset{\approx}{I}-n \ \underset{\approx}{M} \ \underset{\approx}{R}(\vec{q})]^{-1} \ \underset{\approx}{M}_E|\vec{q}) \quad . \tag{6.13}$$

The collective mobility μ_C in (3.20) is given by

$$\vec{\vec{\chi}}_{JE}^{\infty}(0) = n\,\mu_C\,\vec{\vec{1}} \quad . \tag{6.14}$$

The other two tensors have a simple long wavelength limit

$$\vec{\vec{\chi}}_{JM}^{\infty}(0) = \vec{\vec{\chi}}_{ME}^{\infty}(0) = n\,\vec{\vec{1}} \quad , \tag{6.15}$$

as we have used in (3.20).

In analogy to (6.14) we may find a similar expression for the self-mobility μ_S, which is defined from the mean velocity of a single particle of the suspension, when an applied force acts only on that particle. The self-mobility is given by

$$\mu_S\,\vec{\vec{1}} = \lim_{\substack{N\to\infty \\ \Omega\to\infty}} (0|\underset{\approx}{B}_{JE}(1)|0)\big|_{\vec{R}_1=0} \quad . \tag{6.16}$$

This may be expressed as

$$\mu_S\,\vec{\vec{1}} = (6\pi\eta a)^{-1}\,\vec{\vec{1}} + (0|\underset{\approx}{M}_{J}\underset{\approx}{R}_S\left[\underset{\approx}{I}-n\,\underset{\approx}{M}\,\underset{\approx}{R}_S\right]^{-1}\underset{\approx}{M}_E|0) \tag{6.17}$$

where the recurrence operator $\underset{\approx}{R}_S$ is given by

$$n\,\underset{\approx}{R}_S = \underset{\approx}{S}^{(1)} \quad . \tag{6.18}$$

The self-mobility is directly related to the self-diffusion coefficient by the Einstein relation $D_S = k_B T\,\mu_S$.

The above statistical mechanical expressions for the transport coefficients n_{eff}, μ_C and μ_S are exact. There remains the task of evaluating the expressions. Beenakker and Mazur [23,24] have studied the same transport coefficients in a somewhat different framework. They perform an approximate self-consistent calculation and obtain good agreement with experimental results. Our own calculations [25,26] are based on a cluster expansion of the operators $\underset{\approx}{S}(1)$ and $\underset{\approx}{S}(1,2)$ defined in (5.7) and (5.11).

7. EXPLICIT CALCULATIONS

The basic elements which must be calculated are the reaction field operator $\underset{\approx}{S}(1)$ and the short-range connector $\underset{\approx}{S}(1,2)$. We assume that the single body friction kernel and the statistical distribution of particles are fully known.

The probability distribution $P(\vec{R}_1,\ldots,\vec{R}_N)$ is assumed to be normalized to unity and symmetric in the labels $1,\ldots,N$. The partial distribution functions

$$n(1,\ldots,s) = \frac{N!}{(N-s)!} \int\ldots\int d(s+1)\ldots dN\, P(1,\ldots,N) \qquad (7.1)$$

give the probability of finding a configuration of s particles, whatever the configuration of the remaining N-s particles. In practice only the two-particle distribution function $n(1,2)$ is well known. The three-particle distribution function $n(1,2,3)$ is less known. The Kirkwood superposition approximation, by which the three-particle distribution is expressed in terms of the two-particle distribution, works well up to a volume fraction of about 0.3.

From an analysis of the averaged multiple scattering expansion it follows that the operators $\underset{\approx}{S}(1)$ and $\underset{\approx}{S}(1,2)$ may both be expressed in terms of cluster expansions of the form [18]

$$\underset{\approx}{S}(1) = \sum_{\ell=2}^{\infty} \underset{\approx}{S}_\ell(1) \quad , \quad \underset{\approx}{S}(1,2) = \sum_{\ell=2}^{\infty} \underset{\approx}{S}_\ell(1,2) \quad , \qquad (7.2)$$

where $\underset{\approx}{S}_\ell(1)$ and $\underset{\approx}{S}_\ell(1,2)$ are expressed in terms of the solution of a scattering problem involving ℓ particles, averaged over positions with certain well-defined correlation functions. We quote only the two-particle contributions explicitly.

The two-particle contribution to $\underset{\approx}{S}(1)$ is given by

$$\underset{\approx}{S}_2(1) = \int d2\, n(2)g(1,2)\underset{\approx}{N}_{11}(1,2) \quad , \qquad (7.3)$$

where $g(1,2)$ is the pair distribution defined by $n(1,2) = n(1)n(2)g(1,2)$, and $\underset{\approx}{N}_{11}(1,2)$ is a nodal connector defined from the solution of the two-sphere scattering problem. For just two particles the nodal connector may be found from the connector operator matrix $\underset{\approx}{\boldsymbol{V}}$ defined in (4.29) according to $\underset{\approx}{N}_{11}(1,2) = \underset{\approx}{\boldsymbol{V}}_{11}(1,2)$. The two-particle contribution to $\underset{\approx}{S}(1,2)$ is given by

$$\underset{\approx}{S}_2(1,2) = g(1,2)\left[\underset{\approx}{N}_{12}(1,2)-\underset{\approx}{G}_0\right] + h(1,2)\underset{\approx}{G}_0 \qquad (7.4)$$

where $h(1,2) = g(1,2) - 1$ is the pair correlation function and $\underset{\approx}{N}_{12}(1,2)$ is a nodal connector given by $\underset{\approx}{N}_{12}(1,2) = \underset{\approx}{\boldsymbol{V}}_{12}(1.2)$. For $\ell > 2$ the expressions for the cluster operators in (7.2) involve more complicated correlation functions and nodal connectors.

An important contribution to the transport coefficients η_{eff} and μ_C is made by a virtual overlap term. The overlap term arises from the last term in (7.4) which yields a contribution

$$\underset{\approx}{S}_{ov}(1,2) = -\theta(2a-|\vec{R}_1-\vec{R}_2|)\underset{\approx}{G}_0 \quad . \tag{7.5}$$

This is the only contribution to $\underset{\approx}{S}(1,2)$ where the last but one scatterer overlaps with particle 1. In (6.12) the overlap term contributes

$$\vec{S}_{ov}^{(2)}(\vec{q};\vec{s},\vec{s}') = - \int\limits_{R<2a} \vec{G}_0(\vec{s}-\vec{s}'-\vec{R})e^{i\vec{q}\cdot\vec{R}} \, d\vec{R} \quad . \tag{7.6}$$

We abbreviate

$$\vec{K}(\vec{r},\vec{q}) = \int\limits_{R<2a} \vec{T}(\vec{r}-\vec{R})e^{i\vec{q}\cdot\vec{R}} \, d\vec{R} \quad . \tag{7.7}$$

For $\vec{q} = 0$ the integral may be evaluated fairly easily and yields

$$\vec{K}(\vec{r};0) = \frac{1}{\eta}\left[\frac{4}{3}a^2 - \frac{2}{15}r^2\right]\vec{T} + \frac{1}{15\eta}\vec{r}\vec{r} \qquad \text{for } r < 2a \quad . \tag{7.8}$$

If all other contributions to the recurrence operator in (6.10) are neglected, then one finds from (6.6) and (6.9)

$$\eta_{eff} = \eta\left[1 + \frac{5}{2}\frac{\phi}{1-\phi}\right] \qquad \text{(Saitô)} \quad . \tag{7.9}$$

This expression was first derived by Saitô [27] by use of arguments similar to those used by Lorentz to find the local field in the theory of dielectrics. Thus the Saitô-formula (7.9) is the analogue of the Clausius-Mossotti formula in the theory of dielectrics. It may be cast in the form

$$\frac{\eta_{eff}-\eta}{\eta_{eff}+\frac{3}{2}\eta} = \phi \qquad \text{(Saitô)} \quad . \tag{7.10}$$

Note that (7.9) predicts a divergence of the effective viscosity at a volume fraction $\phi = 1$. Experimentally the effective viscosity of a suspension diverges at $\phi_V \approx 0.65$, in the neighborhood of dense packing [28].

Corrections to the Saitô-formula may be cast in the form [26]

$$\frac{\eta_{eff} - \eta}{\eta_{eff} + \frac{3}{2}\eta} = \phi\,[1 + S] \quad , \tag{7.11}$$

where the correction S must be calculated from the remaining contributions to the recurrence operator (6.10). It is an important
task for theory to show how these correction terms shift the sin-
gularity to a new position ϕ_V. If just two-body terms $\underset{\approx}{S}_2(1)$ and
$\underset{\approx}{S}_2(1,2)$ are taken into account [26], then one finds the singularity at the rather low value $\phi_V \approx 0.364$. This is far too low and
shows that at high density it is essential to take at least three-
body, and probably even a class of many-body contributions into
account.

If we make the same overlap approximation in (6.13), then we
find for the collective mobility [25]

$$\mu_C \approx (6\pi\eta a)^{-1}\,[1-5\phi] \qquad (\text{overlap}) \tag{7.12}$$

The pair terms (7.3) and (7.4) yield the low density expansion [29,30]

$$\mu_C = (6\pi\eta a)^{-1}[1-6.546\phi+0(\phi^2)] \quad . \tag{7.13}$$

This shows a sharp decline of the collective mobility with increasing volume fraction. If the higher order terms are neglected,
then one finds from (7.13) that the collective mobility vanishes
at the low volume fraction $\phi_C \approx 0.15$. It is hard to see how the
neglected terms can shift this significantly. Cichocki and I have
suggested [25] that the theory predicts a dynamical phase transition at a low volume fraction. The prediction is conditional on
the assumption that one may calculate the averages by use of the
unchanged hard sphere distribution function. The change of distribution function will probably prevent the transition to occur at
small volume fraction in the low-frequency collective mobility. It
is an intriguing question whether the high-frequency collective
mobility actually vanishes at low volume fraction.

The virtual overlap term does not contribute in the calculation
of the self-mobility μ_S. A calculation based on the approximation
$\underset{\approx}{S}(1) \approx \underset{\approx}{S}_2(1)$ shows [25] that μ_S vanishes at a volume fraction
$\phi_S \approx 0.42$. This is probably too low. The self-consistent theory
of Beenakker and Mazur [23], as well as experimental data [31,32],
indicate a higher value. The transition point is expected to lie

higher than the glass transition, which is characterized by the
vanishing of the low-frequency or long-time self-diffusion coefficient.

The calculation of the transport coefficients n_{eff}, μ_C and μ_S
at high frequency, i.e. for unchanged distribution function, is
far from complete. For low density there are exact results, but at
high density further analysis is necessary.

8. EVOLUTION OF THE DISTRIBUTION FUNCTION

For the calculation of the frequency dependence of the transport coefficients the time evolution of the distribution function
must be taken into account. Provided the particles are sufficiently small this time evolution is dominated by Brownian motion. On
the timescale seen in light scattering experiments, typically 10^{-4}
sec, the momentum distribution may be regarded to be nearly Maxwellian [33]. On this timescale it therefore suffices to consider
the time-dependent configuration space distribution function
$P(\vec{R}_1,\ldots,\vec{R}_N,t)$. For Brownian particles the time-evolution of this
distribution function is governed by the N-body Smoluchowski equation. In formulating this equation we must account for hydrodynamic as well as direct interactions between particles.

We consider first the translational sphere velocities as given
by the creeping flow equations, when the spheres are subject to
applied forces $\{\vec{E}_j\}$, to zero torques and to an incident flow
$(\vec{v}_0(\vec{r})$, $p_0(\vec{r}))$, which satisfies the homogeneous equations (2.1).
The velocity of sphere j is

$$\vec{U}_j = \sum_{k=1}^{N} \vec{\mu}_{jk}^{tt}(1,\ldots,N) \cdot \vec{E}_k + \int \vec{C}_j(\vec{r};1,\ldots,N) \cdot \vec{v}_0(\vec{r})d\vec{r} \quad , \quad (8.1)$$

where the mobility tensor $\vec{\mu}_{jk}^{tt}$ is the same as in (2.19) and $\vec{C}_j(\vec{r})$
is a convection kernel which may be expressed in terms of a mul-
tiple scattering expansion. It is related to the kernel $\underset{\approx}{T}_{JM}(1,\ldots,N)$
by

$$\vec{C}_j(\vec{r}';1,\ldots,N) = \sum_{k=1}^{N} \int \vec{T}_{JM,jk}(\vec{r},\vec{r}';1,\ldots,N)d\vec{r} \quad . \quad (8.2)$$

The N-body Smoluchowski equation reads in abbreviated notation [34]

$$\frac{\partial P}{\partial t} + \underset{\sim}{\nabla} \cdot [\underset{\approx}{C} \cdot \vec{v}_0 \, P] + \underset{\sim}{\nabla} \cdot [\underset{\approx}{\mu}^{tt} \cdot \underset{\sim}{E} \, P]$$

$$= \underset{\sim}{\nabla} \cdot \underset{\approx}{D} \cdot [\underset{\sim}{\nabla} P + \beta(\underset{\sim}{\nabla}\Phi)P] \quad , \quad (8.3)$$

where the 3Nx3N diffusion matrix is given by

$$\vec{D}_{jk}(1,\ldots,N) = k_B T \, \vec{\mu}_{jk}^{tt}(1,\ldots,N) \quad . \tag{8.4}$$

Furthermore $\beta = 1/k_B T$ and $\Phi(1,\ldots,N)$ is the potential of direct interactions. For zero incident flow and in the absence of applied forces the equation has the equilibrium solution

$$P_{eq}(1,\ldots,N) = \exp[-\beta\Phi(1,\ldots,N)]/Q(\beta) \quad , \tag{8.5}$$

where $Q(\beta)$ is the normalization factor.

For an incident flow and applied forces which oscillate at frequency ω one may apply linear response theory to (8.3) and obtain the change of distribution function [34] at frequency ω. Once the distribution function has been found one may evaluate the frequency-dependent transport coefficients from the averaged multiple scattering expansion as described above. Reliable explicit calculations have been performed at zero frequency for the lowest order density correction to the single particle results [35,36]. Attempts have been made to extend the theory to higher density [37,38], but much work remains to be done.

REFERENCES

1) M. Smoluchowski, Proc. 5th Intern. Cong. Math. 2 (1912) 192.

2) C.W. Pyun and M. Fixman, J. Chem. Phys. 41 (1964) 937.

3) G.K. Batchelor, J. Fluid Mech. 52 (1972) 245.

4) B.U. Felderhof, Physica 82A (1976) 596.

5) P. Nozières, Physica 147A (1987) 219.

6) B.U. Felderhof, Physica 153A (1988) 217.

7) G.K. Batchelor, Ann. Rev. Fluid Mech. 6 (1974) 227.

8) R.J. Elliott, J.A. Krumhansl, and P.L. Leath, Rev. Mod. Phys. 46 (1974) 465.

9) J. Happel and H. Brenner, Low Reynolds Number Hydrodynamics (Noordhoff, Leyden, 1973).

10) P. Mazur and W. van Saarloos, Physica 115A (1982) 21.

11) B.U. Felderhof, Physica 151A (1988) 1.

12) P. Mazur and D. Bedeaux, Physica 76 (1974) 235.

13) S.R. de Groot, The Maxwell equations, Studies in Statistical
Mechanics, Vol. IV, eds. J. de Boer and G.E. Uhlenbeck
(North-Holland, Amsterdam, 1969).

14) B. Cichocki, B.U. Felderhof and R. Schmitz, PhysicoChem. Hyd.
10 (1988) 383.

15) B.U. Felderhof and R.B. Jones, Physica 146A (1987) 404.

16) B.U. Felderhof and R.B. Jones, Physica 93A (1978) 457.

17) M.S. Wertheim, Mol. Phys. 25 (1973) 211.

18) B. Cichocki and B.U. Felderhof, J. Stat. Phys. 51 (1988) 57.

19) A. Einstein, Ann. Phys. 19 (1906) 289, 34 (1911) 591.

20) L.L. Foldy, Phys. Rev. 67 (1945) 107.

21) M. Lax, Rev. Mod. Phys. 23 (1951) 287, Phys. Rev. 85 (1952) 621.

22) B.U. Felderhof and B. Cichocki, J. Stat. Phys. 55 (1989) 1157.

23) C.W.J. Beenakker and P. Mazur, Physica 120A (1983) 388, 126A
(1984) 349.

24) C.W.J. Beenakker, Physica 128A (1984) 48.

25) B. Cichocki and B.U. Felderhof, Physica 154A (1989) 213.

26) B. Cichocki, B.U. Felderhof, and R. Schmitz, Physica 154A
(1989) 233.

27) N. Saitô, J. Phys. Soc. Japan 5 (1950) 4, 7 (1952) 447.

28) J.C. van der Werff, C.G. de Kruif, C. Blom and J. Mellema,
Phys. Rev. 39A (1989) 795, 3696.

29) G.K. Batchelor, J. Fluid Mech. 52 (1972) 245.

30) B. Cichocki and B.U. Felderhof, J. Chem. Phys. 89 (1988) 1049.

31) A. van Veluwen, H.N.W. Lekkerkerker, C.G. de Kruif and A. Vrij,
Faraday Disc. Chem. Soc. 83 (1987) 59.

32) R.H. Ottewill and N.S.J. Williams, Nature 325 (1987) 232.

33) P.N. Pusey and R.J.A. Tough, in Dynamic Light Scattering and
Velocimetry: Applications of Photon Correlation Spectroscopy,
ed. R. Pecora (Plenum, New York, 1981).

34) B.U. Felderhof and R.B. Jones, Physica 119A (1983) 591, 146A
(1987) 404.

35) G.K. Batchelor, J. Fluid Mech. 83 (1976) 97, 131 (1983) 155,
137 (1983) 467.

36) B. Cichocki and B.U. Felderhof, J. Chem. Phys. 89 (1988) 3705.

37) W.B. Russel and A.P. Gast, J. Chem. Phys. 84 (1986) 1815.

38) N.J. Wagner and W.B. Russel, Physica 155A (1989) 475.

FUNDAMENTAL PROBLEMS IN STATISTICAL MECHANICS VII
H. van Beijeren, Editor
© *Elsevier Science Publishers B.V., 1990*

STATISTICAL MECHANICS OF POLYMERS

Bernard NIENHUIS

Instituut voor theoretische fysica, Universiteit van Amsterdam, Valckenier-
straat 65, 1018 XE Amsterdam, Netherlands

The description of configurational statistics of long linear polymers in solution
is presented and discussed. Emphasis is placed on scaling and universal properties. Many of the results follow from a formal equivalence with spin models
in ferromagnetism. Some of the more recent developments for two-dimensional
systems, based in part on a relation with models for crystal surfaces, are briefly
reviewed.

1   Introduction

Polymers exist in an enormous variety both in nature and artificial materials. In order to

understand the properties of these large molecules and of the materials made from them,

it is necessary to have some insight in their configurational statistics. In these lectures we

will discuss some developments in the understanding of the static macroscopic properties

of large linear polymers in solution. The fact that this can be discussed in general terms

at all, without specifying the chemical nature of the molecules, is due to a great measure

of universality. A large number of macroscopic properties of chain molecules turns out to

depend only on a limited number of parameters. This fact is much to the advantage of the

theorist, because it allows a great freedom to model the real physical system by abstract

constructs. The theorist will therefore often tailor his models not so much to match the

real system accurately, as to allow the mathematical techniques at his disposal to carry

him as far as possible. While this procedure is legitimate it is still necessary to verify that

the essential ingredients of the physical system are still present in the theoretical model.

The attempts to include what are believed to be these essential ingredients in the theory

form the guide through the topics of these lectures. They should be viewed primarily as

an introduction into a field that deserves to be treated more extensively.

We will discuss in these lectures only chain polymers, which consist of a linear repetition

sequence of a identical molecular groups. These repeat units, or monomers, may be single

atoms such as sulphur, but usually consist of whole groups of atoms. In polymethylene

for instance the backbone of the chain is formed by a sequence of carbon atoms, each of which is coupled to two hydrogen atoms, and the monomer is $-CH_2-$. Polystyrene has the same carbon chain as a skeleton, but now of each four hydrogen atoms one is replaced by a hydrogenated carbon ring, so that the repeat unit is $-CH_2CHC_6H_5-$.

2 Ideal chain models

On the scale of atoms the spatial structure of a polymer molecule is determined almost completely by its chemical bonds. Though thermal and quantum mechanical fluctuations affect this local structure, it is still reasonable to view the skeleton of chemical bonds of a large molecule as locally rigid. When however the molecule is viewed at a larger scale of say thousands of monomers, the small local fluctuations accumulate to an almost complete lack of correlation between such widely separated chemical bonds. On sufficiently large scale the chain can for most practical purposes be described as a flexible string, of which all spatial configurations are equally probable. This brings us to the simplest model: the ideal chain. Consider the polymer as a sequence of identical units, each of which is already large compared to the monomer. Number these units along the chain and let the position of the k-th unit be $\vec{r}_k$, and the distance between two consecutive units $\vec{a}_k = \vec{r}_k - \vec{r}_{k-1}$. The simplest result is obtained, when it is assumed that the $\vec{a}_k$ are identically distributed and totally uncorrelated. In that case the mean square of the end-to-end distance of a chain of L units is clearly

$$\left\langle \vec{r}_L^{\,2} \right\rangle = \left\langle \sum_{k=1}^{L} \vec{a}_k^{\,2} \right\rangle = L \left\langle a^2 \right\rangle . \tag{1}$$

Therefore the root mean square of this distance, denoted as R_L grows as L^ν with $\nu = 1/2$. Note that in this power law relation the coefficient $\langle a^2 \rangle$ depends on details of the model, in particular how the local configurations are distributed, but the power is equal to $1/2$ irrespective of all this. The shape of the entire distribution q of the end-to-end distance is Gaussian for the ideal chain

$$q\left(\vec{r}_L\right) = \left(2\pi \left\langle a^2 \right\rangle L\right)^{-d/2} \exp\left(-\frac{\vec{r}_L^{\,2}}{2 \left\langle a^2 \right\rangle L}\right), \tag{2}$$

again irrespective of the distribution of the individual $\vec{a}_k$.

Among the ingredients to the configurational distribution that have been ignored so far, is the residual stiffness. The chain is treated as infinitely flexible on sufficiently large scale. In order to see if this is justified we now allow each pair of consecutive units to be correlated and compute how this affects the large scale behavior. Suppose the distribution function ρ_L of the entire configuration factorizes in functions depending on two consecutive $\vec{a}_k$ only, symbolically

$$\rho_L\left(\vec{a}_1, \vec{a}_2, .., \vec{a}_L\right) = \prod_{k=1}^{L-1} \rho_2\left(\vec{a}_k, \vec{a}_{k+1}\right). \tag{3}$$

Furthermore let us assume (for computational convenience) that the length of the $\vec{a}_k$ does not fluctuate, but takes a single value $|a|$. Then assuming isotropy in space we find

$$\int \vec{a}_k \ \rho_2\left(\vec{a}_k, \vec{a}_{k+1}\right) \ d\vec{a}_{k+1} = p \ \vec{a}_k \tag{4}$$

with $p < 1$. And as a consequence $\left\langle \vec{a}_k \cdot \vec{a}_{k+m} \right\rangle = p^m \, |a|^2$. From this one can readily derive the root-mean-square end-to-end distance

$$R_L^2 = |a|^2 \left(L + 2p\frac{L - 1 - Lp + p^L}{(1 - p)^2}\right). \tag{5}$$

From inspection of this result it is clear that the presence of the correlation p changes the coefficient of the square root dependence of R_L on large L and moreover gives a correction for relatively small L, specifically $L < L_p = -1/\ln p$, resulting in a linear dependence in this regime. This retrospectively justifies the earlier simpler model where correlations were completely ignored. However, the result also confirms that the asymptotic power law depencence only sets in on scales larger than the persistence length L_p, below which the stiffness of the chain makes it behave like a hard rod. Before we proceed to incorporate other interactions, note that the results so far are independent of spatial dimensionality d.

3   The self-avoiding walk

Another effect that has thus far been ignored in these treatments, is the fact that real polymers occupy a finite volume, that cannot be penetrated by other molecules or by other segments of the same molecule. This so called excluded-volume interaction is of an essentially different nature than the local stiffness, as it immediately affects segments which are far separated in terms of the molecular sequence. While the exact incorporation of this effect is very difficult, its consequences can be shown in good approximation by a surprisingly simple argument due to Flory[1]. Suppose a chain of length $L \gg 1$ extends over an end-to-end distance R_L in a space of d dimensions. Formally the partition sum Z_L can be written as the product

$$Z_L = z^L \, R_L^{d-1} \, \exp\left(-\frac{R_L^2}{2a^2 L}\right) \, \left\langle \exp\left(\frac{-U}{kT}\right) \right\rangle, \tag{6}$$

where the average is taken over all ideal chain configurations, and the excluded-volume effect is written in terms of a penetration energy U. The Gaussian factor gives the number of ideal chain configurations (2) of L units, extending over a distance of R_L. The parameter z is the number of configurations for a single unconstrained unit. It will be clear that all difficulty now resides in the computation of the average. To overcome this a mean-field like approximation will be made, replacing U by $\langle U \rangle$, and in the same spirit approximating $\langle U \rangle$ as if the polymer is evenly distributed over the sphere with radius R_L. Within this crude approximation the energy is replaced by the volume of this sphere multiplied by the square of the density within it.

$$\left\langle \exp\left(\frac{-U}{kT}\right) \right\rangle \sim \exp\left(-cst.L^2 \, R_L^{-d}\right). \tag{7}$$

The value of R_L that will actually be assumed can now be found by maximizing Z_L with respect to R_L. This is done by locating the zero's of

$$\frac{\partial \ln Z_L}{\partial \ln R_L} \sim (d-1) \, - \, R_L^2 \, L^{-1} \, + \, R_L^{-d} \, L^2, \tag{8}$$

in which the model-dependent positive coefficients are omitted. We assume that $R_L = L^\nu$ for some unknown ν. In the limit $L \to \infty$ one of the two positive terms will be negligeable and the other must balance with the one negative term. The two cases, expressed in the unknown ν now read

$$\begin{aligned} 0 \; &< \; 2\nu - 1 \; = \; 2 - d\nu \\ 0 \; &= \; 2\nu - 1 \; > \; 2 - d\nu. \end{aligned} \tag{9}$$

Solving the equality for ν and interpreting the inequality as a limit on the domain of validity, we obtain

$$\nu = \frac{3}{2+d} \quad for \quad d \le 4$$

$$and \tag{10}$$

$$\nu = 1/2 \quad for \quad d \ge 4.$$

This famous result turns out to be correct in one, two, four and more dimensions, but appears in (numerically small) error only in three dimensions, this in spite of the fact that the argument is far from rigorous. Comparison with the ϵ-expansion[2] shows, however, that also the results for non-integer dimension below four are not exact. Besides the surprising accuracy of these exponents, this argument gives a clear indication that the excluded volume effect is capable to change the critical exponent from their classical value. The optimized partition sum within this approximation irrespective of dimension, is given by

$$Z_L = z^L L^{(d-1)\nu} \exp(-cL^{2\nu-1}),\tag{11}$$

where z and c are model dependent, and ν is given by (10).

4 Relation to ferromagnetism

The statistical mechanics of polymers in solution has made significant progress through a formal relation with the theory of the critical point in spin models due to De Gennes[3] and Des Cloizeaux[4] . This relation is presented in this and utilised in the next section. It will be shown that the thermodynamic functions of polymers in solution, are formally equivalent to those of an n-component spin model or n-vector model in the limit $n \to 0$. For conceptual convenience the spin model is defined on the lattice and as a consequence the equivalent model for polymers takes into account only configurations on the lattice, but we emphasize that the equivalence itself is not thus restricted, but is valid also in the continuum.

The spin model is defined by the Hamiltonian

$$-\frac{\mathcal{H}}{kT} = \sum_{<i,j>} K \, \vec{s}_i \cdot \vec{s}_j + \sum_m H \, \hat{e} \cdot \vec{s}_m \; .\tag{12}$$

The spins $\vec{s}$ are n-component vectors which assume values on the surface of the sphere with squared radius n. The first sum is over nearest-neighbor pairs of sites of the lattice. The magnetic field has strength H and the direction of the unit vector $\hat{e}$. The partition sum

$$Z = \int \prod_i \frac{d\vec{s}_i}{\Omega_n} \prod_{<j,k>} \exp(K \, \vec{s}_j \cdot \vec{s}_k) \prod_m \exp(H \, \hat{e} \cdot \vec{s}_m)\tag{13}$$

in which Ω_n is the surface of the n-dimensional sphere, can be expanded in powers of the exchange coupling K and the field H. This is done by expansion of the exponentials and of the products. Before the integration over the spins has been performed, each term in this expansion is a product of powers of the various $K\,\vec{s}_j\cdot\vec{s}_k$ and $H\,\hat{e}\cdot\vec{s}_m$. The computation of the partition sum now involves three operations, the integration of the spin variables, the summation over spin components implicit in the scalar products, and the summation over the various terms in the expansion.

The integration over the spin variables can be readily performed by virtue of the fact that the integrand factorizes completely into products of components of the same spin, such as

$$\int \frac{d\vec{s}}{\Omega_n} = 1$$
$$\int \frac{d\vec{s}}{\Omega_n} s^\alpha = 0$$
$$\int \frac{d\vec{s}}{\Omega_n} s^\alpha s^\beta = \delta_{\alpha,\beta}$$

(14)

the spin components being indicated by upper Greek indices. The first of these simply defines the quantity Ω_n to be the surface of the hypersphere. The second follows from reflection symmetry. The third involves the same symmetry, and is normalized defining the radius of the sphere to be $n^{1/2}$. Now, bearing in mind that we intend to take the formal limit $n \to 0$, we note that higher products are of higher order in n

$$\int \frac{d\vec{s}}{\Omega_n} \prod_{j=1}^{2k-1} s^{\alpha_j} = 0$$
$$\int \frac{d\vec{s}}{\Omega_n} \prod_{j=1}^{2k} s^{\alpha_j} = O(n^{k-1}),$$

(15)

for $k \geq 1$. Again the odd product follows from reflection symmetry of the sphere. To verify the result of the even product note that the integral can only be non-zero when the indices are at least pairwise equal. The spherical symmetry then requires that the result be a constant times the sum over all pairings of indices of the product of Kronecker δ's of the pairs. The constant can be found from the observation that the known integral

$$\int \frac{d\vec{s}}{\Omega_n} \left|\vec{s}\right|^{2k} = n^k$$

(16)

can be expressed in the same even products over spin components.

In the limit $n \to 0$ the terms that will survive integration over the spin variables, involve either none or two factors of each spin. The bookkeeping of these terms is facilitated by a diagram expansion, keeping track of the powers of $K\,\vec{s}_j\cdot\vec{s}_k$, denoted by a bond between

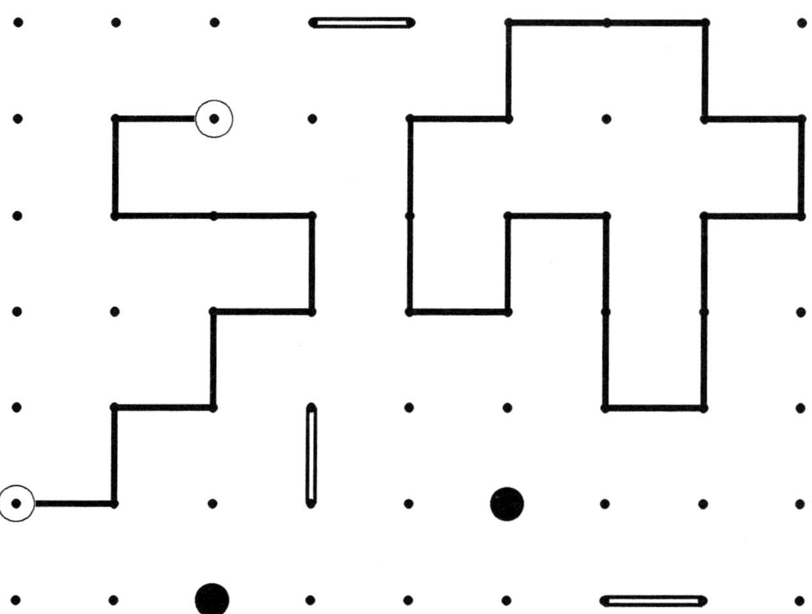

Figure 1: An example of a diagram in the expansion in powers of K and H

sites j and k, and of $H\,\hat{e}\cdot\vec{s}_m$, marked as a circle on site m. An example of such diagrams on the square lattice is given in figure 1. In the diagrams corresponding with the surviving terms in the expansion, each site is either covered by two circles (marked as a filled circle in fig. 1), connected by two bonds, connected by one bond and covered by one circle, or not marked at all. Therefore the diagram consists entirely of the following disconnected parts; i) continuous chains of bonds terminating in circled site at both ends; ii) continuous closed polygons of bonds; iii) single lattice edges marked by two bonds; and iv) single sites marked by a filled circle. To obtain the contributing factor from the chains and loops, we must repeatedly apply the rule

$$\int \frac{d\vec{s}}{\Omega_n} (\vec{a}\cdot\vec{s})(\vec{s}\cdot\vec{b}) = (\vec{a}\cdot\vec{b}) \tag{17}$$

for given vectors $\vec{a}$ and $\vec{b}$. Thus the loops and double bonded edges can be reduced to the integral $\int d\vec{s}\,/\Omega_n(\vec{s}\cdot\vec{s}) = n$, and the chains and dotted sites to $(\hat{e}\cdot\hat{e}) = 1$. Thus we see that each loop and double bonded edge contributes a factor n to the diagram. In the limit $n \to 0$ therefore, the only surviving diagrams consist entirely of chains and of dotted sites. Each of these contributes a factor H^2 to the weight of the diagram, and the chains collectively contribute the factor K^L, where L is the total number of bonds of all the chains taken together. Thus the partition sum of the spin model in this limit can be written

$$Z(K,H) = \sum_{chains} H^{2N} K^L, \tag{18}$$

where the sum is over all possible diagrams consisting of self- and mutually avoiding chains of bonds on the lattice, N is the number of chains and L their combined length, measured in bonds. In this notation chains of length zero are allowed, and represented by the dotted sites. If these for any reason would not be desired they can be trivially summed out, for each given configuration of larger chains, such that the resulting partition sum is still given by (18) after a rescaling with a factor $(1 + H^2/2)$ per site, and renormalizing $K \to K/(1 + H^2/2)$ and $H^2 \to H^2/(1 + H^2/2)$.

At this point it will be obvious that the chain diagrams in the partition sum are to be interpreted as configurations of polymers on the lattice. The lattice must now be viewed as an approximation, and we will restrict the inferences from the equivalence to properties that are universal, i.e. lattice independent. The lattice distance here plays the role of two distinct physical distances. (i) The minimal separation between units, or thickness of the polymer, due to the fact that no two chain segments may occpuy the same site. (ii) The persistence length L_p, below which scale the stiffness of the chain is noticable, due to the fact that that no weight variation is associated with the corners of the chain. Though there is no reason in principle that these two intrinsic lengths are similar, the predictions of the model will only be used for scales much larger than the lattice constant. These results then apply to physical systems on scales much that are large relative to both the polymer thickness and the persistence length.

Note that $Z(K,H)$ is not the partition sum of a given set of polymers but is like a generating function, or a grand canonical partition sum in which both the number of polymers and the total number of monomers is allowed to fluctuate. On first sight eq. (18) may look like the proper description of an ensemble resulting from equilibrium polymerization, but this is a subtle deception. During polymerization in equilibrium there is nothing to prevent the formation of closed ring polymers, which on the lattice would be represented by polygons. This difference, innocent as it may look, does affect the universal properties such as scaling exponents. Therefore (18) will first be viewed rather as a generating function of polymer partition sums, and later also as no more than an approximation to the ensemble of a large number polymers in solution.

5 Scaling relations

In this section we will utilize the equivalence between polymer statistics and spin models to derive various properties and relations in the field of polymers from existing known properties of the critical point in spin models. Without further discussion, it will be taken for granted that the spin model even in the limit $n \to 0$ has a continuous phase transition at zero field ($H = 0$) and some finite critical value $K = K_c$. Furthermore it is presumed that various thermodynamic functions such as the susceptibility and the specific heat have algebraic singularities, which may be derived from the singular part of the free energy per site

$$f(\tau, H) \sim |\tau|^{d\nu} F(H|\tau|^{-\Delta}), \tag{19}$$

where $\exp(-\tau) = K/K_c$. Here F is some universal but unknown function of the scaling variable $H|\tau|^{-\Delta}$, still dependent on the sign of τ. Besides this scaling contribution, the free energy also has a regular contribution, which is analytic at the critical point, and corrections to scaling, which are more weakly singular, and numerically subordinate to (19). From the partition sum (18) it follows directly that the the free energy in zero field H vanishes identically, in particular, for $\tau > 0$ and as $x \to 0$ the function $F(x)$ vanishes quadratically in x. It is tempting to claim a similar behavior for $\tau < 0$, but that is beyond the radius of convergence of the expansion of the partition sum (18). It is believed, that in the limit $H \to 0$, taken after the thermodynamic limit, the free energy has a non-zero value for $\tau < 0$.

Because the power of H in $Z(K, H)$ (18) is equal to twice the number of chains in the contributing diagram, the single polymer can be recovered from the zero field susceptibility of the spin model.

$$\chi = \frac{1}{V} \frac{\partial^2 \ln Z(K, H)}{\partial H^2}\bigg|_{H=0} = \sum_L Z_L K^L, \tag{20}$$

where V is the volume of the system, and Z_L is the partition sum of a single polymer of length L starting at the origin. Since χ diverges algebraically when the critical point is approached, we write

$$\chi \sim \tau^{-\gamma}, \tag{21}$$

as $\tau \to 0$, with $\gamma = 2\Delta - d\nu$. From this it follows that

$$Z_L \sim K_c^L \, L^{\gamma-1}, \tag{22}$$

asymptotically for large L. This result can be verified by substituting it in (20), and replacing the sum by an integral. The form is similar to that obtained from Flory's mean field theory (11), in that it consists of an exponential and an algebraic factor. The 'stretched' exponential, i.e. the exponential with a power less than one in the exponent, in (11) does not greatly affect the similarity since it corresponds to weaker confluent singularities in χ, which are not taken into account here. A more serious discrepancy is the independence of the exponent γ on the end-to-end exponent ν as we will see later.

For the partition sum of a single closed polymer we need to relax the limit $n \to 0$, and take the derivative of $Z(K, H)$ with respect to n at both $n = 0$ and $H = 0$.

$$\frac{1}{V} \left. \frac{\partial \ln Z(K,H)}{\partial n} \right|_{H=0,n=0} = \sum_L C_L \, K^L. \tag{23}$$

For other values of n the free energy for the spin model near the critical point has a singular contribution $f \sim \tau^{d\nu}$. However, since the partition sum itself in the limit $n \to 0$ is trivially unity, the free energy must vanish, and the singular term has a coefficient that vanishes linearly in n, so that in this case $\partial f / \partial n \sim \tau^{d\nu}$. Therefore the partition sum of the single closed polymer of length L behaves like

$$C_L \sim K_c^L \, L^{-1-d\nu}. \tag{24}$$

This quantity, though not of immediate physical relevance, will be used later in the discussion of correlation functions.

Now we will turn to the shape of the isolated chain polymer in solution. The susceptibility of the spin model can be written as the spatial sum of the correlation function

$$\chi = \sum_{\vec{r}} q\left(\vec{r}\right), \tag{25}$$

where $q(\vec{r}_j - \vec{r}_k) = \langle \vec{s}_j \cdot \vec{s}_k \rangle$. Then clearly

$$q\left(\vec{r}\right) = \sum_L K^L Q_L\left(\vec{r}\right),\tag{26}$$

where $Q_L(\vec{r})$ is the number of chain configurations of length L, with end points in the origin and $\vec{r}$. From renormalization theory we know that

$$q\left(\vec{r}\right) \sim \tau^{d\nu-\gamma} S(r\tau^{\nu}),\tag{27}$$

sufficiently close to the critical point. When it is assumed that also $Q_L(\vec{r})$ has a scaling form, i.e. its $\vec{r}$-dependence is entirely through $\vec{r}$ multiplied by some power of L, it can be derived that for sufficiently large r and L

$$Q_L\left(\vec{r}\right) \sim K_c^{-L} L^{\gamma-d\nu-1} P\left(rL^{-\nu}\right).\tag{28}$$

In retrospect (28) can be verified by substitution in (26) and comparing zero-th and second moments in $\vec{r}$ with those of (27). The scaling form (28) serves to show that the spatial distribution of polymers of any sufficiently great length L is given by this one function P. Even without saying anything about P, except that its second moment exists, it is evident that the exponent ν is the same as that of the root mean square of the end-to-end distance

$$R_L \sim L^{\nu}.\tag{29}$$

Before we discuss the behavior of $P(x)$ we should distinguish various regimes in physical distance r. When $r \sim 1$, the lattice of the model plays a dominant role, and the scaling form (28) does not apply. When $r \sim L$ the polymer is close to completely stretched, and again the lattice makes itself noticable. The form (28) is valid only in the scaling regime $1 \ll r \ll L$, and here we expect the scaling function $P(x)$ to be universal for all polymers in good solvent. Though $P(x)$ has not been calculated in general, some insight has been gained into its short range[3] and long range[5] extremes, $1 \ll r \ll L^{\nu}$ and $L^{\nu} \ll r \ll L$.

In the short range limit the two ends of the polymer approach each other closely, so that the entire configuration looks much like a closed ring polymer which is briefly interrupted. Therefore, the number of such confgurations in this limit is

$$Q_L(\vec{r}) \sim L\, C_L\, D(r),\tag{30}$$

where C_L is given by (24). The function D gives the number of ways the origin can be connected to $\vec{r}$ via a relatively short chain, in the presence of a very much longer chain already connecting these points. When (24) and (28) are substituted in (30), it is seen that

$$P(x) \sim x^{(\gamma-1)/\nu}, \quad for \ x \ll 1 \tag{31}$$

In the long range limit it has been observed empirically[6], from model calculations, that $P(x)$ behaves as a stretched exponential

$$P(x) \sim \exp\left(-x^\delta\right), \tag{32}$$

where δ is appreciably larger than one. While only plausible theoretical arguments have been put forward for this behavior, the value of the exponent δ can be related to other exponents, once the form itself has been granted. The argument[5] follows from the observation that the correlation function of the spin model, at $K < K_c$ decays exponentially for sufficiently large r

$$q\left(\vec{r}\right) \sim \exp\left(-r\tau^\nu\right), \tag{33}$$

In this expression (26) is substituted for $q(\vec{r})$ and subsequently (28) for Q_L. It is reasonable to assume that for $r \gg \tau^{-\nu}$ the summation is dominated by the long range behavior (32) of P and the exponent δ must satisfy

$$\sum_L L^{\gamma-d\nu-1} \exp\left(-\tau L - r^\delta L^{-\delta\nu}\right) \sim \exp\left(-r\tau^\nu\right). \tag{34}$$

Notice that the summand behaves smoothly as function of L, assumes a maximum value for some large L, and decays rapidly at both the large and small L side. Therefore the sum can be estimated as the value of the largest term, multiplied by the number of terms of the same order of magnitude. Keeping only the exponential dependence on r and suppressing the algebraic corrections, which already have been ignored in the hypothesis (32), we find

$$\exp\left(-r^{\delta/(\delta\nu+1)} \tau^{\delta\nu/(\delta\nu+1)}\right) \sim \exp(-r\tau^\nu). \tag{35}$$

Therefore $\delta = 1/(1 - \nu)$. In Fisher's[5] original argument for this relation, he also takes into account factors which depend on r and L only as a power. However, the main result concerns the value of δ, and the algebraic factors require a treatment more elaborate than they receive. For this reason only the essentials of the argument, sufficient for δ are repeated here.

So far we have been concerned only with the description of a single or isolated polymer in solution. In principle the partition sum $Z(K, H)$ itself describes a grand canonical ensemble of a polymer solution. The number of polymers is controled by H, and the total length of the polymers by K. We now proceed to derive the form of the osmotic pressure of the solution as a function of the polymer length and density, following Des Cloizeaux[4] and De Gennes[3] .

The ostmotic pressure is given by the volume derivative of the total free energy

$$\frac{\Pi}{kT} = \frac{\partial}{\partial V} \ln Z = \frac{\ln Z}{V} = f(\tau, H). \tag{36}$$

The average polymer density is readily found from the ensemble as

$$C_p = \langle N \rangle = \frac{H}{2} \frac{\partial f(\tau, H)}{\partial H}. \tag{37}$$

And likewise the monomer density is

$$C_m = \langle L \rangle = \frac{\partial f(\tau, H)}{\partial \tau}. \tag{38}$$

Taking the familiar scaling form of the free energy (19), similar forms for the osmotic pressure, the polymer and monomer density readily follow. In the expression of the osmotic pressure the variables τ and H can then be eliminated in favour of C_p and C_m, resulting in

$$\frac{\Pi}{kT} \sim C_p \, G(C_m^{d\nu} C_p^{1-d\nu}), \tag{39}$$

where G is some other scaling function, which can be expressed in F. At sufficiently low densities the osmotic pressure is proportional to C_p, consistent with the scaling form (39), and following directly from the known small H behavior of the free energy. This ideal gas behavior is to be expected when the polymers are well separated. However, at large densities, when the polymers deeply penetrate each others domains, one expects that the osmotic pressure is a function solely of the monomer density. When this plausible assumption is confronted with (39), it follows that

$$\frac{\Pi}{kT} \sim C_m^{d\nu/(d\nu-1)}. \tag{40}$$

It is readily verified that the crossover takes place when the polymer density is of the order of R_L^{-d} where the geometrical size of the polymers $R_L \sim L^\nu$, and the average length is estimated as $L = C_m/C_p$.

A word of caution is in order. For the partition sum (18) to be useful for the description of a prepared set of polymers of given lengths, be it polydisperse or monodisperse, it must be shown that the average variation in the length of each polymer is small relative to the average length itself. We may take for the squared variation of the length

$$\frac{\langle L^2 \rangle - \langle L \rangle^2}{\langle N \rangle^2} = \frac{\frac{\partial^2 f(\tau,H)}{\partial \tau^2}}{\left[\frac{H}{2}\frac{\partial f(\tau,H)}{\partial H}\right]^2}, \tag{41}$$

and for the length itself $\langle L \rangle / \langle N \rangle$. When these quantities are expressed in τ and H by substitution of the scaling form of f, it is seen that in that in the limit of very long polymers, the thus estimated length fluctuations are as large or larger than the length itself. Of course it were better if we could compute $\langle L/N \rangle$ and $\langle L^2/N^2 \rangle - \langle L/N \rangle^2$, being the true average length and variation. However, while no better estimates are available, we must admit that the above results for the osmotic pressure of a polymer solution, may not apply to the physical situation.

6   Discussion of exact results in two dimensions

In the previous section we have seen that many of the properties of the polymer in solution, in suitable limits can be expressed in scaling functons and a few exponents such as γ and ν. Both the scaling functions and the value of the exponents are universal, i.e. dependent on dimension only. This universality allows a lot of freedom to the theorist who wants to calculate them, as almost any reasonable model will give the same exponents and scaling functions. Nevertheless, the universal quantities in three dimensions have not been calculated, except in approximation. In two dimensions which is of considerably less, but still some[8] experimental interest, the exponents have been determined exactly. In this section we give these derivations in outline only. The original arguments are quite elaborate and are given extensively in a recent review[9].

Previously we have seen that a model of chains on the lattice, is equivalent to the limit $n \to 0$ of an n-component isotropic spin model. This equivalence is valid in any dimension. In two dimensions, another equivalence has brought the theory some steps forward.

This equivalence is with the so called solid-on-solid (SOS) models[9], invented to describe equilibrium fluctuations the surface of a crystal. In these models the crystal surface is represented by a configuration of discrete height variables on a two-dimensional lattice. The lattice being two of the spatial coordinates of the crystal surface, and the heights representing the third. In the case of interest the height variables assume integer values, and those of two neighboring sites may differ by at most one unit. The configurations of the model can thus be represented by domain walls between areas that differ in height by one unit. If the height variables sit on a triangular lattice, the domain walls consist of edges of the honeycomb lattice and are naturally prevented from crossing. The domain walls in the SOS model now form closed polygons, and have an intrinsic orientation associated with the sign of the difference of heights of the interior and the exterior. Open chains are introduced by admitting screw dislocations into the model, where a mismatch of heights allows a domain wall to end. The closed polygon configurations are suppressed by a subtle association of a phase factor with the curvature of the domain walls, so that after summation over the orientation of the domain walls, each closed polygon has gotten some fixed factor, which is then tuned to zero.

The reason that the equivalence with SOS models is useful, is that renormalization group arguments have shown that they have a critical regime, at which a very extensive family of critical exponents all can be expressed in a single albeit unknown parameter. In this case, as in many others, some of these exponents were known[9] from the exact solution of the related eight-vertex model. This knowledge suffices to fix the one unknown parameter, and thus the exponents follow, including those of the polymer problem.

$$\nu = 3/4, \quad \gamma = 43/32. \tag{42}$$

These methods using renormalization and asymptotic equivalence with SOS models is commonly known as Coulomb-gas technique, the Coulomb gas being simply another disguise the SOS models may wear. Later the same arguments were used to compute many other exponents, associated with polymers consisting of long linear segments and a given number of branch points[10]. Similarly by arguments such as those used in these lectures for the short range behavior of the end-to-end correlation function, properties of correlations functions of the polymer describing the probability of close approaches distant segments of the same polymer have been computed.

More recently these results have been confirmed[11,12] by an entirely different approach. It is based on Bethe-Ansatz calculations applied directly to the honeycomb chain model. So far, this method has led to no new results for the polymer in good solvent, but has given a rigorous basis to some exponents already known. Even though this method is at least as elaborate as the Coulomb-gas technique, this is a satisfactory development. In the earlier method an equally elaborate exact solution of a related model has been used as an ingredient anyway. The Bethe-Ansatz calculation of the honeycomb model has been followed since, by a similar calculation for the square lattice[14,15]. This gave a solid confirmation of the same exponents for another lattice, in agreement with universality.

This latter calculation also yielded results for another universality class of polymer solutions, the so called Θ-solution, which signals the onset of phase separation between polymer and solvent when the solvent does not dissolve the polymer so well. The situation for this problem is not quite so satisfactory as for the so called good solvent, as there are different candidates for this universality class in the literature. One model, closely akin to the honeycomb model[16] for the good solvent, has yielded the exact exponents $\nu = 4/7$ and $\gamma = 8/7$ by means of Coulomb gas techniques. The other model is the just mentioned square lattice model, approached by Bethe-Ansatz methods[15], which even though it is soluble, has so far resisted extraction of the exact value of the exponent. Numerically[14] it is already clear from the value for $\nu = 0.526 \pm 0.003$ that this model represents a different universality class, though the value for $\gamma = 1.15 \pm 0.01$ may well be the same. It may be expected that this discrepancy will be resolved in the near future. Likewise, the rapid development of understanding of exactly soluble models, and of principles like conformal invariance gives hope that many more of the universal quantities concerning two dimensional polymer solutions will eventually be computed.

References

[1] P. Flory, Priciples of polymer chemistry (Cornell University Press, Ithaca NY, 1971)

[2] K. G. Wilson and J. Kogut, Phys. Rep. 12C (1974) 75

[3] P.G. de Gennes, Scaling concepts in polymer physics (Cornell University Press, Ithaca NY, 1979)

[4] J. des Cloizeaux, J. Physique 36 (1975) 281.

[5] M.E. Fisher, J. Chem. Phys. 44 (1966) 616.

[6] C. Domb, J. Gillis and G. Wilmers Proc. Phys. Soc. (London) 85 (1965) 625.

[7] D.S. McKenzie, Phys. Rep. 27 (1976) 35.

[8] R. Vilanove and F. Rondelez, Phys. Rev. Lett. 45 (1980) 1986

[9] B. Nienhuis, Coulomb Gas formulation of two-dimensional phase transitions, in: Phase transitions and critical phenomena, Vol 11. eds. C. Domb and J.L. Lebowitz (Academic Press, London, 1987), pp. 1-53.

[10] B. Duplantier, Phys. Rev. Lett. 57 (1986) 941.

[11] R.J. Baxter, J. Phys. A19 (1986) 2821, and J. Phys. A20 (1987) 5241.

[12] M.T. Batchelor and H.W.J. Blöte, Phys. Rev. Lett. 61 (1988) 138, and Phys. Rev. B39 (1989) 2391.

[13] R.J. Baxter, Exactly solved models in statistical mechanics (Academic Press, London, 1982).

[14] H.W.J. Blöte and B. Nienhuis, J. Phys. A22 (1989) 1415.

[15] M.T. Batchelor, B. Nienhuis and S.O. Warnaar, Phys. Rev. Lett. 62 (1989) 2425.

[16] B. Duplantiers and H. Saleur, Nucl. Phys. B290 (1987) 291, and Phys. Rev. Lett. 47 (1986) 3179.

FUNDAMENTAL PROBLEMS IN STATISTICAL MECHANICS VII
H. van Beijeren, Editor
© *Elsevier Science Publishers B.V., 1990*

DYNAMICAL PHASE TRANSITIONS IN SPIN MODELS AND AUTOMATA

Bernard DERRIDA

Service de Physique Théorique[†] de Saclay

91191 Gif-sur-Yvette Cédex France

Systems in statistical mechanics are usually defined as collections of N objects which interact through an Hamiltonian $\mathcal{H}$. Then one can study the dynamics of these systems by introducing dynamical rules which are compatible with the Hamiltonian $\mathcal{H}$.

One can consider more general systems (automata, neural network models, spin models with non symmetric interactions) which are defined only by dynamical rules without reference to any Hamiltonian. The aim of these lectures is to describe some properties of such systems and a few methods which can be useful to study them. The outline of these lectures is the following:

1. INTRODUCTION

1.A. Examples

1.B. Attractors and Valleys

1.C. Mean field models with exactly solvable dynamics

2. RANDOM NETWORKS OF AUTOMATA

2.A. Kauffman model

2.B. Distribution of activities

2.C. Effect of noise

2.D. Finite dimensional case and damage spreading

3. SPIN MODELS

3.A. Distance method in presence of noise

3.B. The ferromagnetic case

3.C. Distances in spin glasses

4. CONCLUSION

[†] Laboratoire de L'institut de Recherche Fondamentale du Commissariat à l'Energie Atomique

1. INTRODUCTION

Let us start with a few examples of dynamical rules for the time evolution of spin models or automata.

1.A. Examples :

1.A.1. System governed by an Hamiltonian : the Ising Model.

The system consists of N Ising spins $S_i = \pm 1$ which are located on a regular lattice. The energy $\mathcal{H}(\mathcal{C})$ of a spin configuration

$$\mathcal{C} = \{S_1, S_2, ..., S_N\} \tag{1}$$

is given by

$$\mathcal{H}(\mathcal{C}) = -\sum_{i,j} J_{ij}\ S_i S_j - \sum_i h_i\ S_i \tag{2}$$

where J_{ij} is the interaction between spin i and spin j and h_i is the local field on site i.

To describe the properties of this system in thermal equilibrium at temperature T, one writes that each configuration is occupied with a probability $P_{\text{eq}}(\mathcal{C})$

$$P_{\text{eq}}(\mathcal{C}) = Z^{-1}\exp(-\mathcal{H}(\mathcal{C})/T) \tag{3}$$

where Z is the partition function

$$Z = \sum_{\mathcal{C}} \exp(-\mathcal{H}(\mathcal{C})/T) \tag{4}$$

There are several ways of introducing dynamical rules for this system. Let me first describe the so called sequential heat bath dynamics.

- *The sequential heat bath dynamics*

One starts with some initial spin configuration $\mathcal{C}_0$. Then to make the system evolve from a configuration $\mathcal{C}_t$ at time t to a configuration $\mathcal{C}_{t+\Delta t}$ at time $t + \Delta t$ (usually one chooses $\Delta t = 1/N$ so that each spin is updated on average once per unit time) one goes through the following steps :

1. Choose a site i at random $1 \leq i \leq N$

2. Compute the probability $p_i(t)$

$$p_i(t) = \frac{1}{2} + \frac{1}{2}\tanh\left(\sum_j \frac{J_{ij}S_j(t) + h_i}{T}\right) \tag{5}$$

3. Choose a random number $z_i(t)$ uniformly distributed between 0 and 1

4. Update the spin S_i according to the following rule

$$S_i(t + \Delta t) = \text{sign}\left(p_i(t) - z_i(t)\right) \tag{6}$$

and leave the other spins unchanged $S_j(t + \Delta t) = S_j(t)$ for $j \neq i$. One can repeat this procedure again and again. We see that each configuration $\mathcal{C}_{t+\Delta t}$ depends on the previous configuration $\mathcal{C}_t$ through (5) and on some noise (the choice of the site i to update and the random number $z_i(t)$)

$$\mathcal{C}_{t+\Delta t} = F(\mathcal{C}_t, \text{Noise}_t) \tag{7}$$

This is an example of *stochastic dynamics*.

The probability $P_t(\mathcal{C})$ of finding the system in a configuration $\mathcal{C} = \{S_i\}$ at time t satisfies the following Master equation :

$$P_{t+\Delta t}(\mathcal{C}) = \frac{1}{N}\sum_{i=1}^{N}\left[\frac{1}{2} + \frac{S_i}{2}\tanh\left(\sum_j \frac{J_{ij}S_j + h_i}{T}\right)\right][P_t(\mathcal{C}) + P_t(\mathcal{C}_i)] \tag{8}$$

where $\mathcal{C}_i$ is the configuration obtained from $\mathcal{C}$ by flipping the spin S_i. One can easily check that the equilibrium (3) is a fixed point of the Master equation (8). In doing that one can notice that it is essential for $P_{\text{eq}}(\mathcal{C})$ to be the fixed point of (8) that the interactions J_{ij} are symmetric

$$J_{ij} = J_{ji} \tag{9}$$

The matrix (8) which gives $P_{t+\Delta t}(\mathcal{C})$ as a function of the $P_t(\mathcal{C}')$ satisfies the conditions of the Perron Frobenius theorem[1] (all elements are positive or zero and there is a path from any $\mathcal{C}$ to any $\mathcal{C}'$). This implies that there is a single eigenvalue 1 for which the eigenvector is $P_{\text{eq}}(\mathcal{C})$ and that all the other eigenvalues have a modulus less than 1. Therefore, one knows that

$$P_t(\mathcal{C}) \longrightarrow P_{\text{eq}}(\mathcal{C}) \qquad \text{as} \quad t \longrightarrow \infty. \tag{10}$$

There exist other dynamical rules[2] than the sequential heat bath dynamics (for example the Metropolis algorithm) which lead to the same equilibrium (3). In order to speed up the computer simulation, it is tempting to update several spins simultaneously. Then one can show that the heat bath dynamics always lead to the equilibrium (3) provided that the spins which are updated simultaneously do not interact directly (i.e. i and j can be updated in parallel if $J_{ij} = 0$).

- *Parallel heat bath dynamics*

In the full parallel dynamics all the spins are updated at each time step. So one starts with some initial configuration C_0 and the configuration C_{t+1} is obtained from C_t by

$$S_i(t+1) = \text{sign}\left[\frac{1}{2} + \frac{1}{2}\tanh\left(\sum_j \frac{J_{ij}S_j(t) + h_i}{T}\right) - z_i(t)\right] \tag{11}$$

where the $z_i(t)$ are N random numbers uniformly distributed between 0 and 1.

Here again, one can write a Master equation for $P_t(C)$:

$$P_{t+1}(C) = \sum_{C'}\prod_{i=1}^{N}\left[\frac{1}{2} + \frac{S_i}{2}\tanh\left(\sum_j \frac{J_{ij}S'_j + h_i}{T}\right)\right] P_t(C') \tag{12}$$

and using the Perron Frobenius theorem[1], one knows that in the long time limit

$$P_t(C) \longrightarrow \tilde{P}_{\text{eq}}(C) \tag{13}$$

because all the eigenvalues of (12) except 1 have a modulus less than 1.

Because all the spins are updated simultaneously, the equilibrium $\tilde{P}_{\text{eq}}$ is no longer given by (3) but by[3]

$$\tilde{P}_{\text{eq}}(C) = \text{Constant}\prod_{i=1}^{N} e^{h_i S_i/T}\cosh\left(\sum_j \frac{J_{ij}S_j}{T} + \frac{h_i}{T}\right) \tag{14}$$

Here again, to check that (14) is the fixed point of (12), one needs to use the symmetry (9) of the interactions.

1.A.2. Systems without Hamiltonian

- *Neural network models; non symmetric spin glasses*

The simplest neural network models are defined as systems of N neurons $S_i(t) = \pm 1$ which interact through synapses J_{ij}. One can choose for neural network models the same stochastic dynamics as for spin models :

$$S_i(t + \Delta t) = \text{sign} \left[\frac{1}{2} + \frac{1}{2}\tanh \left(\sum_j \frac{J_{ij}S_j(t) + h_i}{T} \right) - z_i(t) \right] \tag{15}$$

with $\Delta t = 1/N$ for sequential dynamics and $\Delta t = 1$ for parallel dynamics and where $z_i(t)$ is a random number uniformly distributed between 0 and 1. The only difference with spin models is that the interactions are in general non symmetric

$$J_{ij} \neq J_{ji} \tag{16}$$

Although the time evolution of $P_t(\mathcal{C})$ is still given by (8) or (12), and that we know from the Perron Frobenius theorem that $P_t(\mathcal{C}) \longrightarrow P_{\text{eq}}(\mathcal{C})$, the equilibrium is no longer given by (4) or (14) and in general one does not know the expression giving $P_{\text{eq}}(\mathcal{C})$ in terms of the J_{ij} and the h_i.

So for systems with non symmetric interactions, the study of the equilibrium does not look simpler than the study of the non equilibirum.

- Networks of automata

Automata are other examples of systems wich evolve without Hamiltonian. In the simplest cases, they are defined as systems of N spins $S_i = \pm 1$ with K input sites $j_1(i), ..., j_K(i)$ and a Boolean function f_i associated to each site i. Then the system evolves according to the following rule

$$S_i(t + 1) = f_i(S_{j_1(i)}(t), ..., S_{j_K(i)}(t)) \tag{17}$$

Since the functions f_i and the input sites $j_1(i)...j_K(i)$ of each site i are fixed, the system is deterministic. Therefore (17) defines a map in phase space

$$\mathcal{C}_{t+1} = F(\mathcal{C}_t) \tag{18}$$

An example of automata is clearly the case of spin models or neural networks with parallel dynamics at zero temperature. Then (17) takes the following form

278   *B. Derrida*

$$S_i(t + 1) = \text{sign}\left(\sum_j J_{ij} S_j(t) + h_i \right) \tag{19}$$

1.B. Attractors and valleys

In the above examples, we have seen two kinds of dynamics : deterministic dynamics and stochastic dynamics:

1.B.1. Deterministic dynamics (Ising model or neural networks with parallel dynamics at zero temperature, automata).

Then, the configuration C_{t+1} depends only on the previous configuration

$$C_{t+1} = F(C_t) \tag{20}$$

For deterministic dynamics (eq.(20)), the time evolution always ends up by becoming periodic. This is because phase space is finite (2^N different configurations). After a time $t > 2^N$, the system must have visited twice the same configuration, say at times t_1 and t_2

$$C_{t_1} = C_{t_2}$$

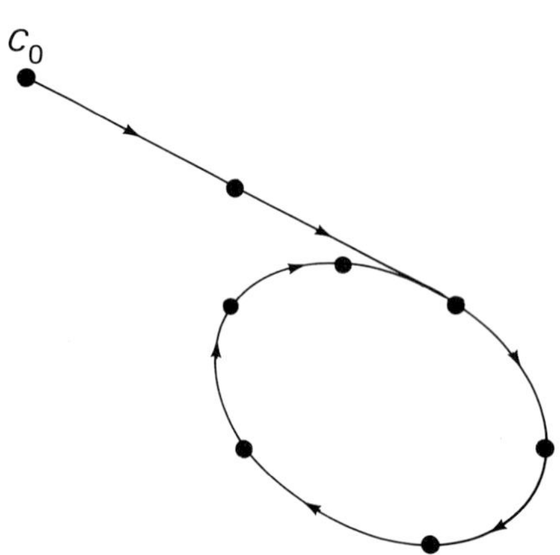

FIGURE 1

The trajectory of a configuration in phase space

Since the dynamics are deterministic, one has for any later time τ

$$C_{t_1+\tau} = F^\tau(C_{t_1}) = F^\tau(C_{t_2}) = C_{t_2+\tau} \tag{21}$$

There are in general (and even for finite N) several periodic attractors : phase space is broken into the *basins of attraction* (that we will call *valleys*) of these different attractors. The only a priori restriction on the period is that it is less than 2^N.

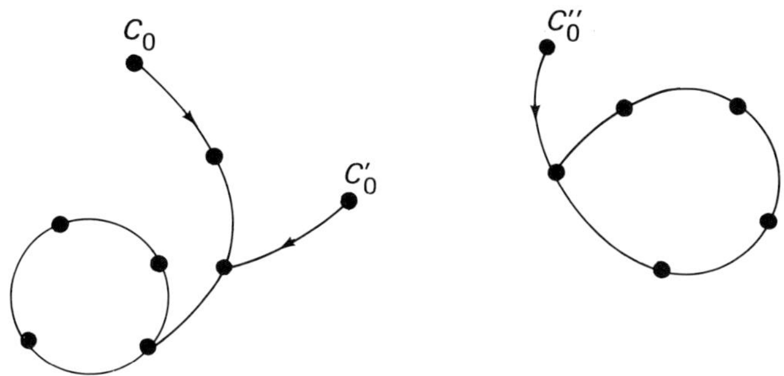

FIGURE 2
Two randomly chosen configurations can either belong to the
same valley (C_0 and C_0') or to distinct valleys (C_0 and C_0'')

It is probably in the case of deterministic dynamics that the difference between systems which evolve according to an Hamiltonian (zero temperature parallel dynamics with symmetric interactions $J_{ij} = J_{ji}$) and systems which evolve without Hamiltonian (non symmetric J_{ij}, automata) is the most apparent. For systems which evolve according to an Hamiltonian, one can introduce an energy function (often called a Lyapunov function) which decreases with time. The consequence is that the cycle length of Hamiltonian systems is always very small. We are now going to show that it is at most 2 for spin models at zero temperature with parallel dynamics and symmetric interactions ($J_{ij} = J_{ji}$).

The spins evolve according to

$$S_i(t + 1) = \text{sign} \left(\sum_j J_{ij} S_j(t) \right) \tag{22}$$

If one defines $E(t)$ by

$$E(t) = - \sum_{ij} J_{ij} S_i(t) S_j(t + 1) \tag{23}$$

One can easily see (using the symmetry $J_{ij} = J_{ji}$) that there are two ways of writing $E(t)$

$$E(t) = -\sum_i g_i(t)S_i(t+1) = -\sum_i g_i(t+1)S_i(t) \tag{24}$$

where

$$g_i(t) = \sum_j J_{ij}S_j(t) \tag{25}$$

Because the dynamics (22) can be written as

$$S_i(t+2) = \text{sign}(g_i(t+1)) \tag{26}$$

One can see from (24) that

$$E(t+1) = -\sum_i |g_i(t+1)| \tag{27}$$

by comparing (24) and (27) one gets that

$$E(t+1) - E(t) = -\sum_i |g_i(t+1)| \left[1 - S_i(t)S_i(t+2)\right] \tag{28}$$

So $E(t)$ is a decreasing function of time. Because phase space is finite, it exists a time t_0 such that $E(t_0 + 2) = E(t_0)$ which implies from (28) that

$$S_i(t_0 + 2) = S_i(t_0) \qquad \forall \ i \tag{29}$$

This means that the system has reached a cycle 2.

One should notice that the same problem with non symmetric J_{ij} ($J_{ij} \neq J_{ji}$) can have any cycle length. The difference between symmetric and non symmetric interactions is that with non symmetric J_{ij}, $E(t)$ cannot be written in the two forms as in (24).

1.B.2. Stochastic dynamics (Ising model or neural networks at non zero temperature)

For stochastic dynamics the configuration $\mathcal{C}_{t+1}$ depends on the previous configuration $\mathcal{C}_t$ and on some stochastic variable Noise_t

$$\mathcal{C}_{t+1} = F\left(\mathcal{C}_t, \text{Noise}_t\right) \tag{30}$$

Usually stochastic dynamics allow a finite system to go from any configuration to any other configuration. This means that the phase space of a finite system consists of a single valley. It is only in the thermodynamic limit ($N \longrightarrow \infty$) that one can observe well defined valleys.

For finite systems at non zero temperature one can only define quasi attractors and the number of these attractors depends on the time scale. To illustrate this fact let us take a simple example : the infinite range ferromagnetic Ising model which is governed by the following Hamiltonian

$$\mathcal{H} = -\frac{1}{N} \sum_{i<j} S_i S_j \tag{31}$$

As usual with infinite range models one has to normalize the interactions with N in order to keep a temperature scale independent of N.

To describe the time evolution of this system, one could write the Master equation (8). However, because the system is infinite range one can write directly a Master equation for the total magnetization M of the system :

$$\begin{aligned}
P_{t+1/N}(M) &= \frac{N+M+2}{2}\left[\frac{1}{2} - \frac{1}{2}\tanh\left(\frac{M+1}{NT}\right)\right] P_t(M+2) \\
&+ \frac{N+M}{2}\left[\frac{1}{2} + \frac{1}{2}\tanh\left(\frac{M-1}{NT}\right)\right] P_t(M) \\
&+ \frac{N-M}{2}\left[\left(\frac{1}{2} - \frac{1}{2}\tanh\left(\frac{M+1}{NT}\right)\right)\right] P_t(M) \\
&+ \frac{N-M+2}{2}\left[\left(\frac{1}{2} + \frac{1}{2}\tanh\left(\frac{M-1}{NT}\right)\right)\right] P_t(M-2) \tag{32}
\end{aligned}$$

One can diagonalize this matrix for finite N (here $N = 100$). If one calls $(\lambda_i)^{1/N}$ its eigenvalues (one introduces this $1/N$ because $\Delta t = 1/N$), one gets for the largest eigenvalues the temperature dependence shown on figure 3 :

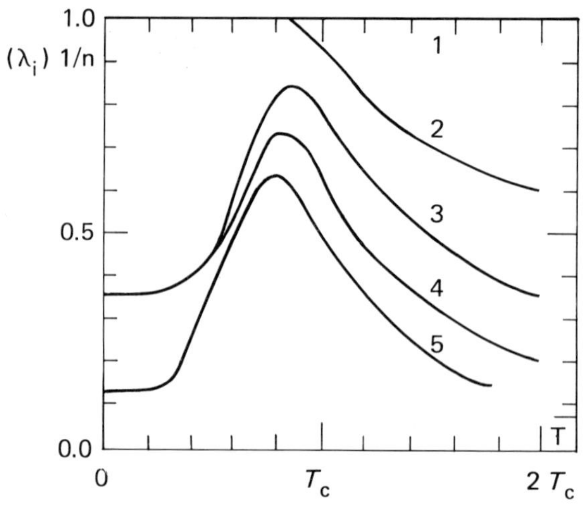

FIGURE 3

The eigenvalues of the Master Equation (32) versus temperature

We see that there is always an eigenvalue 1 corresponding to the normalization $\sum_M P_t(M) = 1$. For $T > 1$, the other eigenvalues are far from 1 : there is a single valley and the system is paramagnetic.

For $T < 1$, the second eigenvalue λ_2 is very close to 1 (it would be degenerate to 1 only for $N = \infty$). This means that for time scales t such that $(\lambda_2/\lambda_1)^t$ close to 1, the system behaves as if there were two attractors. It is only for times such that $(\lambda_2/\lambda_1)^t \ll 1$ that one observes a single valley.

We see with this example that for stochastic dynamics at non zero temperature the number of attractors depends on the time scale. In the limit $t \longrightarrow \infty$, for a finite system, one always has a single valley. However for short enough times (if $(\lambda_2/\lambda_1)^t$ is close enough to 1), the system behaves as if there were several attractors.

1.C. Mean fields models

It is possible to build a family of models with non symmetric interactions for which the dynamics can be calculated analytically. These models are defined as systems of N Ising spins $S_i(t) = \pm 1$ which evolve in time. For each site i, K input sites $j_1(i), ..., j_K(i)$ are chosen at random among the N sites. Then the spins evolve according to

$$S_i(t+1) = \text{sign}\left[\frac{1}{2} + \frac{1}{2}\tanh\left(\sum_j \frac{J_{ij}S_j(t) + h_i}{T}\right) - z_i(t)\right] \qquad (33)$$

where in (33) the sum over j runs over the K input sites of site i and $z_i(t)$ is as above a random number uniformly distributed between 0 and 1.

The only reason[14,40] which makes these models exactly soluble is that the K inputs of each site are chosen at random among the N sites. This implies that in the limit $N \longrightarrow \infty$, the K^t sites which belong to the tree of ancestors of a given site i are all different (this is true as long as $t \ll \log N$).

Since to compute the value of a spin $S_i(t)$ at time t from the initial condition $\{S_j(0)\}$, only the K^t ancestors of site i are involved and because with probability 1, there is no loop in this tree of ancestors, at each time step, the inputs $j_1(i)...j_K(i)$ of each site i are not correlated (one assumes that the spins $S_i(0)$ are not correlated at time $t = 0$). Therefore if one define $m_i(t)$ by

$$m_i(t) = \langle S_i(t)\rangle \qquad (34)$$

where in (34), the average $\langle \ \ \rangle$ means an average over the initial condition $\{S_i(0)\}$ and over the history (i.e. the $z_i(t)$), one can deduce from (33) :

$$m_i(t+1) = \sum_{\sigma_1 = \pm 1} \cdots \sum_{\sigma_K = \pm 1} \prod_{r=1}^{K} \frac{1 + m_{j_r}(t)\sigma_r}{2}\tanh\left(\sum_j \frac{J_{ij_r}\sigma_r + h_i}{T}\right) \qquad (35)$$

Equation (35) expresses the fact that the ancestors $S_{j_r}(t)$ of site i are not correlated at time t. This is a consequence of the fact that the tree of ancestors has no loop and the spins are not correlated at $t = 0$. The equation (35) was first obtained in the case of the Kauffman model[7] for which the time evolution (33) is replaced by a random Boolean function. It is however much more general since its validity depends only on the absence of correlations in the initial condition and of the fact that the inputs of each site are chosen at random.

In order to illustrate these equations, let us discuss two examples :

1.C.1. The ferromagnetic case

- It is the special case where the K interactions J_{ij} of each site i are equal

$$J_{ij} = J \qquad (36)$$

Then by choosing the initial condition with $m_i(t)$ independent of i, one sees from (35) that at any later time t, the solution remains uniform :

$$m_i(t) = m(t) \tag{37}$$

and that the magnetization $m(t)$ evolves according to

$$m(t+1) = \sum_{n=0}^{K} \binom{K}{n} \left(\frac{1+m(t)}{2}\right)^n \left(\frac{1-m(t)}{2}\right)^{K-n} \tanh\left(\frac{(2n-K)J+h}{T}\right) \tag{38}$$

It is important to notice that the fact that the inputs $j_1(i)...j_K(i)$ are chosen at random is essential to derive (38). This model is clearly a model with non symmetric interactions since when a site j is an ancestor of a site i, it is very unprobable for large N that i is an ancestor of j.

The corresponding model with symmetric interactions which would be a usual tree with $J_{ij} = J_{ji} = J$ is a much more difficult problem for which one does not know the analytic expression of the magnetization as a function of time.

From (38), one sees that in zero field ($h = 0$), the system can either be in a paramagnetic phase (if $m = 0$ is the attractive fixed point) or in a ferromagnetic phase (if there is a pair of attractive fixed points $\pm m^*$ with $m^* > 0$). The transition temperature T_c below which the system is ferromagnetic can be computed from (38) by linearizing around $m = 0$ and one finds that T_c is solution of

$$1 = \sum_{n=0}^{K} \binom{K}{n} \frac{(2n-K)}{2^K} \tanh\left(\frac{(2n-K)J}{T_c}\right) \tag{39}$$

1.C.2. Chaotic phase in a non symmetric spin glass[6,8]

Consider again a system a N spins $S_i(t) = \pm 1$ with K random inputs $j_1(i)...j_K(i)$ associated to each site i. Let us describe the dynamical properties of this system in the case of parallel dynamics at zero temperature (deterministic dynamics)

$$S_i(t+1) = \text{sign}\left(\sum_j J_{ij} S_j(t) + h_i\right) \tag{40}$$

where the interaction J_{ij} and the local fields h_i are randomly distributed according to Gaussian distributions

$$\rho(J_{ij}) = \frac{1}{\sqrt{2\pi}}\exp - \left(\frac{J_{ij}^2}{2}\right) \tag{41}$$

$$\rho(h_i) = \frac{1}{\sqrt{2\pi\Delta}}\exp - \left(\frac{h_i^2}{2\Delta}\right) \tag{42}$$

We are going to show that there exists a critical value Δ_c of Δ such that for $\Delta < \Delta_c$ the system has a chaotic behavior. Let us compare two configurations $\{S_i(t)\}$ and $\left\{\tilde{S}_i(t)\right\}$ which evolve according to the same rules (40) :

$$S_i(t+1) = \text{sign}\left(\sum_j J_{ij}S_j(t) + h_i\right)$$

$$\tilde{S}_i(t+1) = \text{sign}\left(\sum_j J_{ij}\tilde{S}_j(t) + h_i\right) \tag{43}$$

To do so, we can define the distance $d(t)$ between these 2 configurations

$$d(t) = \frac{1}{2N}\sum_{i=1}^{N}\left|S_i(t) - \tilde{S}_i(t)\right| \tag{44}$$

This distance is the fraction of spins which are different in the two configurations. One can show that the time evolution of this distance is given by

$$d(t+1) = \sum_{n=0}^{K}\binom{K}{n}[1 - d(t)]^{K-n}[d(t)]^n \frac{2}{\pi}\tan^{-1}\left(\sqrt{\frac{n}{K - n + \Delta}}\right) \tag{45}$$

Formula (45) can easily be understood. At time t, a given site i has a probability $\binom{K}{n}(1-d)^{K-n}d^n$ of having n inputs different in the two configurations. Then for such a site, one can write (44) as

$$S_i(t+1) = \text{sign}(X + Y)$$

$$\tilde{S}_i(t+1) = \text{sign}(X - Y) \tag{46}$$

where X and Y are Gaussian variables of width $\Delta + K - n$ and n. Then the probability that $S_i(t+1) \neq \tilde{S}_i(t+1)$ is given by

$$\mathrm{Prob}\left(S_i(t+1) \neq \tilde{S}_i(t+1)\right) = \mathrm{Prob}(|X| < |Y|) = \frac{2}{\pi}\tan^{-1}\left(\sqrt{\frac{n}{K-n+\Delta}}\right) \qquad (47)$$

To obtain (47), we have used the fact that X and Y are sums of independent random variables. This is justified because the inputs $S_{j_1}(t)...S_{j_K}(t)$ are uncorrelated.

From (45), we see that $d = 0$ is always a fixed point. Clearly, if two configurations are identical, they remain identical for ever. When Δ varies, this fixed point becomes unstable for $\Delta < \Delta_c$ where Δ_c is solution of

$$1 = K \frac{2}{\pi} \tan^{-1}\left(\sqrt{\frac{1}{K-1+\Delta_c}}\right) \qquad (48)$$

and a new attractive fixed point d^* appears. Since $d = 0$ is unstable, one can conclude that two configurations which are initially very close have trajectories which diverge implying that the system is chaotic.

It is interesting to notice that for these simple mean field models, the distance between 2 configurations converges to a fixed value d^* independent of their initial distance.

2. NETWORKS OF RANDOM AUTOMATA

In this chapter we are going to consider a single class of models : networks of random automata. The system consists of N spins $S_i(t) = \pm 1$. Each spin S_i receives inputs from K other sites $j_1(i)...j_K(i)$ and evolves according to a random Boolean function f_i of K variables

$$\tilde{S}_i(t+1) = f_i\left(S_{j_1(i)}(t), ..., S_{j_K(i)}(t)\right) \qquad (49)$$

There are 2^{2^K} different Boolean functions of K variables : for example for $K = 1$, there are 4 Boolean functions g_1, g_2, g_3 and g_4 :

σ	$g_1(\sigma)$	$g_2(\sigma)$	$g_3(\sigma)$	$g_4(\sigma)$
$-$	$+$	$+$	$-$	$-$
$+$	$+$	$-$	$+$	$-$

Here, we choose for each site i a random Boolean f function with a probability $p^n(1-p)^{2^K-n}$ where n is the number of times that the function takes the value -1. (In the above example g_1 would be chosen with probability $(1 - p)^2$, g_2 and g_3 with probability $p(1 - p)$ and g_4 with probability p^2). For $p = 1/2$, all the Boolean functions have equal weights.

In this chapter we will first consider the mean field case (Sections A, B and C) for which the inputs are $j_1(i)...j_K(i)$ of each site i are chosen at random among the N sites. Then we will discuss the finite dimensional case where the inputs are the neighbors of a site on a regular lattice (Section D).

2.A. The Kauffman model

The Kauffman model was introduced as a model for cell differentiation. It describes a system of N genes $S_i = \pm 1$ (+1 if the gene is on and -1 if it is off). The activity of a gene i is influenced by K other genes $(j_1(i), j_2(i)...j_K(i))$.Each gene is influenced by other genes which are usually far away and it is a reasonable approximation to assume that $j_1(i)...j_K(i)$ are chosen at random among the N genes. Then another simplifying assumption is to take a discrete time and random Boolean functions f_i to make evolve the gene activities.

So in the Kauffman model the $S_i(t)$ evolve according to (49) with random Boolean functions f_i and random inputs $j_1(i)...j_K(i)$.

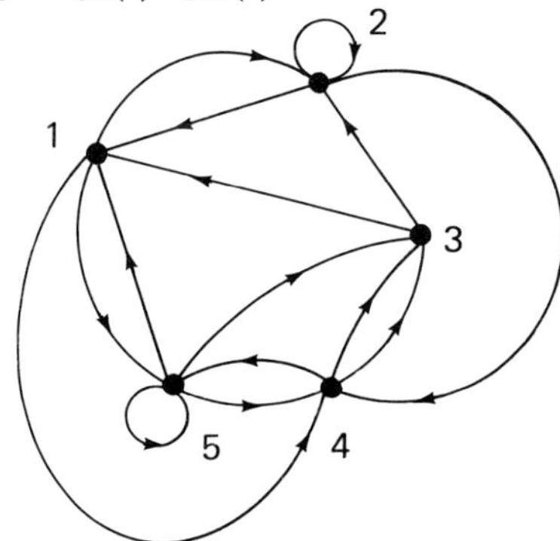

i	$j_1(i)$	$j_2(i)$	$j_3(i)$
1	2	3	5
2	3	2	1
3	4	4	5
4	2	5	1
5	1	5	4

FIGURE 4

Example of the connections in the Kauffman
model for $N = 5$ sites and $K = 3$

The first results[9-11] obtained by Kauffman concerned the periods of the attractors. Kauffman studied numerically how the period of an attractor depends on the system size N. He found two regimes when $p = 1/2$: for $K \leq 2$

$$\text{Period} \sim N^x \tag{50}$$

whereas for $K > 2$, the period increases exponentially with N

$$\text{Period} \sim e^{N\alpha} \tag{51}$$

Up to now, there does not exist any exact analytical calculation of the period except for $K = 1$ [12,13]. In particular the value of α (in eq.(51)) as a function of K has not yet been calculated analytically (see the conclusion for an attempt).

Because this model belongs to the class of mean field models described in 1.C. (the inputs are chosen at random), there are some quantities for which the time evolution can be calculated exactly. As in example 1.C.2., one can calculate the time evolution of the distance $d(t)$ between two configurations $\{S_i(t)\}$ and $\left\{\tilde{S}_i(t)\right\}$:

$$d(t) = \frac{1}{2N} \sum_{i=1}^{N} \left| S_i(t) - \tilde{S}_i(t) \right|. \tag{52}$$

By using exactly the same reasoning as in 1.C.2., one obtains[14,15] that

$$d(t+1) = 2p(1-p)\left(1 - (1-d(t))^K\right) \tag{53}$$

Here again, the probability that a spin i has all its inputs the same in the two configurations is $(1-d(t))^K$. For the spin i to be different at time $t+1$, first we need that not all the inputs are the same (probability $= 1-(1-d)^K$ and then that f gives two different outputs (probability$=2p(1-p)$). From (54) we see that there is a critical line

$$K_c(p) = \frac{1}{2p(1-p)} \tag{54}$$

For $K < K_c$, the only (attractive) fixed point is $d = 0$. So for any initial distance $d(0)$, the final distance vanishes. The system forgets completely its initial condition since different initial configurations always end up by becoming identical. By choosing $\left\{\tilde{S}_i(0)\right\} = \{S_i(1)\}$, we see that for almost all i one has in the limit $t \longrightarrow \infty$

$$S_i(t) = \tilde{S}_i(t) = S_i(t+1) \tag{55}$$

and therefore (almost) all spins are fixed in time. Since almost all spins are fixed in time, one can say that the system is in *its frozen phase*.

For $K > K_c$, the fixed point $d = 0$ becomes unstable and there appears a new attractive fixed point $d^* \neq 0$. Two initial conditions, even if they are very close, remain different. In the long time limit their distance $d(t)$ does not depend on their initial distance and converges to d^*, the attractive fixed point of (53). Since close trajectories in phase space have the tendency to diverge, one can say that $K > K_c(p)$ corresponds to the *chaotic phase*.

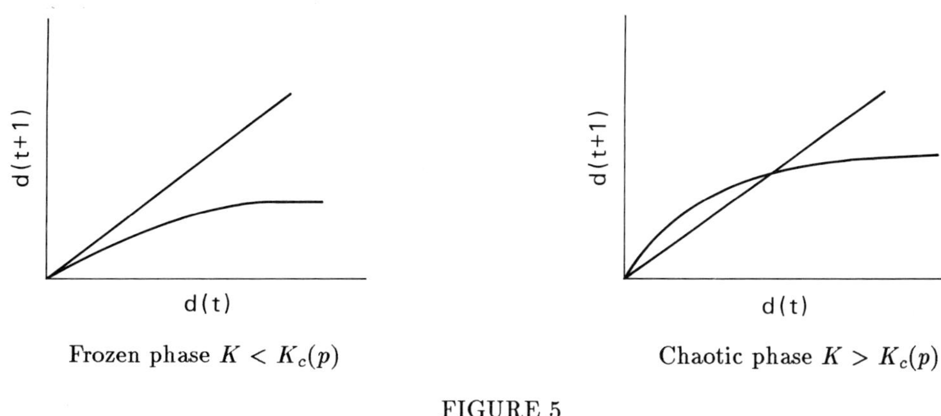

Frozen phase $K < K_c(p)$ Chaotic phase $K > K_c(p)$

FIGURE 5

2.B. Distribution of activities in the Kauffman model[7,16]

There are other quantities which can be computed exactly for the Kauffman model. If one defines $m_i(t)$ as

$$m_i(t) = \langle S_i(t) \rangle \tag{56}$$

where $\langle \ \rangle$ means an average over an ensemble of initial conditions and if the spins are chosen to be uncorrelated in the initial condition, i.e. the probability of starting with some initial condition $\{S_i\}$ is

$$\prod_{i=1}^{N} \left(\frac{1 + m_i(0)S_i}{2} \right) \tag{57}$$

we have seen in 1.C. that almost all the $m_i(t)$ will satisfy (in the limit $N \longrightarrow \infty$)

$$m_i(t+1) = \sum_{\sigma_1 = \pm 1} \cdots \sum_{\sigma_K = \pm 1} \prod_{r=1}^{K} \left(\frac{1 + m_{j_r}(t)\, \sigma_r}{2} \right) f_i(\sigma_1, ..., \sigma_r) \tag{58}$$

From (58) one can write a recursion[7] for the distribution of activities $P_t(m)$ which is defined by

$$P_t(m) = \frac{1}{N} \sum_{i=1}^{N} \delta(m - m_i(t))$$ (59)

In the limit $t \longrightarrow \infty$, $P_t(m)$ converges to a stationnary distribution $P_\infty(m)$ which is the solution of an integral equation[7]. For the Kauffman model one finds two sorts of shapes :

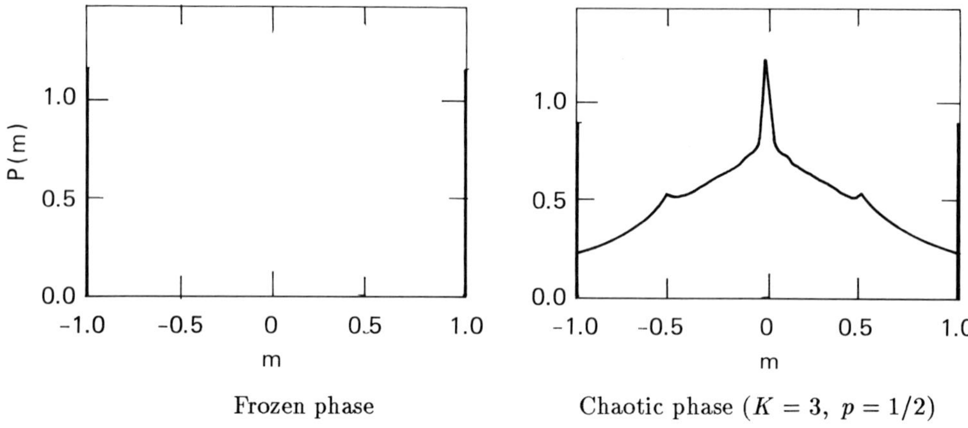

Frozen phase   Chaotic phase ($K = 3$, $p = 1/2$)

FIGURE 6

In the frozen phase $K < K_c(p)$, $P_\infty(m)$ consists of the sum of two delta functions at $+1$ and at -1 : almost all the spins have a value independent of their initial conditions.

For $K > K_c(p)$, in addition to the two delta functions at $m = \pm 1$, one has a continuous part in $P_\infty(m)$.

There is another definition of $P_\infty(m)$ which gives exactly the same distribution. If one defines m_i as the time average of $S_i(t)$ for a fixed initial condition

$$m_i = \lim_{t_0 \longrightarrow \infty} \frac{1}{t_0} \sum_{t=1}^{t_0} S_i(t)$$ (60)

and if $P(m)$ is the distribution of these m_i

$$P(m) = \frac{1}{N} \sum_{i=1}^{N} \delta(m - m_i)$$ (61)

one can show[7] that $P(m)$ satisfies exactly the same integral equation as $P_\infty(m)$ and therefore has the shapes given in Figure 6. The delta functions at $m = \pm 1$ correspond to the spins which do not move whereas the continuous part in $P(m)$ represents the moving spins.

2.C. The effect of noise

For all the automata models defined by (49), one can be interested to study the effect of an external random noise. To do so, one replaces the time evolution (49) by a stochastic rule :

$$
\begin{aligned}
S_i(t+1) &= f_i\left(S_{j_1(i)}(t), ..., S_{j_K(i)}(t)\right) \quad \text{with probability} \quad \frac{1}{2} + \frac{1}{2}\tanh\frac{1}{T} \\
S_i(t+1) &= -f_i\left(S_{j_1(i)}(t), ..., S_{j_K(i)}(t)\right) \quad \text{with probability} \quad \frac{1}{2} - \frac{1}{2}\tanh\frac{1}{T}
\end{aligned}
\tag{62}
$$

The parameter T in (62) plays the role of a temperature. Clearly in the zero temperature limit, one recovers the deterministic case (49).

To implement (62), one chooses at each time step N random numbers $z_i(t)$ uniformly distributed between 0 and 1 and one updates $S_i(t+1)$ by

$$
S_i(t+1) = \text{sign}\left(\frac{1}{2} + \frac{1}{2}\tanh\left(\frac{1}{T}\right) f_i\left(S_{j_1}(t), ..., S_{j_K}(t)\right) - z_i(t)\right)
\tag{63}
$$

We see that as in other stochastic cases, a configuration $C_{t+1} = \{S_i(t+1)\}$ depends on the previous configuration $C_t = S_i(t)$ and on the thermal noise $\text{Noise}_t = z_i(t)$.

$$
C_{t+1} = F(C_t, \text{Noise}_t)
\tag{64}
$$

So the configuration C_t of the system at time t depends on its initial condition C_0 and on the thermal noise between times 0 and $t-1$. To study the respective effects of the initial condition and of the thermal noise, one can measure the distance $d(t)$ between two configurations (52), but there are several distances which can be measured:

2.C.1. Uncorrelated noise

First one can measure the distance between two configurations which are subjected to uncorrelated noises i.e.

$$
\begin{aligned}
S_i(t+1) &= \text{sign}\left[\frac{1}{2} + \frac{1}{2}\tanh\frac{1}{T} f_i\left(\{S_j(t)\}\right) - z_i(t)\right] \\
\tilde{S}_i(t+1) &= \text{sign}\left[\frac{1}{2} + \frac{1}{2}\tanh\frac{1}{T} f_i\left(\{\tilde{S}_j(t)\}\right) - \tilde{z}_i(t)\right]
\end{aligned}
\tag{65}
$$

For the Kauffman model and again because the inputs of each site i are uncorrelated, one can obtain an exact expression for the time evolution of $d(t)$. One finds that for $p = 1/2$

$$d(t + 1) = \frac{1}{2} \left(1 - (1 - d(t))^K \tanh^2 \frac{1}{T} \right) \tag{66}$$

In the limit $T \longrightarrow 0$, one recovers (53) (for $p = 1/2$). There is not much to say on $d(t)$ in this case because there is always an attractive fixed point of (66) which changes smoothly with T without any trace of a phase transition.

2.C.2. Same noise

Another way of defining the distance between two configurations consists of using the same noise for the two configurations

$$z_i(t) = \tilde{z}_i(t) \tag{67}$$

One can again calculate the time evolution of the distance $d(t)$ in this case and one finds for $p = 1/2$

$$d(t + 1) = \left(1 - (1 - d(t))^K \right) \frac{1}{2} \tanh \frac{1}{T} \tag{68}$$

$d = 0$ is always a fixed point : two identical configurations remain identical if they are subjected to the same thermal noise. As for uncorrelated noise one recovers (53) in the limit $T \longrightarrow 0$.

One can see in (68) that there is a dynamical phase transition at a temperature T_c given by

$$\frac{K}{2} \tanh \frac{1}{T_c} = 1 \tag{69}$$

For $T > T_c$, $d = 0$ is an attractive fixed point and two different initial configurations always become identical. The system forgets quickly its initial condition.

For $T < T_c$, $d = 0$ becomes an unstable fixed point and a new attractive fixed point $d^* \neq 0$ appears. The trajectories of two initially close configurations (if their initial distance is non zero) diverge and their distance converges to d^*.

2.C.3. Damage spreading

Even if the initial distance is zero (for example if the two configurations differ initially by a finite number of spins), there is a finite probability that in the long time limit the two configurations end up at the distance d^*. If one defines the survival probability $P(T)$ that two configurations which differ initially by a single spin are still different in the long time limit, one can show that $P(T) \neq 0$ for $T < T_c$ and that $P(T)$ is the non zero solution of

$$P(T) = 1 - \exp\left(-\frac{K}{2}P(T)\tanh\frac{1}{T}\right) \qquad (78)$$

This expression can be understood by writing a recursion for $Q_m(t)$, the probability that 2 configurations which differ initially by m spins have become identical before or at the tth time step. For $p = 1/2$, the recursion has the following form

$$Q_m(t+1) = (c_m)^N + \sum_{n \geq 1}^{N} \frac{N!}{n!(N-n)!}(1 - c_m)^n c_m^{N-n} Q_n(t) \qquad (79)$$

where c_m is the probability that a randomly chosen site will be identical at time $t+1$ knowing that the two configurations differ at time t by m spins

$$c_m = \left(1 - \frac{m}{N}\right)^K + \left(1 - \left(1 - \frac{m}{N}\right)^K\right)\left(\frac{1}{2} + \frac{1}{2}(1 - \tanh\frac{1}{T})\right) \qquad (80)$$

For large N and m finite, (79) becomes

$$\begin{aligned}
Q_m(t+1) = {} & \exp\left(-\frac{mK}{2}\tanh\frac{1}{T}\right) \\
& + \sum_{n \geq 1}\left(\frac{mK}{2}\tanh\frac{1}{T}\right)^n \frac{1}{n!}\exp\left(-\frac{mK}{2}\tanh\frac{1}{T}\right)Q_n(t)
\end{aligned} \qquad (81)$$

If at time t and for finite m, we choose the following form of the $Q_m(t)$

$$Q_m(t) = [a(t)]^m \qquad (82)$$

where $a(t)$ is a parameter, one gets from (81), that $Q_m(t+1)$ keeps this m dependence with $a(t+1)$ given by

$$a(t+1) = \exp\left[-\frac{K}{2}(1 - a(t))\tanh\frac{1}{T}\right] \qquad (83)$$

In the limit $t \longrightarrow \infty$, $a(t)$ converges to a fixed point a^* (with $a^* = 0$ if $T > T_c$ and $a^* \neq 0$ if $T < T_c$). It is then clear that

$$P(T) = 1 - a^* \tag{84}$$

since $1 - a^*$ is the probability that two configurations which differ by a single spin will never meet and this leads to (78).

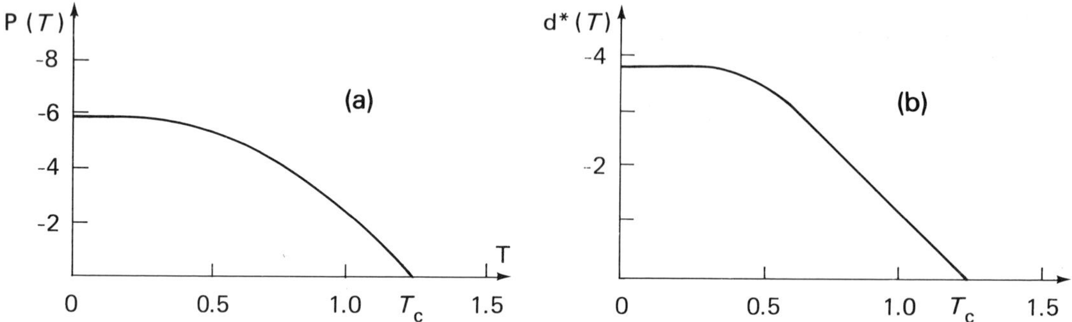

FIGURE 7a and 7b
The survival probability $P(T)$ and the distance $d^*(T)$ versus temperature
for $K = 3$ and $p = 1/2$. They both vanish at a dynamical phase transition
$T_c \simeq 1.243$ solution of (69).

2.D. Random automata in finite dimension

For the Kauffman model (deterministic version as in sections 2.A and 2.B) the evolution of the distance $d(t)$ and the distribution of local magnetisations can be calculated exactly. This is because the Kauffman model is a mean field model (Section 1.C.).

One can wonder what would be the effect of correlations when the system has some spatial structure. The simplest way of generalizing the Kauffman model to a finite dimensional model is to put the spins $S_i(t)$ on a regular lattice (of coordination number K) and to consider that the inputs $j_1(i)...j_K(i)$ are the neighbors of site i on the lattice. So the model in finite dimension is a system of N spins $S_i(t)$ on a regular lattice. For each i:

(1) the inputs $j_1(i)...j_K(i)$ are the nearest neighbors of i on the lattice

(2) the function f_i is a random Boolean function of K variables.

Since the number K of inputs is fixed, we can only vary the parameter p (introduced at the beginning of section 2) to see both the frozen and the chaotic phases. Because of the symmetry $p \longrightarrow 1 - p$, one needs only to study the range $0 \le p \le .5$.

For these finite dimensional systems, no analytical work has been done so far and all the known results have been obtained by numerical simulations. Let me describe here some results obtained for the 2d square lattice[17−23].

- The distance

The first quantity one can measure is the distance $d(t)$ between two different configurations C_t and $\tilde{C}_t$ as defined in equation (52). An important difference with the Kauffman model discussed in section 2.A. is that in finite dimension for $0 < p \le .5$, the distance $d(\infty)$ in the long time limit depends on the initial distance $d(0)$.

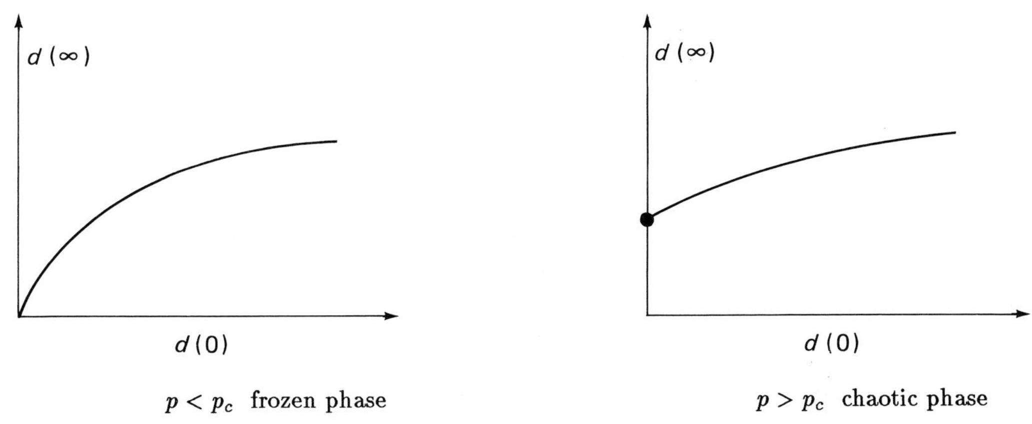

FIGURE 8

The distance $d(t)$ for $t \longrightarrow \infty$ versus the initial distance

The transition between the chaotic and the frozen phase can nevertheless be seen by looking at how $d(\infty)$ depends on $d(0)$. When $d(0) \longrightarrow 0$, one observes[17] a threshold at

$$p_c \simeq .26 \tag{85}$$

(More recent numerical simulations[19] give a higher estimate $p_c \simeq .29$).

For $p < p_c$ (the frozen phase), $d(\infty)$ vanishes as $d(0) \longrightarrow 0$ whereas for $p > p_c$ (the chaotic phase) $d(\infty)$ has a non zero limit as $d(0) \longrightarrow 0$. The behavior of $d(\infty)$ as a function of $d(0)$ is very reminiscent of the magnetisation as a function of the magnetic field h for a ferromagnet. Above T_c, m vanishes as $h \longrightarrow 0$ whereas below T_c, m has a finite limit in the limit $h \longrightarrow 0$.

- The damage spreading

In the chaotic phase, a small difference has the tendency to spread over the whole system. This means that if one considers two configurations which differ at $t = 0$ by a single spin, they have a finite probability of differing by a macroscopic number of spins at time $t = \infty$. Critical exponents related to this problem of damage spreading have been calculated numerically[19,23].

Figure 9 obtained by AU Neumann illustrates the damage spreading phenomenon in the case of the 2d Kauffman model. One starts for a sample of 100×100 spins with 2 configurations which are identical except for a small square of 10×10 at the center. After many time steps, one sees that the damage has spread over the whole system in the chaotic phase $p > p_c$ whereas it remains localized in the frozen phase $p < p_c$

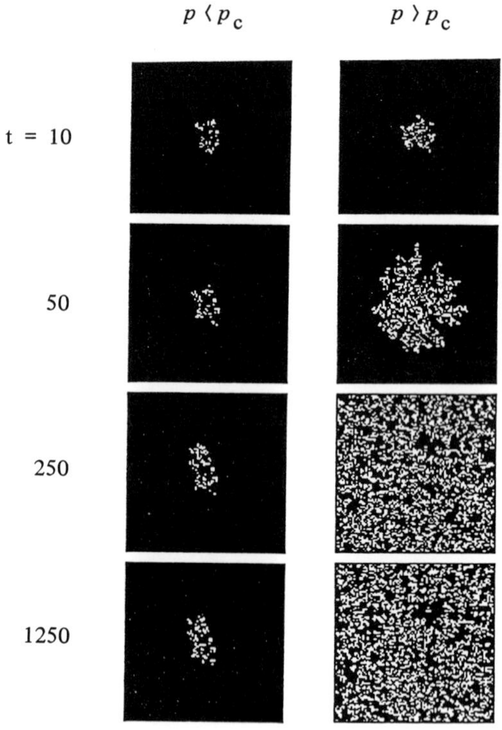

FIGURE 9

Damage spreading for the 2d Kauffman model (figure by A.U. Neumann)

- The distribution of activities

One can also measure the distribution of local magnetisations $P(m)$ as defined by equations (60) and (61)[22]. One starts with a random initial configuration and one measures for each site i the magnetisation m_i averaged over a long (but finite $\sim 1000 steps$) time for a large system.

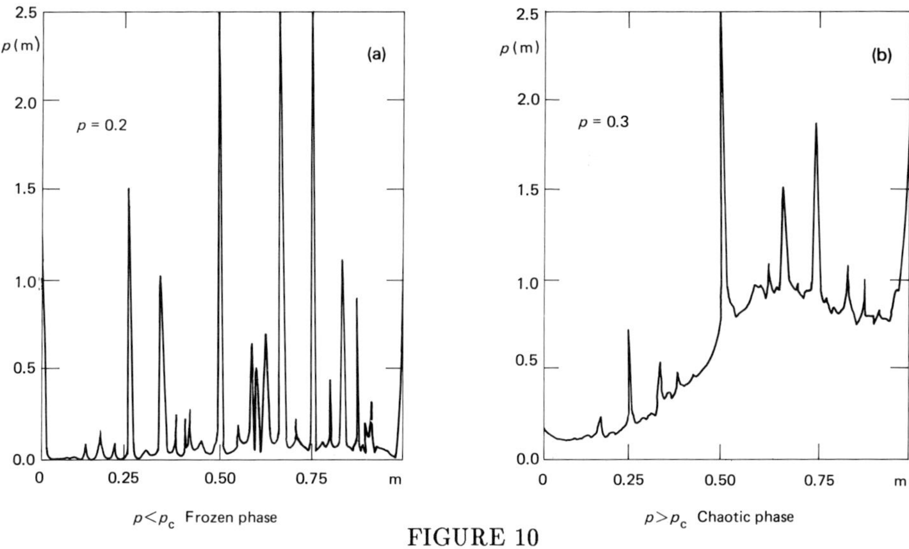

FIGURE 10

$P(m)$ versus m for the 2 dimensional case

For $p < p_c$ (frozen phase), $P(m)$ is a sum of delta functions at simple rational values of m. This corresponds to either spins which remain fixed for ever (after a few time steps) or to spins which belong to small clusters which have a periodic behavior with a short period.

In the chaotic phase ($p > p_c$) we see that in addition to the delta functions at simple rational values of m, there is a continuous part in $P(m)$. This continuous part is due to the presence of an infinite cluster of spins with no period (i.e. an infinite period).

We have seen that the distance $d(t)$ and the distribution of magnetisations $P(m)$ have more structure in the finite dimensional case than in the Kauffman model discussed in section 2.A, $(d(\infty)$ depends on $d(0) and$ $P(m)$ has an infinite number of delta functions at all rational values of m). There is nevertheless a clear numerical evidence for a transition from the frozen to the chaotic phase (the fact that $d(\infty)$ does not vanih as $d(0) \longrightarrow 0$ and the presence of a continuous part in $P(m)$ are two clear signatures of the chaotic phase).

The effect of noise on automata in finite dimension has been studied numerically[24] and the results are very similar to those shown in Figure 8 (for the mean field case).

3. SPIN MODELS

3.A. Distance method in presence of noise

For deterministic dynamics, we have seen in section 2 that one can observe dynamical phase transitions between a chaotic and a frozen phase by measuring the distance between two configurations which evolve according to the same rules.

For stochastic dynamics, we can recover these dynamical phase transitions provided that we use the same thermal noise to update the two configurations (2.C).

One can then try to see how these dynamical phase transitions are related to usual equilibrium phase transitions of systems which possess an Hamiltonian[25].

Consider a system of N Ising spins $S_i = \pm 1$ for which the Hamiltonian $\mathcal{H}$ is

$$\mathcal{H} = -\sum_{i,j} J_{ij} S_i S_j \tag{86}$$

Two configurations $\{S_i(t)\}$ and $\left\{\tilde{S}_i(t)\right\}$ which evolve according to the heat bath dynamics are subjected to the same thermal noise (same random number $z_i(t)$)

$$
\begin{aligned}
S_i(t + \Delta t) &= \text{sign}\left[\frac{1}{2} + \frac{1}{2}\tanh\left(\sum_j \frac{J_{ij} S_j(t)}{T}\right) - z_i(t)\right] \\
\tilde{S}_i(t + \Delta t) &= \text{sign}\left[\frac{1}{2} + \frac{1}{2}\tanh\left(\sum_j \frac{J_{ij} \tilde{S}_j(t)}{T}\right) - z_i(t)\right]
\end{aligned}
\tag{87}
$$

and we measure the distance $d(t)$:

$$d(t) = \frac{1}{2N} \sum_{i=1}^{N} \left| S_i(t) - \tilde{S}_i(t) \right| \tag{88}$$

In the limit $T \longrightarrow \infty$, each time that a spin i is updated, it becomes identical in the two configurations. Therefore

$$d(t) = d(0) e^{-t} \tag{89}$$

So at high temperature the two configurations subjected to the same noise attract each other and one expects $d(t)$ to vanish in the long time limit.

At low temperature, if the system has several low temperature phases, there is always the possibility that each configuration falls into a different valley and therefore one expects that $d(t)$ remains non zero, at least for far enough initial conditions to allow the two configurations to fall into different valleys.

One can then wonder whether the only mechanism for two configurations (subjected to the same noise) to remain different (in the limit $t \longrightarrow \infty$) is the appearence of infinite barriers between valleys. The distance method has already been used for several systems and two sorts of effects have been found which give a non vanishing distance in the long time limit:

1. The two configurations belong to two valleys in phase space separated by high energy barriers, like in the case of ferromagnets[25−27].

2. The chaotic character of the dynamics makes two initially close trajectories diverge (automata, spin glasses, neural networks)[25,6].

We are now going to compare the dynamical phase transition observed by looking at the distance (between two configurations subjected to the same noise) and the equilibrium phase transition for two systems: First the ferromagnetic Ising model for which we will show that the dynamical phase transition coincides with the phase transition at equilibrium. Then we will consider the case of spin glasses and see how the equilibrium phase transition could be related to the dynamical properties of the distance.

3.2 The ferromagnet

When one compares two configurations subjected to the same thermal noise at a temperature T, one can measure two quantities: the survival probability $P(T)$, which is the probability that the two configurations are still different (after a long time t) and the average distance $D(T) = \langle d \rangle$ between these two configurations (where the average is done over the cases for which the two configurations are still different).

- *The mean field ferromagnet*

For the mean field model defined by (31)

$$\mathcal{H} = -\frac{1}{N} \sum_{i<j} S_i S_j \tag{90}$$

one can calculate[28] analytically the survival probability $P(T)$ and the distance $D(T)$ for sequential or parallel dynamics.

The survival probability depends on how the two initial conditions are chosen. For sequential dynamics, one finds that

$$P(T) = \frac{2}{\pi}\,\tan^{-1}\left[\left(2\,\frac{1-T}{T}\right)^{1/2}\right] \tag{91}$$

when the two initial configurations C_0 and $\tilde{C}_0$ are opposite (C_0 is random) and

$$P(T) = \frac{2}{\pi}\,\tan^{-1}\left[(1-T)^{1/2}\right] \tag{92}$$

when the two initial conditions C_0 and $\tilde{C}_0$ are independent random configurations.

When the two configurations remain different in the long time limit, one can calculate their distance $D(T)$ and one finds[28] that this distance is equal to the magnetization $m(T)$

$$D(T) = m(T) \tag{93}$$

which is the fixed point ($m > 0$) of

$$m = \tanh\left(\frac{m}{T}\right) \tag{94}$$

For this mean field model, we see that above $T_c(=1)$, the two configurations meet quickly, so $P(T)$ vanishes. Below T_c, there is a finite probability $P(T)$ that one configuration falls into the $+$ phase and the other in the $-$ phase: When the temperature changes, the relative effects of the initial conditions and of the thermal noise changes. A higher temperature increases the probability that the two configurations fall into the same valley.

The fact that the distance $D(T)$ is exactly equal to the magnetization is more surprising. One expects that if one configuration falls into the $+$ phase and the other in the $-$ phase, their distance is comparable to $m(T)$. But the exact equality (93) is not a priori obvious. We will see below that the equality (93) holds in finite dimension for any system which contains only ferromagnetic bonds.

- The barrier heights

Before doing that, let me mention that the comparison of two or more configurations subjected to the same noise can be used in ferromagnetic systems to measure barrier heights.

For any finite system, two configurations subjected to the same noise will ultimately meet. One can define the time τ_2 for two configurations to meet. In the low temperature phase, this time τ_2 increases with the system size because, with a finite probability, each configuration is in a

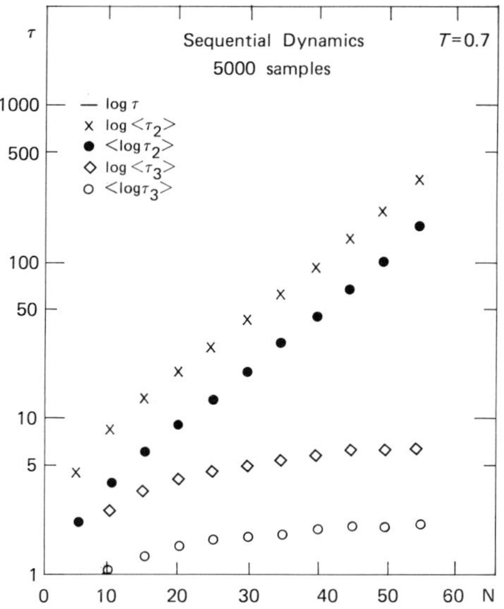

FIGURE 11

Sequential dynamics : in the ferromagnetic phase ($T = 0.7 < T_c = 1$), the time τ_2 for two configurations to meet as a function of N. The slope of τ_2 is a measure of the barrier height. The time τ_3 that three configurations remain different is much shorter because there are only two valleys.

different phase and the time for them to meet is the time for one of the two configurations to jump from one phase to the other.

Fig.11 gives τ_2 as a function of the system size for the mean field model (90). We see that τ_2 increases exponentially with N and the slope of log τ_2 versus N gives a measure of the free energy barrier between the two phases[28].

One can compare more than 2 configurations. The time τ_3 shown in figure 1 represents the time for two among 3 configurations to meet. This time is much shorter because, the mean field ferromagnet possesses only 2 valleys. Therefore, three configurations cannot remain different for very long. Of course, τ_3 could increase with N for systems (like the 3 Potts model) which have at least three low temperature phases.

In the case of automata, with noise (Section 2.B), the times τ_2, τ_3, τ_4... can be measured and one finds that they all increase exponentially with N in the chaotic phase[29].

- *The finite dimensional case*

The fact that the distance $D(T)$ is equal to the magnetization can be understood for any

Ising model with ferromagnetic interactions. The derivation of this result was explained to me by J.L. Lebowitz[30].

Let us choose for the two initial configurations $\{S_i(0)\}$ and $\left\{\tilde{S}_i(0)\right\}$, two configurations which satisfy

$$S_i(0) \leq \tilde{S}_i(0) \qquad \forall i \tag{95}$$

Because the bonds are ferromagnetic, one can check from the dynamical rule (87), that at any later time, this property remains valid

$$S_i(t) \leq \tilde{S}_i(t) \tag{96}$$

Therefore the distance $d(t)$ defined by (88) becomes

$$d(t) = \frac{1}{2N} \sum_{i=1}^{N} \left(\tilde{S}_i(t) - S_i(t) \right) = \frac{1}{2} \left[\tilde{m}(t) - m(t) \right] \tag{97}$$

We see that the distance is related to the magnetization. In particular if the two configurations fall into the same phase, $d(t) \longrightarrow 0$ because $m(t)$ and $\tilde{m}(t)$ converge to the same value, whereas if one configuration falls into the $+$ phase and the other in the $-$ phase, $d(t)$ is equal to the spontaneous magnetization.

This analogy between quantities measured by comparing two configurations and magnetic properties can be extended to correlation functions[27].

3.C. Distances in the spin glass problem

One can study the time evolution of the distance between pairs of configurations subjected to the same thermal noise in all kinds of systems. In this section, I will describe what is observed in the case of spin glasses.

- *The Sherrington Kirkpatrick model*[31,32]

The Sherrington Kirkpatrick model (SK model) plays the role in spin glass theory of the mean field model (90) discussed in the previous section for the ferromagnet.

The Hamiltonian of the SK model is

$$\mathcal{H} = -\frac{1}{\sqrt{N}} \sum_{i<j} J_{ij} S_i S_j \tag{98}$$

where in (98) the J_{ij} are random variables ($J_{ij} = \pm 1$ or J_{ij} Gaussian with $\langle J_{ij}^2 \rangle = 1$).

For any distribution of interactions J_{ij} which satisfies $\langle J_{ij}^2 \rangle = 1$, the transition temperature T_c of the SK model defined by (98) is

$$T_c = 1 \tag{99}$$

To compare the time evolution of these two configurations subjected to the same thermal noise, one can consider the following three situations.

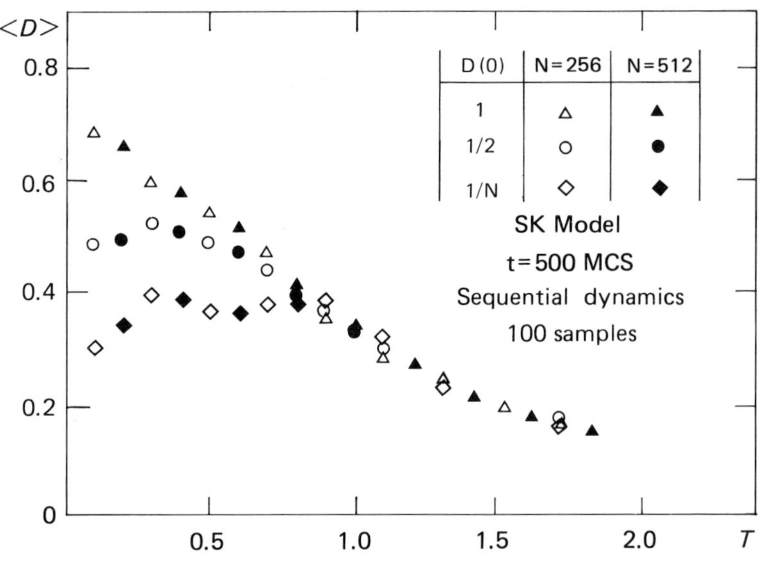

FIGURE 12

Sherrington Kirkpatrick model : The distance $D(T)$ versus temperature T for three initial distances; opposite initial conditions ($d(0) = 1$), uncorrelated initial conditions ($d(0) = 1/2$), identical initial conditions except for one spin ($d(0) = 1/N$).

a) The two initial configurations C_0 and $\tilde{C}_0$ are opposite, C_0 being random ($S_i(0) = -\tilde{S}_i(0)$ for all i). Therefore $d(0) = 1$. (The data corresponding to this case will be represented by triangles in the figures).

b) The two initial configurations C_0 and $\tilde{C}_0$ are random and independent: $d(0) = 1/2$ (squares in the figures).

c) The two initial configurations C_0 and $\tilde{C}_0$ differ by a single spin: $d(0) = 1/N$ (diamonds in the figures).

For each of these three initial conditions, one can measure the survival probability $P(T)$ and the distance $D(T)$. One finds that the survival probability $P(T)$ is 1 at all temperature (for large enough systems) for each of the three initial situations. The results[33] for $D(T)$ are shown in figure 12.

We see that the distance $D(T)$ does not vanish at any temperature. Numerical calculations performed at higher temperature indicate that $D(T)$ would vanish at $T = \infty$ and that $D(T) \sim 1/T^2$.

So the temperature where $D(T)$ vanishes is not related to the phase transition in the SK model. However, as it can be seen in Fig.12, one observes two regions for the distance. A high temperature regime where $D(T)$ does not depend on the initial distance and a low temperature regime where $D(T)$ presents remanence effects and does depend on the initial distance. The temperature which separates these two regimes seems to be close to $T_c = 1$. However the data of Fig.12 are not of good enough quality to be certain that the equilibrium transition temperature $T_c = 1$ is the place where the remanence effects in the distance $D(T)$ start to appear.

- *The 3 dimensional spin glass*

The same calculation can be repeated for finite dimensional systems[25]. In 3d with $\pm J$ bonds, one finds that below a temperature $T_1 \simeq 4J$, the survival probability $P(T) \neq 0$. For initial conditions a and b, one finds that $P(T) = 1$ (as for the automata in presence of noise of Section 2.C) whereas $P(T)$ is a function of temperature for the case c (two initial configurations which differ by a single spin) which is very similar to what was obtained for automata (Fig.7a). So the survival probability $P(T)$ in the case of 3d spin glasses has in all cases a similar temperature dependence as in the automata model.

The distance $D(T)$ obtained for the 3d $\pm J$ spin glass is given in figure 13. We see clearly 3 regimes :

- A high temperature regime $T > T_1$ where the distance $D(T)$ vanishes.

- An intermediate regime $T_2 < T < T_1$ where the distance $D(T)$ is non zero but is independent of the initial distance.

- A low temperature regime $T < T_2$ where the distance $D(T)$ depends on the initial distance.

From figure 13, one can estimate that $T_1 \simeq 4.$ and $T_1 \simeq 1.5$. These estimates may have some dependence on the system size and on the time t of the simulation and a good finite size scaling analysis[26] would be necessary to improve these estimates.

The main question about these results is how to compare them with what is already known in the 3d spin glass[34,35]. Unlike in the ferromagnetic case, these does not exist here any proof that

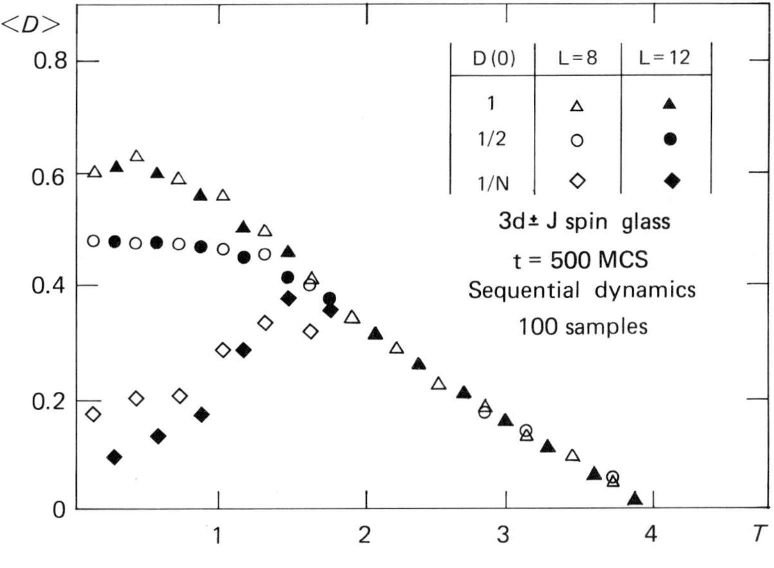

FIGURE 13

The distance $D(T)$ versus temperature (as in Fig.12) for the 3d $\pm J$ spin glass $(N = L^3)$.

the dynamical phase transitions associated to the distance (between two configurations subjected to the same noise) should coincide with equilibrium phase transitions. So one can only compare the numerical data.

In his extensive numerical simulations on the 3d $\pm J$ spin glass[35], Ogielsky obtained that there are 3 regimes for 3d $\pm$ spin glass: a high temperature region $T > T_1^*$ where the spin autocorrelation function $q(t)$ decreases exponentially with time. An intermediate region $T_2^* < T < T_1^*$ where $q(t)$ has a slow decay (stretched exponential). A low temperature phase $T < T_1^*$ where $q(t)$ does not decay to zero. His estimates for T_1^* and T_2^* were $T_1^* \simeq 4.5$ (because T_1^* is probably the transition temperature of the ferromagnet) and $T_2^* \simeq 1.2$.

We see that the values of $T_1 \simeq 4.$ and $T_2 \simeq 1.5$ are rather close to $T_1^* \simeq 4.5$ and $T_2^* \simeq 1.2$ by Ogielsky. The quality of the data for the determination of T_1 and T_2 is however not good enough at the moment to be certain that they are the same.

Since the lower temperature T_2 seems to be close (in the 3d case and for the SK model) to the equilibrium phase transition, it was tempting to study the effect of a magnetic field h. The numerical data obtained so far indicate that $T_2(h)$ decreases[36] with h for the SK model whereas it increases in the 3d case[37].

4. CONCLUSION

These lectures describe a few models and methods which have been developed recently in the study of the dynamics of spin models and automata. There is a special emphasis on the distance method which consists of comparing the time evolution of two configurations.

For deterministic models, this method gives a good way of finding the phase boundary between a frozen and a chaotic phase (section 2).

For stochastic systems, one finds that this method gives dynamical phase transitions (section 3) but, except in the ferromagnetic case, one does not know yet how these dynamical phase transitions are related to the equilibrium phase transitions[38,39]. It would be, of course, very interesting to understand better the relation between the dynamical phase transitions (section 3) and the equilibrium ones.

Another challenging question is the calculation of the periods in deterministic models. The models discussed in sections 1.C and 2. are exactly soluble mean field models[40] in the sense that in the limit $N \longrightarrow \infty$, there are some quantities (local magnetizations, distances,...) which can be calculated exactly. Much less is known on these models for finite N. In particular, there does not exist at present any analytic way of computing the period of these mean field models.

To conclude these lectures, I would like to mention an attempt in that direction. The formula I obtained for the period of a Kauffman model is in the limit $N \longrightarrow \infty$

$$\frac{\log(\text{Period})}{N} = \frac{1}{2} \int P(m) dm \, \log\left(\frac{2}{1+m^2}\right) \tag{100}$$

where $P(m)$ is the distribution of activities discussed in section 2.2. Let me explain briefly how this formula was obtained and what assumption was made to derive it. Consider a random map[41]

$$C' = F(C) \tag{101}$$

in a phase space of 2^N configurations. For each $\mathcal{C}$, one chooses its image $\mathcal{C}' = F(\mathcal{C})$ at random according to a given probability $p(\mathcal{C})$. Then one can show[42] that the typical period of this random map is

$$\text{Period} \sim \left(\sum_{\mathcal{C}} (p(\mathcal{C}))^2 \right)^{-1/2} \tag{102}$$

Now let us identify a configuration $\mathcal{C}$ in phase space with a spin configuration $\{S_i\}$ and choose for $p(\mathcal{C})$

$$p(\mathcal{C}) = \prod_{i=1}^{N} \left(\frac{1 + m_i S_i}{2} \right) \tag{103}$$

Then from (102) one gets

$$\text{Period} \sim \left[\frac{1}{2^N} \prod_{i=1}^{N} (1 + m_i^2) \right]^{-1/2} \tag{104}$$

which gives (100) in the limit $N \longrightarrow \infty$.

So (100) is exact for the random map model defined by (103). This random map model has the same local magnetizations as the Kauffman model. If one assumes that the correlations between the spins do not change the period, then one concludes that the exponential growth of the period in the Kauffman model would be given by (100). At present, I do not know whether this assumption (which neglects the correlations as one usually does in mean field models) is justified for the mean field models discussed in section 1.C and 2.

These lectures have for main content the result of collaborations which came out from very pleasant and stimulating discussions with H. Flyvbjerg, O. Golinelli, A.U. Neumann, Y. Pomeau, D. Stauffer and G. Weisbuch.

REFERENCES

1) Gant Macher (1959) Matrix Theory Vol. II, p.53, Chelsea Publishing Company N.Y.

2) K. Binder (1984) Applications of the Monte Carlo Method in Statistical Physics (Berlin : Springer Verlag).

3) P. Peretto, Biol. Cybern. 50, 51 (1984).

4) E. Bienenstock, F. Fogelman Soulié and G. Weisbuch, Disordered Systems and Biological Organisation (Heidelberg : Springer-Verlag).

5) E. Goles, Comportement dynamique des réseaux d'automates, Thèse Grenoble 1985.

6) B. Derrida, J. Phys. A20, L721 (1987).

7) B. Derrida, H. Flyvbjerg, J. Phys. A20, L1107 (1987).

8) K.E. Kürten, J. Phys. A21, L615 (1988); Phys. Lett. A129, 157 (1988).

9) S.A. Kauffman, J. Theor. Biol. 22, 437 (1969).

10) S.A. Kauffman, Physica D10, 145 (1984).

11) A.E. Gelfand and C.C. Walker, Ensemble Modelling, (M. Dekker) (1984).

12) H. Flyvbjerg and N.J. Kjaer, J. Phys. A21, 1695 (1988).

13) H.J. Hilhorst and M. Nijmeijer, J. Physique 48, 185 (1987).

14) B. Derrida and G. Weisbuch, J. Physique 47, 1297 (1986).

15) B. Derrida and Y. Pomeau, Europhys. Lett. 1, 45 (1986).

16) H. Flyvbjerg, J. Phys. A21, L509 (1988).

17) B. Derrida, D. Stauffer, Europhys. Lett. 2, 739 (1986).

18) G. Weisbuch, D. Stauffer, J. Physique 48, 11 (1987).

19) D. Stauffer, Phys. Mag. B56, 901 (1987).

20) L. de Arcangelis, J. Phys. A20, L369 (1987).

21) L. de Arcangelis, D. Stauffer, J. Physique 48, 1881 (1987).

22) B. Derrida, Les Houches (1986) "Chance and Matter", J. Souletie, J. Vannimenus and R. Stora eds.,

23) S. Stölzle, J. Stat. Phys. 53, 995 (1988).

24) L. de Arcangelis, A. Coniglio, Europhys. Lett. 7, 113 (1988).

25) B. Derrida, G. Weisbuch, Europhys. Lett. 4, 657 (1987).

26) A.U. Neumann, B. Derrida, J. de Physique 49, 1647 (1988).

27) A. Coniglio, L. de Arcangelis, H.J. Herrmann, N. Jan, Europhys. Lett. 8, 315 (1989).

28) O. Golinelli, B. Derrida, J. Physique 49, 1663 (1988).

29) O. Golinelli, B. Derrida, J. Physique 50, 1587 (1989).

30) J.L. Lebowitz, private communication.

31) D. Sherrington, S. Kirkpatrick, Phys. Rev. Lett. 35, 1972 (1975); Phys. Rev. B17, 4384 (1978).

32) M. Mezard, G. Parisi, M. Virasoro, "Spin Glass Theory and Beyond" 1987, World Scientific Publishing Co.

33) B. Derrida, to appear in Phys. Reports.

34) K. Binder, A.P. Young, Rev. Mod. Phys. 58, 801 (1986).

35) A. Ogielsky, Phys. Rev. B32, 7384 (1985).

36) B. Derrida, unpublished.

37) L. de Arcangelis, H.J. Herrmann, A. Coniglio, to appear J. Phys. A, (1989).

38) O. Martin, J. Stat. Phys. 41, 249 (1985).

39) L. de Arcangelis, A. Coniglio, H.J. Herrmann, Europhys. Lett. 8, 315 (1989).

40) B. Derrida, E. Gardner, A. Zippelius, Europhys. Lett. 4, 167 (1987).

41) B. Derrida, H. Flyvbjerg, J. Physique 48, 971 (1987).

42) C. Beck, Phys. Lett. A136, 121 (1989).

FUNDAMENTAL PROBLEMS IN STATISTICAL MECHANICS VII
H. van Beijeren, Editor
© *Elsevier Science Publishers B.V., 1990*

A THEORY OF AGING IN SPIN GLASSES

H.J. HILHORST

Laboratoire de Physique Théorique et Hautes Energies,
Université de Paris-Sud, Bât. 211, 91405 Orsay, France

Recent experimental data on aging in spin glasses can be explained by the interplay of two types of characteristic lengths : (i) a time dependent domain size, and (ii) an overlap length associated with each pair of nearby equilibrium states.

1. INTRODUCTION

Slow relaxation processes in nature have received ample attention over the last many years. These lectures are about a special class of such processes, namely those that involve aging phenomena. To our knowledge no general definition of aging has been given, and we shall therefore attempt one.

It may happen that a physical system is approaching its final equilibrium state so slowly that experimentally it looks like being in thermodynamic equilibrium. That the system still evolves in time becomes clear when one applies an external force to it : the response then appears to vary with time. The system is said to be <u>aging</u>. The term comes from materials science (see e.g. Ref. 1), where the physical mechanisms which cause aging are usually related to slow changes in the system's structure on a mesoscopic length scale. These do not involve much free energy and are therefore not easily detectable by thermodynamic measurements.

We shall be concerned exclusively with aging associated with magnetic relaxation in spin glasses. Spin glasses are magnetic materials that are said to be disordered because of the simultaneous occurrence of competing ferro- and antiferromagnetic interactions. They have a critical temperature T_c, typically in the range of 10 to 50 K. However, unlike ferromagnets, even for temperatures T below T_c they magnetize only in response to a magnetic field. Whereas for $T > T_c$ the magnetic response is immediate, for $T < T_c$ it becomes extremely slow.

Aging in spin glasses was discovered by Lundgren et al.[2] in 1983 in CuMn, and has since been found and studied in several spin glass materials, in particular by the group in Uppsala, and by Ocio et al.[3] and Refregier et al.[4] at Saclay. Aging phenomena can be observed in zero or small magnetic fields, typically $\lesssim 10$ G. In too high fields they disappear. They do not occur to the same degree in all spin glasses [5]. Moreover, aging in magnetic relaxation is not limited to spin glasses and has recently been observed [6,7] in some of the high T_c superconducting oxides.

2. MAGNETIC RELAXATION IN SPIN GLASSES

In magnetic relaxation experiments in spin glasses the experimental control parameters are

- the temperature T ;
- the magnetic field H ;

- an "infinitesimal" probing field $h_o e^{i\omega t}$

The observed quantities are

   - the magnetization M ;
   - the susceptibility χ.

More rarely one also carries out measurements of energy changes in the system.

In a traditional <u>field jump experiment</u> one places a sample in a field, cools it to a "working" temperature below T_c, cuts the field, and monitors the relaxation of the magnetization, M(t). Typically the relaxation curve is very flat, and it may happen that within the experimentally available time window   (10 s to 10^5 s) M(t) relaxes by an amount of, say, no more than 5 % of its initial value.

Much effort has been spent on a precise determination of the analytic form of the relaxation curve. Among the proposed behaviors have been

$$M(t) \sim \frac{1}{\log(t/\tau_o)} \; , \tag{1a}$$

$$M(t) \sim \left(\frac{\tau_o}{t}\right)^a \; , \quad a > 0 \tag{1b}$$

$$M(t) \sim e^{-(t/\tau_o)^p} \; , \quad 0 < p < 1 \tag{1c}$$

multiplicative combinations of these, or still more complicated forms. The "stretched exponential" or Kohlrausch law (1c) has come to enjoy a great deal of popularity. No firm theoretical basis exists, however, for any of these analytic forms. The fundamental theoretical approach would be to write down and solve a master equation for the probability distribution P(s,t) over the spin configurations s, assuming a set of spin interactions and flip rates. Such a procedure, however, is too complicated and has not led to results that can be meaningfully compared to experiment.

In a more phenomenological approach one expresses the relaxation curve M(t) as a weighted sum of exponentials,

$$M(t) = M(0) \int_0^\infty d\Delta \, p(\Delta) \, e^{-t/\tau_\Delta} \tag{2}$$

The physical input here is that one considers τ_Δ as the typical time needed for the system to cross a free energy barrier Δ, and writes the Arrhenius law

$$\tau_\Delta = \tau_o \, e^{\Delta/k_B T} \tag{3}$$

with τ_o a microscopic time. The problem of determining the unknown decay curve M(t) is thereby replaced with the problem of determining the unknown weight function $p(\Delta)$, and obviously not much has been gained. Suitable choices of $p(\Delta)$ will reproduce any of the three behaviors (1).

3. AGING AT A CONSTANT TEMPERATURE AND MAGNETIC FIELD

The discovery of aging in 1983 by Lundgren et al.[2] had the merit of providing for a much larger variety of experimental data in need of an interpretation. We shall describe here a few out of the many aging phenomena discovered and choose a pedagogical rather than a chronological order.

3.1. The simplest aging phenomenon

Suppose one takes a sample in <u>zero</u> magnetic field and quenches it, at a time $t = 0$, to a temperature T below T_c (typically $T \approx 0.8\ T_c$). One then probes the system with an "infinitesimal" field $h(t) = h_o\ e^{i\omega t}$ (typically $\omega \approx 10^2$ Hz) in order to determine its dynamic susceptibility at that frequency. It is then found [8,4] that the imaginary part χ'' of χ varies with time : it is written as $\chi''(\omega\ ;\ t)$. Obviously the concept of a time and frequency dependent susceptibility makes sense only if the t dependence is slow on the scale ω^{-1}. Typically $\chi''(\omega\ ;\ t)$ decays to a constant value $\chi''_{eq}(\omega)$ on a time scale of minutes.

3.2. Explanation

The explanation [9,10] is as follows. Let us consider, for definiteness, Ising spins. In the low temperature phase the spins s_i are frozen in, i.e. fluctuate around the nonzero average values

$$< s_i > = m_i \quad \text{(all i),} \quad \text{or} \quad < s_i > = - m_i \quad \text{(all i)} \qquad \text{(4 a,b)}$$

(In zero field there is no net magnetization density : the volume average of the m_i vanishes). Eq. (4) defines two possible spin glass orderings. Just after the quench, at time $t = 0^+$, the spins point in random directions. With time a pattern of domains will begin to appear, each domain being an ordered region of type (4a) or one of type (4b). Two hypotheses are basic to what follows :

(1) the typical linear domain size $l(t)$ increases with time t ;

(2) the typical magnetic relaxation rate $\tau^{-1}(l)$ of a spin in a domain of size l decreases with l (this being due to the stronger coherence of spins in a larger domain).

The precise analytic forms of $l(t)$ and $\tau(l)$ are of secondary importance. For definiteness we shall suppose, however, that

$$l(t) \sim a_o\ \left(\frac{t}{\tau_o}\right)^p \quad , \qquad \text{(5 a)}$$

$$\tau(l) \sim \tau_o\ \left(\frac{l}{a_o}\right)^z \quad , \qquad \text{(5 b)}$$

where $p > 0$ and $z > 0$, and a_o and t_o are a microscopic length and time, respectively. Substitution of (5 a) in (5 b) then shows that at a time t after the quench the typical magnetic relaxation time is given by

$$\tau(t) \sim \tau_o\ \left(\frac{t}{\tau_o}\right)^{pz} \qquad \text{(6)}$$

(more precisely : $\tau^{-1}(t)$ is the instantaneous magnetic relaxation <u>rate</u> a time t after the quench).

The concept of a time-dependent relaxation spectrum was used already by Lundgren et al.[11]. Once it has been accepted, it is not hard to calculate the time-dependent dynamic susceptibility. Koper and Hilhorst [9] show that the above picture leads to

$$\chi''(\omega\,;t) - \chi''_{eq}(\omega) \sim t^{-pz}\,, \tag{7}$$

i.e. a decay towards equilibrium as observed in the experiment [2].

4. AGING IN THE PRESENCE OF MAGNETIC FIELD JUMPS. THE PROBLEM

The reasoning so far is rather simple-minded and straightforward, and could also apply to, say, an antiferromagnet in a field (see Section 9). There is, however, a class of experiments which reveals the necessity to extend this theory in a very nontrivial way. This can be illustrated by any experiment in which, following the quench, the sample is exposed to still further changes (jumps or continuous variations) of either temperature or magnetic field. One of these is the field jump experiment, which we shall consider more carefully now.

A preliminary remark is that for the field values that are used the thermal energy of the spins is typically a factor $\sim 10^4$ larger than their magnetic energy. Consequently the picture of up and down domains remains very nearly valid. At first sight the hypotheses of Section 3 then seem to allow for a qualitative understanding of the $M(t)$ curves in a field jump experiment. If the field H is cut at time $t_w > 0$ (the quench having taken place at $t = 0$), then $M(t)$ will relax as

$$\frac{M(t)}{H} \sim \chi_{eq}\, e^{-(t-t_w)/\tau(t)} \tag{8 a}$$

$$= \chi_{eq}\, \exp\left[-\left(\frac{t-t_w}{\tau_0} \right)\left(\frac{t}{t_w} \right)^{-pz} \right] \tag{8 b}$$

Here χ_{eq} is the equilibrium dc susceptibility. For different waiting times t_w (but at the same fixed small enough value of H), Eq. (8 b) gives a set of curves that are in satisfactory agreement with experiment [2,3,4] provided one chooses $pz \approx 1$.

The problem arises if one increases H at fixed t_w. Eq. (7 b) predicts that the relaxation curves for $M(t)/H$ should be independent of H. Experimentally [12] such independence (in a sample of Cd Cr$_{1.7}$ In$_{0.3}$ S$_4$) is indeed observed in the magnetic field range until ~ 10 G. However, for $H \gtrsim 20$ G the relaxation becomes much faster. This effect cannot be explained on the basis of the nonlinearity in the equilibrium M(H) curve, which is still almost straight at $H = 20$ G. It is a true dynamic nonlinearity. The problem is how to explain it.

5. RANDOMNESS AS A FUNDAMENTAL PROPERTY OF THE EQUILIBRIUM SPIN GLASS

5.1. Temperature dependence of spin-spin correlations

We have to recall first a few properties of the equilibrium spin glass. The essential point is that, below T_c, the nature of the equilibrium spin glass state changes randomly with temperature. In order to explain what is meant by this, we consider the spin-spin correlation function at temperature T, $\langle s_i\, s_j \rangle \equiv g_{ij}(T)$. Since the frozen-in averages $\langle s_i \rangle = \pm m_i$ of Eq. (4) vary randomly as a function of i, obviously $g_{ij}(T)$ is also a random function of i and of j. This is not very surprising. The important point is that $g_{ij}(T)$ <u>depends randomly also on T</u> ! I.e., for the same sample, g_{ij} (T + ΔT) is a function that is very different from g_{ij} (T). The degree of similarity between g_{ij} (T) and g_{ij} (T + ΔT) is conveniently expressed by <u>an overlap length</u> $l_{\Delta T}^{eq}$. This is a quantity associated with a pair of equilibrium states of the same sample. It has the property that

$$g_{ij} \ (T) \ \approx \ g_{ij} \ (T + \Delta T) \qquad\qquad \text{for} \ |i\text{-}j| \ \lesssim \ l_{\Delta T}^{eq}$$

$$g_{ij} \ (T) \ \text{and} \ g_{ij} \ (T + \Delta T) \quad \text{uncorrelated} \ \ \text{for} \ |i\text{-}j| \ \gg \ l_{\Delta T}^{eq} \tag{9}$$

For a definition see e.g. Ref. 18. Obviously $l_{\Delta T}^{eq}$ must increase to infinity when $\Delta T \to 0$.

The reason for this random temperature dependence was exposed by Bray and Moore [14] and by Fisher and Huse [15]. At a given temperature, the stability of long-range spin glass order implies that only small length scale thermal excitations are possible ; large-scale fluctuations are suppressed by increasingly high energy barriers. However, when the equilibrium temperature is changed by an amount ΔT, the situation in some sense the reverse : the large-scale spin coherence will be disrupted more easily than the short-distance order. In fact, <u>no matter how small ΔT is, the long-range order associated with the equilibrium state at temperature T will become unstable beyond some distance</u> $l_{\Delta T}^{eq}$. Bray and Moore [14] show that

$$l_{\Delta T}^{eq} \ \sim \ a_o \left(\frac{J}{\Delta T}\right)^{x_T} \qquad (\Delta T \ \text{small}) \tag{10}$$

where J is a typical value of the spin interaction energy and x_T an exponent which the authors relate to the properties of the thermal excitations.

5.2. Field dependence of spin-spin correlations

One should expect that, for small magnetic fields H, the spin correlation functions depend randomly also on H. Hence an overlap length $l_{\Delta H}^{eq}$ is associated with each pair of equilibrium spin glass states (at the same temperature) but in magnetic fields that differ by an amount ΔH. In analogy to (10) we put

$$l_{\Delta H}^{eq} \ \sim \ a_o \left(\frac{J}{\mu_o \Delta H}\right)^{x_H} \qquad (\Delta T \ \text{small}) \tag{11}$$

where μ_o is the spin's magnetic moment and x_H a positive exponent (Remark : If the magnetic field destroys the spin glass phase, then there is a finite equilibrium correlation length ξ_H which tends to zero for $H \to 0$. In the small fields H that we consider, we assume ξ_H to be larger than any other length in the problem, so that we are effectively in a spin glass phase).

The incompatibility of the long-distance correlations even in nearby equilibrium states of a spin glass must have consequences for the dynamics. We turn to these consequences now.

6. AGING IN THE PRESENCE OF FIELD JUMPS. EXPLANATION OF DYNAMIC NONLINEARITIES

When in the field jump experiment the field is cut, the time-dependent domain size will have reached the value $l(t_w)$. This quantity refers to domains in the sense of the equilibrium ordering at field H, which overlaps with the zero field equilibrium only up to length scales l_H^{eq}. Therefore, if the waiting time t_w has been so long that $l(t_w) > l_H^{eq}$, the domains formed before the field cut and whose size exceeds $l(t_w)$, will be incompatible with the zero-field equilibrium state to be formed after t_w. Hence the relaxation of the magnetization will set out at a time scale which is the minimum of $\tau(l(t_w))$ and $\tau(l_{\Delta H}^{eq})$. Calling this scale $\tau_{jump}(H)$ and using (5 b), (6) and (11) this gives

$$\tau_{jump}(H) \sim t_o \min\left[\left(\frac{t_w}{t_o}\right)^{p\,z}, \left(\frac{J}{\mu_o H}\right)^{zx_H}\right] \tag{12}$$

For small field jumps H this is independent of H (linear regime) ; for larger jumps it begins to get shorter (dynamical nonlinearities), in agreement with the experimental observations [12,13].

Suppose that one accepts the various power laws that have been postulated above. Since x_H is known from independent arguments [14], and the product pz from the linear regime (see Section 4), a study of the crossover to the nonlinear regime should in principle lead to values for p and z individually. An analysis of data by Alba et al. [12] has led Koper and Hilhorst [9] to the estimates

$$p \approx 0.5 \qquad\qquad z \approx 1.8 \tag{13}$$

7. AGING IN AN ARBITRARY TIME-DEPENDENT FIELD

In the previous section only two equilibrium states played a role : the ones in field H and in field zero. Aging effects were reducible to the interplay between two lengths : the time dependent domain size $l(t)$ with respect to ordering in one of the states, and the overlap length l_H^{eq} between the two equilibrium states.

Clearly more complicated experimental situations may arise, for example (i) when the field makes several steps $H_1 \to H_2 \to ... \to H_n$ not necessarily all of the same sign, or (ii) when H varies continuously, e.g. as $H(t) = H_o e^{i\Omega t}$, with an amplitude H_o which (in contrast to the amplitudes of the probing fields used to measure the ac susceptibility) is not infinitesimal. In such cases one can find the behavior of the magnetization by generalizing the reasoning of the preceding sections, as has been shown in Ref. 9. An interesting situation arises when $H(t)$ does not settle down with time to a constant value. Then the system cannot approach an equilibrium state of any kind closer than up to a length scale that will depend on $H(t)$. For $H(t) = H_o e^{i\Omega t}$ one should expect [19], possibly after a transient effect as discussed in Section 3, a periodically time-dependent nonequilibrium state in which one can again measure $\chi(\omega;t)$ for $\omega \gg \Omega$.

8. AGING IN THE PRESENCE OF TEMPERATURE VARIATIONS

8.1. Temperature jumps

Experiments have also been carried out in which, during the waiting time, temperature jumps were performed. Since the equilibrium state varies randomly with T as well as with H, one should expect that, in principle, temperature jumps have the same (linear and nonlinear) effects on domain growth as do magnetic field jumps, the only difference being that the role of $l_{\Delta H}^{eq}$ is taken over by

$l_{\Delta T}^{eq}$. It indeed appears experimentally that a crossover from a linear to a nonlinear regime takes place for ΔT in the range from 0.1 to 1 K.

Temperature jumps have, however, a very important complication associated with them that is absent for field jumps, and that is clearly brought out by the experiments [4,16]. This is that the time scale of (supposedly) thermally activated phenomena such as the growth and the breaking-up of domains, is proportional to $\exp\left(\frac{E_o}{k_B T}\right)$, where E_o is a microscopic energy. Hence when applying the heuristic arguments of the previous sections to situations with nonconstant temperature [9] one should also take this variation of the time scale into account. As a consequence, whereas there is a perfect symmetry between (small) magnetic field jumps ΔH and $-\Delta H$, there is no such symmetry between ΔT and $-\Delta T$. This asymmetry has been beautifully demonstrated in experiments [4,16] with several temperature jumps.

8.2. Continuous temperature variations

As in the case of continuous magnetic field variations, a continuum of equilibrium states is involved. A special case has been considered by Koper and Hilhorst [17] within the context of a model calculation. We shall come back to this in Section 10.

9. AGING IN AN ANTIFERROMAGNET

This section has no other purpose than the pedagogical one of demonstrating why aging in a spin glass is different from aging in an unfrustrated magnetic material. We consider an Ising antiferromagnet, which is close to a spin glass in that below its critical temperature T_N it is ordered without spontaneous magnetization. When such a system is quenched to below T_N, domains of both ordered phases will start to grow. Since in a pure antiferromagnet one does not expect any slowing down due to a hierarchy of energy barriers, this growth will in general be fast. Again, the magnetic relaxation rate and the ac susceptibility will vary with the typical domain size. If (in a suitably chosen material, or in a Monte Carlo simulation) the size increase takes place on a sufficiently slow time scale, then aging will manifest itself in e.g. a field jump experiment.

This situation is very analogous to aging in spin glasses as described in section 3. The difference appears, however, when one considers the nonlinear regime. An ordered antiferromagnetic domain in equilibrium at a temperature T and in a magnetic field H can adapt, without large-scale rearrangements, to new temperature and field values $T+\Delta T$ and $H+\Delta H$. Hence one expects no dynamic nonlinearities in aging in an antiferromagnet.

10. A "RANDOM EQUILIBRIUM" MODEL

In an attempt to construct a more microscopic model of aging phenomena, Koper and Hilhorst [18] introduced what might be called a "random equilibrium" (REQ) model of a spin glass. This model is essentially one-dimensional, and hence has no phase transition. However, on length scales less than its correlation length it exhibits several of the features of a true spin glass.

The REQ model is thought of as representing a spin glass with short-range random couplings on a d-dimensional lattice (a "bar") of $(Lb) \times b^{d-1}$ sites. One imagines that this system can be renormalized via some real-space method in which blocks of b^d spins are represented by block spin variables s_k ($k = 1,...,L$). The renormalized system is at a temperature T, which is a smooth function of the original temperature, and is supposed to have nearest-neighbor interactions $J_{k,k+1}$ of some typical strength J. The effects of frustration in the original system will cause the precise values of the individual couplings $J_{k,k+1}$ to depend very sensitively on the original temperature, and hence on T. So one writes $J_{k,k+1}(T)$. An exact calculation of these couplings is not feasible. It is replaced with the postulate that $J_{k,k+1}(T)$ takes only the values $\pm J$ in a stochastically temperature-dependent way, with correlation function

$$\overline{J_{k,k+1}(T)J_{k,k+1}(T')} = J^2 e^{-\Gamma |T-T'|} \tag{14}$$

Here Γ is a constant that arises in the renormalization process.

The model thus defined is, at any <u>fixed</u> temperature, nothing but a Mattis spin glass which can be transformed into a ferromagnet. The domain size of the preceding sections is identified with the typical length of a sequence of satisfied bonds. If this system is quenched from a high to a low temperature, the domain size $l(t)$ begins to grow and this growth can be calculated e.g. for Glauber dynamics. One finds $l(t) \sim \sqrt{t}$, which is the model equivalent of Eq. (5 a). (The equilibrium correlation length is supposed to be much larger than any other length in the problem and unattainable on experimental time scales). A temperature <u>variation</u> ΔT, however, has the nontrivial effect of randomly changing the signs of a fraction $\Gamma \Delta T$ of all bonds, thereby reducing the domain size. This is what makes the REQ model interesting.

This model reproduces many of the aging effects associated with temperature jumps in real spin glasses. Other situations can also be studied within its context. In particular, the case of continuous temperature variations has been considered in Ref. 17. For a linear variation, $T(t) = T_o + \dot{T} t$, it was shown that on a time scale such that $\dot{T} t \ll T_o$ the system approaches a stationary state with a domain size l_{stat} given by

$$l_{stat}(\dot{T}) \sim (\Gamma \dot{T})^{-\frac{1}{3}} \tag{15}$$

In the REQ model this manifests itself in an ac susceptibility depending on the rate of temperature change $\dot{T}$ as

$$\chi_{stat}(\omega; \dot{T}) \sim (\Gamma \dot{T})^{-\frac{1}{3}} \chi_{eq}(\omega) \tag{16}$$

in the appropriate range of values of $\dot{T}$. Analogous behavior should be expected in real spin glasses, although the power $\frac{1}{3}$ may not be the same. In the presence not of temperature but of magnetic field changes a similar effect can be studied.

REFERENCES

1) L.C.E. Struik, Physical aging in amorphous polymers and other materials (Elsevier Scient. Publ.1978).

2) L. Lundgren, P. Svedlindh, P. Nordblad and O. Beckman, Phys. Rev. Lett. 51 (1983) 911.

3) M. Ocio, M. Alba and J. Hammann, J. de Physique Lett. 46 (1985) L 1101.

4) P. Refregier, E. Vincent, J. Hammann and M. Ocio, J. de Physique 48 (1987) 1533.

5) N. Bontemps and R. Orbach, Phys. Rev. B37 (1988) 4708.

6) K.A. Müller, M. Takashige and J.G. Bednorz, Phys. Rev. Lett. 58 (1987) 1143.

7) C. Rossel, Y. Maeno and I. Morgenstern, Phys. Rev. Lett. 62 (1989) 681.

8) P. Svedlindh, P. Granberg, P. Nordblad, L. Lundgren and H.S. Chen, Phys. Rev. B35 (1987) 268.

9) G.J.M. Koper and H.J. Hilhorst, J. de Physique 49 (1988) 429.

10) D.S. Fisher and D.A. Huse, Phys. Rev. B38 (1988) 373.

11) L. Lundgren, P. Nordblad, P. Svedlindh and O. Beckman, J. Appl. Phys. 57 (1985) 3371.

12) M. Alba, E. Vincent, J. Hammann and M. Ocio, J. Appl. Phys. 61 (1987) 4092.

13) P. Nordblad, L. Lundgren and L. Sandlund, Europhys. Lett. 3 (1987) 235.

14) A.J. Bray and M.A. Moore, Phys. Rev. Lett. 58 (1987) 57.

15) D.S. Fisher and D.A. Huse, Phys. Rev. 56 (1986) 1601.

16) P. Nordblad, P. Svedlindh, L. Sandlund and L. Lundgren. Phys. Lett. A120 (1987) 475.

17) G.J.M. Koper and H.J. Hilhorst, Physica, to appear.

18) G.J.M. Koper and H.J. Hilhorst, Physica, A155 (1989) 431.

19) M. Nifle and H.J. Hilhorst, unpublished.

FUNDAMENTAL PROBLEMS IN STATISTICAL MECHANICS VII
H. van Beijeren, Editor
© *Elsevier Science Publishers B.V., 1990*

LINEAR RESPONSE THEORY FOR CELLULAR AUTOMATA FLUIDS

Matthieu H. ERNST

Institute for Theoretical Physics, University of Utrecht
3508 TA Utrecht, The Netherlands

1   Introduction

Lattice gas cellular automata are models of non-equilibrium fluids, that have received considerable attention during the last few years [1] -[10]. The basic paper in the field is a review by Frisch et al [11]. On the *microscopic scale* these models consist of many interacting particles living on the sites of a regular space lattice with a small set of allowed velocities and with over-simplified dynamics. Time is integer-valued. On the *macroscopic space and time scale*, large compared to a lattice distance and to a single time step respectively, the dynamic equations have a structure similar to the Navier-Stokes equations of fluid dynamics.

One may consider CA-fluids as practical approximations for simulating physical systems, for which the microscopic time evolution can be computed exactly without any round off errors. The solutions to the nonlinear hydrodynamic equations under various physical boundary conditions can be studied efficiently by construction of computing machines specialized to implement the simple dynamics of cellular automata. Then one has appropriate tools to study all sorts of convective instabilities, transport properties and mesoscopic phenomena. They can model multi-component and multi-phase systems, phase separation, transition to pattern formation, Rayleigh-Taylor and Saffman-Taylor instabilities, reaction systems, capillary flow in porous media, surface tension, rheology of suspensions, etc.[8].

Here we consider a complimentary use of CA-fluids as bona fide statistical mechanical models to study the relationship of macrodynamics to its underlying microdynamics for systems with many degrees of freedom. Thus we take seriously the lattice gas cellular automata (LGCA's) as models for a fluid, and apply standard methods of non-equilibrium statistical mechanics [12] to investigate the conditions under which hydrodynamic equations can be expected to apply and the relationship of the parameters in these equations to the mycrodynamics.

The main goal of these lecture notes is to study the decay of small deviations from equilibrium in CA-fluids and their transport properties. Irreversible thermodynamics provides the phenomenological description of fluids not too far from equilibrium. Then the transport fluxes are linearly related to the thermodynamic driving forces, such as gradients in the temperature and fluid velocity, through the *constitutive relations* that define the transport coefficients [13]. Here we will extend the method of Dufty and the author [14] to include temperature dependence and derive the Green-Kubo relations for the viscosities and heat conductivity of CA-fluids. We will therefore have to develope proper concepts of local equilibrium thermodynamics, local temperature and flow velocity in non-equilibrium CA-fluids.

Essential in constructing cellular automata that qualify as non-equilibrium fluids are the conservation laws, the invariances and the separation of time scales between microscopic and macroscopic dynamics. As to the conservation laws, mometum conservation is vital in the existence of a flow velocity and sound waves. The macroscopic equation for energy balance is less vital for the study of problems involving convection. Frequently energy conservation is dropped as a basic ingredient in the microscopic collision rules. Such *athermal* models would not have a temperature. The formalism developed here is general anough to include all desired conservation laws, as well as the undesired spurious ones. It appears that practically all LGCA's have spurious conservation laws, as artifacts of the discrete space-time structure [15]-[18]. These spurious conservation laws will generate additional hydrodynamic equations containing spurious transport coefficients. The Green-Kubo expressions for these transport coefficients have been derived in [14,16]. There exist several cases in the literature where properties of CA-fluids can only be explained if spurious conservation laws are taken into account [15,16,19]. The spurious conservation laws will not be considered any further in these lecture notes.

A second important point are the invariances or symmetries of the LGCA's versus those of continuous fluids. Ofcourse the discrete translational symmetries in space and time disappear in the coarse grained description on large spatial and temporal scales. However, the lack of isotropy and Galilei invariance in LGCA's are serious problems, that have consequences at the macroscopic level. The difference between the discrete rotational symmetries of Bravais lattices and the isotropy of continuous fluids manifests itself here in the invariant forms of the fourth rank viscosity tensor. Only an isotropic viscosity tensor yields the typical structure of the Navier-Stokes equations of fluid dynamics, as will be discussed in section 5. For instance, on a triangular lattice the viscosity tensor has the isotropic form with two physical (shear and bulk) viscosities, whereas on a lattice with square or cubic symmetry there are in general three different types of viscosity coefficients. For the same reason a cubic crystal has three different elastic coefficients, whereas an isotropic solid, such as Debije's elastic continuum, only has two. Therefore only very special lattices, such as the triangular lattice in two dimensions, are suitable to study the Navier-Stokes equations.

Next we discuss Galilei invariance. The equations of motion in continuous systems are invariant under the group of Galilei transformations. This invariance property is the basic reason why the nonlinear convective term $\vec{\nabla} \cdot \rho \vec{u}\vec{u}$ in the Navier-Stokes equation has this simple form, where $\vec{u}$ denotes the flow velocity and ρ the mass density of the fluid. LGCA's with their very limited set of allowed velocity states lack Galilei invariance. The reason is that adding the same constant velocity to all particles of the CA-fluid will in general generate configurations containing inadmissible velocity states. The nonlinear convective term $\vec{\nabla} \cdot g(\rho)\rho \vec{u}\vec{u}$ in CA-fluids contains the non-Galilean factor $g(\rho)$ that differs from unity. We refer to the literature [11] for a discussion of this problem. It is only briefly touched upon in section 5, as these lecture notes are focussing on *linear deviations* from equilibrium.

The plan of the notes is as follows. We start in section 2 with a description of the lattice gas models, their microdynamics and conservation laws. Section 3 introduces the basic concepts of equilibrium and non-equilibrium statistical mechanics appropriate for CA-fluids. Section 4 presents the time correlation function method leading to the Green-Kubo formulae for the CA-fluids. Section 5 discusses the general form of the fluid dynamic equations and the effects of the lack of isotropy and Galilei invariance. In section 6 the general frame work is applied to derive explicit formulae not only for the viscosities, but also for the heat conductivity in temperature dependent fluids. In section 7 the Green-Kubo formulae are evaluated in the Boltzmann approximation. Finally we conclude with a discussion.

2 Simple models of nonequilibrium fluids

A lattice gas cellular automaton or CA-fluid consists of a collection of N indistinguishable point particles, living on the V sites $\vec{r} = r_1\vec{e}_1 + r_2\vec{e}_2 + \ldots$ of a d-dimensional Bravais lattice and having periodic boundary conditions. Here $\{\vec{e}_i\}$ with $i = 1, 2, \ldots d$ is the set of basis vectors spanning the unit cell. The total volume of the unit cell and system is v_0 and $v_0 V$ respectively. For example, on the square lattice $\vec{e}_1 = (1, 0)$, $\vec{e}_2 = (0, 1)$ and $v_0 = 1$ and on the triangular lattice $\vec{e}_1 = (1, 0)$, $\vec{e}_2 = \frac{1}{2}(1, \sqrt{3})$ and $v_0 = \frac{1}{2}\sqrt{3}$. At each site there exists a set of b allowed velocity states $(i = 0, 1, \ldots, b-1)$, connecting the site with its nearest neighbors, next nearest neighbors, etc. The model is also referred to as a b-bits model. Rest particles, having $\vec{c}_0 = 0$, are also allowed in certain models. In each state, specified by $\{\vec{c}_i, \vec{r}\}$, at most one particle is allowed. So the particles obey the *Fermi exclusion rule*. The state of the whole system at time t is described by the set of occupation numbers $n(\cdot) = \{n(\vec{c}_i, \vec{r}, t)\}$, where $n(\vec{c}_i, \vec{r}, t)$ equals 1 or 0, if the state is respectively occupied or empty at time t, or more precisely at time $t - 0$.

The time evolution consist out of a collision step followed by a propagation step. The collisions are local and occur at integer time values $t = 1, 2, \ldots$, in which the precollisional velocities at $t - 0$ change into postcollisional ones at $t + 0$. In the subsequent propagation step the particle moves from its state at $\{\vec{c}_i, \vec{r}\}$ at time $t + 0$ to the state $\{\vec{c}_i, \vec{r} + \vec{c}_i\}$ at time $t + 1 - 0$. The dynamics can be expressed formally by the time evolution equation

$$n(\vec{c}_i, \vec{r} + \vec{c}_i; t + 1) = n(\vec{c}_i, \vec{r}; t) + I(\vec{c}_i \mid n(\vec{r}; t)). \qquad (2.1)$$

The shift from $\vec{r}$ to $\vec{r} + \vec{c}_i$ represents the free streaming. The collision term I can only take the values $0, +1, -1$ and depends in a nonlinear fashion on the occupation numbers $n(\vec{c}_i, \vec{r}; t)$ at site $\vec{r}$ with $i = 0, 1, 2, \ldots, b-1$. Beyond this general form no further restrictions are required on the collision term except that it supports a set of conservation laws. The collision rules for the two basic models [11], the HPP model and the FHP model, are shown in the upper right block of figure 1a and in figure 2 respectively. The 4-bits HPP model, introduced by Hardy, de Pazzis and Pomeau [20] is defined on the square lattice, and the 6- or 7-bits FHP model, introduced by Frisch, Hasslacher and Pomeau [21] is defined on the triangular lattice. The 7-bits version contains a rest particle with $\vec{c}_0 = 0$, as shown in figure 2.

As an illustration the explicit form of the nonlinear collision term $I_i = I(\vec{c}_i \mid n)$ is constructed for the HPP model:

$$I_i(n) = n_{i+1}n_{i-1}\overline{n}_i\overline{n}_{i+2} - n_in_{i+2}\overline{n}_{i+1}\overline{n}_{i-1}. \tag{2.2}$$

Here we use the short hand notation $n_i = n(\vec{c}_i, \vec{r}; t)$ for the occupation number of a particle and $\overline{n}_i = 1 - n_i$ for the occupation number of a hole with i = 1,2,3,4 (mod 4). Also note that a "collision" of two, three or four indistinguishable unlabeled particles, in which velocities are simply exchanged, is not to be considered a collision.

The conservation laws at the microscopic level determine number and structure of the slow hydrodynamic variables, the macroscopic conservation laws and the fluid dynamic equations. For the existence of a macroscopic fluid velocity $\vec{u}(\vec{r}, t)$ momentum conservation is clearly essential. To have a continuity equation number conservation must hold. Ofcourse, for systems in *thermal equilibrium* energy conservation is the most important conservation law. Frequently energy conservation at the microscopic level is not imposed (athermal models). For example, energy is trivially conserved in the collisions in the left part of figure 1 and in figure 2a,b,c and not conserved at all in figure 2e. To account for energy conservation and true thermal effects one only needs to extend the set of allowed velocities, as illustrated in figure 1. As an example we consider a thermal model with energy conservation, introduced by d'Humières et al [22]. Allowed velocities are equal to the nearest neighbor vectors with $\mid \vec{c}_i \mid = 1$ and to the next nearest neighbor ones with $\mid \vec{c}_i \mid = \sqrt{2}$. There is also a rest particle with $\mid \vec{c}_i \mid = 0$. The collision rules between particles of equal speed are the same as in the HPP model. In addition one allows energy exchanging collisions, illustrated in the right part of figure 1, with nontrivial conservation laws for total number, momentum and energy. On the square and cubic lattice the model is respectively 9-bits and 19-bits.

The time evolution of LGCA's may be *deterministic* or *stochastic*. In a stochastic version of the FHP model collisions (a) and (c) in figure 2 occur at each time step with equal probability; in a deterministic version of the same model one might choose at odd/even times the right/left turning alternative in (a) and (c) of figure 2 respectively.

The structure of the collision term in the dynamic equation (2.1) is rather similar to that in the Boltzmann equation. The last term on the right hand side of (2.2) represents the loss term due to the direct collisions and the first term represents the gain term due to the restituting collisions. The factors $\overline{n} = 1 - n$ ensure that multiple collisions can only occur if the appropriate post-collisional states are empty, i.e. occupied by "holes". This is a consequence of the Fermi exclusion rule. An important feature that distinguishes a lattice gas fluid from a continuous fluid is the occurrence of multiple collisions where n particles ($n = 2, 3, 4, \ldots b$) are simultaneously present at the same site.

Our general derivation applies to an unspecified set of conservation laws. In several of

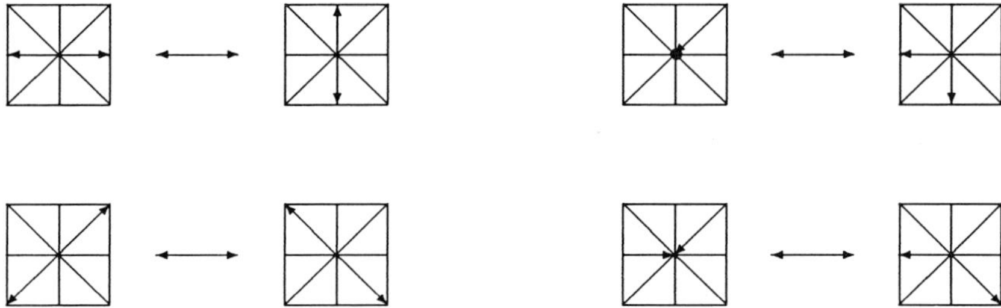

Figure 1: Collision rules for a LGCA with square or cubic symmetry and non-trivial energy conservation (see reference [23]). If only particles of unit speed are allowed, the HPP model is recovered.

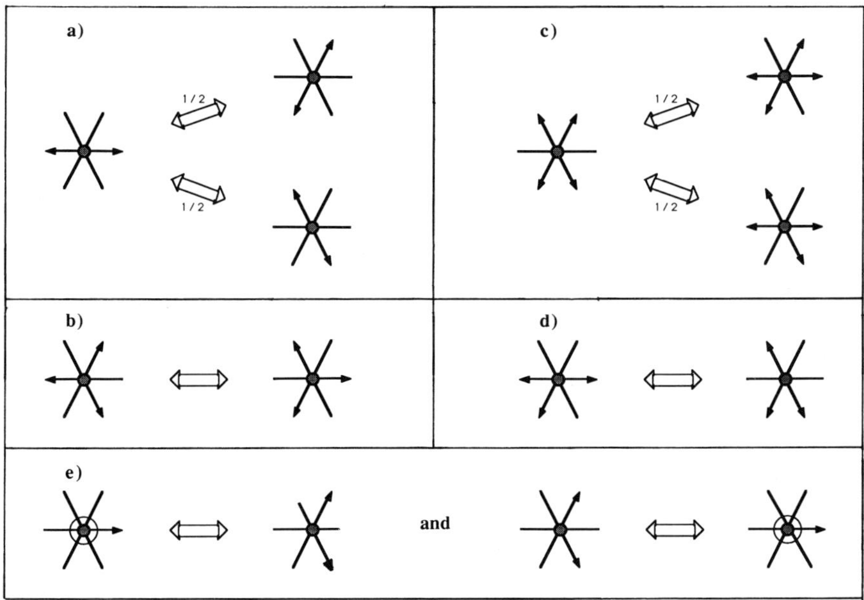

Figure 2: Collision rules for the FHP models: (a) head-on collision with two equiprobable outcomes; (b) triple collision; (c) dual of head-on collision under particle-hole exchange; (d) head-on collision with spectator; (e) binary collisions involving one rest particle (represented by a circle)

the explicit examples we include the energy among the conserved quantities. If there exist additional spurious conservation laws, such as the staggered momenta that are appearing in the 6- or 7-bits FHP model of figure 2 [15,17] and in the thermal model of figure 1 [18], they can be incorporated in a straightforward manner in the present formalism, as has been shown in [16,14]. An obvious defect of the HPP model is the existence of many unphysical conservation laws, referred to as line invariants. For instance, the total momentum parallel to any lattice row is a constant of the motion. Here we only include the standard set of conservation laws and we study the local conserved densities or hydrodynamic fields,

$$A(\vec{r}, t) = \sum_c a(\vec{c}) n(\vec{c}, \vec{r}; t), \tag{2.3}$$

corresponding to the so called *collisional invariants*, where $a(\vec{c}) = \{1, \vec{c}, \frac{1}{2}c^2, \ldots\}$, which are conserved in binary, triple, quadruple,... collisions. More explicitly we have the local number, momentum and energy densities,

$$
\begin{aligned}
\rho(\vec{r}, t) &= \sum_c n(\vec{c}, \vec{r}; t) \\
\vec{g}(\vec{r}, t) &= \sum_c \vec{c}\, n(\vec{c}, \vec{r}; t) \\
\epsilon(\vec{r}, t) &= \sum_c \tfrac{1}{2}c^2 n(\vec{c}, \vec{r}; t),
\end{aligned}
\tag{2.4}
$$

where volumes are measured in units of the volume v_0 of the unit cell. The total number of particles N, the total momentum of the system $\vec{P}$ and the total energy H are constants of the motion. The corresponding densities change only through free streaming, but not through collisions. To derive the local conservation laws from the microdynamic equations (2.1), we observe that the collision term in (2.1) satisfies the relation

$$\sum_c a(\vec{c}) I(\vec{c}\,|\,n) = 0. \tag{2.5}$$

The local microscopic conservation laws take the form

$$\sum_c a(\vec{c}) \left[n(\vec{c}, \vec{r} + \vec{c}; t + 1) - n(\vec{c}, \vec{r}; t) \right] = 0. \tag{2.6}$$

In most of our analysis the details of how the dynamics for $n(\vec{c}, \vec{r}; t)$ is generated and the structure of the lattice are unimportant.

3   Total and local equilibrium

The equilibrium state of a *continuous fluid*, in which energy is the only conserved quantity, can be described by a grand canonical ensemble $\rho_0(\Gamma) \sim \exp(\nu N - \beta H)$ where H is the Hamiltonian, β the inverse temperature and ν/β the chemical potential. If the total momentum is also conserved, then the corresponding grand ensemble is $\rho_0(\Gamma) \sim \exp[\nu N - \beta H + \vec{\gamma} \cdot \vec{P}]$ where $\vec{\gamma}$ is proportional to the average velocity of the system.

In general, if there is a set of conserved quantities $\boldsymbol{A}_0 = \{N, H, \vec{P}, \ldots\}$ with conjugate thermodynamic variables $\boldsymbol{b} = \{\nu, -\beta, \vec{\gamma}, \ldots\}$, the probability density in the grand ensemble is $\rho_0(\Gamma) \sim \exp[\boldsymbol{b} \cdot \boldsymbol{A}_0]$. The same holds for the ensemble density $\rho_0(n(\cdot))$ of a lattice gas of indistinguishable particles with Fermi exclusion. In this case

$$\rho_0(n(\cdot)) = \exp[-Q + \boldsymbol{b} \cdot \boldsymbol{A}_0] \tag{3.1}$$

and the conserved quantities are

$$\begin{aligned}
N &= \sum_{rc} n(\vec{c}, \vec{r}; t) \\
\vec{P} &= \sum_{rc} \vec{c}\, n(\vec{c}, \vec{r}; t) \\
H &= \tfrac{1}{2} \sum_{rc} c^2\, n(\vec{c}, \vec{r}; t).
\end{aligned} \tag{3.2}$$

Here the energy H and the conjugate inverse temperature β may have to be *excluded* from the pairs $\{\boldsymbol{A}_0, \boldsymbol{b}\}$, if H is either not conserved at all (see figure 2e) or essentially the same as N (single speed models). The spurious conservation laws existing in many LGCA's and their conjugate thermodynamic fields may have to be *included* in the exponent of the grand ensemble (3.1). This can be done in a systematic manner [18], but will not be considered here. Furthermore, we only consider systems where the total momentum $\vec{P}$ vanishes on the average in equilibrium. Consequently, its conjugate field $\vec{\gamma}$ is set equal to zero. In doing so one keeps the fluctuations of the total momentum in the description.

In the equilibrium case with or without energy conservation statistical mechanics of LGCA's is very similar to that for the ideal Fermi gas in the occupation number representation. The simplest case is the grand canonical ensemble for a system at rest,

$$\rho_0(n(\cdot)) = \exp[-Q + \nu N - \beta H] = \prod_{rc} \left[\frac{e^{(\nu - \frac{1}{2}\beta c^2)n(\vec{c}, \vec{r})}}{1 + e^{\nu - \frac{1}{2}\beta c^2}} \right]. \tag{3.3}$$

The ensemble density is normalized to unity so that

$$Q = \ell n \left\{ \sum_N \sum_{n(\cdot)} {}' \exp\left[\sum_{rc} (\nu - \tfrac{1}{2}\beta c^2) n(\vec{c}, \vec{r}) \right] \right\} = V \sum_c \ell n \left\{ 1 + e^{\nu - \frac{1}{2}\beta c^2} \right\}. \tag{3.4}$$

The prime indicates that the summation over occupation numbers is constraint to have exactly N particles. We note that the summand does not contain a factor $1/N!$ because the particles are indistinguishable. The distribution function or average occupation of the single particle state $\{\vec{c}, \vec{r}\}$ in equilibrium is

$$f(c) = \langle n(\vec{c}, \vec{r}) \rangle = [1 + e^{-\nu + \frac{1}{2}\beta c^2}]^{-1}. \tag{3.5}$$

It has the form of the Fermi distribution. The most important thermodynamic quantities are,

$$\langle N \rangle = \rho V = \partial Q / \partial \nu = V \sum_c f(c)$$
$$\langle H \rangle = \epsilon V = -\partial Q / \partial \beta = V \sum_c \tfrac{1}{2} c^2 f(c)$$
$$Q = \beta p v_0 V = -V \sum_c \ln(1 - f(c))$$
$$S = -\langle \ln \rho_0(n(\cdot)) \rangle = -V \sum_c \{ f(c) \ln f(c) + (1 - f(c)) \ln(1 - f(c)) \} \quad , \qquad (3.6)$$

where S is the entropy, $p = p(\beta, \nu)$ is the thermodynamic pressure and v_0 the volume of the unit cell. The equation of state $p(\beta, \rho)$ is obtained by eliminating ν in favor of ρ using the first equation in (3.6). We also note that the kinetic pressure p_K, as defined through the virial theorem or through the momentum flux density,

$$p_K v_0 = \sum_c c_x^2 f(c) = \tfrac{2}{d} \epsilon, \qquad (3.7)$$

differs from the thermodynamic pressure p. In the low density limit p approaches the ideal gas law, $p v_0 = \rho k_B T$. In the literature [23] a kinetic temperature has been introduced through the relation $p_K v_0 = \rho k_B T_K$. However this temperature is very different from the thermodynamic temperature, as illustrated in figure 4. All thermodynamic quantities have the same form as in the ideal Fermi gas with the exception that the sum over velocities is restricted to a small set of b different values. If the $\vec{c}$-summation in both expressions for the pressure could be replaced by a $\vec{c}$-integration over three-dimensional velocity space, then the system would be Galilei-invariant and one verifies by partial integration that thermodynamic and kinetic pressure would be equal. Figure 3 (isotherms) and figure 4 (isochores) show plots of the equation of state for the kinetic pressure, which is actually the caloric equation of state, in the temperature dependent two-dimensional 9-bits model of figure 1, where $v_0 = 1$. The pressure at zero temperature ($\theta \equiv \exp(-\tfrac{1}{2}\beta)$) is,

$$p_K = \epsilon = \tfrac{1}{2}(\rho - 1)H(\rho - 1)H(5 - \rho) + (\rho - 3)H(\rho - 5)H(9 - \rho), \qquad (\theta = 0) \qquad (3.8)$$

where $H(x)$ is the unit step function. At infinite temperature $p_K v_0 = \rho c_0^2$, and in particular for the 9-bits model,

$$p_K = \epsilon = \tfrac{2}{3}\rho, \qquad (\theta = 1). \qquad (3.9)$$

The isotherms at finite temperature are located between these two limiting cases. The isotherm in figure 3 is at $\theta = 0.2$. Figure 4 shows the isochores for the 9-bits model at $\rho = \tfrac{1}{2}, 3, 7, 9$. The isochore for the completely filled lattice (where $\rho = b$) gives a temperature-independent pressure, $p_K = \tfrac{2}{d} \epsilon = b c_0^2$. Here c_0 is the *speed of sound*, defined through

$$c_0^2 = \tfrac{1}{b} \sum_c c_x^2 = \tfrac{1}{bd} \sum_c | \vec{c} |^2 . \qquad (3.10)$$

Next, we consider fluctuations $\delta a = a - \langle a \rangle$, which are typical for particles obeying the Fermi exclusion principle. For the fluctuation in the occupation number of a single state one finds

$$\langle(\delta n)^\ell\rangle = \partial^{\ell-1} f(c)/\partial \nu^{\ell-1}.$$

(3.11)

This can be written out in terms of distribution functions:

$$
\begin{aligned}
\langle(\delta n)^2\rangle &= f(c)(1 - f(c)) \equiv \kappa(c) \\
\langle(\delta n)^3\rangle &= f(c)(1 - f(c))(1 - 2f(c)) \\
\langle(\delta n)^4\rangle &= f(c)(1 - f(c))(1 - 6f(c) + 6f^2(c)).
\end{aligned}
$$

(3.12)

The basic correlation functions in CA-fluids are,

$$
\begin{aligned}
\langle \delta n(\vec{c}, \vec{r})\delta n(\vec{c}', \vec{r}')\rangle &= \langle(\delta n)^2\rangle \delta_{cc'}\delta_{rr'} \\
\langle \delta n(\vec{c}, \vec{r})\delta n(\vec{c}', \vec{r}')\delta n(\vec{c}'', \vec{r}'')\rangle &= \langle(\delta n)^3\rangle \delta_{cc'}\delta_{cc''}\delta_{rr'}\delta_{rr''}.
\end{aligned}
$$

(3.13)

In case the energy is not conserved one has to set $\beta = 0$ in (3.1) and (3.5). Consequently the average occupation number is independent of $\vec{c}$ and equals the reduced density

$$f = \langle N\rangle/bV = \rho/b.$$

(3.14)

This leads to a significant simplification. For instance, $p_K v_0 = \rho c_0^2$. Except at low densities, also here the kinetic pressure differs from the thermodynamic pressure $p v_0 = -b c_0^2 \ell n(1 - f)$, as given by the logarithm of the grand partition function (3.6), where β has been replaced by c_0^{-2} to use units with the same dimension. This concludes the discussion of the equilibrium properties of lattice gas cellular automata in a uniform state of equilibrium.

Next we consider the case of a non-uniform equilibrium state, in which the thermodynamic fields $\boldsymbol{b}(\vec{r}) = \{\nu(\vec{r}), \vec{\gamma}(\vec{r}), -\beta(\vec{r}), \ldots\}$ vary with position over distances containing many sites. The ensemble density for this state of local equilibrium is obtained by replacing the exponent in (3.1) by a sum over sites

$$\rho_\ell(n(\cdot)) = \exp[-Q_\ell + \sum_r \boldsymbol{b}(\vec{r})\cdot\boldsymbol{A}(\vec{r})] = \rho_0(n(\cdot))\frac{\exp[\sum_r \delta\boldsymbol{b}(\vec{r})\cdot\delta\boldsymbol{A}(\vec{r})]}{\langle\exp[\sum_r \delta\boldsymbol{b}(\vec{r})\cdot\delta\boldsymbol{A}(\vec{r})]\rangle}.$$

(3.15)

Here $\delta\boldsymbol{b}(\vec{r}) = \boldsymbol{b}(\vec{r}) - \boldsymbol{b}$ is the deviation from uniformity in the conjugate fields and $\boldsymbol{A}(\vec{r}) = \{\rho(\vec{r}), \vec{g}(\vec{r}), \epsilon(\vec{r}) \ldots\}$ stands for the set of local conserved densities or hydrodynamic fields (2.4). Consequently local equilibrium does not change through collisions, but changes only through streaming. The local equilibrium analog of (3.5) is ofcourse

$$f_\ell(\vec{c}) = \langle n(\vec{c}, \vec{r})\rangle_\ell = [1 + \exp(-\boldsymbol{b}(\vec{r})\cdot\boldsymbol{a}(\vec{c}))]^{-1}.$$

(3.16)

Local equilibrium provides a mechanism for defining thermodynamic variables $\boldsymbol{b}(\vec{r}, t)$ outside thermal equilibrium [12], such as the local chemical potential $\nu(\vec{r}, t)$, the local field $\vec{\gamma}(\vec{r}, t)$,

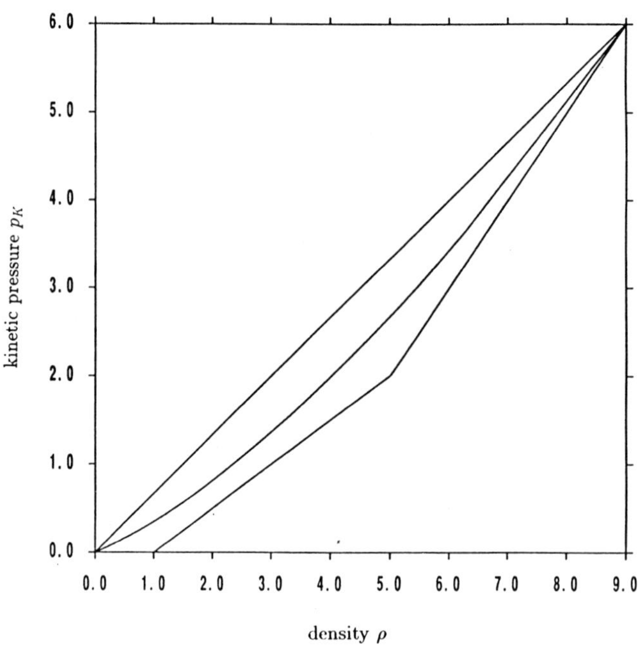

Figure 3: Isotherms of the kinetic pressure versus average occupation per site ρ at $\theta = \exp(-\frac{1}{2}\beta) = 0, 0.2, 1$ in the two-dimensional 9-bit model of figure 1 with non-trivial energy conservation.

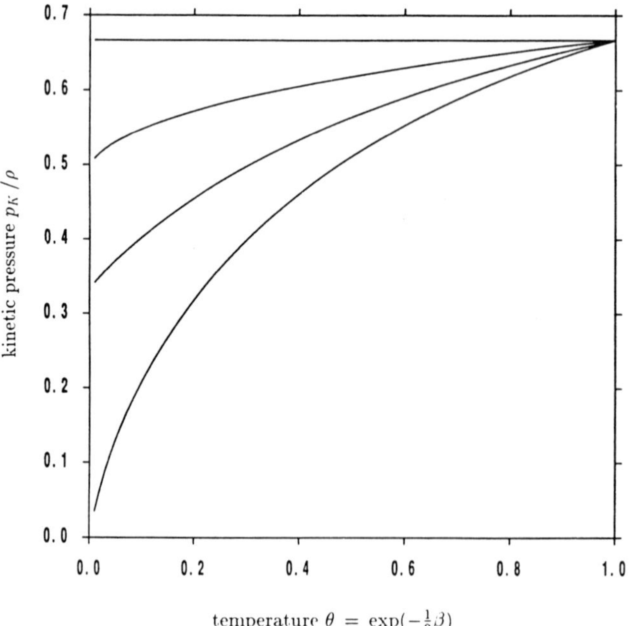

Figure 4: Isochores for the kinetic pressure versus θ at $\rho = \frac{1}{2}, 3, 6, 9$ for the same 9-bits model.

conjugate to the momentum density, and the local temperature $-\beta(\vec{r}, t)$, if energy is conserved. Suppose the conserved densities in some nonequilibrium state are

$$\langle \boldsymbol{A}(\vec{r}, t)\rangle_{ne} \equiv \boldsymbol{a}(\vec{r}, t) = \{\rho(\vec{r}, t), \rho(\vec{r}, t)\vec{u}(\vec{r}, t), \epsilon(\vec{r}, t)\ldots\}, \tag{3.17}$$

where the brackets denote an average of the phase function $\boldsymbol{A}(\vec{r})$ over a nonequilibrium ensemble ρ_{ne}. Then we can define the nonequilibrium local conjugate fields $\boldsymbol{b}(\vec{r}, t)$, appearing in (3.15), by imposing the condition, $\boldsymbol{a}(\vec{r}, t) = \langle \boldsymbol{A}(\vec{r})\rangle_\ell$, to be valid at any instant of time. More explicitly, the quantities $\nu(\vec{r}, t)$, $\beta(\vec{r}, t)$ and $\vec{\gamma}(\vec{r}, t)$ are defined through the relations,

$$\begin{aligned}
\rho = \rho_\ell(\nu, \beta, \vec{\gamma}) &= \sum_c f(\vec{c}, \vec{r}; t) \\
\rho\vec{u} = \vec{g}_\ell(\nu, \beta, \vec{\gamma}) &= \sum_c \vec{c} f(\vec{c}, \vec{r}; t) \\
\epsilon = \epsilon_\ell(\nu, \beta, \vec{\gamma}) &= \sum_c \tfrac{1}{2}c^2 f(\vec{c}, \vec{r}; t).
\end{aligned} \tag{3.18}$$

In these lecture notes we will be mainly concerned with small deviations from equilibrium, in which case we can expand the exponents in (3.15) in powers of $\delta\boldsymbol{b}(\vec{r})$. For instance, the quadratic terms in this expansion will determine the nonlinear convective terms in the Euler equations, containing the so called *non-Galilean factor* caused by the lack of Galilei invariance of the LGCA's [11]. It will be calculated at the end of section 5. However, we restrict ourselves mainly to linear deviations, since our primary goal is to derive Green-Kubo expressions for the linear transport coefficients in the Navier-Stokes equations.

For small deviations from equilibrium the local fields $\delta\boldsymbol{b}(\vec{r}, t)$ can be solved explicitly from (3.18) in terms of $\delta\boldsymbol{a}(\vec{r}, t) = \boldsymbol{a}(\vec{r}, t) - \boldsymbol{a}$ with $\boldsymbol{a} = \langle \boldsymbol{A}_0\rangle/V$. In order to do so it is convenient to linearize (3.15) around uniform equilibrium

$$\rho_\ell = \rho_0[1 + \tfrac{1}{V}\sum_k \boldsymbol{A}_k{}^\dagger \cdot \boldsymbol{b}_k + \ldots]. \tag{3.19}$$

In this notation $\boldsymbol{A}_k$ and $\boldsymbol{b}_k$ denote column vectors. The row vectors are the Hermitean adjoints. Fourier transforms are defined as

$$\boldsymbol{A}_k(t) = \sum_r e^{-i\vec{k}\cdot\vec{r}}\delta\boldsymbol{A}(\vec{r}, t) \tag{3.20}$$

and similar definitions for $\boldsymbol{b}_k(t)$ and $\boldsymbol{a}_k(t)$ in terms of the deviations $\delta\boldsymbol{b}(\vec{r}, t)$ and $\delta\boldsymbol{a}(\vec{r}, t)$. The vectors $\vec{k} = k_1\vec{e}_1^{\,*} + k_2\vec{e}_2^{\,*}$ are restricted to the first Brillouin zone of the reciprocal lattice, where $\{\vec{e}_i^{\,*}, i = 1, 2, \cdots d\}$ is the set biorthogonal to $\{\vec{e}_i, i = 1, 2, \cdots d\}$ satisfying $\vec{e}_j^{\,*} \cdot \vec{e}_i = \delta_{ij}$ and hence $\vec{k} \cdot \vec{r} = \sum_{i=1}^d k_i r_i$. In a Cartesian coordinate system, appropriate for square and cubic lattices, the sets $\{\vec{e}_i\}$ and $\{\vec{e}_i^{\,*}\}$ coincide.

It follows from (3.19) that to linear order in the deviations

$$\boldsymbol{a}_k(t) = \langle \boldsymbol{A} \mid \boldsymbol{A}\rangle \cdot \boldsymbol{b}_k(t) = \boldsymbol{\chi} \cdot \boldsymbol{b}_k(t), \tag{3.21}$$

where the inner product is defined as

$$\langle C \mid D \rangle = \tfrac{1}{V}\langle C_k D_k{}^\dagger \rangle. \tag{3.22}$$

The relation (3.21) defines the *local temperature* $-\beta_k(t)$, chemical potential $\nu_k(t)$ and the local field $\vec{\gamma}_k(t)$ in terms of the average number density $\rho_k(t)$, momentum density $\rho\vec{u}_k(t)$ and energy density $\epsilon_k(t)$. According to irreversible thermodynamics [13] the gradients of these conjugate fields are the thermodynamic driving forces X_b featuring in the *constitutive relations* $J_a = L_{ab}X_b$, where J_a are the irreversible fluxes and L_{ab} the linear transport coefficients. The matrix L_{ab} is positive definite because irreversible entropy production is positive. The currents and linear transport coefficients will be calculated in the next section. The elements of the susceptibility matrix χ are the thermodynamic derivatives. They can be calculated using (3.13) and (3.22) and are independent of $\vec{k}$ due to lack of spatial correlations in the equilibrium state.

$$\chi = \langle A \mid A \rangle = \sum_c a(\vec{c})a(\vec{c})\kappa(c) \tag{3.23}$$

where the weight $\kappa(c) = f(c)(1 - f(c))$. If *energy is conserved* the relation (3.21) becomes

$$\begin{aligned}
\rho_k &= \chi_{\rho\rho}\nu_k - \chi_{\rho\epsilon}\beta_k \\
\epsilon_k &= \chi_{\epsilon\rho}\nu_k - \chi_{\epsilon\epsilon}\beta_k \\
\rho\vec{u}_k &= \chi_{\vec{g}\vec{g}} \cdot \vec{\gamma}_k \equiv \rho\chi_u\vec{\gamma}_k
\end{aligned} \tag{3.24}$$

The susceptibilities can also be written in terms of thermodynamic derivatives,

$$\begin{aligned}
\chi_{\rho\rho} &= (\partial\rho/\partial\nu)_\beta \\
\chi_{\rho\epsilon} &= -(\partial\rho/\partial\beta)_\nu = (\partial\epsilon/\partial\nu)_\beta \\
\chi_{\epsilon\epsilon} &= -(\partial\epsilon/\partial\beta)_\nu \\
\rho\chi_u &= \tfrac{1}{d}\sum_c c^2\kappa(c) = \tfrac{2}{d}\chi_{\epsilon\rho} \\
&= \tfrac{2}{d}(\partial\epsilon/\partial\nu)_\beta = (\partial p_K v_0/\partial\nu)_\beta.
\end{aligned} \tag{3.25}$$

If the thermodynamic pressure p and kinetic pressure p_K were equal (which is not the case!), standard thermodynamic relations would yield $\chi_u = 1/\beta$ and $\vec{u}_k = \vec{\gamma}_k/\beta$. This is the relation that applies in a continuous fluid. In athermal CA-fluids without energy conservation equations (3.25) simplifies to

$$\begin{aligned}
\rho_k &= \chi_n\nu_k \\
\rho\vec{u}_k &= \chi_{\vec{g}\vec{g}} \cdot \vec{\gamma}_k = \rho\chi_u\vec{\gamma}_k.
\end{aligned} \tag{3.26}$$

To avoid confusion with the temperature dependent susceptibilities above we introduce here

$$\begin{aligned}
\chi_{\rho\rho} &\equiv \chi_n = bf(1 - f) \\
\rho\chi_u &= b\kappa c_0^2 = \rho c_0^2(1 - f),
\end{aligned} \tag{3.27}$$

where we have used (3.12) and (3.10) with $f = \rho/b$ the reduced density. To formally solve (3.21) for $\boldsymbol{b}_k$ we introduce the microscopic fluctuations $\boldsymbol{B}_k$ as linear combinations of $\boldsymbol{A}_k$'s through the relation,

$$\boldsymbol{A}_k = \langle \boldsymbol{A} \mid \boldsymbol{A} \rangle \cdot \boldsymbol{B}_k. \tag{3.28}$$

By multiplying (3.28) on the right with $\boldsymbol{A}_k$ one verifies that the fluctuations $\boldsymbol{A}_k$ and $\boldsymbol{B}_k$ are *reciprocal* in the sense

$$\langle \boldsymbol{B} \mid \boldsymbol{A} \rangle = \langle \boldsymbol{A} \mid \boldsymbol{B} \rangle = \boldsymbol{1}. \tag{3.29}$$

Multiplying (3.28) with $\boldsymbol{A}_k$ yields the inverse of the susceptibility matrix,

$$\langle \boldsymbol{B} \mid \boldsymbol{B} \rangle = \langle \boldsymbol{A} \mid \boldsymbol{A} \rangle^{-1} = \chi^{-1}. \tag{3.30}$$

Before closing this section we note that the complete set of conserved densities $\boldsymbol{A}_k(t)$ determines the hydrodynamic subspace of slow excitations. In the next section we need the projection operator $\mathcal{P}$, defined as

$$\mathcal{P} =\mid \boldsymbol{A} \rangle \cdot \langle \boldsymbol{B} \mid =\mid \boldsymbol{B} \rangle \cdot \langle \boldsymbol{A} \mid . \tag{3.31}$$

It projects any dynamical variable onto the hydrodynamic subspace, or stated differently, it determines the local equilibrium contribution of that variable up to linear order in the deviations from equilibrium.

4 Correlation function method

4.1 Onsager regression hypothesis

The hydrodynamic equations are the equations of motion for the averages of the local conserved densities $\boldsymbol{a}(\vec{r}, t)$. These equations contain transport coefficients. The goal of this section is to derive exact expressions for the linear transport coefficients in the form of Green-Kubo relations. To study such linear transport coefficients nonlinear terms in the flow field and other nonlinear deviations from total equilibrium, may be neglected. The presentation given here follows closely the work of Dufty and the author [14], but has been extended here to include thermal effects, whenever applicable. There exist several other derivations of Green-Kubo expressions for transport coefficients in LGCA's, based on fluctuating hydrodynamics [24,27], on the Chapman-Enskog method [25,26,28] and on the projection operator technique [16].

Suppose the system of interest is described at the initial time $t = 0$ by a local equilibrium ensemble where the initial average density and flow field deviate only slightly from total equilibrium. The same holds for the thermodynamic parameters $\delta \boldsymbol{b}(\vec{r}, 0)$. By averaging $\boldsymbol{A}_k$ over this initial ensemble we obtain the linear response to small initial deviations from equilibrium [12],

$$\boldsymbol{a}_k(t) = V^{-1}\langle \boldsymbol{A}_k(t)\boldsymbol{A}_k{}^{\dagger}(0)\rangle \boldsymbol{b}_k(0) = \boldsymbol{G}(\vec{k},t)\boldsymbol{b}_k(0) = \boldsymbol{G}(\vec{k},t)\boldsymbol{\chi}^{-1}\boldsymbol{a}_k(0). \tag{4.1}$$

This expression relates the conjugate fields $\boldsymbol{b}_k(0)$ linearly to the initial densities $\boldsymbol{a}_k(0)$ for large space and time scales. Hence, the right hand side of (4.1) provides exactly the same information as the equilibrium correlation functions of the conserved densities. This is the gist of Onsager's regression hypothesis.

In terms of the inner product (3.22) the hydrodynamic correlation functions are

$$\boldsymbol{G}(\vec{k},t) = \langle \boldsymbol{A}(t) \mid \boldsymbol{A}(0)\rangle = \langle \boldsymbol{A}(0) \mid \boldsymbol{A}(-t)\rangle, \tag{4.2}$$

where we have used stationarity of the equilibrium ensemble. This holds for stochastic as well as for deterministic dynamics. In the following analysis a linear equation for $\boldsymbol{G}(\vec{k},t)$ is obtained with the exact coefficients of the evolution matrix identified to order k^2 in terms of correlation functions of the conserved variables and their fluxes. The result has the form,

$$\left[\partial_t + ik\boldsymbol{\Omega} + k^2\boldsymbol{\Lambda}(\hat{k}) + \cdots\right]\boldsymbol{G}(\vec{k},t) = 0. \tag{4.3}$$

The time derivative occurs because this equation is valid only for large times in which a continuum limit applies. Also, the dots on the left side indicate terms of higher order in k. By identifying this exact evolution equation with the equations of fluid dynamics, one obtains the Green-Kubo relations for the transport coefficients. Navier-Stokes order hydrodynamics results from retaining only terms up through order k^2. The matrix $\boldsymbol{\Omega}$ describes the reversible (Euler) dynamics while $\boldsymbol{\Lambda}$ describes Navier-Stokes order dissipation. The elements of $\boldsymbol{\Lambda}$ are the transport coefficients. The Fourier transform of (3.17), $\boldsymbol{a}_k(t)$, also satisfies (4.3) on account of (4.1). This basic result is Onsager's regression hypothesis, stating that the long time decay of long wavelength components of the equilibrium fluctuations follows the macroscopic laws. The correlation function method is therefore an efficient way to obtain exact expressions for $\boldsymbol{\Omega}$ and $\boldsymbol{\Lambda}$ without explicit consideration of the nonequilibrium state.

4.2 Green-Kubo relations

The analysis is straightforward and makes only use of the conservation laws and stationarity. The conservation law (2.6) can be written in the more suggestive form, using a forward difference

$$\Delta_t \boldsymbol{A}_k(t) \equiv \boldsymbol{A}_k(t+1) - \boldsymbol{A}_k(t) = -\boldsymbol{I}_k(t+1). \tag{4.4}$$

Here the flux $\boldsymbol{I}_k$ is defined by

$$\boldsymbol{I}_k(t) = \sum_c \left[e^{i\vec{k}\cdot\vec{c}} - 1\right]\boldsymbol{a}(\vec{c})\hat{n}_k(\vec{c},t) = \sum_c \left[i\vec{k}\cdot\vec{c} - \tfrac{1}{2}(\vec{k}\cdot\vec{c})^2 + \cdots\right]\boldsymbol{a}(\vec{c})\hat{n}_k(\vec{c},t), \tag{4.5}$$

and $\hat{n}_k(\vec{c}, t)$ is the Fourier transform of $\delta n(\vec{c}, \vec{r}; t)$. We first derive an identity, needed below and valid for conserved densities only. By applying (4.4) to both sides of the last equality in (4.2) and using stationarity we obtain,

$$\Delta_t G(\vec{k}, t) = -\langle I \mid A(-t-1)\rangle = \langle A \mid I(-t)\rangle. \tag{4.6}$$

It is important to stress that the present derivation does not use microscopic reversibility or invertibility of the equations of motion, and therefore applies to both deterministic and stochastic dynamics. After taking Laplace transforms the above identity becomes

$$\langle I \mid \widetilde{A}(s)\rangle = -\langle A \mid \tilde{I}(s)\rangle. \tag{4.7}$$

Here we have introduced a discrete Laplace transform, which is slightly different for densities and currents

$$\begin{aligned}
\widetilde{A}_k(s) &= \sum_{t=0}^{\infty} e^{-st} A_k(-t-1) \\
\tilde{I}_k(s) &= \sum_{t=0}^{\infty} e^{-st} I_k(-t).
\end{aligned} \tag{4.8}$$

We continue with the Laplace transform of the conservation law (4.4), to which we apply the projectors $\mathcal{P}$, defined in (3.31), and $Q = 1 - \mathcal{P}$. The result is,

$$\begin{aligned}
(1 - e^{-s})\mathcal{P}\widetilde{A}(s) - \mathcal{P}\tilde{I}(s) &= \mathcal{P}A = A \\
(1 - e^{-s})Q\widetilde{A}(s) - Q\tilde{I}(s) &= QA = 0.
\end{aligned} \tag{4.9}$$

The first equation, written out in component form, reads

$$(1 - e^{-s})\langle A \mid \widetilde{A}(s)\rangle - \langle A \mid \tilde{I}(s)\rangle = \langle A \mid A\rangle = \chi, \tag{4.10}$$

where use has been made of (3.23). With the help of identity (4.7) and the relation $\mathcal{P} + Q = 1$ the second term on the left hand side can be rewritten as,

$$- \langle A \mid \tilde{I}(s)\rangle = \langle I \mid \mathcal{P}\widetilde{A}(s)\rangle + \langle I \mid Q\widetilde{A}(s)\rangle = \langle I \mid B\rangle\langle A \mid \widetilde{A}(s)\rangle + \langle I \mid Q\widetilde{A}(s)\rangle. \tag{4.11}$$

The last term can be eliminated in favor of the current-current correlation function after multiplying (4.9) with I. This yields formally

$$[1 - e^{-s} + M(\vec{k}, s)]\langle A \mid \widetilde{A}(s)\rangle = \langle A \mid A\rangle \tag{4.12}$$

with the M-matrix given by

$$\begin{aligned}
M(\vec{k}, s) &= \langle I \mid B\rangle + \langle I \mid Q\tilde{I}(s)\rangle[(1 - e^{-s})\langle A \mid \widetilde{A}(s)\rangle]^{-1} \\
&\simeq \left[\langle I \mid A\rangle + \langle I \mid Q\tilde{I}(s)\rangle\right]\langle A \mid A\rangle^{-1}.
\end{aligned} \tag{4.13}$$

To obtain the second equality we have taken the *long wave length limit* and retained in $\boldsymbol{M}(\vec{k}, s)$ only terms at least of $\mathcal{O}(k^2)$. Since the current-current correlation is already of $\mathcal{O}(k^2)$ on account of (4.5), the term $[\cdots]$ on the first line may be replaced by its dominant small-k behavior, $\langle \boldsymbol{A} \mid \boldsymbol{A} \rangle$. The k-expansion of the one but last term in (4.13) follows from (4.5) and (3.13) as

$$\langle \boldsymbol{I} \mid \boldsymbol{A} \rangle = ik\langle \boldsymbol{J} \mid \boldsymbol{A} \rangle - \tfrac{1}{2}k^2\langle \boldsymbol{J} \mid \boldsymbol{J} \rangle + \ldots, \tag{4.14}$$

where the flux $\boldsymbol{J}$ may still depend on the direction of $\vec{k}$, but not on its magnitude,

$$\boldsymbol{J}_k(t) = \sum_{rc} \hat{k} \cdot \vec{c}\,\boldsymbol{a}(\vec{c})\hat{n}_0(\vec{c}, t) = \sum_{rc} \hat{k} \cdot \vec{c}\,\boldsymbol{a}(\vec{c})\delta n(\vec{c}, \vec{r}; t). \tag{4.15}$$

The matrix $\boldsymbol{M}(\vec{k}, s)$ then becomes

$$\boldsymbol{M}(\vec{k}, s) = ik\boldsymbol{\Omega} + k^2 \left[\langle \boldsymbol{J} \mid Q\tilde{\boldsymbol{J}}(s) \rangle - \tfrac{1}{2}\langle \boldsymbol{J} \mid \boldsymbol{J} \rangle \right] \langle \boldsymbol{A} \mid \boldsymbol{A} \rangle^{-1}. \tag{4.16}$$

In the last equation the Euler matrix $\boldsymbol{\Omega}$ has been introduced, which satisfies the following relations,

$$\boldsymbol{\Omega}\boldsymbol{\chi} = \langle \boldsymbol{J} \mid \boldsymbol{A} \rangle = \langle \boldsymbol{A} \mid \boldsymbol{J} \rangle = \boldsymbol{\chi}\boldsymbol{\Omega}^\dagger = \sum_c (\hat{k} \cdot \vec{c})\boldsymbol{a}(\vec{c})\boldsymbol{a}(\vec{c})\kappa(c). \tag{4.17}$$

To obtain the equations of fluid dynamics we need the behavior of (4.12) on large spatial and temporal scales, where $\vec{r}$ and t can be treated as continuous variables and where $k \longrightarrow 0$ and $s \longrightarrow 0$. In this limit $1 - e^{-s}$ in (4.12) may be replaced by $s - \tfrac{1}{2}s^2 + \ldots$, corresponding to $\partial_t + \tfrac{1}{2}\partial_t^2 + \ldots$ in time language. Here one may eliminate $\tfrac{1}{2}s^2$ with the help of the relation $s \longrightarrow ik\boldsymbol{\Omega}$ and make the replacement

$$-\tfrac{1}{2}s^2 \longrightarrow \tfrac{1}{2}k^2\boldsymbol{\Omega}^2 = \tfrac{1}{2}k^2\langle \boldsymbol{J} \mid \mathcal{P}\boldsymbol{J} \rangle\langle \boldsymbol{A} \mid \boldsymbol{A} \rangle^{-1}, \tag{4.18}$$

where we have used (4.17). It is correct to $\mathcal{O}(k^2)$-terms included. This procedure actually amounts to eliminating the second time derivative of $\boldsymbol{G}(\vec{k}, t) = \langle \boldsymbol{A}(0) \mid \boldsymbol{A}(-t) \rangle$ with the help of the Euler equations, $(\partial_t + ik\boldsymbol{\Omega})\boldsymbol{G}(\vec{k}, t) = 0$. Similar things are done in constructing the normal solution [29] to the Boltzmann equation with the help of the Chapman-Enskog method or in the multiple time scale formalism [31]. The resulting equations for the hydrodynamic correlation functions then take the form

$$[s + ik\boldsymbol{\Omega} + k^2\boldsymbol{\Lambda}(\hat{k}, s)]\langle \boldsymbol{A} \mid \widetilde{\boldsymbol{A}}(s) \rangle = \langle \boldsymbol{A} \mid \boldsymbol{A} \rangle, \tag{4.19}$$

where the relaxation matrix $\boldsymbol{\Lambda}$ is given by

$$\boldsymbol{\Lambda}(\hat{k}, s) = \boldsymbol{L}(\hat{k}, s)\boldsymbol{\chi}. \tag{4.20}$$

Finally the matrix of transport coefficients is given by the small-s limit of the Green-Kubo relations,

$$L(\widehat{k},s) = \langle \boldsymbol{J} \mid Q\widetilde{\boldsymbol{J}}(s)\rangle - \tfrac{1}{2}\langle \boldsymbol{J} \mid Q\boldsymbol{J}\rangle = \sum_{t=0}^{\infty}{}^{*}e^{-st}\boldsymbol{\varphi}(t). \tag{4.21}$$

The asterisk on the summation sign indicates that the term with $t = 0$ has only a weight $\tfrac{1}{2}$. This is a consequence of the discreteness of time. In the literature on CA-fluids the transport coefficients are frequently written as [11], [16],[24] -[28],

$$L(\widehat{k},s) = \sum_{t=0}^{\infty} e^{-st}\boldsymbol{\varphi}(t) + \boldsymbol{L}^{P} \tag{4.22}$$

with the *propagation part* of the transport matrix defined as

$$\boldsymbol{L}^{P} = -\tfrac{1}{2}\boldsymbol{\varphi}(0) = -\tfrac{1}{2}\langle \widehat{\boldsymbol{J}} \mid \widehat{\boldsymbol{J}}\rangle. \tag{4.23}$$

In equations above we have introduced the current correlation function

$$\boldsymbol{\varphi}(t) = \langle \boldsymbol{J} \mid Q\boldsymbol{J}(-t)\rangle = \langle \widehat{\boldsymbol{J}}(t) \mid \widehat{\boldsymbol{J}}\rangle = \langle \boldsymbol{J}(t) \mid \boldsymbol{J}\rangle - \langle \boldsymbol{J} \mid \boldsymbol{A}\rangle\langle \boldsymbol{A} \mid \boldsymbol{A}\rangle^{-1}\langle \boldsymbol{A} \mid \boldsymbol{J}\rangle. \tag{4.24}$$

Note that the subtracted term on the second line is constant in time. It contains the *subtracted flux*, defined with the help of the projector $Q = 1 - \mathcal{P}$,

$$\widehat{\boldsymbol{J}} \equiv Q\boldsymbol{J} = \boldsymbol{J} - \boldsymbol{A}\langle \boldsymbol{A} \mid \boldsymbol{A}\rangle^{-1}\langle \boldsymbol{A} \mid \boldsymbol{J}\rangle = \boldsymbol{J} - \boldsymbol{A}\boldsymbol{\Omega}^{\dagger}. \tag{4.25}$$

The subtracted flux $\widehat{\boldsymbol{J}}$ is the component of $\boldsymbol{J}$ that is orthogonal to the conserved variables in the sense that $\langle \widehat{\boldsymbol{J}} \mid \boldsymbol{A}\rangle = 0$.

Equations (4.19) and (4.21) are the primary results of this section. They are exact expressions for the transport matrix in the equations of motion for the hydrodynamic correlation functions.

If equation (4.19) is transformed back to time language the dissipation term will in general involve memory effects of the form $\int_0^t d\tau \phi(\tau)G(\vec{k},t-\tau)$. However, if the correlation function $\phi(t)$ decays sufficiently fast on the hydrodynamic time scale, then $\boldsymbol{\Lambda}(\widehat{k},s)$ and $\boldsymbol{L}(\widehat{k},s)$ may be replaced by their small-s limits, $\boldsymbol{\Lambda}(\widehat{k})$ and $\boldsymbol{L}(\widehat{k})$ respectively. Then the hydrodyamic correlation functions $\boldsymbol{G}(\vec{k},t) = \langle \boldsymbol{A}(t) \mid \boldsymbol{A}\rangle$ obey for long times and long wave lengths the *instantaneous* linear equations (4.3). Schmitz and Dufty [30] have recently shown that $\boldsymbol{L}(\widehat{k})$ is a non-negative matrix that can be properly identified with dissipation. One implication is that diagonal elements $L_{aa}(k)$ of the transport matrix are positive.

Explicit evaluation of Euler matrix $\boldsymbol{\Omega}$ and susceptibility matrix $\langle \boldsymbol{A} \mid \boldsymbol{A}\rangle$ in (4.17) and (3.23) requires specification of the LGCA. Evaluation of the Green-Kubo relations $\boldsymbol{L}(\widehat{k},s)$ for the transport coefficients is more difficult and involves a detailed analysis of the many body dynamics.

The primary differences in the form of the Green-Kubo relations obtained here, from those for continuous fluids [12] are due to the discrete sum over time instead of an integration, and the fact that only one half of the $t = 0$ contribution occurs. These are effects of the discrete space-time that persist at the macroscopic level. The existence of hydrodynamic equations in Navier-Stokes order requires that $L(\widehat{k}, s)$ remain finite as $s \longrightarrow 0$. For fluid type models $L(\widehat{k}, 0)$ is finite for dimensionality $d > 2$. At $d = 2$ typical matrix elements of $L(\widehat{k}, s)$ show a weak divergence proportional to $ln(s)$ as $s \longrightarrow 0$, leading to long time tails $\sim t^{-1}$ in the current-current correlation functions, as recently reviewed in [32].

5 Fluid dynamics for LGCA's

Based om the discussion of section 4.1 the hydrodynamic correlation functions $G(\vec{k}, t)$ decay according to the same laws as small macroscopic deviations from equilibrium. So we obtain from (4.1) and (4.2) the relaxation equations for fluid excitations,

$$[\partial_t + ik\Omega + k^2 \Lambda(\widehat{k})]a_k(t) = 0, \tag{5.1}$$

provided the small-s-limit of $\Lambda(\widehat{k}, s)$ exists. A convenient alternative form, involving the gradients of the conjugate fields (3.21) is obtained with the help of (4.17) and (4.20). The resulting equation has the same form as the linearized Navier-Stokes equation,

$$\partial_t a_k + ik\langle J \mid A\rangle b_k(t) + k^2 L(\widehat{k})b_k(t) = 0. \tag{5.2}$$

It may also be written in the form of a macroscopic conservation law,

$$\partial_t a_k + ikj_k^\ell + ikj_k^D = 0, \tag{5.3}$$

where the current is the sum of a *local equilibrium current*,

$$j_k^\ell = \langle J \mid A\rangle b_k, \tag{5.4}$$

and a *dissipative current*

$$j_k^D = -ikL(\widehat{k})b_k. \tag{5.5}$$

The local equilibrium current corresponds to the Euler part of the fluid dynamic equations; the dissipative current to the Navier-Stokes part. The dissipative currents have the standard form of the *linear constitutive relations*, $J_a = L_{ab}X_b$, of irreversible thermodynamics [13] with transport coefficients expressed in terms of Green-Kubo formulae (4.21).

First we carry out some general simplifications. If the microscopic current J_a itself is conserved, then the projected current $\widehat{J}_a$ vanishes identically on account of (4.25). This is the case in equation (5.1) for $a = \rho$, where $J_{k\rho} = \vec{g}_k$. Consequently $L_{\rho a} = L_{a\rho} = 0$ for $a = \{\rho, \vec{g}, \epsilon\}$

and the continuity equation (5.3) with $a = \rho$ has no dissipative current. Furthermore, the gradient of the chemical potential $ik\nu_k$ does not act as a thermodynamic driving force, at least not in a single component system. The only admissible driving forces are the gradient of the local temperature $ik_\alpha T_k$ with $T = (k_B\beta)^{-1}$ and the gradient of the conjugate flow field $\gamma_{k\alpha}$, which is proportional to the gradient of the fluid velocity $ik_\alpha u_{k\beta}$ on account of (3.24) or (3.26). The dissipative current j_{ka}^D for $a = g_\alpha$ is the stress tensor $P_{k\alpha\beta}^D$ and for $a = \epsilon$ the heat current $q_{k\alpha}$.

According to *Curie's law* [13] a vectorial current, such as $q_{k\alpha}$, does not couple to the tensorial driving force $ik_\alpha u_{k\beta}$. Similarly a tensorial current, such as $P_{k\alpha\beta}^D$, does not couple to the vectorial force $ik_\alpha T_k$. The relevant transport coefficients entering in (5.5) for $a = \epsilon$ and $a = g_\alpha$ are defined through the relations,

$$\begin{aligned}
q_{k\alpha} &= -v_0\lambda_{\alpha\beta}(ik_\beta T_k) \\
P_{k\alpha\beta}^D &= -v_0\eta_{\alpha\beta\gamma\delta}(ik_\gamma u_{k\delta}),
\end{aligned} \tag{5.6}$$

where the heat conductivity $\lambda_{\alpha\beta}$ and the viscosity $\eta_{\alpha\beta\gamma\delta}$ are numerical tensors, respectively of rank two and four. These tensors have symmetry properties dictated by the invariances of the underlying lattice. A continuous fluid is isotropic, whereas a CA-fluid is invariant under a discrete group of rotations and reflections. A tensor is said to be isotropic if it is invariant under all symmetry transformations of the continuous rotatation group. A tensor with square, hexagonal, cubic, ... symmetry is only invariant under the discrete transformation group that leaves the corresponding Bravais lattice invariant.

From the above definitions one can deduce that second rank tensors on all lattices with inversion symmetry are isotropic, i.e.

$$T_{\alpha\beta} = I\delta_{\alpha\beta}, \tag{5.7}$$

where the invariant $I = \frac{1}{d}T_{\alpha\alpha}$. For the same reason tensors of odd rank vanish identically. This proves in fact Curie's law for lattice gases, because the coupling coefficients mentioned above are numerical tensors of rank three.

The most general *isotropic* tensor of rank four is constructed from direct products of Kronecker delta functions,

$$T_{\alpha\beta\gamma\delta} = I_1\delta_{\alpha\beta}\delta_{\gamma\delta} + I_2\delta_{\alpha\gamma}\delta_{\beta\delta} + I_3\delta_{\alpha\delta}\delta_{\beta\gamma} \tag{5.8}$$

with three independent scalars I_p. If the system has only *cubic symmetry*, there is an additional term $I_4\delta_{\alpha\beta\gamma\delta}^{(4)}$. The invariant cubic tensor $\delta^{(4)}$ is defined as

$$\delta_{\alpha\beta\gamma\delta}^{(4)} = \begin{cases} 1 & \text{if } \alpha = \beta = \gamma = \delta \\ 0 & \text{elsewhere.} \end{cases} \tag{5.9}$$

In systems with discrete symmetries the only lattice in two dimensions where fourth rank tensors have the isotropic form (5.8) is the triangular lattice, on which the FHP model is defined.

In four dimensions the FCHC (face centered hypercubic) lattice also yields the isotropic form (5.8). In three dimensions no lattice exists with the required symmetry [31]. To obtain three-dimensional models with the proper isotropic symmetry one uses an FCHC lattice that is only one lattice unit wide in the fourth dimension [11].

If in addition the fourth rank tensor is symmetric in, say, the first pair of indices, then $I_2 = I_3$ and it can always be written as

$$T_{\alpha\beta\gamma\delta} = I\{\delta_{\alpha\gamma}\delta_{\beta\delta} + \delta_{\alpha\delta}\delta_{\beta\gamma} - \tfrac{2}{d}\delta_{\alpha\beta}\delta_{\gamma\delta}\} + I'\delta_{\alpha\beta}\delta_{\gamma\delta}. \tag{5.10}$$

If the tensor has cubic symmetry and is symmetric in the index pair (α, β) we may write equation (5.10) as,

$$T_{\alpha\beta\gamma\delta} = I\left(\delta_{\alpha\gamma}\delta_{\beta\delta} + \delta_{\alpha\delta}\delta_{\beta\gamma} - 2\delta^{(4)}_{\alpha\beta\gamma\delta}\right) + I'\delta_{\alpha\beta}\delta_{\gamma\delta} + I''(\delta^{(4)}_{\alpha\beta\gamma\delta} - \tfrac{1}{d}\delta_{\alpha\beta}\delta_{\gamma\delta}). \tag{5.11}$$

Because the stress tensor $P^D_{\alpha\beta}$ in (5.6) is symmetric in (α, β) (conservation of total angular momentum), the viscosity tensor of a cubic symmetric LGCA can have, in principle, three independent scalar viscosities, where I corresponds to the shear viscosity η, I' to the bulk viscosity ζ and I'' to an unphysical ϑ-viscosity, typical for a CA-fluid with cubic symmetry. The representation (5.11) is convenient because the viscosity coefficients η, ζ, ϑ, defined in this manner, are *non-negative*, as will be shown in the next section with the help of the Green-Kubo formulae. In this case the fluid dynamic equations do not have the isotropic fluid symmetry of the Navier-Stokes equations, unless the two terms containing $\delta^{(4)}$ cancel, i.e. unless the *isotropy condition*,

$$\vartheta = 2\eta, \tag{5.12}$$

is satisfied. However, a viscosity tensor with hexagonal symmetry is automatically isotropic, it has at most two independent non-negative scalar viscosities and leads automatically to fluid dynamic equations with the proper isotropic fluid symmetry. This is the fundamental reason why the FHP model is an acceptable model for simulating two-dimensional hydrodynamic problems.

Next we return to the local equilibrium current in (5.4). In fact, $j^\ell_{ka}(t)$ is equal to the local equilibrium average $\langle J_{ka}\rangle_\ell$, calculated from the *linearized* local equilibrium distribution (3.19). Thus, it is correct to $\mathcal{O}(\delta)$ in the deviations $\delta\boldsymbol{a}$ and $\delta\boldsymbol{b}$ from equilibrium. The Euler terms are therefore correctly given to $\mathcal{O}(k\delta)$. The $\mathcal{O}(k\delta^2)$-terms will be discussed at the end of this section.

First we derive the explicit form of the fluid dynamic equations (5.2) for CA-fluids with the standard conserved densities (2.4) of number, momentum and energy. Here the macroscopic densities $\boldsymbol{a}_k(t)$ are given through (3.17) and the conjugate fields $\boldsymbol{b}_k(t) = \chi^{-1}\boldsymbol{a}_k(t)$ through (3.21). The nonvanishing inner products in (5.4) can be expressed in the susceptibilities (3.24) using (4.17) with the result,

$$\begin{aligned}
\langle J_\rho \mid A_{g\alpha}\rangle &= \langle J_{g\alpha} \mid A_\rho\rangle = \hat{k}_\alpha\chi_{\epsilon\rho} \\
\langle J_{g\alpha} \mid A_\epsilon\rangle &= \langle J_\epsilon \mid A_{g\alpha}\rangle = \hat{k}_\alpha\tfrac{1}{d}\chi_{\epsilon\epsilon} \equiv \hat{k}_\alpha\tfrac{1}{d}h_K\chi_{\epsilon\rho}.
\end{aligned} \tag{5.13}$$

Here we have introduced the kinetic enthalpy per particle, h_K, to which we return in section 6 in connection with the Green-Kubo formula for the heat conductivity. With the help of (3.24) the linearized fluid dynamic equations for the CA-fluid become

$$\partial_t \rho_k + ik_\alpha \rho u_{k\alpha} = 0$$
$$\partial_t \rho u_{k\alpha} + ik_\alpha v_0 p_k + v_0 \eta_{\alpha\beta\gamma\delta} k_\beta k_\gamma u_{k\delta} = 0$$
$$\partial_t \epsilon_k + ik_\alpha \rho h_K u_{k\alpha} + v_0 \lambda k^2 T_k = 0, \tag{5.14}$$

where we have used that the second rank heat conductivity tensor is isotropic. The volume of the unit cell v_0 appears because the number and energy density are measured in these units. On account of (3.24) we have identified the (kinetic) pressure deviation as,

$$v_0 p_k = \tfrac{2}{d}[\chi_{\epsilon\rho}\nu_k - \chi_{\epsilon\epsilon}\beta_k] = \tfrac{2}{d}\epsilon_k \tag{5.15}$$

In athermal CA-fluids the last equation in (5.14) is absent. There we get from (3.26) and (3.27),

$$\langle J_\rho \mid A_{g\alpha} \rangle = \langle J_{g\alpha} \mid A_\rho \rangle = \hat{k}_\alpha c_0^2 \chi_n. \tag{5.16}$$

The equations for the flow field $\vec{u}_k$ are the same as above, but the pressure deviation is given by $v_0 p_k = c_0^2 \rho_k$.

In summary, we conclude that the linearized Navier-Stokes equation for a CA-fluid with an isotropic viscosity tensor has the form,

$$\partial_t \rho \vec{u}_k + i\vec{k} v_0 p_k + v_0 \eta k^2 \vec{u}_k + v_0 \eta_\ell \vec{k}\vec{k} \cdot \vec{u}_k = 0, \tag{5.17}$$

where $\eta_\ell = \eta(1 - \tfrac{1}{d}) + \zeta$ is the longitudinal viscosity. In the case of cubic symmetry the structure is very different. We only quote the special two-dimensional case where both the shear and bulk viscosity vanish identically and ϑ is a positive viscosity coefficient,

$$\partial_t \rho u_x + \nabla_x c_0^2 \delta\rho + \tfrac{1}{2}\vartheta\nabla_x(\nabla_x u_x - \nabla_y u_y) = 0$$
$$\partial_t \rho u_y + \nabla_y c_0^2 \delta\rho - \tfrac{1}{2}\vartheta\nabla_y(\nabla_x u_x - \nabla_y u_y) = 0, \tag{5.18}$$

as follows from (5.6) and (5.11). This is actually the fluid dynamic equation [27], appropriate for the HPP model, as we shall see in the next section.

If one is interested in nonlinear convective terms and other nonlinear contributions of $\mathcal{O}(k\delta^2)$ in the Euler equations, one needs to extend the expansion of the local equilibrium distribution in (3.19) to $\mathcal{O}(\delta^2)$. A simple calculation [22,18] shows,

$$\partial_t \rho u_\alpha + \nabla_\beta T_{\alpha\beta\gamma\delta} u_\gamma u_\delta = -\nabla_\alpha p + \nabla_\beta \eta_{\alpha\beta\gamma\delta} \nabla_\gamma u_\delta, \tag{5.19}$$

where T is also a fourth rank tensor that is symmetric in all four indices,

$$T_{\alpha\beta\gamma\delta} = \tfrac{1}{2}\gamma_0^2 \sum_c c_\alpha c_\beta c_\gamma c_\delta \langle(\delta n)^3\rangle. \tag{5.20}$$

Here $\vec{\gamma} = \gamma_0 \vec{u}$ is determined from (3.24) as $\gamma_0 = 1/\chi_u$, and the third order fluctuation formula is calculated in (3.12). For isotropic systems this tensor has the form (5.8) with $I_1 = I_2 = I_3$ and the nonlinear convection term reduces to $\vec{\nabla} \cdot g(\rho)\rho\vec{u}\vec{u}$ instead of the standard term $\vec{\nabla} \cdot \rho\vec{u}\vec{u}$. This is caused by the lack of Galilei invariance in CA-fluids. The consequences of the non-Galilean factor $g(\rho)$ being different from unity are extensively discussed in the literature [11,31]. They are outside the scope of these lectures that concentrate on linear excitations and transport coefficients.

In lattices with cubic symmetry there is an additional term $I_4 \delta^{(4)}_{\alpha\beta\gamma\delta}$. The nonlinear Euler terms only have fluid symmetry if I_4 vanishes, i.e. if the *isotropy condition* is satisfied,

$$I_4 = \tfrac{1}{d(d-1)} \sum_c [(d+2)c^{(4)} - 3c^4]\langle(\delta n)^3\rangle = 0, \tag{5.21}$$

where $c^{(4)} = \sum_\alpha c_\alpha^4$ is a cubic invariant. For the two-dimensional model of figure 1 the isotropy condition, written out in terms of Fermi distributions, reads

$$f_1(1 - f_1)(1 - 2f_1) = 4f_2(1 - f_2)(1 - 2f_2) \tag{5.22}$$

where $f_i(i = 1, 2)$ refers to $f(c)$ in (3.5) with $c^2 = 1, 2$ respectively. This is ofcourse a severe restriction because the two-dimensional (ρ, θ)-phase space with $\theta = \exp(-\tfrac{1}{2}\beta)$ is reduced to a line, $\theta(\rho)$. Numerical solution of the isotropy condition (5.22) shows two branches with $0 \le \theta \le 1$ and $0 \le \rho \le 9$: one decreasing from $\theta(0) = \tfrac{1}{4}$ to approximately zero at $\rho = 2.6$ and a second branch decreasing from $\theta(4.5) \simeq 1$ to zero as ρ approaches 7. In the remaining ρ-intervals no physically acceptable solutions have been found.

6 Transport coefficients

6.1 Viscosities in athermal CA-fluids

We apply the correlation function method to obtain expressions for the viscosities in athermal lattice gases without energy conservation with hexagonal and cubic symmetries. The HPP and FHP model of section 2 are contained in the description. The slow variables are the local densities $A_{ka}(t) = \sum_c a_a(\vec{c})\hat{n}_k(\vec{c}, t)$ with $a = \{\rho, g_\alpha, \epsilon\}$, which are assumed to form a complete set of conservation laws. The relevant susceptibilities have been calculated in (3.25) and (3.27). By combining equations (4.21)-(4.25) with (5.2) for $a = g_\alpha$ and with (5.14) we can identify

$$v_0 \eta_{\alpha\beta\gamma\delta}(s)u_{k\delta} = \langle J_{\alpha\beta} \mid Q\tilde{J}_{\gamma\delta}(s)\rangle^* \gamma_{k\delta}, \tag{6.1}$$

where the asterisk on the inner product is a short hand for the first line in (4.21). The conjugate field $\vec{u}_k = \chi_u \vec{\gamma}_k$ can be eliminated and we find the Green-Kubo formula for the viscosity tensor, valid for any space lattice with or without energy conservation,

$$\eta_{\alpha\beta\gamma\delta}(s) = \frac{1}{\chi_u v_0}\langle J_{\alpha\beta} \mid Q\tilde{J}_{\gamma\delta}(s)\rangle^* = \frac{1}{\chi_u V}\sum_{t=0}^{\infty}{}^* e^{-st}\langle \hat{J}_{\alpha\beta}(t)\hat{J}_{\gamma\delta}(0)\rangle, \tag{6.2}$$

where $\mathcal{V} = v_0 V$ is the volume of the system. Ofcourse the explicit form of the subtracted current in (4.25) will depend on which quantities are conserved. For CA-fluids without energy conservation the susceptibility χ_u is given by (3.10), where c_0 is the speed of sound. The subtracted microscopic stress can be calculated from (4.25) and (5.16) with the result

$$\hat{J}_{\alpha\beta}(t) = \sum_{rc}(c_\alpha c_\beta - \delta_{\alpha\beta}c_0^2)\delta n(\vec{c},\vec{r};t), \tag{6.3}$$

where c_0 is the speed of sound (3.10). This is in general not a traceless tensor. The symmetry properties of fourth rank tensors have been discussed in (5.8) to (5.11). On lattices with hexagonal symmetry, such as used in the two-dimensional FHP models, the fourth rank viscosity tensor is isotropic, and I and I' correspond respectively to the shear and bulk viscosity η and ζ. By appropriate contractions of tensor indices they can be expressed in terms of scalar invariants. For the shear viscosity we get for instance, $\eta = ((d+2)(d-1))^{-1}[\eta_{\alpha\beta\alpha\beta} - d^{-1}\eta_{\alpha\alpha\beta\beta}(s)]$. The simplest formula is in fact just a tensor element of (6.2),

$$\eta(s) = \eta_{xyxy}(s) = \frac{1}{\chi_u \mathcal{V}} \sum_{t=0}^{\infty} {}^{*} e^{-st}\langle\sigma_{xy}(t)\sigma_{xy}\rangle \tag{6.4}$$

with the microscopic stress tensor,

$$\sigma_{xy}(t) = \sum_{rc} c_x c_y \delta n(\vec{c},\vec{r};t). \tag{6.5}$$

For the bulk viscosity we find similarly,

$$\zeta(s) = \tfrac{1}{d^2}\eta_{\alpha\alpha\beta\beta}(s) = \frac{1}{\chi_u \mathcal{V}} \sum_{t=0}^{\infty} {}^{*} e^{-st}\langle\sigma(t)\sigma(0)\rangle, \tag{6.6}$$

with the projected bulk viscosity current,

$$\sigma(t) = \sum_{rc}(\tfrac{1}{d}c^2 - c_0^2)\delta n(\vec{c},\vec{r};t). \tag{6.7}$$

These expressions represent the viscosities appropriate for d-dimensional lattices with isotropic fourth rank tensors. The propagation part of the transport coefficients is given by one half of the $(t=0)$-term in (6.4) and (6.6). For the 6-bits FHP model one finds for instance, $\eta^P = -\tfrac{1}{4}\sqrt{3}\rho$ and $\zeta^P = 0$. For the 7-bits version of this model the results are $\eta^P = -\tfrac{1}{4}\sqrt{3}\rho$ and $\zeta^P = -\tfrac{1}{14}\sqrt{3}\rho$.

It is apparent from (5.10) and (5.11) that the shear viscosity vanishes in one dimension. The bulk viscosity vanishes in all dimensions for single speed models, since $c^2 = dc_0^2 = 1$ on account of (3.10). Somewhat similar results for the viscosities have been derived by Rivet and Frisch [24,25], Hayot [26] and Zanetti [16].

Next we consider d-dimensional lattices with *cubic symmetry*, where the fourth rank viscosity tensor (5.11) has in general three independent scalar viscosity coefficients: the shear viscosity η, the bulk viscosity ζ and the ϑ-viscosity, typical for cubic symmetry. There are many ways to express the viscosities in cubic invariants or in tensor elements. Convenient forms are $\eta = \eta_{xyxy}$ and $\zeta = d^{-2}\eta_{\alpha\alpha\beta\beta}$, which are again given by (6.4) and (6.6). This implies that $\zeta = 0$ for all single speed models since $c_0^2 = \frac{1}{d}$. For the ϑ-viscosity we obtain,

$$
\begin{aligned}
\vartheta(s) &= \eta_{xxxx}(s) - \eta_{xxyy}(s) \\
&= \frac{1}{2\chi_u V} \sum_{t=0}^{\infty} {}^{*} e^{-st} \langle (J_{xx}(t) - J_{yy}(t))(J_{xx}(0) - J_{yy}(0)) \rangle.
\end{aligned}
\qquad (6.8)
$$

This formula is valid for all lattices with square and cubic symmetry, without energy conservation, independent of the set of allowed velocity states. If such models admit only zero velocities (rest particles) and velocities that are multiples of the $2d$-nearest neighbor lattice vectors, then $c_x c_y = 0$ in (6.3). Consequently $\sigma_{xy} = 0$ and $\eta_{xyxy} = \eta = 0$ on account of (5.11). It is apparent from (5.10) and (5.11) that the shear viscosity and ϑ-viscosity vanish in one dimension. The two-dimensional HPP model, defined on the sqare lattice, is a simple example of a single speed model with these properties. There the viscosity tensor takes the simple form,

$$
\eta_{\alpha\beta\gamma\delta} = \vartheta \left(\delta^{(4)}_{\alpha\beta\gamma\delta} - \tfrac{1}{d}\delta_{\alpha\beta}\delta_{\gamma\delta} \right),
\qquad (6.9)
$$

where ϑ is a positive viscosity coefficient. Substitution of these results in (5.14) yields the fluid dynamic equation (5.18) for the HPP model, derived in section 5. The result for the ϑ-viscosity, derived in reference [27] resembles (6.8), apart from the missing relative weight of one half in front of the $(t = 0)$ term.

6.2   Viscosities in thermal CA-fluids

In models with energy conservation equation (6.2) for the complete viscosity tensor, as well as the separate expressions (6.4) for the shear viscosity, (6.6) for the bulk viscosity and (6.8) for the ϑ-viscosity remain valid. Here the susceptibility χ_u in (3.27) should be replaced by (3.25) and we recall, that if thermodynamic and kinetic pressure were equal (Galilei invariance), the susceptibility would simply be $\chi_u = 1/\beta$. The Green-Kubo formulae for the viscosities in a temperature dependent CA-fluid would then be identical to those for the continuous fluid [12] apart from the propagation viscosity (4.23). Including energy among the conserved quantities has an important effect on the viscosity coefficients of CA-fluids: the *bulk viscosity ζ vanishes* identically, independent of the set of allowed velocity states and independent of the symmetry of the underlying lattice. This is similar to the case of continuous fluids where the bulk viscosity vanishes in the absence of statistical correlations (low density). On a triangular lattice with an appropriately extended set of velocity states, such that non-trivial energy conservation can be built into the collision rules, there would only be a non-vanishing shear viscosity. The sound waves would be damped by heat dissipation only. The reason for the dissappearing bulk viscosity is that the relevant current is the trace of the stress tensor , $J_{\alpha\alpha}(t) = \sum_{rc} c^2 n(\vec{c}, \vec{r}; t)$, as can be deduced from (6.3). It is proportional to the total energy, which is a conserved quantity. Hence, the subtracted current $\hat{J}_{\alpha\alpha} = QJ_{\alpha\alpha} = 0$ for all LGCA's with point particles and energy conservation.

The model of figure 1, defined on square and cubic lattices, has two non-vanishing viscosities, the shear viscosity η and the unphysical viscosity ϑ. Only in cases where it is possible to choose the parameters in the collision rules (assuming that there is sufficient freedom to do so) such that the isotropy condition (5.12), $\vartheta = 2\eta$, can be satisfied, would the cubic tensor (5.11) reduce to the isotropic form (5.10) and would one recover fluid dynamic equations with isotropic symmetry and a non-vanishing shear viscosity [23], if simultaneously the isotropy condition (5.21) for the nonlinear part of the Euler equation can be satisfied.

As a side remark we note that this model has also two spurious conservation laws of staggered momentum, which will introduce additional unphysical equations in the set of fluid dynamic equations [18]. If one drops the restriction of point particles and considers collisions with non-zero impact parameter, then the bulk viscosity may be non-vanishing. Atleast one lattice gas model of this type, the hard hexagon model, has been studied in the literature [33].

6.3 Heat conductivity

In thermal models there exists also dissipation through heat conduction and the Green-Kubo expression for this transport coefficient is contained in the present formalism. By combining (4.21)-(4.25) with (5.2) for $a = \epsilon$ and with (5.14) we can identify,

$$v_0 k^2 \lambda(s) T_k = -k^2 L_{\epsilon\epsilon}\beta_k = -k_\alpha k_\beta \langle J_{\epsilon\alpha} \mid Q\tilde{J}_{\epsilon\beta}(s)\rangle^* \beta_k, \tag{6.10}$$

where the asterisk has the same meaning as in (6.2) and $\beta_k = -T_k/(k_B T^2)$. The Green-Kubo formula for the heat conductivity is given by the small-s limit of

$$\lambda(s) = \frac{1}{dv_0}\langle J_{\epsilon\alpha} \mid Q\tilde{J}_{\epsilon\alpha}(s)\rangle^* = \frac{\beta}{dTV}\sum_{t=0}^{\infty}{}^* e^{-st}\langle \vec{Q}(t)\cdot\vec{Q}(0)\rangle. \tag{6.11}$$

The subtracted energy current can be deduced from (3.24) and (5.13) as

$$\hat{J}_{k\epsilon} = J_{k\epsilon} - h_K \hat{k}\cdot\vec{g}_k \equiv \hat{k}\cdot\vec{Q}, \tag{6.12}$$

where we have introduced

$$\vec{Q}(t) = \sum_{rc} \vec{c}(\tfrac{1}{2}c^2 - h_K)\delta n(\vec{c},\vec{r};t). \tag{6.13}$$

The kinetic enthalpy per particle h_K is defined in (5.13). Subtraction of the enthalpy term $h_K \vec{P}$ makes $\vec{Q}$ orthogonal to the conserved total momentum $\vec{P}$. The propagation part (4.23) of the heat conductivity is given by

$$\lambda^P = \frac{\beta}{2dTV}\langle\vec{Q}\cdot\vec{Q}\rangle = \frac{\beta}{2dTv_0}\sum_c c^2(\tfrac{1}{2}c^2 - h_K)^2\kappa(c), \tag{6.14}$$

where the weight $\kappa(c)$ is defined in (3.12). The Green-Kubo formula for the heat conductivity looks very similar to the one for continuous fluids [12] with one subtle difference, that is a consequence of the lack of Galilei-invariance: the quantity h_K above is the *kinetic enthalpy* per particle and not the thermodynamic enthalpy h. The latter is defined through derivatives of the thermodynamic pressure (3.6),

$$\beta\left(\frac{\partial p v_0}{\partial v}\right)_\beta = \rho \quad \text{and} \quad -\beta\left(\frac{\partial p v_0}{\partial \beta}\right)_v = \epsilon + p v_0 \equiv \rho h, \tag{6.15}$$

where ρ, ϵ and p are defined in (3.6). Consequently $h = (\partial v/\partial \beta)_p$. The kinetic enthalpy h_K was introduced in (5.13) and satisfies

$$h_K = \frac{\chi_{\epsilon\epsilon}}{\chi_{\epsilon\rho}} = -\frac{(\partial\epsilon/\partial\beta)_v}{(\partial\epsilon/\partial v)_\beta} = \left(\frac{\partial v}{\partial\beta}\right)_{p_K} \tag{6.16}$$

where we have used the relation $p_K v_0 = \frac{2}{d}\epsilon$ between kinetic pressure and energy density.

In the preceding discussion the transport coefficient have been cast into the general form of a diagonal element of the transport matrix,

$$L(s) = \frac{1}{V} \sum_{t=0}^{\infty} {}^* e^{-st} \langle J(t) J(0)\rangle \tag{6.17}$$

with a current of the general form,

$$J_a(t) = \sum_{rc} w_a(c)\delta n(\vec{c}, \vec{r}; t). \tag{6.18}$$

The currents for the different transport coefficients are given by

$$\begin{aligned}
w_\eta(c) &= c_x c_y \\
w_\zeta(c) &= \tfrac{1}{d}c^2 - c_0^2 \\
w_\vartheta(c) &= c_x^2 - c_y^2 \\
w_\lambda(c) &= c_x(\tfrac{1}{2}c^2 - h_K)
\end{aligned} \tag{6.19}$$

According to the discussion at the end of section 4 the transport coefficients are non-negative, because they have been expressed as diagonal elements of a transport matrix. Analytic evaluation of the Green-Kubo formulae is straight forward in the Boltzmann approximation, as will be discussed in the next section.

7   Boltzmann approximation

In this section we evaluate the Green-Kubo expressions in the Boltzmann approximation. The Boltzmann approximation or Stosszahlansatz is the assumption that positions and velocities of particles before collision are uncorrelated. This implies that recollisions are neglected and that the probability for the occurrence of an ℓ-tuple collision at site $\vec{r}$ is simply given by the product of ℓ single particle distribution functions, defined as $f(\vec{c},\vec{r};t) = \langle n(\vec{c},\vec{r};t)\rangle_{ne}$. The Boltzmann equation for the CA-fluids follows then from (2.1) by replacing $\langle I_c(n)\rangle_{ne}$ by $I_c(f)$, i.e. by replacing the average of a product of occupation numbers by the product of their averages, $f = \langle n\rangle_{ne}$. The result is

$$f(\vec{c},\vec{r}+\vec{c};t+1) = f(\vec{c},\vec{r};t) + I_c(f(\vec{r};t)). \tag{7.1}$$

We note that the collision term vanishes if f is replaced by the total or local equilibrium distribution, given in (3.5) and (3.16) respectively.

How is this approximation implemented in evaluating Green-Kubo formulae? The basic idea of the method is as follows. Correlated collisions, such as ring collisions, require at least two time steps. Therefore the value of the time correlation function $\Gamma(t)$ after one time step is entirely determined by uncorrelated collisions. Hence the relation $\Gamma(1) = (1 - \Omega)\Gamma(0)$ defines the linearized Boltzmann collision operator Ω. By assuming that this relation holds between any consecutive time steps one obtains the Boltzmann approximation,

$$\Gamma(t) = (1 - \Omega)\Gamma(t - 1) = (1 - \Omega)^t\Gamma(0). \tag{7.2}$$

To make this more quantitative the transport coefficient in (6.17) for $s = 0$ is expressed in terms of Γ as,

$$L = \sum_{t=0}^{\infty} {}^{*}\sum_{cc'} w(c)\Gamma_{cc'}(t)w(c') \tag{7.3}$$

with summation convention for repeated indices. The asterisk indicates that the term at $t = 0$ only counts half. The basic time correlation function Γ is defined as,

$$\Gamma_{cc'}(t) = \frac{1}{V}\sum_{rr'}\langle \delta n(\vec{c},\vec{r};t)\delta n(\vec{c}',\vec{r}';0)\rangle. \tag{7.4}$$

On account of (3.13) its initial value is given by,

$$\Gamma_{cc'}(0) = \kappa(c)\delta_{cc'}. \tag{7.5}$$

With the help of the microdynamic equation (2.1) we obtain,

$$\Gamma_{cc'}(1) - \Gamma_{cc'}(0) = \sum_{r}\langle I_c(n)\delta n_{c'}(0)\rangle, \tag{7.6}$$

where all occupation numbers in I_c refer to site $\vec{r}$ and time $t = 0$. Let $\delta n = n - \langle n\rangle = n - f$ be the fluctuation around total equilibrium, then the collision operator can be written as,

$$I_c(n) = I_c(f) - \Omega_{cc'}\delta n_{c'} - \Omega_{cc'c''}\delta n_{c'}\delta n_{c''} + \cdots \tag{7.7}$$

The first term on the right hand side vanishes on account of (7.1). In a b-bits model the last term on the right hand side contains a product of at most b factors δn_c, each of which refers to a *different velocity state*. Substitution of (7.7) in (7.6) shows that only the linear term $\Omega_{cc'}\delta n_{c'}$ is non-vanishing, yielding

$$\Gamma_{cc'}(1) - \Gamma_{cc'}(0) = -\Omega_{cc'}\kappa(c'). \tag{7.8}$$

Combining this with (7.5) and writing it in matrix form gives (7.2) for $t = 1$. Next we insert (7.2) into (7.3), sum the geometric series and obtain an expression for the transport coefficient in the Boltzmann approximation,

$$L_0 = w(c)\left(\tfrac{1}{\Omega} - \tfrac{1}{2}\right)_{cc'}\kappa(c')w(c'). \tag{7.9}$$

The subtracted term, $L^P = -\tfrac{1}{2}\sum_c w^2(c)\kappa(c)$, represents the propagation part of the transport coefficient (see for instance (6.14)).

One way to proceed would be to calculate the eigenfunctions and eigenvalues of the $b \times b$ collision matrix,

$$\Omega_{cc'}\kappa(c')\psi(c') = \lambda\kappa(c)\psi(c). \tag{7.10}$$

The eigenfunctions with eigenvalue $\lambda = 0$ are the collisional invariants, $\psi(c) = a(c) = \{1, \vec{c}, \tfrac{1}{2}c^2, \ldots\}$. The inverse matrix Ω^{-1} in (7.9) does exist, since the b-vector $w(c)$ belongs to the orthogonal complement of the null space. This can be seen from equations (4.25), (6.17) and (6.19), yielding the orthogonality condition,

$$\langle \hat{J} \mid A \rangle = \sum_c w(c)a(c)\kappa(c) = 0. \tag{7.11}$$

Explicit evaluation of L_0 requires ofcourse a complete specification of the collision rules. As an illustration we apply the Boltzmann approximation to two simple examples: the ϑ-viscosity of the two-dimensional HPP fluid, defined on the square lattice, and the the bulk viscosity for a one-dimensional CA-fluid. Starting with the HPP fluid we construct the collision matrix Ω_{ij} with $(i,j) = \{1,2,3,4\}$ from equations (7.7) and (2.2). The result is,

$$\Omega_{ij}\delta n_j = f(1 - f)\{\delta n_i + \delta n_{i+2} - \delta n_{i+1} - \delta n_{i-1}\}. \tag{7.12}$$

The current

$$w(c_j) = c_{jx}^2 - c_{jy}^2 = (1, -1, 1, -1), \tag{7.13}$$

as given by (7.9) and (6.19), is itself an eigenvector of Ω with eigenvalue $\lambda = 4f(1 - f)$. Insertion of these results in (6.8) and (6.19) gives the ϑ-viscosity for the HPP model,

$$\vartheta_0 = \frac{1}{2\chi_u}L_0 = \tfrac{1}{4}\rho\left(\tfrac{1}{\lambda} - \tfrac{1}{2}\right)\sum_c w^2(c) = \frac{1}{1-f} - 2f \tag{7.14}$$

with χ_u given in (3.27). We also note that the viscosity is strictly positive, $\vartheta_0 \geq 1$ for the physical range of reduced densities, $0 \leq f \leq 1$.

As a second example we consider the bulkviscosity of a one-dimensional fluid type cellular automaton, with five velocities per site ($c = 0, \pm 1, \pm 2; b = 5$), as introduced by d'Humières et al [28] with possible collisions

$$\begin{aligned}
(-1) + (+1) &\Longleftrightarrow (+2) + (-2) \\
(0) + (-1) &\Longleftrightarrow (+1) + (-2) \\
(0) + (+1) &\Longleftrightarrow (-1) + (+2).
\end{aligned} \tag{7.15}$$

As a small variation on the original model we impose the constraint that no collisions occur if more than two particles are present a one site. The model has two collisional invariants ($a = 1, c$). The fluid velocity has only a longitudinal component and supports sound waves, propagating with a sound speed c_0, given by $c_0^2 = \frac{1}{5}\sum_c c^2 = 2$, and damped by the bulk viscosity, given in (6.6) and (6.7). The function $w(c)$ in (6.19) is here $w(c) = c^2 - 2$. Equation (7.9) then takes the form,

$$\zeta_0 = \frac{1}{\chi_u}L_0 = \tfrac{1}{10}\rho w(c)\left(\tfrac{1}{\Omega} - \tfrac{1}{2}\right)_{cc'} w(c') \tag{7.16}$$

where we have used the susceptibility in (3.27). It is a straight forward extension of (2.2) to construct the collision term $I_i(n)$ and its linearized form Ω_{ij} with $(i,j) = (2,1,0,-1,-2)$ from the collision rules in (7.15). The result is

$$\Omega_{ij} = f(1-f)^2 \begin{pmatrix} 2 & -2 & -1 & 0 & 1 \\ -2 & 3 & 0 & -1 & 0 \\ -1 & 0 & 2 & 0 & -1 \\ 0 & -1 & 0 & 3 & -2 \\ 1 & 0 & -1 & -2 & 2 \end{pmatrix} \tag{7.17}$$

The eigenvectors and eigenvalues $\lambda_j = f(1-f)^2\overline{\lambda}_j$ are,

$$\begin{aligned}
\psi_0 &= (\ 1,\ 1,\ 1,\ 1,\ 1) & (\overline{\lambda}_0 = 0) \\
\psi_1 &= (\ 2,\ 1,\ 0,\ -1,\ -2) & (\overline{\lambda}_1 = 0) \\
\psi_2 &= (\ 0,\ 1,\ -2,\ 1,\ 0) & (\overline{\lambda}_2 = 2) \\
\psi_3 &= (\ 1,\ -2,\ 0,\ 2,\ -1) & (\overline{\lambda}_3 = 5) \\
\psi_4 &= (3,\ -2,\ -2,\ -2,\ 3) & (\overline{\lambda}_4 = 5)
\end{aligned} \tag{7.18}$$

In the present case the current is a linear combination of two eigenfunctions,

$$w(c) = (2,-1,-2,-1,2) = \tfrac{1}{3}\psi_2 + \tfrac{2}{3}\psi_4. \tag{7.19}$$

Combining this with (7.16) gives the Boltzmann value for the bulk viscosity in the present model,

$$\zeta_0 = \tfrac{1}{10}\rho \sum_c w(c) \left[\tfrac{1}{3}(\tfrac{1}{\lambda_2} - \tfrac{1}{2})\psi_2 + \tfrac{2}{3}(\tfrac{1}{\lambda_4} - \tfrac{1}{2})\psi_4 \right] = \frac{3}{2(1-f)^2} - \tfrac{7}{2}f \qquad (7.20)$$

where $f = \tfrac{1}{5}\rho$ is the reduced density, $(0 \le f \le 1)$.

Suppose we would like to apply the Boltzmann approximation to the Green-Kubo formula for the viscosities in the 6- or 7-bits FHP model, defined on the triangular lattice. Here there are more velocity states per site with many binary, triple and quadruple collisions, for which the analogs of (7.15) have to be given. From there one constructs the $b \times b$ collision matrix Ω and determines its eigenvalues and eigenfunctions. The explicit results for the viscosities in these models have been calculated by d'Humières and Lallemand [34].

As mentioned in section 4, the transport coefficients in one- and two-dimensional models do not exist in the long time limit because of very long memory effects , the so called *long time tails*. In two dimensions the transport coefficients grow as $\ell n(t)$ and in one dimension as t^α with α in the range 0.3 to 0.5 [18,32]. The transport coefficients L_0 calculated here in the Boltzmann approximation, are the *short time* or *bare* transport coefficients. In most two-dimensional systems the effects of the divergences on the transport coeffiients are too small to be detectable in computer simulations. However, the long time tail in the velocity correlation function of the two-dimensional FHP model, which is proportional to $1/t$, has been measured in computer simulations by Frenkel and collaborators over time intervals of up to 500 time steps over a large range of densities. There is almost perfect agreement with the predictions from mode coupling theory [32], [35]-[38]. The coefficients in the long time tails over the intermediate time intervals of the computer experiments are entirely determined by the bare transport coefficients.

How to extend kinetic theory for CA-fluids beyond the approximation of uncorrelated collisions is clear, at least in a formal sense, from the fluctuation expansion in (7.6) and (7.7), iterated for t time steps. One easily developes a graphical representation of the terms in this fluctuation expansion, using standard methods of statistical mechanics [39]. However, the diagrams have still to be analyze in order to be able to select the most important ones and to develope resummation techniques to take them into account.

8 Conclusions

In these lectures we have developed nonequilibrium statistical mechanics of cellular automata fluids, in close parallel with linear response theory for continuous fluids. The main message conveyed in these lecture notes is that lattice gas cellular automata with b velocity states may be considered as bona fide, although extremely simplified, statistical mechanical models of nonequilibrium systems with many degrees of freedom, that are able to model many aspects of hydrodynamics and transport properties of fluids.

1. *Thermodynamic equilibrium* can be described by the grand canonical ensemble, as dis-

cussed in section 3. The formalism can handle both athermal CA-fluids without energy conservation, as well as thermal ones with energy conservation and hence with a temperature $T = (k_B\beta)^{-1}$. A CA-fluid has strong similarities with the ideal Fermi-gas in the occupation number representation because of the Fermi exclusion rule for double occupancy of single particle states $\{\vec{c}, \vec{r}\}$. The average occupation number is the Fermi-Dirac distribution $f(c)$ in (3.5) which contains a temperature β and a chemical potential ν/β. In the high temperature limit the thermal models approach the athermal ones and the Fermi distribution $f = \rho/b$ becomes energy independent. The average occupation per site ρ in a b-bits model is restricted to $0 \leq \rho \leq b$. In lattice gases the *virial theorem* does not hold, i.e. the thermodynamic pressure p in (3.6) obtained from the partition function, differs from the kinetic pressure p_K in (3.7), or equivalently, from the average momentum current. The same is in fact true for a one-component plasma with a fixed neutralizing back ground charge. The equation of state for the kinetic pressure, $v_0 p_K(\rho, \beta) = \frac{2}{d}\epsilon(\rho, \beta)$, which is in fact the caloric equation of state, is shown in figure 3 (isotherms) anf figure 4 (isochores) for the square LGCA of figure 1.

2. Figure 1 also illustrates that a *kinetic temperature*, defined through $\epsilon(\rho, \beta) \equiv \frac{1}{2}d\rho k_B T_K$, is quite different from the thermodynamic temperature $T = (k_B\beta)^{-1}$. The same comment applies ofcourse to an ideal Fermi gas. In the high temperature limit $(T \longrightarrow \infty)$ the kinetic temperature approaches a constant, $k_B T_K \longrightarrow c_0^2$, where c_0 is the speed of sound (3.10) in athermal fluids. At low temperature, $(T \longrightarrow 0)$, the kinetic temperature is a function $T_K(\rho)$ of the density, given through (3.8) for the square lattice gas of figure 1.

3. *Local equilibrium* concepts, which form the basis of irreversible thermodynamics [13], make it possible to define local temperature $T(\vec{r}, t)$ and local chemical potential outside thermal equilibrium. This is done through the caloric equation of state, $\epsilon = \epsilon(\rho, \beta)$ [12], defined here through (3.18) or in linearized form through (3.21). Exactly the same temperature definition is followed in the derivation of Green-Kubo formulae for continuous fluids [12]. The local equilibrium ensemble is then used to derive the nonlinear hydrodynamic or *Euler equations* of the ideal fluid *without dissipation*. It is ofcourse the kinetic pressure that enters in the Euler equations of an ideal CA-fluid, as an average of the microscopic momentum current. The energy balance equation contains the kinetic enthalpy per particle, h_K, given by the fluctuation formula (6.16). It differs from the thermodynamic enthalpy $\rho h = \epsilon + p v_0 = (\partial \nu / \partial \beta)_p$ and also from $\epsilon + p_K v_0 = \frac{d+2}{d}\epsilon$. In a continuous fluid all three quantities are identical. The same enthalpy h_K enters the Green-Kubo formula for the heat conductivity (6.11) in thermal CA-fluids. Also in section 5 we discuss how the nonlinear terms in the Euler equations are obtained.

4. The *linear transport* coefficients L_{ab}, such as viscosity and heat conductivity, are defined in irreversible thermodynamic through the constitutive relations, $J_a = L_{ab}X_b$, where J_a and X_b are respectively the thermodynamic fluxes (heat and particle current, stress tensor) and driving forces (gradients in local temperature, chemical potential and local flow velocity). The linear response theory of section 4 for CA-fluids then yields Green-Kubo formulae (6.11) for L_{ab}, analogous to those for continuous fluids. They are discrete time sums over time correlation functions of so-called subtracted fluxes with the restriction that the first term at $t = 0$ has a relative weight $\frac{1}{2}$. Alternatively, if the initial term is given weight unity, then there appears

a negative contribution of half the term at $t = 0$, which has been called the "propagation" part of the transport coefficient. It is a vestige of the discrete space-time of the LGCA that persists at the macroscopic level. The derivation applies to CA-fluids with and without energy conservation, to point particles with deterministic or stochastic dynamics, to single speed and multi-speed models (including rest particles), and to lattices for which the macroscopic equations may or may not have fluid isotropy. The only condition on the dynamics is the existence of local conservation laws (2.6) and an equilibrium state.

5. *Isotropy* of fourth rank tensors is a required condition to obtain fluid dynamic equations with the proper fluid symmetries. The fluid dynamic equations for CA-fluids, defined on a triangular lattice, such as the FHP models, and on the quasi 4-dimensional FCHC lattice have the isotropic symmetry of the Navier-Stokes equations. LGCA's with square and cubic symmetry give fluid dynamic equations (5.18) that do not possess the symmetry of the Navier-Stokes equations (5.17). The nonlinear terms (5.19) in the Euler equations are isotropic, if the isotropy condition (5.21) is satisfied, as shown by d'Humières at al [22]. The viscous terms in the Navier-Stokes equations are isotropic, if the special cubic viscosity and the shear viscosity satisfy the condition $\vartheta = 2\eta$. Both viscosities are positive.

6. The viscosity tensor (6.2) of thermal and athermal CA-fluids has been expressed as a Green-Kubo formula, independent of the lattice symmetry. It contains the susceptibilities χ_u, given by (3.25). It reduces in the high temperature limit to the value $\chi_u = c_0^2(1 - f)$ of the athermal models. In the hypothetical case that the thermodynamic and kinetic pressure would be equal (as is the case in continuous fluids), then $\chi_u = 1/\beta$. For comparison we also recall from remark 2 that the kinetic temperature approaches $k_B T_K \longrightarrow c_0^2$ as $T \longrightarrow \infty$. In the case of square or cubic symmetry there are in general three positive viscosities, η, ζ, ϑ, respectively given in (6.4), (6.6) and (6.8). In case of isotropic symmetry, where $\vartheta = 2\eta$, only shear and bulk viscosity remain.

7. The bulk viscosity *vanishes* identically in all *thermal* CA-fluids, as shown in section 6.2. This result is independent of the lattice structure and independent of the set of allowed velocity states. The reason is that the trace of the microscopic stress tensor is a conserved quantity.

8. To evaluate the Green-Kubo formulae for transport coefficients in LGCA's one needs to develope generalized kinetic equations. For fluid-type LGCA's no such generalizations exist beyond the Boltzmann equation. It is straightforward to evaluate the Green-Kubo formulae in the Boltzmann approximation. This leads to the general result (7.9), where Ω is the linearized Boltzmann collision operator for these models. The transport coefficients of CA-fluids measured in computer simulations [25] agree reasonably well with the values obtained from the Boltzmann approximation. In two-dimensional models, where transport coefficients don't exist, these Boltzmann values are usually referred to as bare or short time transport coefficients.

9. In the Boltzmann approximation the positivity of the transport coefficients can be verified from the explicit results, derived in section 7. It is a consequence of the positive definiteness of the Boltzmann collision operator, which is the same as the linearized form of the H-theorem. However, Schmitz and Dufty [30] have obtained the stronger result, that the transport matrix , given by the Green-Kubo formula (4.21), is a positive definite matrix, sothat

its diagonal elements are positive. By generalizing (7.20) one sees that a *sufficient condition* for the positivity of transport coefficients is that *all non-vanishing* eigenvalues of the Boltzmann collision operator (7.10) satisfy the inequality $\mathcal{R}e\lambda \leq 2$. Although this is not a necessary condition, the inequality is satisfied for all cases that we have investigated. A example less trivial than (7.14) or (7.18) is the 9-bits model of figure 1, where the eigenvalues depend on density and temperature. It would even hold if the different collisions in figure 1 are given arbitrary weights $W_i (i = 1, 2, 3, 4)$ with $0 \leq W_i \leq 1$.

Acknowledgements

It is a pleasure to thank S. Das for the numerical work needed to produce the isotherms and isochores in figures 3 and 4 and for numerically solving the isotropy condition at the Euler level.

References

[1] G.D. Doolen, Ed., *Lattice gas methods for partial differential equations*, (Addison-Wesley Publ. Co., 1989)

[2] Proceedings of the workshop on *Large nonlinear systems*, Complex Systems 1, No. 4 (1987)

[3] R. Monaco, Ed., *Discrete kinetic theory, lattice gas dynamics and foundations of hydro-dynamics*, (World Scientific, Singapore, 1989)

[4] M. Mareschal, Ed., *Microscopic simulation of complex flows*, (Plenum, New York, 1990)

[5] G.D. Doolen, Ed. Proceedings of the workshop on *Lattice gas methods for PDE's*, Los Alamos, September 1989, to appear in *Physica* D (1990)

[6] B. Hasslacher, Discrete fluids, *Los Alamos Science* 15 (special issue) 175, 211 (1988)

[7] T.Shimomura, G.D. Doolen, B.Hasslacher and C. Fu, in Ref.1, p.3

[8] J.P. Boon, to appear in *Physicalia* 11 (1989)

[9] S. Wolfram, *Rev. Mod. Phys.* 55, 601, (1984)

[10] S. Wolfram, Ed., *Theory and application of cellular automata*,(World Scientific, Singapore, 1986)

[11] U. Frisch, D. d'Humières, B. Hasslacher, P.Lallemand, Y. Pomeau and J.P. Rivet, *Complex Systems* 1, 649 (1987) and Ref. 1, p. 75

[12] J.A. McLennan, *Introduction to nonequilibrium statistical mechanics*, (Prentice Hall, USA, 1989)

[13] S.R. de Groot and P. Mazur, *Non-equilibrium thermodynamics*, (North Holland Publ, Amsterdam, 1962)

[14] M.H. Ernst and J.W. Dufty, *J. Stat. Phys.* 58, 57 (1990)

[15] L.P. Kadanoff, G. R. McNamara and G. Zanetti, in Ref.1, and *Phys. Rev.* A40, 4527 (1989)

[16] G. Zanetti, *Phys. Rev.* A40, 1539 (1989)

[17] D. d'Humières, Y.H. Qian and P. Lallemand, in Ref.1, p. 102

[18] M.H. Ernst, in: *Liquids, freezing and the glass transition*, Les Houches Session LI, 1989, D. Levesque, J.P. Hansen and J. Zinn-Justin, Eds. (Elsevier Science Publ, 1990)

[19] T.Naitoh, M.H. Ernst and J.W. Dufty, to be published

[20] J. Hardy, Y. Pomeau and O. de Pazzis, *J. Math. Phys.* 14, 1746 (1973)

[21] U. Frisch, B. Hasslacher and Y. Pomeau, *Phys. Rev. Lett.* 56, 1505 (1986)

[22] D. d'Humières, P.Lallemand and U. Frisch, *Europhys. Lett.* 2, 291 (1986)

[23] S. Chen, M. Lee, K.H. Zhao and G.D. Doolen, *Physica* D 37, 42 (1989)

[24] U. Frisch and J.P. Rivet, in Ref.1, p. 437

[25] J.P. Rivet,in Ref.1, p.399

[26] F.Hayot, *Physica* 28D, 210 (1987)

[27] J. Hardy, O. de Pazzis and Y. Pomeau, *Phys. Rev.* A13, 1949 (1976)

[28] D. d'Humières, P. Lallemand and Y.H. Qian, *C.R. Acad. Sci. Paris II* 308, 585 (1988) and preprint ENS (1989)

[29] J.R. Dorfman and H. van Beijeren, in *Statistical mechanics, Part B: Time dependent processes*, B.J. Berne, Ed., (Plenum Press, New York, 1977) p. 65

[30] R.Schmitz and J.W. Dufty, private communication

[31] S. Wolfram, *J.Stat. Phys.* 45, 471 (1986) and in Ref.1, p. 19

[32] M.H. Ernst,in Proceedings of the workshop on *Lattice gas methods for PDE's*, Los Alamos, September 1989, to appear in *Physica* D (1990)

[33] M.E. Colvin, A.J.C. Ladd and B.J. Alder, *Phys. Rev. Lett.* 61, 381 (1988)

[34] D. d'Humières and P.Lallemand, in Ref.1, p. 297

[35] D. Frenkel and M.H. Ernst, *Phys. Rev. Lett.* 63, 2165 (1989)

[36] M. van der Hoef and D. Frenkel, *Phys. Rev.* A (1989), submitted

[37] D. Frenkel, in *Fundamental Problems in Statistical Mechanics* VII, Ed. H. van Beijeren, (North Holland Publ. Co, Amsterdam, 1990)

[38] M. van der Hoef, in Ref. 5

[39] Shang-Keng Ma, *Modern theory of critical phenomena*, Frontiers in Physics 46, (W. Benjamin, Inc. 1976),chapt XIV

FUNDAMENTAL PROBLEMS IN STATISTICAL MECHANICS VII
H. van Beijeren, Editor
© Elsevier Science Publishers B.V., 1990

KINETIC THEORY OF DENSE GASES AND LIQUIDS

Henk van Beijeren

Instituut voor Theoretische Fysica, Princetonplein 5, P.O.Box 80.006,
3508 TA Utrecht, The Netherlands

1 INTRODUCTION

The temporal behaviour of a dilute gas is described satisfactorily by the Boltzmann equation. Formally this equation can be obtained by using cluster expansion methods to close the *BBGKY*-hierarchy equations and going to the low density limit [1]. If the ensemble describing the initial state satisfies certain asymptotic factorization properties, related to Boltzmann's Stosszahlansatz, the Boltzmann equation results. Lanford has given a rigorous derivation of the Boltzmann equation for a hard sphere system under these conditions [2], but his method so far only works up till a finite fraction of the mean free time between collisions, mainly by lack of existence theorems for solutions of the Boltzmann equation. The need for an ensemble of systems to derive the Boltzmann equation implies that this equation describes *the average time behaviour* of a large class of macroscopically indistinguishable, but microscopically different systems. It is for this average behaviour that Boltzmann's celebrated *H*-theorem holds and the approach of entropy to its equilibrium value is monotonic. Individual systems typically exhibit small deviations from this average behaviour, which can be incorporated into the Boltzmann equation by adding fluctuating terms to it [3]. For dense gases and liquids no such unambiguous procedures are available for obtaining kinetic equations of general validity. When compared to the dilute gas two major complications arise: the ubiquitous presence of many-particle interactions and the fact that the energy density is not determined by the one-particle distribution function alone, but in addition requires the pair distribution function. Hence, even on hydrodynamic time scales the one-particle distribution function does not suffice to describe the state of the fluid, in contrast to the situation in a dilute gas. A special case for which these complications are absent is that of particles interacting through a hard core potential. These collisions never involve more than two particles and potential energy of interparticle interaction is absent. For a system of smooth hard spheres Enskog derived his famous Enskog equation [4]. Under appropriate initial conditions this equation becomes exact for very short times, but its main applications are for long times, where it is only an approximation. For many purposes it is a very good approximation at not too high densities, for example the transport coefficients of the hard sphere fluid are predicted correctly within 5% for densities up to 3/4 of the solidification density, but especially at densities close to solidification the discrepancies between the predictions of the Enskog theory and molecular dynamics (MD) results for the hard sphere system become quite large [5]. To our understanding these discrepancies are the consequence of dynamical correlations between colliding particles, that means correlations between the velocities of such particles, resulting from previous collisions in the system. A way to account for these correlations is by applying *mode coupling theories*. This can be done either within the framework of kinetic theory [6] or in more phenomenological ways, e.g. by using fluctuating hydrodynamics [7].

However, even for lower densities, where these corrections are not needed, a valid criticism remains that real molecules are not hard spheres, hence one would like to extend Enskog's theory so as to be applicable to more realistic potentials. An early try at this was made by Davis, Rice and Sengers (DRS) [8], who considered systems of particles interacting through a square-well potential. For this system all collisions are still instantaneous, so involving pairs of particles only, but a non-vanishing interparticle potential energy density is present. DRS designed a kinetic equation for the one-particle distribution function alone, which has the shortcoming of not satisfying conservation of energy. Recently Karkheck et al. [9] used the formal tool of maximization of entropy to close the hierarchy equations for the square-well system and thus derived a generalization of the DRS-equation. This consists of two coupled equations, for the one-particle distribution function and the potential energy respectively, and satisfies all conservation laws. An interesting feature of these equations is that out of equilibrium there are two different temperature fields [10], conjugated to kinetic and potential energy density respectively. This is a kinetic effect and does not play a major role in the hydrodynamic equations, which are of the usual form, with transport coefficients that can be derived explicitly from the kinetic equations. Only the bulk viscosity differs from that obtained in the DRS theory.

In the sequel I will briefly discuss the Enskog equation in its standard and revised form, next derive the square-well kinetic equations, discuss the H-theorem for these equations and compare the resulting transport coefficients to MD results. Next I will consider the eigenvalues and eigenfunctions of the corresponding linearized equations and make a comparison to the corresponding quantities for linearized Enskog theory and to MD results.

2   Kinetic theory of the square-well fluid

2.1   The Enskog equation

Before embarking on the kinetic theory for a square-well fluid I first want to briefly discuss the hard sphere model. At moderate to relatively high densities the time evolution in this model is described approximately by the Enskog equation. This is a closed kinetic equation for the one-particle distribution function $f^{(1)}(\vec{r}, \vec{v}, t)$ describing the probability density of finding at time t a particle with velocity $\vec{v}$ and position $\vec{r}$. As starting point for deriving the Enskog equation one can take the exact first $BBGKY$-hierarchy equation (see section 2.2), which for the hard sphere system assumes the form

$$\left(\frac{\partial}{\partial t} + \vec{v}_1 \cdot \frac{\partial}{\partial \vec{r}_1} + \frac{1}{m}\frac{\partial}{\partial \vec{v}_1} \cdot \vec{F}^{ext}(\vec{r}_1, \vec{v}_1, t)\right) f^{(1)}(\vec{r}_1, \vec{v}_1, t)$$

$$= \int d\vec{r}_2 \int d\vec{v}_2 \, \delta(r_{12} - \sigma) \mid \vec{v}_{12} \cdot \hat{r}_{12} \mid \left\{ \Theta(\vec{v}_{12} \cdot \vec{r}_{12}) f^{(2)}(\vec{r}_1, \vec{r}_2, \vec{v}_1^{\,*}, \vec{v}_2^{\,*}, t) \right.$$

$$\left. - \Theta(-\vec{v}_{12} \cdot \vec{r}_{12}) f^{(2)}(\vec{r}_1, \vec{r}_2, \vec{v}_1, \vec{v}_2, t) \right\} \tag{1}$$

Here $\vec{F}^{ext}$ is an external force field, which may depend on position and velocity as well as explicitly on time; $\Theta(x)$ is the unit step function, which equals zero for $x < 0$ and unity for $x \geq 0$; $\vec{v}_{12} = \vec{v}_1 - \vec{v}_2$ and $\vec{r}_{12} = \vec{r}_1 - \vec{r}_2$; $r_{12} = \mid \vec{r}_{12} \mid$ and $\hat{r}_{12} = \vec{r}_{12}/r_{12}$. Further $\vec{v}_1^{\,*}$ and $\vec{v}_2^{\,*}$ are the precollisional velocities giving rise to postcollisional velocities $\vec{v}_1$ and $\vec{v}_2$ in a collision with relative position $\vec{r}_{12}$. Their explicit forms follow from the conservation law $\vec{v}_1^{\,*} + \vec{v}_2^{\,*} = \vec{v}_1 + \vec{v}_2$ and the collision law $\vec{v}_{12}^{\,*} = \vec{v}_{12} - 2(\vec{v}_{12} \cdot \hat{r}_{12})\hat{r}_{12}$. Finally, $f^{(2)}(\vec{r}_1, \vec{r}_2, \vec{v}_1, \vec{v}_2, t)$ is the pair distribution function describing the probability density of finding two different particles with coordinates $\vec{r}_1, \vec{v}_1$ and $\vec{r}_2, \vec{v}_2$ respectively at time t. The Enskog equation is obtained by expressing this pair distribution function in terms of $f^{(1)}$ through the *closure relation*

$$\delta(r_{12} - \sigma)\Theta(-\vec{v}_{12} \cdot \vec{r}_{12}) f^{(2)}(\vec{r}_1, \vec{r}_2, \vec{v}_1, \vec{v}_2, t) \;=\; \delta(r_{12} - \sigma)\Theta(-\vec{v}_{12} \cdot \vec{r}_{12}) f^{(1)}(\vec{r}_1, \vec{v}_1, t)$$

$$f^{(1)}(\vec{r}_2, \vec{v}_2, t)\, g(\vec{r}_1, \vec{r}_2, t) \tag{2}$$

The prefactors $\delta(r_{12} - \sigma)\,\Theta(-\vec{v}_{12} \cdot \vec{r}_{12})$ have been added to emphasize that (2) is required to hold for precollisional configurations only. The function $g(\vec{r}_1, \vec{r}_2, t)$ is a purely positional pair correlation function, determined entirely by the density field $n(\vec{r}, t)$, which follows from $f^{(1)}$ as

$$n(\vec{r}, t) = \int d\vec{v}\, f^{(1)}(\vec{r}, \vec{v}, t), \tag{3}$$

hence g is a functional of the one-particle distribution function. For the precise form of this functional different choices are possible. In standard Enskog Theory (SET) it is defined as

$$g^{SET}(\vec{r}_1, \vec{r}_2, t) = g\left(r_{12} \mid n\left(\frac{\vec{r}_1 + \vec{r}_2}{2}, t\right)\right) \tag{4}$$

where $g(r \mid n)$ denotes the pair correlation function for two spheres at distance r in a uniform equilibrium system at density n.

In Revised Enskog Theory (RET) [11] on the other hand $g(\vec{r}_1, \vec{r}_2, t)$ is defined as the pair correlation function in a non-uniform equilibrium ensemble in which the average local density equals $n(\vec{r}, t)$ as given by[1] (3). This pair correlation function may be specified either by its Mayer expansion,

$$g^{RET}(\vec{r}_1, \vec{r}_2, t) = 1 + \int d\vec{r}_3\, f(r_{13})f(r_{23})n(\vec{r}_3, t) + \ldots, \tag{5}$$

with $f(r)$ the Mayer function of the hard sphere fluid, or by the functional relation [12]

$$n(\vec{r}_1)n(\vec{r}_2)g^{RET}(\vec{r}_1, \vec{r}_2) = \delta n(\vec{r}_1)/\delta(\beta\mu(\vec{r}_2)) - n(\vec{r}_1)\delta(\vec{r}_1 - \vec{r}_2) \tag{6}$$

---

[1]For physically realizable $f^{(1)}(\vec{r}, \vec{v}, t)$ one can always find an external potential in which the equilibrium density field is of the form $n(\vec{r}, t)$.

The Boltzmann equation for a hard sphere system is recovered by setting $g(\vec{r}_1, \vec{r}_2, t) = 1$ in (2) and by replacing $f^{(1)}(\vec{r}_2, \vec{v}_2, t)$ with $f^{(1)}(\vec{r}_1, \vec{v}_2, t)$. So there are two notable differences:

a. In the Enskog equation the probability for two hard spheres to be right next to each other (and therefore their probability to collide) is enhanced by the presence of other spheres. The enhancement factor is assumed to be $g(\vec{r}_1, \vec{r}_2, t)$, its equilibrium value as a functional of $n(\vec{r}, t)$.

b. In contrast to the Boltzmann equation the Enskog equation does not ignore the difference in position between the colliding particles. Therefore collisional transfer contributions to the transport of energy and momentum are taken into account, whereas they are absent from the Boltzmann equation.

The importance of this can be understood in the following way:

Collisional transfer is the instantaneous transport of momentum or energy from the center of one colliding sphere to that of the other in a binary collision. These transfers take place over distances proportional to σ. This is to be contrasted with convective transport, in which momentum and energy are transported as particles carry them along during free flight. If you follow the track of one particle you see that each collisional transfer in which the particle is involved, is followed by a convective transport over a distance of the order l_{mf}, the mean free path between collisions. Hence the ratio between the amounts of convective transport and collisional transfer is proportional to l_{mf}/σ. As a consequence convective transport dominates at low densities, where $l_{mf}/\sigma \gg 1$, whereas at high densities $l_{mf}/\sigma \ll 1$ and collisional transfer is the dominating mode of transport.

Now, why does one have the distinction between SET and RET? Basically this is for historical reasons. Enskog did not know how to define the pair correlation function for non-uniform systems and for symmetry reasons proposed [4] his prescription (4), which is invariant under the exchange of densities in $\vec{r}_1$ and $\vec{r}_2$, and coincides with the RET-prescription up to terms of order $(\nabla n)^2$. As a consequence, through the Navier-Stokes level (second order in the spatial gradient) SET and RET yield identical hydrodynamic equations and transport coefficients, but beyond this level differences show up.

In the presence of a non-uniform constant external potential the equilibrium distribution function is a stationary solution of (2) combined with (5) or (6) (RET), but not for (2) and (4) (SET) [13]. Furthermore, RET can be readily generalized to mixtures [14], while satisfying the requirements of irreversible thermodynamics, especially the validity of Onsager's reciprocal relations. This cannot be done for SET. Hence, I think RET definitely is preferable to SET, although neither theory is exact.

2.2 The square-well hierarchy equations

For a fluid of particles interacting through a continuous differentiable potential of the form

$$\Phi(\vec{r}_1 \ldots \vec{r}_N) = \sum_{i<j} \phi(r_{ij}) \tag{7}$$

the Liouville equation can be written in the form [14]:

$$\left\{ \frac{\partial}{\partial t} + \sum_{i=1}^{N} \left(\vec{v}_i \cdot \frac{\partial}{\partial \vec{r}_i} + \frac{1}{m} \frac{\partial}{\partial \vec{v}_i} \cdot \vec{F}^{ext}(\vec{r}_i, \vec{v}_i) \right) \right\} \rho(x^N, t)$$

$$= \sum_{i<j} \frac{1}{m} \frac{\partial \phi(r_{ij})}{\partial \vec{r}_{ij}} \cdot \left(\frac{\partial}{\partial \vec{v}_i} - \frac{\partial}{\partial \vec{v}_j} \right) \rho(x^N, t) \tag{8}$$

Here x^N stands for $x_1, x_2, \ldots x_N$ and x_i for $\vec{r}_i, \vec{v}_i$ and the density in phase space ρ is defined as a function of velocities rather than momenta. $\partial/\partial\vec{v}_i$ acts on ρ as well as on $\vec{F}^{ext}$. If one requires ρ to be symmetric under the exchange of particle indices, which is physically reasonable, the *BBGKY* hierarchy equations for the *s*-particle distribution functions are obtained by integrating (8) over all but s particles. All terms of the forms $\vec{v}_i \cdot \frac{\partial}{\partial \vec{r}_i} \rho$, $\frac{1}{m} \cdot \vec{F}_i \frac{\partial}{\partial \vec{v}_i} \rho$ and $\frac{1}{m} \frac{\partial \phi(r_{ij})}{\partial \vec{r}_{ij}} \cdot \frac{\partial}{\partial \vec{v}_i} \rho$, with $i > s$, vanish on performing these integrations. Introducing the *s*-particle distribution functions as

$$f^{(s)}(x^s, t) = N(N-1)\ldots(N-s+1) \int dx_{s+1}\ldots dx_N \rho(x^N, t) \tag{9}$$

one obtains the hierarchy equations as:

$$\left\{ \frac{\partial}{\partial t} + \sum_{i=1}^{s} \left[\left(\vec{v}_i \cdot \frac{\partial}{\partial \vec{r}_i} + \frac{1}{m} \frac{\partial}{\partial \vec{v}_i} \cdot \vec{F}^{ext}(\vec{r}_i, \vec{v}_i) \right) - \sum_{j=i+1}^{s} \frac{1}{m} \left(\frac{\partial \phi(r_{ij})}{\partial \vec{r}_{ij}} \cdot \left(\frac{\partial}{\partial \vec{v}_i} - \frac{\partial}{\partial \vec{v}_j} \right) \right) \right] \right\}$$

$$f^{(s)}(x^s, t) = \int dx_{s+1} \sum_{i=1}^{s} \frac{1}{m} \frac{\partial \phi(r_{is+1})}{\partial \vec{r}_{is+1}} \cdot \frac{\partial f^{(s+1)}(x^{s+1}, t)}{\partial \vec{v}_i} \tag{10}$$

This set of equations expresses the time derivatives of the *s*-particle distribution functions in terms of the $s + 1$-particle distribution functions and therefore is closed only at the N-particle level where one recovers the Liouville equation. In order to extract a tractable kinetic equation one has to find a closure relation, expressing a low order distribution function $f^{(s+1)}$ for the interacting configurations occurring on the right hand side of (10) in terms of the lower order distribution functions $f^{(1)} \ldots f^{(s)}$. For instance, the Boltzmann equation is obtained by expressing $f^{(2)}$ for collisional configurations as a product of $f^{(1)}$'s just preceding the collision [1].

For the square well potential (see Fig. 1),

$$\phi(r) = \begin{array}{ll} \infty & r < \sigma \\ -\epsilon & \sigma \leq r < R \\ 0 & R \leq r, \end{array} \tag{11}$$

$\partial\phi(r)/\partial\vec{r}$ is not well-defined. Instead of continuously the velocities of the particles change instantaneously at collisions, bot at the hard core and at the square-well edge. The Liouville equation is replaced by a so-called pseudo-Liouville equation, of the form [15]

$$\left\{ \frac{\partial}{\partial t} + \sum_{i=1}^{N} \vec{v}_i \cdot \frac{\partial}{\partial \vec{r}_i} + \frac{1}{m} \frac{\partial}{\partial \vec{v}_i} \cdot \vec{F}^{ext}(\vec{r}_i, \vec{v}_i) \right\} \rho(x^N, t) = \sum_{i<j} \overline{T}_{ij} \rho(x^N, t) \tag{12}$$

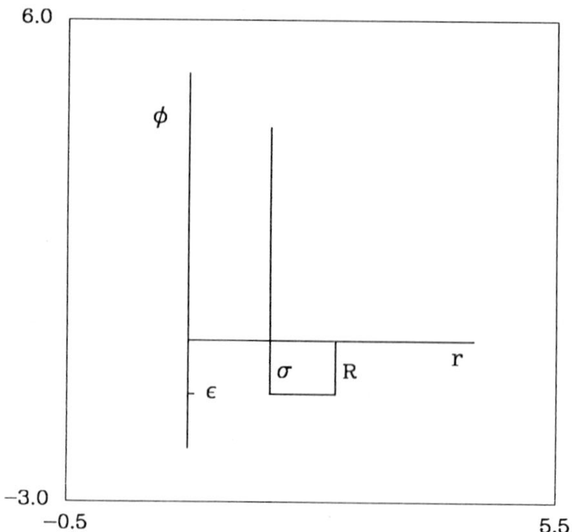

Figure 1: The square well potential

The *binary collision operator* $\overline{T}_{ij}$ consists of two parts: a real part which, if the conditions for a collision between particles i and j are satisfied, replaces $\vec{v}_i$ and $\vec{v}_j$ by their precollisional values, and a virtual part which, under the same conditions, leaves $\vec{v}_i$ and $\vec{v}_j$ unchanged[2]. The latter contribution may be considered to correct the free streaming contribution in case a collision between i and j takes place. For the square-well potential four different types of collisions may be distinguished (see Fig. 2): (1) a collision at the hard core, (2) a collision entering the square well, (3) a collision leaving the square well, and (4) a collision at the inner side of the square-well edge, which happens if the radial relative velocity is too small for the particles to escape from the well. All four types of collision contribute a real and a virtual term to $\overline{T}_{ij}$, which can be written as [9,16]

$$\overline{T}_{ij} = \sum_{p=1}^{4} \overline{T}_{ij}{}^{(p)}$$

$$\overline{T}_{ij}{}^{(p)} = \lim_{t \to 0+} S^0_{-t}\delta(r_{ij} - R_p)\,|\,\vec{v}_{ij} \cdot \hat{r}_{ij}\,|\,\left\{\Theta^r_p(\vec{v}_{ij} \cdot \hat{r}_{ij})b^{(p)}(\hat{r}_{ij}) - \Theta^v_p(\vec{v}_{ij} \cdot \hat{r}_{ij})\right\} S^0_{-t} \qquad (13)$$

Here, $R_1 = \sigma$ and $R_2 = R_3 = R_4 = R$. Further,

$$\Theta^r_1(x) = \Theta^v_1(-x) = \Theta^r_2(x) = \Theta(x);$$
$$\Theta^v_2(x) = \Theta(x - v_{esc}); \; \Theta^r_3(x) = \Theta(v_{esc} - x); \; \Theta^v_3(x) = \Theta(-x);$$
$$\Theta^r_4(x) = \Theta^v_4(-x) = \Theta(-x)\Theta(v_{esc} + x),$$

---

[2]This argument is somewhat subtle. The time evolution of the distribution functions is described by backward streaming. Therefore the conditions for, say, a hard core collision between particles i and j are $r_{ij} = \sigma$ and $\vec{v}^*_{ij} \cdot \hat{r}_{ij} < 0$. In a virtual collision $\vec{v}_{ij}$ does not change, so one has the condition $\vec{v}_{ij} \cdot \hat{r}_{ij} < 0$

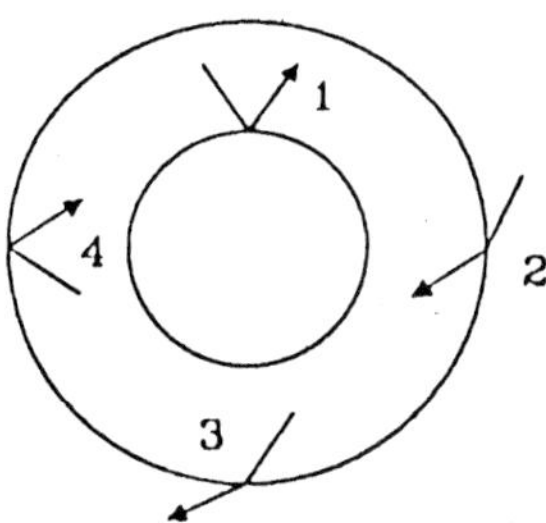

Figure 2: The four different types of collisions

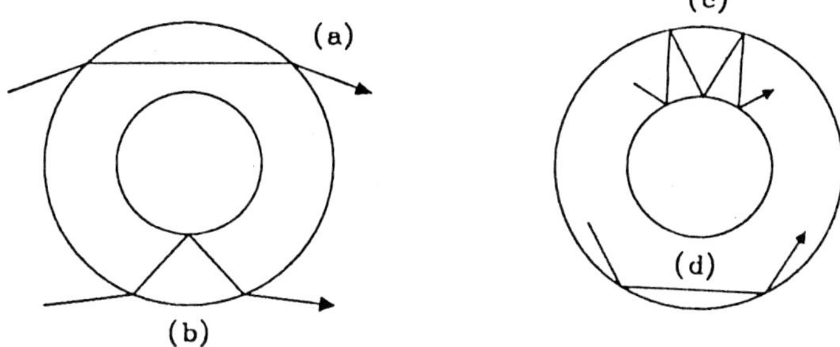

Figure 3: The four possible types of partial collision sequences for a square-well pair.

with $v_{esc} = (4\epsilon/m)^{1/2}$ the escape velocity from the square well. S^0_{-t} is the free streaming operator, which lets the particles stream backward without colliding during a time t. The insertion of the infinitesimal free streaming operators is needed to define the action of $\overline{T}_{ij}$ on functions that are discontinuous at $r_{ij} = \sigma$ or R. The operators $b^{(p)}$ describe the change in $\vec{v}_i$ and $\vec{v}_j$ under a collision of type p. They are most concisely specified by giving their action on the relative velocity $\vec{v}_{ij}$, as they all leave the center of mass velocity $\vec{V}_{ij} = (\vec{v}_i + \vec{v}_j)/2$ unchanged.

Specifically:

$$
\begin{aligned}
b^{(1)}(\hat{\sigma})\vec{v}_{ij} &= b^{(4)}(\hat{\sigma})\vec{v}_{ij} = \vec{v}_{ij} - 2(\vec{v}_{ij}\cdot\hat{\sigma})\hat{\sigma} \\
b^{(2)}(\hat{\sigma})\vec{v}_{ij} &= \vec{v}_{ij} + \hat{\sigma}\left\{\sqrt{(\vec{v}_{ij}\cdot\hat{\sigma})^2 + v^2_{esc}} - (\vec{v}_{ij}\cdot\hat{\sigma})\right\} \\
b^{(3)}(\hat{\sigma})\vec{v}_{ij} &= \vec{v}_{ij} - \hat{\sigma}\left\{\sqrt{(\vec{v}_{ij}\cdot\hat{\sigma})^2 - v^2_{esc}} + (\vec{v}_{ij}\cdot\hat{\sigma})\right\}
\end{aligned}
\tag{14}
$$

Notice that I have defined as a collision a single passage through the square-well edge or rebound from the hard core or the inside of the square-well edge. However, at low densities a full two-body collision consists of a sequence of either entering, hard-core and leaving collision (Fig. 3a) or just an entering and a leaving collision (Fig. 3b).

364　　　　　　　　　　　　　　　　　*H. van Beijeren*

In addition, there are bound states, characterized by either an alternating sequence of type 1 and type 4 collisions (Fig. 3c), or just a chain of type 4 collisions (Fig. 3d). The partial collisions of one full collision sequence are fully correlated: once the entering collision takes place the eventual hard core collision and the leaving collision are completely determined. At high density this is not true any more. As soon as the mean free path becomes small compared to the square-well width, the particle velocities in subsequent partial collisions between the same pair of particles become largely uncorrelated because of the large number of collisions with other particles happening in between. In the theory to be presented here, as well as in *DRS* theory, subsequent partial collisions will be assumed to possess no velocity correlations at all, hence the validity of these theories a priori will be restricted to relatively high densities.

The hierarchy equations for the square well fluid are obtained from (12), again by integrating over the coordinates of $N - s$ particles, with the result[3]

$$\left\{ \frac{\partial}{\partial t} + \sum_{i=1}^{s} \left[\vec{v}_i \cdot \frac{\partial}{\partial \vec{r}_i} + \frac{1}{m} \frac{\partial}{\partial \vec{v}_i} \cdot \vec{F}^{ext}(\vec{r}_i, \vec{v}_i) - \sum_{j=1+1}^{s} \overline{T}_{ij} \right] \right\} f^{(s)}(x^s, t)$$

$$= \sum_{i=1}^{s} \int dx_{s+1} \overline{T}_{is+1} f^{(s+1)}(x^{s+1}, t) \tag{15}$$

Note that the hard sphere fluid is contained in this as a special case, namely the limit $\epsilon \longrightarrow 0$, in which $\overline{T}^{(2)}$, $\overline{T}^{(3)}$ and $\overline{T}^{(4)}$ all become zero.

2.3　Closure of the hierarchy. Maximization of entropy

A practical tool for obtaining closure of the hierarchy equations is the procedure of *maximization of entropy*. For example, if one wants to derive a closed equation for $f^{(1)}(x, t)$ alone, as one has in RET, one constructs the ensemble ρ^{ME} that maximizes entropy, defined as

$$S = -k_B \int dx^N \rho(x^N) \log \left(N! h^{3N} \rho(x^N) \right), \tag{16}$$

under the constraint that

$$\int dx_1 \ldots dx_N \sum_{i=1}^{N} \delta(x - x_i) \rho^{ME}(x^N, t) = f^{(1)}(x, t) \tag{17}$$

This determines ρ^{ME} as a functional of $f^{(1)}$, from which by integration $f^{(2)}$ can be obtained as a functional of $f^{(1)}$ as well, so closure is attained at the level of the first hierarchy equation. This closure is by no means exact, since ρ^{ME} in certain ways only is a poor approximation of the real non-equilibrium ensemble. However, for the hard sphere system this scheme leads to *RET*, which has amply proved its value in spite of its approximative character, hence it seems reasonable using maximization of entropy also in other cases for obtaining closure relations.

For the square-well fluid closure at the level of $f^{(1)}$ alone is not satisfactory, because $f^{(1)}$ does not suffice to describe the energy density in the system. On the other hand, closure at the level of $f^{(2)}$, though possible in principle, gives rise to unwieldy kinetic equations. A

---

[3]To be precise one has to replace $\vec{v}_i \cdot \frac{\partial}{\partial \vec{r}_i}$ by $\lim_{t \to 0+} S^0_{-t} \vec{v}_i \cdot \frac{\partial}{\partial \vec{r}_i} S^0_{-t}$. The infinitesimal free streaming operators are inserted for similar reasons as in (14).

reasonable compromise seems extending the constraint on $f^{(1)}$ by a constraint on the local potential energy density $U(\vec{r}, t)$, which is defined as

$$
\begin{aligned}
U(\vec{r}, t) &= \int dx^N \sum_{i=1}^{N} \left[\delta(\vec{r} - \vec{r}_i) \tfrac{1}{2} \sum_{j \neq i} \phi(r_{ij}) \right] \rho(x^N, t) \\
&= \tfrac{1}{2} \int d\vec{v}_1 dx_2 \phi(r_{12}) f^{(2)}(x_1, x_2, t)
\end{aligned}
\tag{18}
$$

(so it contains only *internal* potential energy density).

This will give rise to a pair of coupled kinetic equations for $f^{(1)}$ and U respectively.

The ensemble ρ^{ME} can be constructed by the method of undetermined Lagrange multipliers, leading to [9]

$$
\begin{aligned}
\delta \int dx^N \rho \Bigg\{ \log \rho + \int dx \lambda(x, t) \sum_{i=1}^{N} \delta(x - x_i) \\
+ \int d\vec{r} \, \beta(\vec{r}, t) \sum_{i=1}^{N} \delta(\vec{r} - \vec{r}_i) \tfrac{1}{2} \sum_{j \neq i} \phi(r_{ij}) \Bigg\} = 0
\end{aligned}
\tag{19}
$$

where δ denotes a variation and $\lambda(x, t)$ and $\beta(\vec{r}, t)$ are the Lagrange multiplier fields conjugated to $f^{(1)}(x, t)$ and $U(\vec{r}, t)$ respectively. Working out (19) one obtains

$$
\int dx^N \delta\rho \left\{ \log \rho + 1 + \sum_i \left[\lambda(x_i, t) + \tfrac{1}{2}\beta(\vec{r}_i, t) \sum_{j \neq i} \phi(r_{ij}) \right] \right\} = 0,
\tag{20}
$$

where the variations $\delta\rho$ may be treated as independent of each other for all ρ. The solution of this equation is of the form

$$
\rho^{ME}(x^N, t) = \frac{1}{N!(h/m)^{3N} Z} \, \Pi_{i=1}^{N} e^{-\lambda(x_i, t)} \exp\left(- \sum_{i<j} \frac{\beta(\vec{r}_i, t) + \beta(\vec{r}_j, t)}{2} \phi(r_{ij}) \right)
\tag{21}
$$

Here Z is a normalization factor, depending on $\lambda(x, t)$ and $\beta(\vec{r}, t)$. The factor $N! h^{3N}$ has been put in for reasons of elegance; in the case of equilibrium, where β is uniform and $\lambda = \tfrac{1}{2}\beta m v^2$, Z reduces to the canonical partition function. The Lagrange multiplier fields $\lambda(x, t)$ and $\beta(\vec{r}, t)$ are determined by the constraints (17) and (18).

Notice that the only velocity dependence of ρ^{ME} is contained in the factors $\exp -\lambda(x_i, t)$, in other words: The velocities of different particles are entirely uncorrelated. For systems out of equilibrium this feature is unrealistic: already for low densities it is known that, even if velocities are uncorrelated initially, they cannot remain so. If the velocities of particles which are about to collide are uncorrelated, as assumed in Boltzmann's Stosszahlansatz, the velocities of particles that just have collided consequentially are strongly correlated. In fact, as we saw already, for closing the hierarchy equations only the two particle correlation function for precollisional configurations is needed and for these the factorization property of the velocity dependence may be considered a generalization of the Stosszahlansatz, which is often called the assumption of molecular chaos. The maximization of entropy procedure, as applied here, leads to a closure relation in which this assumption holds before each partial collision. Therefore, as argued before, it can only be used at fairly high densities.

From the factorization property for the velocities it follows that $f^{(1)}$ can be writen in the form

$$f^{(1)}(x,t) = \frac{n(\vec{r},t)e^{-\lambda(x,t)}}{\zeta(\vec{r},t)(h/m)^3} \tag{22}$$

where I introduced the local fugacity,

$$\zeta(\vec{r},t) = (h/m)^{-3} \int d\vec{v}\, e^{-\lambda(x,t)} \tag{23}$$

It is a functional of $n(\vec{r},t)$ and $\beta(\vec{r},t)$, which may be characterized by its Mayer expansion. The latter assumes the same form as in a non-uniform equilibrium system [17], but the Mayer functions now are defined as

$$f(\vec{r}_1,\vec{r}_2) = e^{-\beta_{12}\phi(r_{12})} - 1 \tag{24}$$

with $\beta_{12} = \frac{1}{2}(\beta(\vec{r}_1) + \beta(\vec{r}_2))$. Similarly one finds that $f^{(2)}$ can be expressed as

$$f^{(2)}(x_1,x_2,t) = f^{(1)}(x_1,t)f^{(1)}(x_2,t)g(\vec{r}_1,\vec{r}_2,t) \tag{25}$$

where g, through the elimination of ζ, can be expressed as a functional of $n(\vec{r},t)$ and $\beta(\vec{r},t)$. For g too the same Mayer expansion holds as in non-uniform equilibrium, with the Mayer function defined by (24). Through (18) the potential energy density $U(\vec{r},t)$ may now be expressed as a functional of $n(\vec{r},t)$ and $\beta(\vec{r},t)$ and inversion of this relation[4] will express $\beta(\vec{r},t)$ as a functional of $U(\vec{r},t)$ and $n(\vec{r},t)$.

Inserting (25) into the first hierarchy equation, one obtains the following kinetic equation for $f^{(1)}(x,t)$:

$$\left\{\frac{\partial}{\partial t} + \vec{v}_1 \cdot \frac{\partial}{\partial \vec{r}_1} + \frac{1}{m}\frac{\partial}{\partial \vec{v}_1} \cdot \vec{F}^{ext}(x_1)\right\} f^{(1)}(x_1,t) = \int dx_2 \overline{T}_{12}$$
$$f^{(1)}(x_1,t)f^{(1)}(x_2,t)g(\vec{r}_1,\vec{r}_2,t) \tag{26}$$

This equation looks very similar to the Enskog equation, but it still contains the independent field $\beta(\vec{r},t)$ in the pair correlation function. From the second hierarchy equation one obtains, using (18), (21) and the property

$$\int d\vec{v}_i d\vec{v}_j \overline{T}_{ij} \ldots = 0, \tag{27}$$

the kinetic equation for $U(\vec{r},t)$ as

$$\left(\frac{\partial}{\partial t} + \frac{\partial}{\partial \vec{r}_1} \cdot \vec{u}(r_1,t)\right) U(\vec{r}_1,t) = \frac{\epsilon}{2}\int d\vec{v}_1 \int dx_2 \mid \vec{v}_{12} \cdot \hat{r}_{12} \mid g(\vec{r}_1,\vec{r}_2,t)f^{(1)}(x_1,t)$$
$$f^{(1)}(x_2,t)\left[\Theta(\vec{v}_{12}\cdot\hat{r}_{12} - v_{esc})\delta(r_{12} - R^-) - \Theta(-\vec{v}_{12}\cdot\hat{r}_{12})\delta(r_{12} - R^+)\right] \tag{28}$$

where the local average velocity $\vec{u}(\vec{r},t)$ is defined through

$$\int d\vec{v}\, \vec{v}\, f^{(1)}(\vec{r},\vec{v},t) = n(\vec{r},t)\vec{u}(\vec{r},t) \tag{29}$$

---

[4]Alternatively one has the option of switching to $\beta(\vec{r},t)$ and $f^{(1)}(x,t)$ as independent variables instead of $U(\vec{r},t)$ and $f^{(1)}(x,t)$.

The δ-functions $\delta(r_{12} - R^-)$ and $\delta(r_{12} - R^+)$ imply that the pair correlation function has to be evaluated just inside respectively outside the square-well edge. This distinction is important because this function exhibits a jump at $r_{12} = R$: from (21) and (11) it follows that

$$g(\vec{r}_1, \vec{r}_1 + R^-\hat{r}_{21}, t) = e^{\beta_{12}\epsilon}g(\vec{r}_1, \vec{r}_1 + R^+\hat{r}_{21}, t) \tag{30}$$

In conclusion, the kinetic equations for the square-well fluid consist of (26) and (28) together with the equations expressing g and U as functionals of n and β.

2.4   The H-theorem

One of the attractive features of the maximization of entropy procedure is that it almost automatically provides one with an H-theorem stating the monotonic increase, or at least the non-decrease of entropy. The general argument goes as follows: Consider an ensemble of systems that at time zero is of the form $\rho^{ME}(x^N)$. Its entropy is given initially by (16) with $\rho = \rho^{ME}$. After an infinitesimal time δt the exact time evolution will have evolved ρ to $\rho(x^N, \delta t)$, which in general will not be of the form ρ^{ME} any more. Since the closure relations deriving from approximating $\rho(x^N, t)$ by $\rho^{ME}(x^N, t)$ are satisfied exactly at $t = 0$ (as $\rho(x^N, 0) = \rho^{ME}(x^N, 0)$) the kinetic equations deriving from these are also exact at $t = 0$, hence $f^{(1)}(x, \delta t)$ and $U(\vec{r}, \delta t)$ (or in general those independent functions that determine the constraints on ρ^{ME}) as derived from the kinetic equations, coincide with the same functions as obtained from the ensemble $\rho(x^N, \delta t)$. As a consequence of Liouville's theorem the entropy obtained from $\rho(x^N, \delta t)$ according to (17) is the same as the entropy at $t = 0$. However, since $\rho^{ME}(x^N, \delta t)$ has the same $f^{(1)}(x, \delta t)$ and $U(\vec{r}, t)$ as $\rho(x^N, \delta t)$ and at the same time maximizes entropy for those $f^{(1)}$ and U, we may conclude that the entropy $S^{ME}(\delta t)$ that follows from it, satisfies the inequality

$$S^{ME}(\delta t) \geq S(0) = S^{ME}(0)$$

As this inequality holds for arbitrary initial $f^{(1)}$ and U, we may conclude that the kinetic equations give rise to a monotonically non-decreasing entropy.

Harder to prove is a strong H-theorem, i.e., a theorem stating that the entropy strictly increases except under very special conditions. For the square-well system this proof has been given in ref. [18]. Although straightforward it is technically rather involved, so I won't repeat it here. The outcome is that dS/dt is strictly positive unless $f^{(1)}$ is of the Maxwellian form

$$f^{(1)}(x) = n(\vec{r}) \left(\frac{m}{2\pi k_B T}\right)^{3/2} \exp -m \mid \vec{v} - \vec{u} \mid^2 /2k_B T \tag{31a}$$

and at the same time

$$\beta = \frac{1}{k_B T} \tag{31b}$$

For equilibrium solutions of the kinetic equations (26) and (28), satisfying $\partial f^{(1)}/\partial t = \partial U/\partial t = 0$, one has to require in addition that either there is no external potential and $n(\vec{r})$ is uniform, or, if there is an external potential such that $\vec{F}^{ext}(\vec{r}, \vec{v}) = -d\Phi(\vec{r})/d\vec{r}$, $\vec{u} = 0$ and $n(\vec{r})$ satisfies the equation

$$\frac{\partial}{\partial \vec{r}_1} \left[\log n(\vec{r}_1) \; + \; \beta\Phi(\vec{r}_1)\right] = \int d\vec{r}_2 \hat{r}_{12}\left[e^{\beta\epsilon}\delta(r_{12} - \sigma^+) + (1 - e^{\beta\epsilon})\right.$$

$$\left.\delta(r_{12} - R^+)\right]n(\vec{r}_2)g(\vec{r}_1, \vec{r}_2), \tag{32}$$

which can be interpreted as the condition of uniform chemical potential.

2.5 Hydrodynamic equations. Kinetic and potential temperature

From the kinetic equations (26) and (28) the hydrodynamic equations may be derived by applying the Chapman–Enskog procedure. This requires the expansion of $f^{(1)}(x, t)$ and $\beta(\vec{r}, t)$ as

$$f^{(1)}(x, t) = f_0^{(1)}(x, t) + f_1^{(1)}(x, t) + \ldots \tag{33a}$$

$$\beta(x, t) = \beta_0(x, t) + \beta_1(x, t) + \ldots \tag{33b}$$

where $f_n^{(1)}$ and β_n are the quantities that are solved for in the n-th step of the Chapman-Enskog procedure. The kinetic equations (26) and (28) have to be rewritten somewhat. Setting the external force in (26) equal to zero for simplification I obtain

$$\left(\frac{\partial}{\partial t} + \vec{v}_1 \cdot \frac{\partial}{\partial \vec{r}_1}\right) f^{(1)}(x_1, t) - \Delta J\left(\left\{f^{(1)}(x, t), \beta(\vec{r}, t)\right\}\right) = J_0\left(f^{(1)}(x_1, t), \beta(\vec{r}_1, t)\right) \tag{34}$$

$$\partial U(\vec{r}_1, t)/\partial t + \partial/\partial \vec{r}_1 \cdot (\vec{u}(\vec{r}_1, t)U(\vec{r}_1, t)) \; - \; \Delta K\left(\left\{\beta(\vec{r}, t), n(\vec{r}, t)\right\}\right)$$

$$= \; K_0\left(\beta(\vec{r}_1, t), n(\vec{r}_1, t)\right) \tag{35}$$

Here $J_0\left(f^{(1)}(x_1, t), \beta(\vec{r}_1, t)\right)$ and $K_0\left(\beta(\vec{r}_1, t), n(\vec{r}_1, t)\right)$ are defined as the right hand side of (26) and (28) respectively, in which $f^{(1)}(x_2, t)$ is replaced by $f^{(1)}(\vec{r}_1, \vec{v}_2, t)$ and g by the uniform equilibrium pair correlation function at density $n(\vec{r}_1, t)$ and temperature $(k_B\beta(\vec{r}_1, t))^{-1}$. ΔJ and ΔK are the corrections to these approximations, which can be expanded in terms of the spatial derivatives of $f^{(1)}(x_1, t)$ and $\beta(\vec{r}_1, t)$. The curly brackets denote functional dependence on the enclosed fields. In the hydrodynamic regime, where gradients are felt over macroscopic distances only and time changes occur on macroscopic time scales, the left hand sides of (34) and (35) typically are very small compared to the right hand sides. Therefore, in the zeroth order of the Chapman-Enskog expansion the left hand sides are neglected entirely and from the right hand sides one obtains the equations

$$J_0\left(f_0^{(1)}(x_1, t), \beta_0(\vec{r}_1, t)\right) = 0$$

$$K_0\left(\beta_0(\vec{r}_1, t), n(\vec{r}_1, t)\right) = 0 \tag{36}$$

For each $\beta_0(\vec{r}, t)$ these equations are satisfied by the local Maxwellian distribution

$$f_0^{(1)}(x, t) = n(\vec{r}, t)\left(\beta_0(\vec{r}, t)m/2\pi\right)^{3/2} \exp -\beta_0(\vec{r}, t)m \mid \vec{v} - \vec{u}(\vec{r}, t) \mid^2 /2 \tag{37}$$

An essential requirement of the Chapman-Enskog procedure is that, up to the level of approximation attained, $f_0^{(1)}$ describes the density $n(\vec{r}, t)$ and the local velocity $\vec{u}(\vec{r}, t)$ exactly. Moreover, together with the equation of state

$$U_0(\vec{r}, t) = U_0\left(\beta_0(\vec{r}, t), n(\vec{r}, t)\right), \tag{38}$$

expressing the local potential energy density as the same function of local temperature and particle density as in uniform equilibrium, it exactly describes the local energy density,

$$
\begin{aligned}
e(\vec{r}, t) &= U_0(\vec{r}, t) + \int d\vec{v}\tfrac{1}{2}m \mid \vec{v} - \vec{u}(\vec{r}, t) \mid^2 f^{(1)}(x, t) \\
&= e(\beta_0(\vec{r}, t), n(\vec{r}, t)) \tag{39}
\end{aligned}
$$

In other words, the functions $f_1^{(1)}, f_2^{(1)}$ etc. together with β_1, β_2 etc., do not contribute to the local densities of particles, momentum and *total* energy. Notice that the pair correlation function g_0 occurring in J_0 and K_0 in (36) is to be evaluated as a function of $n(\vec{r}, t)$ and $\beta_0(\vec{r}, t)$ and thereby differs from g, occurring in J_0 and K_0 in (34) and (35), which is the same function, however of the arguments $n(\vec{r}, t)$ and $\beta(\vec{r}, t)$.

The terms on the left hand sides of (34) and (35) acting upon f_0 and β_0, together with the right hand side terms that are linear in $f_1^{(1)}$ and β_1, yield the first order equations. These can be written in the form

$$
\begin{aligned}
\left\{ \left(\frac{\partial}{\partial t}\right)_1 + \vec{v}_1 \cdot \frac{\partial}{\partial \vec{r}_1} \right\} f_0^{(1)}(x_1, t) &- \Delta J_1\left(\{f_0^{(1)}, \beta_0\}\right) \\
&= \Lambda(\vec{r}_1, t) f_1^{(1)}(x_1, t) + \epsilon f_\beta^{(1)}(x_1, t)\beta_1(\vec{r}_1, t) \tag{40}
\end{aligned}
$$

$$
\left(\frac{\partial}{\partial t}\right)_1 U_0(\vec{r}_1, t) + \frac{\partial}{\partial \vec{r}_1} \cdot (\vec{u}(\vec{r}_1, t)U_0(\vec{r}_1, t)) = \frac{\partial K_0\left(\beta_0(\vec{r}_1, t), n(\vec{r}_1, t)\right)}{\partial \beta_0(\vec{r}_1, t)}\beta_1(\vec{r}_1, t) \tag{41}
$$

with

$$
\begin{aligned}
f_\beta^{(1)}(x_1, t) &= \epsilon^{-1}\partial J_0\left(f_0^{(1)}(x_1, t), \beta_0(\vec{r}_1, t)\right) / \partial \beta_0(\vec{r}_1, t) \\
&= g_0(R^+, t) f_0^{(1)}(x_1, t) \int dx_2 \mid \vec{v}_{12} \cdot \hat{r}_{12} \mid f_0^{(1)}(\vec{r}_1, \vec{v}_2, t)\delta(r_{12} - R) \\
&\quad \left[\Theta(\vec{v}_{12} \cdot \vec{r}_{12}) - e^{\beta_0(\vec{r}_1, t)\epsilon}\Theta(-\vec{v}_{12} \cdot \vec{r}_{12} - v_{esc})\right] \tag{42}
\end{aligned}
$$

In (40) ΔJ_1 is obtained by expanding the fields $n(\vec{r}, t)$, $\vec{u}(\vec{r}, t)$ and $\beta(\vec{r}, t)$ around $\vec{r}_1$ and retaining only those terms that are linear in the gradients of these fields. The corresponding term K_1 in (41) vanishes by virtue of symmetry properties. Λ is obtained by substituting (33a) and (33b) for the arguments of J_0 in (34) and keeping only those terms that are linear in $f_1^{(1)}$, while β is replaced by β_0. The terms that are linear in β_1 while $f^{(1)}$ is replaced by $f_0^{(1)}$ are given explicitly as the last term in (40). The index 1 was attached to $\partial/\partial t$ to remind of the fact that higher order terms in the expansion scheme yield additional contributions to the time derivatives of $f_0^{(1)}$ and U_0 [18].

As in the solution of the Boltzmann equation the time derivatives in (40) and (41) can be eliminated through the imposition of solvability conditions. Applying the operations $\int d\vec{v}_1$ or $\int d\vec{v}_1 \vec{v}_1$ to the right hand side of (40) yields zero, hence the same must hold for applying them

370   *H. van Beijeren*

to the left hand side. Seen from a physical viewpoint these solvability conditions yield the hydrodynamic equations for particle and momentum density. Applying the operations $\int d\vec{v}_1$ and $\int d\vec{v}_1 \, \vec{v}_1$ to the first hierarchy equation (see (15)) one obtains the exact but non-closed equations of change,

$$\partial n(\vec{r},t)/\partial t + \partial/\partial \vec{r} \cdot \{n(\vec{r},t)\vec{u}(\vec{r},t)\} = 0 \tag{43}$$

$$\begin{aligned} \partial/\partial t \,\{n(\vec{r},t)\vec{u}(\vec{r},t)\} \; &+ \; \partial/\partial \vec{r} \cdot \{n(\vec{r},t)\vec{u}(\vec{r},t)\vec{u}(\vec{r},t)\} \\ &= \; -\partial/\partial \vec{r} \cdot \int d\vec{v}\, \vec{V}\, \vec{V}\, f^{(1)}(x,t) + \int dx_1 \int dx_2 \vec{v}_1 \overline{T}_{12} \\ & \qquad f^{(2)}(x_1,x_2,t)\delta(\vec{r}-\vec{r}_1) = 0 \end{aligned} \tag{44}$$

where I used again the property $\int d\vec{v}_1 \int d\vec{v}_2 \overline{T}_{12}\ldots = 0$ (eq. (27)) and I introduced the peculiar velocity $\vec{V} = \vec{v} - \vec{u}(\vec{r},t)$. The law of momentum conservation leading to the identity $\int d\vec{v}_1 \int d\vec{v}_2 \,(\vec{v}_2 + \vec{v}_2)\, \overline{T}_{12}\ldots = 0$, can be used to cast the last term in (44) into the form of a divergence:

$$\begin{aligned} &\int dx_1 \int dx_2 \vec{v}_1 \overline{T}_{12} f^{(2)}(x_1,x_2,t)\delta(\vec{r}-\vec{r}_1) \\ &= \tfrac{1}{2}\int dx_1 \int dx_2 (\vec{v}_1 - \vec{v}_2)\overline{T}_{12} f^{(2)}(x_1,x_2,t)\delta(\vec{r}-\vec{r}_1) \\ &= \tfrac{1}{2}\int_0^1 d\lambda \tfrac{\partial}{\partial\lambda} \int d\vec{v}_1 \int dx_2 \vec{v}_1 \overline{T}_{12} f^{(2)}(\vec{r}+(\lambda-1)\vec{r}_{21},\vec{r}+\lambda\vec{r}_{21},\vec{v}_1,\vec{v}_2,t) \\ &= \tfrac{1}{4}\tfrac{\partial}{\partial\vec{r}} \cdot \int_0^1 d\lambda \int d\vec{v}_1 \int dx_2 \vec{r}_{21}\vec{v}_{12}\overline{T}_{12} f^{(2)}(\vec{r}+(\lambda-1)\vec{r}_{21},\vec{r}+\lambda\vec{r}_{21},\vec{v}_1,\vec{v}_2,t) \end{aligned} \tag{45}$$

From (45) and (43) it follows that the equation of state for the local velocity may be written in the form

$$n(\vec{r},t)\partial\vec{u}(\vec{r},t)/\partial t + n(\vec{r},t)\vec{u}(\vec{r},t) \cdot \partial/\partial \vec{r}\, \vec{u}(\vec{r},t) = -\partial/\partial \vec{r} \cdot \overset{\leftrightarrow}{P}(\vec{r},t)/m \tag{46}$$

with the local stress tensor given as

$$\begin{aligned} \overset{\leftrightarrow}{P}(\vec{r},t) \; = \; &\int d\vec{v}\, m\, \vec{V}\, \vec{V} f^{(1)}(x,t) - \frac{m}{4}\int_0^1 d\lambda \int d\vec{v}_1 \int dx_2 \vec{r}_{21}\vec{v}_{12}\overline{T}_{12} \\ &f^{(2)}(\vec{r}+(\lambda-1)\vec{r}_{21},\vec{r}+\lambda\vec{r}_{21},\vec{v}_1,\vec{v}_2,t) \end{aligned} \tag{47}$$

If one expresses $f^{(2)}$ in terms of $f^{(1)}$ and β, and for these quantities substitutes the Chapman-Enskog solution, (46) and (47) reduce to the hydrodynamic equations for the velocity field. The zeroth order form of $\overset{\leftrightarrow}{P}$, required to eliminate $(\partial\vec{u}/\partial t)_1$ from (40), reads

$$\begin{aligned} \overset{\leftrightarrow}{P}_0(\vec{r},t) \; = \; &\left[n(\vec{r},t)/\beta_0(\vec{r},t) + \frac{1}{12}m \int dx_1 \int dx_2 (\vec{r}_{12} \cdot \vec{v}_{12})\overline{T}_{12} \right. \\ &\left. f_0^{(1)}(x_1 t)f_0^{(1)}(\vec{r}_1,\vec{v}_2,t)g_0(\vec{r}_1,\vec{r}_2,t)\delta(\vec{r}-\vec{r}_1) \right] \mathbf{1} \equiv p_0(\vec{r}_1,t)\mathbf{1} \end{aligned} \tag{48}$$

where $\mathbf{1}$ denotes the unit dyadic tensor.

The equation of change for the energy density is obtained by applying $\int d\vec{v}\tfrac{1}{2}mv^2$ to both sides of the first hierarchy equation and adding the equation for the rate of change of potential energy density. Using conservation of energy in the form

$$\int d\vec{v}_1 \int d\vec{v}_2 \left\{ \tfrac{1}{2}mv_1^2 + \tfrac{1}{2}mv_2^2 + \phi(r_{12}) \right\} \overline{T}_{12}\ldots = 0 \tag{49}$$

one finds this equation of change as

$$\frac{\partial}{\partial t}\left\{e(\vec{r},t) + \tfrac{1}{2}\rho(\vec{r},t)u^2(\vec{r},t)\right\} + \partial/\partial\vec{r}\cdot\left\{\vec{u}(\vec{r},t)e(\vec{r},t) + \tfrac{1}{2}u(\vec{r},t)\rho(\vec{r},t)u^2(\vec{r},t)\right\}$$
$$= -(\partial/\partial\vec{r})\cdot\left\{\vec{J}^e(\vec{r},t) + \overset{\leftrightarrow}{P}(\vec{r},t)\cdot\vec{u}(\vec{r},t)\right\} \tag{50}$$

where $\rho(\vec{r},t) = mn(\vec{r},t)$.

Here the energy current density is defined as

$$\vec{J}^e(\vec{r},t) = \int d\vec{v}\,\vec{V}\,\tfrac{1}{2}mV^2 f^{(1)}(x,t) - \tfrac{1}{4}\int_0^1 d\lambda\int d\vec{v}_1\int dx_2 \vec{r}_{21}\tfrac{1}{2}m(V_1^2 - V_2^2)$$
$$\overline{T}_{12}f^{(2)}(\vec{r} + (\lambda-1)\vec{r}_{21}, \vec{r} + \lambda\vec{r}_{21}, \vec{v}_1, \vec{v}_2, t) \tag{51}$$

Introducing the local energy per unit of mass, $\hat{e}(\vec{r},t) = e(\vec{r},t)/\rho(\vec{r},t)$, and using (43) and (46), one may simplify (50) somewhat to

$$\rho(\vec{r},t)\left\{\partial/\partial t + \vec{u}(\vec{r},t)\cdot\partial/\partial\vec{r}\right\}\hat{e}(\vec{r},t) = -\partial/\partial\vec{r}\cdot\vec{J}^e(\vec{r},t) - \overset{\leftrightarrow}{P}(\vec{r},t):\partial/\partial\vec{r}\,\vec{u}(\vec{r},t) \tag{52}$$

where : denotes the contraction of two dyadic tensors, $\overset{\leftrightarrow}{A}:\overset{\leftrightarrow}{B} = \sum_{\alpha\beta} A_{\alpha\beta}B_{\beta\alpha}$. Again, substitution of the closure relation for $f^{(2)}$ into (47) and (51) and employing the Chapman-Enskog type of solution for $f^{(1)}$ and β leads to the hydrodynamic equation for the energy density. The zeroth order solution, needed to eliminate $(\partial\beta_0/\partial t)_1$ in (40), is obtained by setting $\vec{J}_0^e = 0$ and substituting $\overset{\leftrightarrow}{P}_0$ for $\overset{\leftrightarrow}{P}$. Note that under these approximations (43), (46) and (52) reduce to the Euler equations.

Next $f_1^{(1)}$ formally can be solved for from (40) as

$$f_1^{(1)}(x,t) = \Lambda^{-1}(\vec{r},t)\Big[\left\{(\partial/\partial t)_1 + \vec{v}\cdot\partial/\partial\vec{r}\right\}f_0^{(1)}(x,t)$$
$$- \Delta J_1\left(\left\{f_0^{(1)}(x,t),\beta_0(\vec{r},t)\right\}\right) - \epsilon f_\beta^{(1)}(x,t)\beta_1(\vec{r},t)\Big] \tag{53}$$

From the isotropy of Λ in $\vec{V}$ and the form (37) of $f_0^{(1)}$ it follows that $f_1^{(1)}$ is of the general form

$$f_1^{(1)}(x,t) = A(V^2)(\vec{V}\,\vec{V} - 1/3\,V^2\mathbf{1}):\partial/\partial\vec{r}\,\vec{u}(\vec{r},t)$$
$$+ B(V^2)\vec{V}\cdot\partial\log\beta(\vec{r},t)/\partial\vec{r} + H(V^2)\partial/\partial\vec{r}\cdot\vec{u}(\vec{r},t)$$
$$+ K(V^2)\epsilon\beta_1(\vec{r},t) \tag{54}$$

with A, B, H and K implicitly depending on $n(\vec{r},t)$ and $\beta(\vec{r},t)$. Please notice that, in contrast to the case of the Enskog equation, the function $H(V^2)$ is non-zero because $\tfrac{1}{2}mV^2$ is not a zero eigenfunction of $\Lambda(\vec{r},t)$.

The value of $\beta_1(\vec{r},t)$ cannot be obtained by solving (41), because the solvability conditions on (40) and (41) only determine $(\partial/\partial t)_1\,e(\vec{r},t)$, but not $(\partial/\partial t)_1\,U(\vec{r},t)$. Instead $\beta_1(\vec{r},t)$ follows from the condition $e_1(\vec{r},t) = 0$, leading to

$$\int d\vec{v}\tfrac{1}{2}mV^2\left\{H(V^2)\partial/\partial\vec{r}\cdot\vec{u}(\vec{r},t) + K(V^2)\epsilon\beta_1(\vec{r},t)\right\} + \frac{\partial U_0(\vec{r},t)}{\partial\beta_0(\vec{r},t)}\beta_1(\vec{r},t) = 0 \tag{55}$$

or

$$\beta_1(\vec{r},t) = -\left[\frac{\partial U_0(\vec{r},t)}{\partial\beta_0(\vec{r},t)} + \epsilon\int d\vec{V}\tfrac{1}{2}mV^2 K(V^2)\right]^{-1}\int d\vec{v}\tfrac{1}{2}mV^2 H(V^2)\partial/\partial\vec{r}\cdot\vec{u}(\vec{r},t) \tag{56}$$

Notice that corrections to $U(\vec{r}, t)$ resulting from the non-local dependence of $n(\vec{r}_2, t)\, g_0(\vec{r}_1, \vec{r}_2, t)$ on $n(\vec{r}, t)$ and $\beta_0(\vec{r}, t)$, do not enter at first order in the spatial gradient.

From (54) and (56) it is seen that the potential energy temperature

$$T^P(\vec{r}, t) = (k_B \beta(\vec{r}, t))^{-1} \tag{57}$$

and the kinetic energy temperature

$$T^K(\vec{r}, t) = k_B^{-1} \int d\vec{r} \tfrac{1}{2} m V^2 f^{(1)}(\vec{r}, \vec{v}, t) \tag{58}$$

are not identical any more, as they were at the zeroth level of approximation.

The full non-equilibrium temperature $T(\vec{r}, t)$ either may be defined through (38) and (39), i.e.

$$e(\vec{r}, t) = 3/2 n(\vec{r}, t) k_B T(\vec{r}, t) + U(T(\vec{r}, t), n(\vec{r}, t)) \tag{59}$$

where U again is a simple function of its arguments, or by the requirement that to zeroth order ρ^{ME}, as defined in (21), already determines $e(\vec{r}, t)$ correctly. This is equivalent to replacing (59) by a non-local equation of state of the form

$$e(\vec{r}, t) = 3/2 n(\vec{r}, t) k_B T(\vec{r}, t) + U\left(\{T(\vec{r}, t), n(\vec{r}, t)\}\right) \tag{60}$$

There seems to be no compelling argument for preferring either (59) or (60); the definition of temperature in non-equilibrium systems to a certain degree just seems to be a matter of convention. In addition, as is well known and is seen here again, different contributions to the energy density may produce different temperature fields [19].

The Navier-Stokes hydrodynamic equations are obtained by deriving the solvability conditions for $f_2^{(2)}$ and adding these to the Euler equations obtained already. It suffices to give the first order contribution to the stress tensor and the energy current. These are found to be of the form

$$\begin{aligned}
P_{1\alpha\beta}(\vec{r}, t) &= -\eta(\vec{r}, t)\left\{\partial u_\beta(\vec{r}, t)/\partial x_\alpha + \partial u_\alpha(\vec{r}, t)/\partial x_\beta\right. \\
&\quad \left. - 2/3\delta_{\alpha\beta}\partial/\partial\vec{r}\cdot\vec{u}(\vec{r}, t)\right\} - \delta_{\alpha\beta}\kappa(\vec{r}, t)\partial/\partial\vec{r}\cdot\vec{u}(\vec{r}, t)
\end{aligned} \tag{61}$$

$$\vec{J}_1^e(\vec{r}, t) = \lambda(\vec{r}, t)\,(k_B\beta_0(\vec{r}, t))^{-1}\,\partial/\partial\vec{r}\cdot\log\beta_0(\vec{r}, t) \tag{62}$$

The shear viscosity η, the bulk viscosity κ and the coefficient of heat conduction λ can be expressed as

$$\begin{aligned}
\eta(\vec{r}, t) &= \frac{-1}{10}\left[\int d\vec{v}_1\left\{m(\vec{V}_1\vec{V}_1 - \tfrac{1}{3}\vec{V}_1^2\mathbf{1}) - \tfrac{1}{2}\int dx_2\right.\right. \\
&\quad \left. m(\vec{V}_{12}\vec{r}_{12} - \tfrac{1}{3}(\vec{v}_{12}\cdot\vec{r}_{12})\mathbf{1})\overline{T}_{12}\, g_0(\vec{r}_1, \vec{r}_2, t) f_0^{(1)}(\vec{r}_1, \vec{v}_2, t)\right\} \\
&\quad A(V_1^2)(\vec{V}_1\vec{V}_1 - \tfrac{1}{3}V_1^2\mathbf{1}) + \frac{1}{12}\int d\vec{v}_1\int dx_2 \\
&\quad \left. m(\vec{v}_{12}\cdot\vec{r}_{12})\overline{T}_{12}\, g_0(\vec{r}_1, \vec{r}_2, t) f_0^{(1)}(x, t) f_0^{(1)}(\vec{r}_1, \vec{v}_2, t) m(\vec{v}_{12}\cdot\vec{r}_{12})\right]
\end{aligned} \tag{63}$$

$$\kappa(\vec{r},t) \;=\; \frac{-1}{9}\Bigg[\int d\vec{v}_1 \Big\{ mV_1^2 - 3\beta_0^{-1}(\vec{r}_1,t) - \tfrac{1}{2}\int dx_2(\vec{v}_{12}\cdot\vec{r}_{12}) $$

$$\overline{T}_{12}\,g_0(\vec{r}_1,\vec{r}_2,t)f_0^{(1)}(\vec{r}_1,\vec{v}_2,t)\Big\}\Big\{H(V_1^2)+$$

$$+\epsilon K(V_1^2)\int d\vec{v}_1\,\tfrac{1}{2}mV_1^2\,H(V_1^2)\Big[\frac{\partial U_0(\vec{r}_1,t)}{\partial\beta_0(\vec{r}_1,t)} + \epsilon\int d\vec{v}\,\tfrac{1}{2}mV^2 K(V^2)\Big]^{-1}\Big\}$$

$$\frac{1}{8}\int d\vec{v}_1\int dx_2\,m(\vec{v}_{12}\cdot\vec{r}_{12})\overline{T}_{12}\,g_0(\vec{r}_1,\vec{r}_2,t)f_0^{(1)}(x_1,t)f_0^{(1)}(\vec{r}_1,\vec{v}_2,t)$$

$$m(\vec{v}_{12}\cdot\vec{r}_{12})\Bigg] \tag{64}$$

$$\lambda(\vec{r}_1,t) \;=\; \frac{k_B}{3}\Bigg[\int d\vec{v}_1\Big\{\Big(\tfrac{1}{2}\beta_0(\vec{r},t)mV_1^2 - 5/2\Big)\vec{V}_1 - \tfrac{1}{4}\int dx_2\beta_0(\vec{r}_1,t)$$

$$m(V_2^2 - V_1^2)\vec{r}_{21}\,\overline{T}_{12}\,g_0(\vec{r}_1,\vec{r}_2,t)f_0^{(1)}(\vec{r}_1,\vec{v}_2,t)\Big\}\cdot B(V_1^2)\vec{V}_1$$

$$-\frac{1}{8}\int d\vec{v}_1\int dx_2\,\beta_0^2(\vec{r}_1,t)m(V_1^2 - V_2^2)\overline{T}_{12}\,m(V_1^2 - V_2^2)r_{12}^2 f_0^{(1)}(x_1,t)$$

$$f_0^{(1)}(\vec{r}_1,\vec{v}_2,t)\Bigg] \tag{65}$$

where the functions A, B, H and K are defined through the relations

$$A(V^2)(\vec{V}\vec{V} - 1/3)V^2\mathbf{1}) = \Lambda^{-1}(\vec{r}_1,t)\beta_0(\vec{r}_1,t)\Big\{m(\vec{V}\vec{V} - \tfrac{1}{3}V^2\mathbf{1})-$$

$$1/2\int dx_2\,\overline{T}_{12}\,g_0(\vec{r}_1,\vec{r}_2,t)f_0^{(1)}(\vec{r}_1,\vec{v}_2,t)\,m(\vec{r}_{12}\vec{v}_{12} - \tfrac{1}{3}(\vec{r}_{12}\cdot\vec{v}_{12})\mathbf{1}\Big\} \tag{66}$$

$$B(V^2)\vec{V} \;=\; -\Lambda^{-1}\Big\{(\tfrac{1}{2}\beta_0(\vec{r}_1,t)mV_1^2 - 5/2)\vec{V}_1 - \tfrac{1}{2}(1 - P_v)$$

$$\int dx_2\,\overline{T}_{12}\,g_0(\vec{r}_1,\vec{r}_2,t)\,f_0^{(1)}(x_1,t)$$

$$f_0^{(1)}(\vec{r}_1,\vec{v}_2,t)\beta_0(\vec{r}_1,t)(mV_2^2 + \varphi(r_{12}))\vec{r}_{21}\Big\} \tag{67}$$

$$H(V^2) \;=\; \Lambda^{-1}(\vec{r}_1,t)\beta_0(\vec{r}_1,t)\Bigg[\Big\{\frac{\beta_0(\vec{r}_1,t)p_0(\vec{r}_1,t)}{n(\vec{r}_1,t)}$$

$$-\frac{3}{2c_v(\vec{r}_1,t)}\Big(\frac{\partial p(\vec{r}_1,t)}{\partial T(\vec{r}_1,t)}\Big)_{n(\vec{r}_1,t)}\Big\}(mV_1^2 - 3\beta_0^{-1}(\vec{r}_1,t))f_0^{(1)}(x_1,t)$$

$$-\tfrac{1}{2}\int dx_2\,\overline{T}_{12}\,f_0^{(1)}(x_1,t)f_0^{(1)}(\vec{r}_1,\vec{v}_2,t)\,g_0(\vec{r}_1,\vec{r}_2,t)m(\vec{v}_{12}\cdot\vec{r}_{12})\Bigg] \tag{68}$$

$$K(V^2) = \Lambda^{-1}(\vec{r}_1,t)\,f_\beta^{(1)}(x_1,t) \tag{69}$$

In these expressions the inverse of Λ is defined on the space of functions of $\vec{V}_1$ that are orthogonal to 1 and $\vec{V}_1$ under the inner product

$$(f \mid g) = n^{-1}(\vec{r}_1,t)\int d\vec{v}\,f(\vec{V})g(\vec{V})f_0^{(1)}(x_1,t) \tag{70}$$

With the aid of the same inner product the projector P_v is defined as

$$P_v = \beta m\vec{V}\cdot(\vec{V} \mid \tag{71}$$

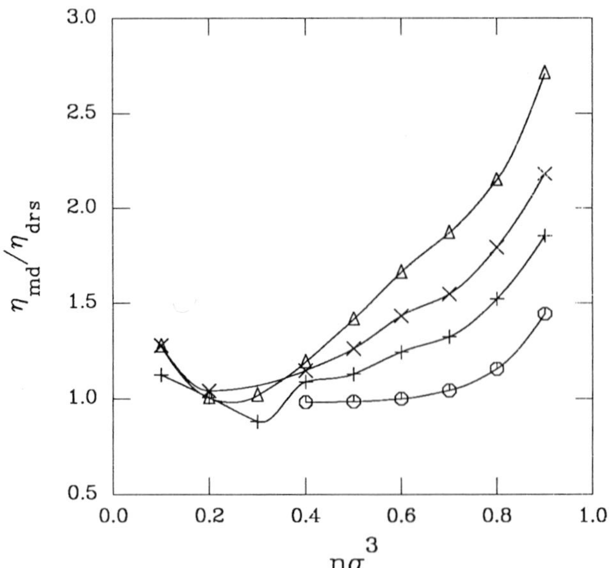

Figure 4: The ratio of MD and theoretical values of the shear viscosity as function of the reduced density $n\sigma^3$. The reduced square well depth $\beta\epsilon$ takes the values 0 (octagons), 0.333 (pluses), 0.5 (crosses) and 0.667 (triangles).

Finally c_v is the specific heat at constant volume per unit of volume.

In fig. 4 the shear viscosity as obtained from (63) is compared to simulation results of J.P.J. Michels [20]. There is reasonable agreement in the density range $0.2 < n\sigma^3 < 0.4$.

For non-vanishing well depths the theory fails at lower densities because of the neghlect ofd correlations between consecutive partial collisions. It also fails at higher densities. This must also be due to correlations between the velocities of colliding particles, but the causes of these correlations are not so obvious as in the low density case. For the coefficients of heat conduction and self diffusion Michels found very similar results [21] and so did Joslin et al. [22] for the bulk viscosity.

3   Linearized Kinetic Theory

For small deviations from equilibrium[5] the kinetic equations (26) and (28) can be linearized around total equilibrium. One sets

$$\begin{aligned}
f^{(1)}(x,t) &= f^{(1)}_{eq}(v)[1 + h(x,t)] \\
U(\vec{r},t) &= U_{eq} + \delta U(\vec{r},t) \\
\beta(\vec{r},t) &= \beta_{eq} + \delta\beta(\vec{r},t)
\end{aligned} \tag{72}$$

with

$$f^{(1)}_{eq}(v) = (\beta_{eq}m/2\pi)^{3/2} n_{eq} e^{-\beta_{eq}mv^2/2}$$

---

[5]For short wave length deviations this need not coincide with small gradients

$$U_{eq} = \frac{1}{2} \int d\vec{r}_2 n_{eq} g(r_{12}) \varphi(r_{12})$$

and $g(r_{12})$ the equilibrium pair correlation function at density n_{eq} and temperature $(k_B \beta_{eq})^{-1}$. The linearized kinetic equations become [23]

$$f_{eq}^{(1)}(v_1)(\partial/\partial t + \vec{v}_1 \cdot \partial/\partial \vec{r}_1) h(x_1, t) = \int dx_2 \overline{T}_{12} f_{eq}^{(1)}(v_1) f_{eq}^{(1)}(v_2)$$

$$\left[g(r_{12}) \left\{ h(x_1, t) + h(x_2, t) \right\} + \int d\vec{r}_3 \left\{ \left. \frac{\delta g(\vec{r}_1, \vec{r}_2)}{\delta n(\vec{r}_3)} \right|_\beta \delta n(\vec{r}_3, t) + \right. \right.$$

$$\left. \left. + \left. \frac{\delta g(\vec{r}_1, \vec{r}_2)}{\delta \beta(\vec{r}_3)} \right|_n \delta \beta(\vec{r}_3, t) \right\} \right] \tag{73}$$

$$\partial/\partial t \, U(\vec{r}_1, t) + U_{eq} \partial/\partial \vec{r}_1 \cdot \vec{u}(\vec{r}_1, t) = \epsilon/2 \int d\vec{v}_1 \int dx_2 \, | \, \vec{v}_{12} \cdot \hat{r}_{12} \, |$$

$$f_{eq}^{(1)}(v_1) f_{eq}^{(1)}(v_2) \Big[\Theta(\vec{v}_{12} \cdot \hat{r}_{12} - v_{esc}) \delta(r_{12} - R^-)$$

$$- \, \Theta(-\vec{v}_{12} \cdot \hat{r}_{12}) \delta(r_{12} - R^+) \Big]$$

$$\left[g(r_{12}) \left\{ h(x_1, t) + h(x_2, t) \right\} + \int d\vec{r}_3 \left\{ \frac{\delta g(\vec{r}_1, \vec{r}_2)}{\delta n(\vec{r}_3)} \delta n(\vec{r}_3, t) + \right. \right.$$

$$\left. \left. + \, \frac{\delta g(\vec{r}_1, \vec{r}_2)}{\delta \beta(\vec{r}_3)} \delta \beta(\vec{r}_3, t) \right\} \right], \tag{74}$$

where $\delta n(\vec{r}, t) = \int d\vec{r} h(x, t)$ and $\delta g/\delta n$ and $\delta g/\delta \beta$ are functional derivatives evaluated in uniform equilibrium. The above equations can be Fourier transformed in space and, after expressing $\delta \beta$ as a linear functional of δU and δn, (74) can be replaced by an equation for the Fourier transformed variable $\hat{\beta}(\vec{k}, t)$. The resulting pair of equations can be expressed in the form

$$\partial/\partial t \begin{pmatrix} \hat{h}(\vec{k}, \vec{v}, t) \\ \hat{\beta}(\vec{k}, t) \end{pmatrix} = L(\vec{k}) \begin{pmatrix} \hat{h}(\vec{k}, \vec{v}, t) \\ \hat{\beta}(\vec{k}, t) \end{pmatrix} \tag{75}$$

The linear operator $L(\vec{k})$ is given explicitly in [23].

For not too large k approximate eigenvalues and eigenfunctions of $L(\vec{k})$ can be found by expanding the function $\hat{h}(\vec{k}, \vec{v}, t)$ in an orthogonal set of polynomials $\Psi_{rlm}(\vec{v})$ of the form

$$\Psi_{rlm}(\vec{v}) = N_{rl} L_{rl}^{l+1/2}(v^2) v^l Y_{lm}(\hat{v}) \tag{76}$$

where Y_{lm} is a spherical harmonic, $L_{rl}^{l+1/2}$ an associated Laguerre, or Sonine, polynomial and N_{rl} a normalization constant. One may introduce phase space functions connected to these Ψ_{rlm} and to $\hat{\beta}$ as

$$\hat{\Phi}_{rlm}(x_1 \cdots x_n) = \frac{1}{\sqrt{N}} \sum_{i=1}^{N} \Psi_{rlm}(\vec{v}_i) e^{i\vec{k} \cdot \vec{r}_i} \tag{77}$$

$$\hat{\Phi}_\beta(x_1 \ldots x_N) = \frac{C_U(k)}{\sqrt{N}} \sum_{i=1}^{N} \left\{ \sum_{j \neq i} \frac{\varphi(r_{ij})}{2} - \int d\vec{r} \left. \frac{\delta U(\vec{r})}{\delta n(\vec{r}_i)} \right|_\beta \right\} e^{i\vec{k} \cdot \vec{r}_i} \tag{78}$$

where C_U can be expressed in terms of k-dependent equilibrium correlation functions between the densities of particles and of potential energy. With the aid of these functions we may define a scalar product as

$$\langle \Psi_a \mid \Psi_b \rangle = \langle \hat{\Phi}_a^* \hat{\Phi}_b \rangle \tag{79}$$

where a runs over rlm and β and the brackets on the right hand side denote an average over the N-particle canonical ensemble. In terms of the Ψ_a $\hat{h}$ and $\hat{\beta}$ may be written as

$$\begin{pmatrix} \hat{h}(\vec{k}, \vec{v}, t) \\ \hat{\beta}(\vec{k}, t) \end{pmatrix} = \sum_{lrm} c_{lrm}(\vec{k}, t)\Psi_{rlm}(\vec{v}) + \hat{\beta}(\vec{k}, t)\Psi_\beta \tag{80}$$

The reason for introducing the scalar product (79) is that under this scalar product the operator L satisfies the symmetry relation

$$L_{ab}(\vec{k}) = \langle \Psi_a \mid L(\vec{k})\Psi_b \rangle = \epsilon_a \epsilon_b L_{ba}^\dagger(\vec{k}) \tag{81}$$

where

$$\epsilon_a = \begin{cases} 1 & \text{for } a = \beta \text{ and for even } l \\ -1 & \text{for odd } l \end{cases} \tag{82}$$

Approximate eigenvalues and eigenfunctions of L are obtained by truncating the infinite matrix L_{ab} at some finite order and constructing the eigenfunctions and eigenvalues of the truncated matrix. For very small k there are five so-called *hydrodynamic modes*, with eigenvalues that tend to zero at least proportionally to k. In the limit $k \longrightarrow 0$ these hydrodynamic modes approach to linear combinations of the following conserved quantities: particle number, energy and momentum. For somewhat larger k-values the analytic continuations (as functions of $\vec{k}$) of these hydrodynamic modes, called extended hydrodynamic modes, still have eigenvalues that are considerably smaller than those of most of the other modes, called the *kinetic modes*, and for still larger k-values, roughly satisfying $k\sigma > 3$, among these extended hydrodynamic modes only the extended heat mode still has an eigenvalue distinctly below the kinetic eigenvalues.

In the present theory, however, a remarkable phenomenon occurs: one non-hydrodynamic mode has an eigenvalue comparable in smallness (for not too small k) to that of the extended heat mode. This mode is called the exchange mode, because it mainly describes the exchange of kinetic and potential energy and vice versa. In fig. 5 the lowest lying eigenvalues as obtained from our theory are compared to spectra obtained by C. Bruin from Delft by fitting MD-data of J.P.J. Michels for correlation functions between particle density, longitudinal momentum density and kinetic and potential energy density to linear combinations of four exponentials with adjustable decay rates.

For the reduced density $n\sigma^3 = 0.8$ and a square well depth satisfying $\beta\epsilon = 0.333$ the predicted heat mode decay rate seems to follow the MD-values very well, whereas MD-values of the exchange mode decay rates, though well below the kinetic eigenvalues, seem to be much larger than the theoretical predictions for these. At $\beta\epsilon = 0.667$ the situation seems reversed, with MD-values for the heat mode decay rate way below the theoretical prediction. One has to be very careful, however, interpreting the fitting procedure. Often several quite different combinations of exponentials can be found to give almost equally good fits.

The limit $\epsilon \longrightarrow 0$ is quite remarkable. Although potential energy density goes to zero, the field $\beta(\vec{r}, t)$ does not. In the limit $\epsilon \longrightarrow 0$ $\beta(\vec{r}, t)$ basically is determined by the number of

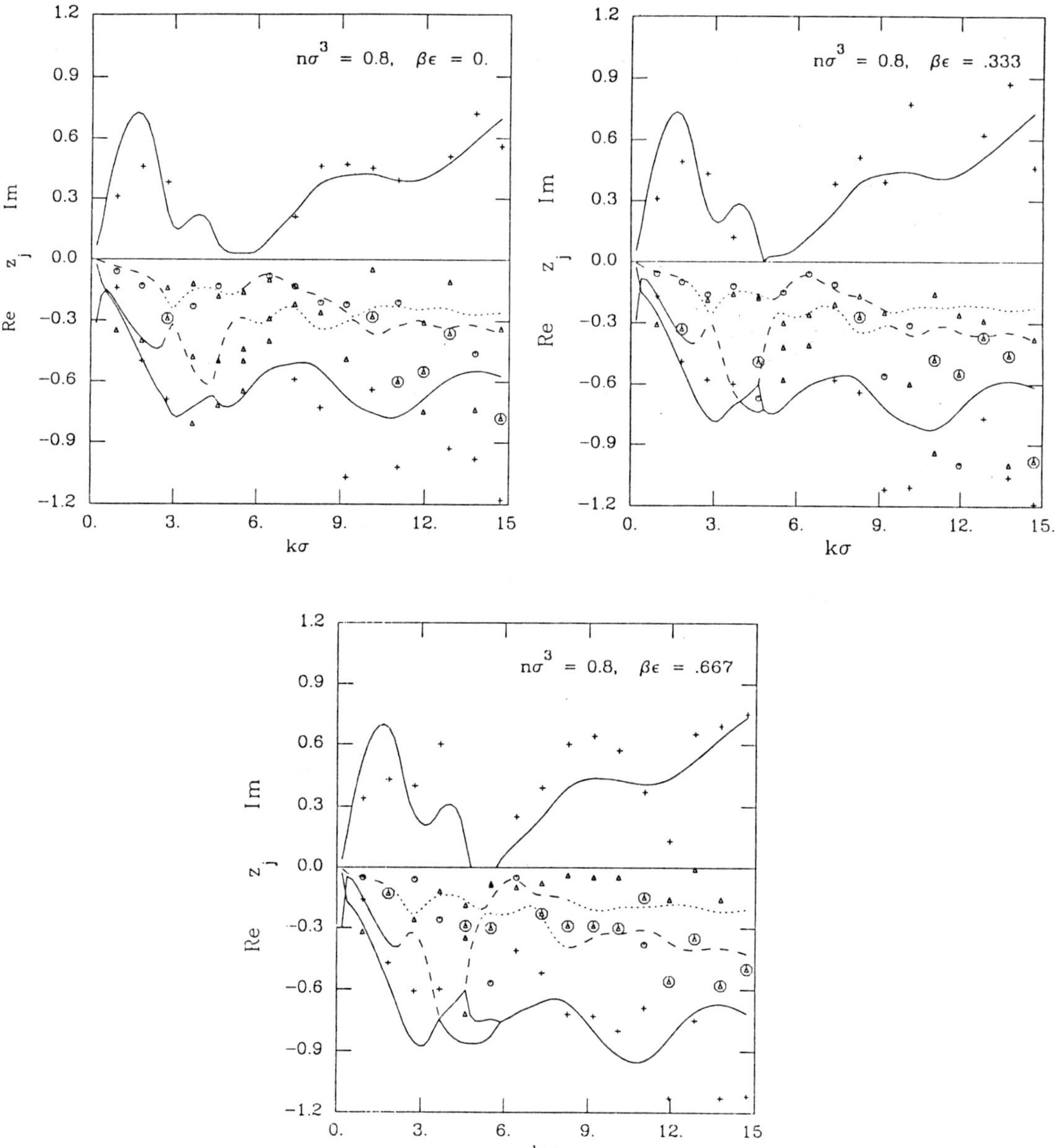

Figure 5: Spectra of the four most slowly decaying eigenmodes of the linearized square-well equation in units t_e^{-1}, where t_e is the mean time between consecutive hard core collisions of the same particle. Theoretical values are presented by lines and MD-results by plotting symbols. Real parts of the eigenvalues are always negative, the magnitude of the imaginary parts of the sound mode eigenvalues is plotted above zero. Dashed lines and triangles represent modes with a large component along β, dotted lines and octagons modes with a large component along the unit function. Some of the MD modes have large components along both β and the unit function. For $\beta\epsilon = 0.667$ a sound propagation gap appears around $\kappa\sigma \approx 0.6$.

particles in the shell $\sigma <| \vec{r}' - \vec{r} |< R$. This implies that β has to diverge as $1/\epsilon$ if one wants to be able to adjust arbitrary (within certain bounds) values for the fields $n(\vec{r}, t)$ and $U(\vec{r}, t)/\epsilon$. The resulting kinetic theory differs from *RET*. Whether or not this is physically satisfactory is unclear.

Discussion

In a restricted range of densities the square well kinetic equation discussed here yields good predictions of transport coefficients for all well depths. For small well depths the theory can be used over a larger density range, although at high density it does not really do better than the Enskog theory, which ignores the well completely [20,21].

As an approximation for realistic potentials the square well potential is not very good. E.g. the dependence of transport coefficients on temperature and density is rather different from that of, say, the Lennard Jones potential. One way to improve on this is by adding a differentiable tail to the square well potential. The maximization of entropy procedure treats such a tail by adding a mean field term to $\vec{F}^{ext}(x_1)$ in (26), which is the average interparticle force exerted on a particle at $\vec{r}_1$, as determined from the pair correlation function $g(\vec{r}_1, \vec{r}_2, t)$. This mean field term does not influence the transport coefficients, yet the latter are changed, because the pair correlation function at the hard core distance is changed by adding the tail potential. In fact a square well step is only needed in this type of theory to produce an irreversible exchange between kinetic and potential energy, and its depth is kept so small that its contribution to the transport coefficient is negligible.

Using Weeks-Chandler-Andersen perturbation theory for choosing effective hard-core diameters and determining the pair correlation function at the hard core distance Karkheck et al. managed to compute transport coefficients for Lennard-Jones systems [24] in good agreement with simulation results and with experimental data for argon.

Summarizing one can say that the Enskog type kinetic equation developed here for the square well potential is not as successful as the Enskog equation for hard spheres. At low densities this is due to the neglect of bound states and of correlations between partial collisions of the same pair of particles, at high densities velocity correlations between colliding particles, resulting from recollisions and steric effects, seem to have an important effect.

The occurrence of an additional slowly decaying mode in the spectrum of the linearized kinetic equation may be confirmed by MD simulations, but this needs a more thorough analysis.

Even if a more complicated kinetic theory will be needed the present one will probably be an essential part of this.
The approximation of a continuous potential by a hard core plus smooth tail, treated by an extension of the square well kinetic theory yields very reasonable results for the transport coefficients, yet is not entirely satisfactory from a theoretical viewpoint. Especially the sensitivity of the results with respect to the choice of the effective hard core diameter is slightly disturbing. However, a more fundamental kinetic theory for dense systems of particles interacting through continuous potentials is not available to date.

Acknowledgements

These notes were largely prepared at the Rockefeller University, where I enjoyed the hospitality of prof. E.G.D. Cohen. Jan Adriaan Leegwater made valuable remarks on the manuscript, prepared the figures and computed the theoretical eigenvalues presented in fig. 5. Jan Michels computed density-density, density-energy and energy-energy time correlation functions by MD-simulations, from which Cok Bruin extracted the eigenvalues appearing as points in fig. 5. Discussions with George Stell and John Karkheck were quite useful. Finally, Sheila Rhyne and Leonie Silkens did a very fine typing job.

References

[1] E.G.D. Cohen, The Boltzmann equation and its generalization to higher densities, in: Fundamental Problems in Statistical Mechanics I, ed. E.G.D. Cohen (North-Holland, Amsterdam 1961) pp. 110-156.

[2] O.E. Lanford, Time evolution of large classical systems, in: Dynamical Systems, Theory and Applications, ed. J. Moser (Lecture Notes in Physics 38, Springer, Berlin, 1975) pp. 1-111.

[3] H. Spohn, Fluctuation theory for the Boltzmann equation, in: Studies in Statistical Mechanics, Vol. 10, eds. E.W. Montroll and J.L. Lebowitz (North-Holland, Amsterdam, 1983) pp. 225-251.

[4] D. Enskog, K. Svenska Akad. Handl. $\underline{63}$ (1921) no. 4, English translation in: S. Brush, Kinetic Theory Vol. 3 (Pergamon, London, 1972).

[5] B.J. Alder, D.M. Gass and T.E. Wainwright, J. Chem. Phys. $\underline{53}$ (1970) 3813.

[6] J.R. Dorfman and E.G.D. Cohen, Phys. Rev. $\underline{A12}$ (1975) 292; I.M. de Schepper, A.F.E.M. Haffmans, H. van Beijeren, Phys. Rev. Lett. $\underline{57}$ (1986) 1715.

[7] T.R. Kirkpatrick, Phys. Rev. $\underline{A32}$ (1985) 3130, T.R. Kirkpatrick and J.C. Nieuwoudt, Phys. Rev. $\underline{A33}$ (1986) 2652 and 2658.

[8] H.T. Davis, S.A. Rice and J.V. Sengers, J. Chem. Phys. $\underline{35}$ (1961) 2210.

[9] J. Karkheck, H. van Beijeren, I.M. de Schepper and G. Stell, Phys. Rev. $\underline{A32}$ (1985) 2517.

[10] H. van Beijeren, J. Karkheck and J.V. Sengers, Phys. Rev. $\underline{A37}$ (1988) 2247.

[11] H. van Beijeren and M.H. Ernst, Physica $\underline{68}$ (1973) 437.

[12] G. Stell in: Phase Transitions and Critical Phenomena, Vol. 5B, eds. C. Domb and M.S. Green (Academic, New York, 1976).

[13] H. van Beijeren, Phys. Rev. Lett. $\underline{51}$ (1983) 1503.

[14] H. van Beijeren and M.H. Ernst, Physica **70** (1973) 225.

[15] M.H. Ernst, J.R. Dorfman, W. Hoegy and J.M.J. van Leeuwen, Physica **45** (1969) 127.

[16] H. Wilbertz, J.P.J. Michels, H. van Beijeren and J.A. Leegwater, J. Stat. Phys. **53** (1988) 1155.

[17] J.K. Percus, The Pair Distribution Function in Classical Statistical Mechanics, in: Classical Fluids, eds. H.L. Frisch and J.L. Lebowitz (Benjamin, New York, 1964) pp. II-33, II-170.

[18] J.R. Dorfman and H. van Beijeren, The kinetic theory of gases, in: Modern Theoretical Chemistry, Vol. 6, Statistical Mechanics B, ed. B.J. Berne (Plenum, New York, 1977).

[19] M.H. Ernst, Physica **32** (1966) 252.

[20] J.P.J. Michels and N.J. Trappeniers, Physica **101A** (1980) 156, **104A** (1980) 243, **107A** (1981) 299.

[21] J.P.J. Michels and N.J. Trappeniers, Physica **107A** (1981) 158, **116A** (1982) 516.

[22] C.G. Joslin, C.G. Gray, J.P.J. Michels and J. Karkheck, The bulk viscosity of a square-well fluid, Mol. Phys., to appear.

[23] J.A. Leegwater, H. van Beijeren and J.P.J. Michels, J. Phys. Cond. M. **1** (1989) 237.

[24] J. Karkheck, G. Stell and J. Xu, Int. J. Thermoph. **10** (1989) 113, J. Chem. Phys. **89** (1988) 5829.

FUNDAMENTAL PROBLEMS IN STATISTICAL MECHANICS VII
H. van Beijeren, Editor
© *Elsevier Science Publishers B.V., 1990*

Computer Simulations in Statistical Physics [1]

D. Frenkel and M. van der Hoef
FOM Institute for Atomic and Molecular Physics
PO Box 41883, 1009 DB Amsterdam, The Netherlands

1 INTRODUCTION

The aim of these lectures is *not* to give a review of computer simulations in statistical physics. The number of applications of computer simulations has grown tremendously over the past two decades to the extent that simulations now probably provide more 'experimental' data to test statistical mechanical theories than real experiments. For this reason the present lectures focus on applications of computer simulations that illustrate the different ways in which computer simulations can be used to gain a better understanding of the complex many-body systems that are studied in statistical physics.

The first question that should be addressed in these lectures is : *Why do computer simulations at all?*. This question is important because the numerical studies that we perform in the context of statistical mechanics are not of the same nature as in, say, material science. In particular, it is usually not our aim to simulate a particular material as accurately as possible. Rather, we tend to study simplified models that contain the essential features of the many-body systems that we wish to understand. We use computer simulations mainly for two reasons:

1. To test theoretical approximations

2. To 'discover' new phenomena, i.e. to gather the building blocks for new theories.

However, computer simulations themselves are no substitute for a theoretical description. On the contrary, the ultimate aim of any simulation should be to make itself redundant: if a simple theory can be constructed that accounts for the computer simulation results, then clearly such a theory is preferable to a computer program.

In these lectures we shall mainly consider the first application of computer simulation. In particular, we shall discuss some recent numerical techniques that enable us to test theories of transport phenomena in simple fluids in great detail. An example of the other application of computer simulations in the exploration of novel phenomena is discussed in considerable detail in a recent review article on computer simulations of liquid crystals [1].

2 LONG-TIME TAILS

2.1 Introduction

In the history of the kinetic theory of fluids, 1969-1970 was a crucial year. In that year Alder and Wainwright [2] published a paper in which they demonstrated the breakdown of the 'Molecular Chaos' assumption. The Molecular Chaos assumption, originally introduced

---

[1]This paper is based on material that has either been published elsewhere or has been submitted for publication.

by Boltzmann as the 'Stoßzahlansatz', states that the collisions experienced by a molecule in a fluid are uncorrelated. One consequence of this assumption is that the velocity auto-correlation function (VACF) of a tagged particle in fluid should decay exponentially. What Alder and Wainwright found is that the VACF of a particle in a moderately dense fluid of hard spheres or hard disks does not decay exponentially but algebraically. These algebraic long-time tails are the consequence of coupling between particle diffusion and shear modes in the fluid.

The Alder-Wainwright simulations caused a complete overhaul of the kinetic theory of dense fluids. The subsequent theoretical analyses of algebraic long-time tails were either based on an extension of kinetic theory [3] or on mode-coupling theory [4]. For a review, see ref. [5] . In the mode-coupling theory by Ernst, Hauge and van Leeuwen [4], it is assumed that the long-time tail is the consequence of coupling between particle diffusion and shear modes in the fluid. To a first approximation the leading term in the long-time tail of the velocity ACF is given by:

$$< v_x(0)v_x(t) > \approx \frac{D-1}{D} \frac{1}{\rho(4\pi(D_0 + \nu_0)t)^{D/2}} \equiv \frac{d_0}{t^{D/2}} , \tag{1}$$

where ρ is the number density, D_0 the 'bare' self-diffusion constant, ν_0 the kinematic viscosity and D the dimensionality.

Following this theoretical work, simulations were performed by Levesque and Ashurst [6] and, most extensively, by Erpenbeck and Wood [7,8] with the aim to verify the validity of eqn. 1. For three-dimensional fluids these simulations are extremely expensive because very long simulations on very large systems must be performed. Even so, the system sizes studied in the simulations of Erpenbeck and Wood were such that it was essential to apply finite-size corrections to the corresponding mode-coupling theory before a meaningful comparison with the simulations could be made. Following such an approach, Erpenbeck and Wood found agreement between their simulation results for the VACF and a finite-size mode-coupling theory for a number of different densities. Nevertheless the statistical accuracy of their data was such that it was not meaningful to verify either the value of the exponent of the algebraic tail or the functional form of the density-dependent tail coefficient independently.

In the case of two-dimensional fluids there is another problem. Ever since the discovery of hydrodynamic tails, it has been realized that a consistent description of mode-coupling effects in a two-dimensional fluid would result in a long-time tail that decays faster than t^{-1}, because in $2D$ the self-diffusion constant itself diverges. In fact, de Schepper and Ernst [9] computed the coefficient (d_1 in eqn. 2) of the first correction to the t^{-1} tail for a system of hard disks. They predicted that this correction should be negative and proportional to $\log(t/t_0)/t$ (where t_0^{-1} is the initial decay rate of the velocity ACF):

$$< v_x(0)v_x(t) > \approx d_0/t + d_1 \log(t/t_0)/t + ... \text{for } t/t_0 \gg 1. \tag{2}$$

However, this prediction is only expected to hold for times that are not too long. Forster et al. [10] argued that as $t \to \infty$, the tail should be renormalized to $1/(t\sqrt{\log t})$.

Thus far it has not been possible to compare these predictions directly with computer simulation data. The reason is that such a comparison requires accurate knowledge of the velocity ACF for very long times (many tens to hundreds of collision times). This requires very long simulations on very large systems (to avoid problems due to spurious correlations caused carried by sound waves in periodic systems). Thus far such calculations have remained beyond the power of presentday computers.

2.2　Lattice gases

In the remainder of this paper we shall indicate how computer simulations on lattice gases can be used to test theoretical predictions about 'long-time tails' in the velocity autocorrelation function of tagged particle in a fluid.

One of the attractive features of lattice gas models is that they are ideally suited to serve as a testing ground for concepts in kinetic theory. The reason is twofold: on the one hand the very simple structure of most lattice-gas models often makes it possible to work out in closed form the consequences of a particular approximation scheme in kinetic theory. On the other hand, lattice-gas models are ideally suited for computer simulation. Thus approximate schemes in kinetic theory can be tried out on lattice gas models before applying them to more realistic models of fluids or solids. Below we show how Lattice Gas Cellular Automata (LGCA's) can be used to verify in great detail the theoretical predictions about long-time tails in the the velocity autocorrelation function of tagged particles.

Lattice gas cellular automata have recently received a lot of attention because these model systems may provide a 'cheap' alternative to simulate the hydrodynamic behaviour of simple fluids [11]. Because of their simple structure which makes it comparatively easy to work out the consequences of a particular approximation scheme, LGCA's are, in principle, ideally suited to serve as a testing ground for concepts in kinetic theory. It is therefore only natural that lattice gas models were considered as promising systems to study the long-time behaviour of the tagged-particle VACF. This approach has been tried by Boon and Noullez [12] and Binder and d'Humières [13] for 2D systems. However, somewhat disappointingly, the statistical accuracy of these (long) numerical simulations is poor. Hence the presence of a long-time tail was hidden in the statistical noise and a special analysis was required to demonstrate that the simulation data are, in fact, compatible with the presence of an algebraic tail of approximately the expected amplitude [14].

However, using a technique that was recently developed by Frenkel [15,16] it is now possible to compute tagged particle VACF's in lattice gases with an efficiency that is a factor 10^6 to 10^{10} higher than the earlier 'brute force' schemes. In the present paper we shall discuss the application of this new technique to the study of long-time tails.

In order to explain the numerical scheme, we first describe its application to lattice Lorentz gases. Thereafter we indicate how the method can be extended to study long-time tails in the velocity autocorrelation function of a tagged particle in a lattice gas cellular automaton.

2.3　Lattice Lorentz gases

Let us first consider a lattice-Lorentz gas with 'bond' disorder. This is a model system where a random walker performs a hopping motion along the nearest-neighbor bonds of a square (2D) or simple-cubic (3D) lattice. The moving particle can jump with equal probability along any of the 4 (or 6) bonds to a neighboring site. However, if the bond is 'broken', the particle remains at its original site until the next trial move. In the bond-disorder model, a finite fraction of the bonds is broken. The broken bonds are distributed randomly over the lattice. Two equivalent versions of this random walk process can be formulated: a) a discrete time version in which a new trial move is attempted every unit time and b) a continuous time version where the waiting times between successive trial moves are Poisson distributed. It is clear that statistical properties of the continuous-time model, such as a time correlation function $C(t)$, can be obtained as a weighted average over all n-step correlations of the corresponding discrete-time model $C_D(n)$:

$$C(t) = \frac{1}{2}C_D(0)\delta(0_+) + \sum_{n=1}^{\infty} C_D(n)\frac{t^n}{n!}\exp(-t), \tag{3}$$

assuming an average rate of one jump per unit time for the continuous time model. It turns out that the theory for transport in lattice Lorentz gases is formulated most compactly for the continuous-time model, whereas the simulations are performed most conveniently for the discrete-time version.

Let us now consider the numerical computation of the velocity correlation function of a lattice-Lorentz gas. The discrete nature of the model leads to the following natural definitions of the 'velocity': the velocity $v_\alpha(t_n)$ of a random walker at discrete time t_n in direction α is $+1(-1)$ if the random walker performed a jump in the $+\alpha(-\alpha)$ direction in time interval n, and is 0 otherwise. With this definition it is easy to see how to compute the velocity auto-correlation function $\phi(t) \equiv\; < \mathbf{v}(0).\mathbf{v}(t_n) >$. In principle this quantity can be computed by considering all possible starting positions of the random walker at time $t = 0$ and then averaging over all possible random walks that this particle can perform in a time interval t_n. The problem with this 'brute force' approach is that very many random walks must be sampled to obtain a reasonable statistical accuracy. To give a specific example: for a 2D lattice-Lorentz gas with a low concentration of broken bonds (say $p=1\%$), the long time tail in the velocity ACF is expected to decay as $\phi(t) \approx p/t^2$. Hence, if we are interested in the long-time behaviour (e.g. $t = 100$), the amplitude of $\phi(t)$ will be $O(10^{-6})$. In order to obtain a signal-to-noise ratio of $O(1)$ we should therefore sample over $O(10^{12})$ independent random walks. Using an efficient random walk procedures [25] such a simulation would require some 10^2 hours of CPU time on a Cyber 205 vector computer.

In fact, a method has been developed to reduce the required computer time by a factor 10^2 [17] , but this technique will not work for lattice-gas cellular automata, to be considered below. Anyway, the approach described in the present paper results in far greater computational gains.

Let us consider the random walk problem that we wish to solve in some detail. In order to compute $< \mathbf{v}(0).\mathbf{v}(t) >$ we could pick a site $\mathbf{X}$ on the lattice at $t = 1/2$ (in order to minimize confusion we compute velocities at integer times and positions at half integer times) pick a point $\mathbf{X}'$ at time $t = t_n - \frac{1}{2}$ that can be reached in a possible random walk in time $t_n - 1$, compute the product $\mathbf{v}(0).\mathbf{v}(t_n)$ and average over: 1) all possible random walks linking $\mathbf{X}(t = \frac{1}{2})$ and $\mathbf{X}'(t = t_n - \frac{1}{2})$, 2) all $\mathbf{X}'$ and 3) all $\mathbf{X}$.

First, consider the computation of the velocity at t_n. If the random walker is at site $\mathbf{X}'$ at time $t_n - \frac{1}{2}$ then it can jump to any of the neighboring sites with equal probability, unless any of the bonds connected to $\mathbf{X}'$ is broken. If all bonds connected to $\mathbf{X}'$ are open, then clearly the average velocity of the random walker in the time interval centered at t_n is 0. Only if one or more bonds are broken can the random walker have a non-zero average velocity in the interval $[t_n - \frac{1}{2}, t_n + \frac{1}{2}]$. Let us denote this velocity by $\mathbf{v}(\mathbf{X}')$. Using this definition, we note that the average velocity of a random walker at time $t = 0$ (i.e. just before its arrival at $\mathbf{X}$ is equal to $-\mathbf{v}(\mathbf{X})$. Clearly, if there are no broken bonds around $\mathbf{X}$, $\mathbf{v}(\mathbf{X}) = 0$ and all random walks starting at $\mathbf{X}$ do not contribute to $< \mathbf{v}(0).\mathbf{v}(t) >$. From now on we restrict our attention to those random walks for which $\mathbf{v}(\mathbf{X}) \neq 0$. The actual number of lattice sites for which $\mathbf{v}(\mathbf{X}) \neq 0$ can easily be counted once the distribution of broken bonds has been specified. The average fraction of sites for which $v_\alpha(\mathbf{X}) \neq 0$. for any direction α is equal to $2p(1 - p)$, where p is the fraction of broken bonds.

Now let us consider the computation of $< \mathbf{v}(0).\mathbf{v}(t) >$. For $t = 0$ the answer is simply $(1 - p)$. For $t = 1$, we obtain $-\overline{v^2(X)}$ where the bar denotes averaging over all lattice sites. In order to compute $< \mathbf{v}(0).\mathbf{v}(t) >$ for $t > 1$, we note that if at time $t - \frac{1}{2}$ a particle is at position $\mathbf{X}$ then after one more time-step, it has a probability $1/2D$ ($D =$ dimensionality) to be on any of the ℓ ($\leq 2D$) connected nearest neighbor sites of $\mathbf{X}$ and a probability $(2D - \ell)/2D$ to be still at site $\mathbf{X}$. The average velocity of this random walker at time $t + 1$ is therefore :

$$\mathbf{v}(t + 1) = \sum_{\text{all } \mathbf{X}' \text{ connected to } \mathbf{X}} \frac{1}{2D}\mathbf{v}(\mathbf{X}') \tag{4}$$

where the original site $\mathbf{X}$ is counted as 'connected' to itself for every broken bond present. We now attribute this average velocity $\mathbf{v}(t + 1)$ to the site $\mathbf{X}$ and denote it by $\mathbf{v}^{(2)}(\mathbf{X})$. Clearly, $< \mathbf{v}(0).\mathbf{v}(2) >= -\mathbf{v}(\mathbf{X}).\mathbf{v}^{(2)}(\mathbf{X})$. Analogously, by iteration we obtain $\mathbf{v}^{(3)}(\mathbf{X})$ from $\mathbf{v}^{(2)}(\mathbf{X})$, and so on.

Note that in this way we compute the contribution to $< \mathbf{v}(0).\mathbf{v}(t) >$ from *all* possible random walks. Hence, once we have specified the distribution of broken bonds on the lattice, the present computation of $< \mathbf{v}(0).\mathbf{v}(t) >$ is *exact*. Statistical errors are only due to the fact that the velocity correlation function should be computed for a large number of independently generated realizations of the bond disorder. The other point to note is that the iteration scheme requires only very simple matrix operations that are eminently vectorizable. To demonstrate the power of the present method, figure 1 shows the long-time tail of the velocity ACF of a $2D$ lattice-Lorentz gas with bond disorder.

Note that the statistical error (points) in figure 1 decreases with increasing time. For the longest times studied in this simulation the statistical error is less than 1 in 10^{10}. The calculation required 13 minutes of CPU-time on a CYBER 205 vector computer. To achieve the same accuracy by straightforward Monte Carlo sampling would require over 10^5 years of CPU time on the same machine. The difference between the statistical error obtained by the present method and those obtained using the Monte Carlo approach [17] is particularly striking for long times. Direct sampling over trajectories yields a statistical error that is essentially independent of time. Hence to compute a t^{-2} long-time tail over a time interval that is 10 times longer required a factor 10^4 (!) more computing time (10^5 in 3D). In contrast, the error in the present matrix-method decreases with time. Hence, to study a 10-times longer correlation interval simple requires a run that is 10 times longer. The relative merit of the present method becomes greater at longer times. In our case the gain with respect to the random walk method was a factor 10^{10}, but the comparison could have been made even more biased by going to longer times.

2.4 Lattice Gas Cellular Automata

Next we shall show how, by a slight extension, the present method can be applied to compute the long-time behaviour of velocity ACF's in lattice-gas cellular automata (LGCA's).

At first sight, the computation of the velocity ACF for a 'particle' in a system of lattice-gas CA's seems very different from the corresponding problem of a random walker in a lattice- Lorentz gas. After all, lattice gas cellular automata are meant to model systems of moving, interacting particles. In particular, LGCA's have been constructed as a highly simplified model for an atomic fluid. In the present paper we shall focus on the properties of LGCA's on a $2D$ triangular lattice (FHP-model [11,21] and on a $3D$ face-centered hyper-cubic (FCHC) lattice.

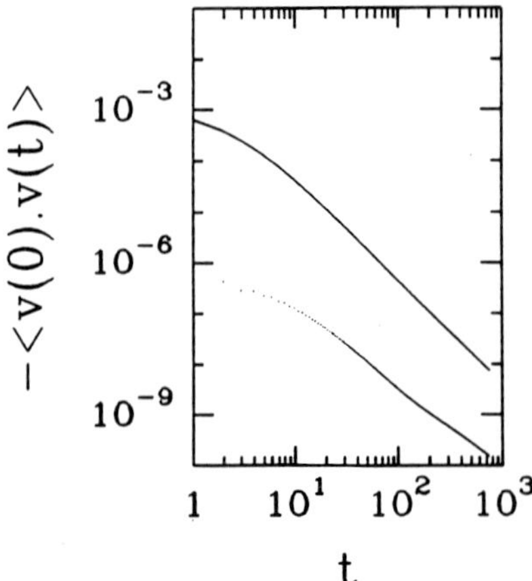

Figure 1: Velocity autocorrelation function of a random walker on a two-dimensional square lattice-Lorentz gas with a fraction of 1% broken bonds. The drawn curve is the velocity correlation function. The dots represent the statistical error.

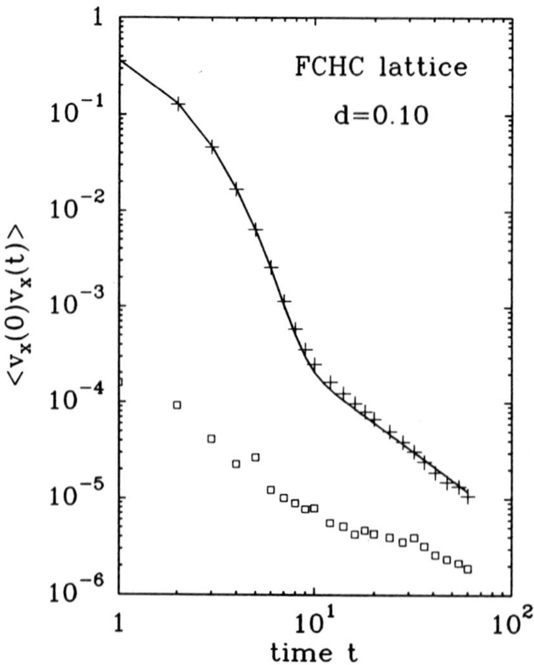

Figure 2: Log-log plot of the normalized velocity autocorrelation function of a 3-dimensional lattice gas cellular automaton on a FCHC lattice for density $d = 0.10$. The density d is in units of average number of particles per link. The time is in units of the time between successive updates of the lattice gas cellular automaton $\Delta t(=1)$. The solid line is the prediction of mode-coupling theory, the crosses are the simulation data. Note that the estimated error (open squares) decreases with increasing t to a value of order 10^{-6}.

In LGCA's, the particles are constrained to move along the bonds joining the lattice sites. No two particles can move along the same bond in the same direction. The state of the lattice is completely specified by indicating which links are occupied and which are empty. This implies that lattice-gas particles are indistinguishable.

The time evolution of the system is governed by the following rules:

1. Propagation: all particles move in one time step (for convenience we choose $\Delta t = 1$) from their initial lattice position (say $\mathbf{X}$) to a new position ($\mathbf{X'} = \mathbf{X} + \mathbf{c}_\alpha$; where $\mathbf{c}_\alpha$ is the velocity of species α). The velocities $\mathbf{c}_\alpha$ are such that at the end of the propagation steps all particles are once more positioned at lattice sites.

2. Collision: the particles at all sites on the lattice undergo a collision that converves the total number of particles and the total momentum at each site. Usually, these local collision rules are deterministic.

Provided that the lattice has a sufficiently high symmetry (e.g. triangular in 2 dimensions) and the collision rules are sufficiently isotropic (for a discussion, see [18]), it can be shown that the equation that governs the time evolution of the distribution function of such a lattice gas becomes equivalent to the Navier-Stokes equation for an incompressible fluid in the limit that the flow velocity is much less than the particle velocity, and all spatial variations in the system occur on a scale that is large compared to the mean free path of the lattice gas particles. In this respect LGCA's model atomic fluids.

When attempting to compute the velocity correlation function of a particle in a lattice gas CA, one is immediately confronted with a conceptual problem. As all lattice gas particles are indistinguishable, the velocity correlation function of 'a particle' is ill defined. As soon as a particle has collided it is no longer possible to identify any of the outgoing particles as the original particles whose velocity ACF we are attempting to compute. To avoid this problem, the particle under consideration must be labeled differently from the rest (say, a 'blue' particle in a sea of 'red' particles). Once the collision rules for all particles have been specified we can then compute the velocity ACF of a single tagged particle. This is the approach that has been pursued by Boon and Noullez for the FHP model [12] and by Binder and d'Humières for the HPP model [13]. As mentioned above, this approach yields poor statistics because we must solve the dynamics of all N lattice-gas particles in order to follow the time-evolution of 1 tagged particle.

An alternative approach that effectively side-steps the problem referred to above has been followed by Colvin, Ladd and Alder [26]. These authors compute not the velocity ACF of a tagged particle but the autocorrelation function of the fluid velocity at a lattice site. This method yields somewhat better statistics and made it possible to observe a long-time tail in the site-velocity ACF. However, the site-velocity ACF does not contain the same mode-coupling contributions as the tagged-particle velocity ACF and is therefore, from a theoretical point of view, of less interest.

Fortunately, the matrix method described in the previous section can be extended in such a way that it becomes possible the compute the velocity ACF of a tagged particle in a lattice-gas cellular automaton with high accuracy. In order to do so we have to impose one restriction on the rules of the lattice gas automaton, namely that the collision rules for a tagged particle with untagged particles result in the occupation of the same output states as in the case of collisions betveen untagged particles. And, most importantly, the tagged particle has equal probability to be in any of the occupied output states. Hence, for the tagged particle the collision rules are stochastic, although for a 'colour-blind' observer,

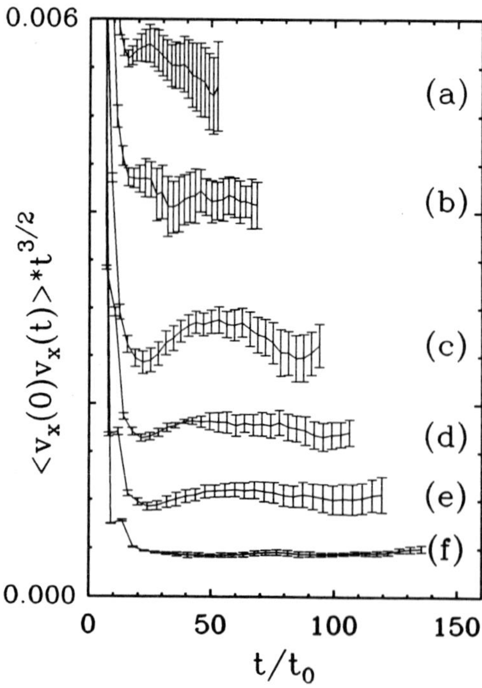

Figure 3: Velocity autocorrelation function multiplied by $t^{3/2}$ for the following densities: (a) $d = 0.2$, (b) $d = 0.3$, (c) $d = 0.5$, (d) $d = 0.6$, (e) $d = 0.7$, (f) $d = 0.8$. The time is in units of mean free time t_0.

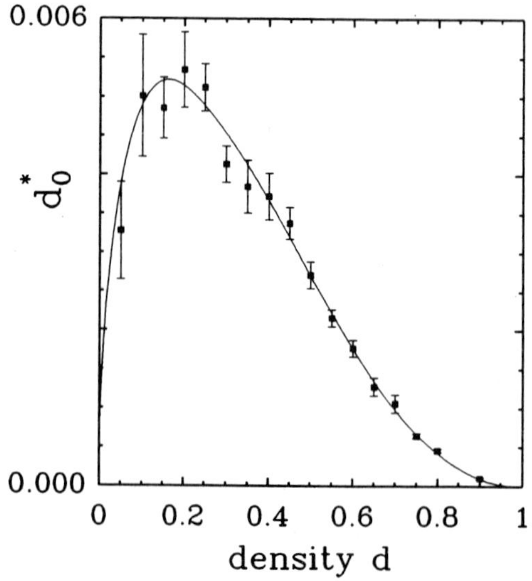

Figure 4: The normalized tail coefficient d_0^* of the tagged-particle VACF for the 3D FCHC system The solid curve is the mode-coupling prediction, the dashed curve is a weighted fit to the simulation data.

the rules are deterministic. Note also that even in 'collisions' that have the same input and output states the tagged particle may still 'collide', i.e. it may change its velocity state.

With these rules, it is obvious that the average velocity of a tagged particle after a collision at site $\mathbf{X}$, depends only on the (colorless) state at that site. In particular, it depends in no way on where the tagged particle was coming from. We can thus define for every (non-empty) site of the lattice the average post collisional velocity that a tagged particle at that site would have at that time:

$$\mathbf{v_X}(t) = \frac{1}{N_{occ}(\mathbf{X})} \sum_{\alpha=1}^{N_{occ}(\mathbf{X})} \mathbf{c}_\alpha \tag{5}$$

where $N_{occ}(\mathbf{X})$ is the total number of particles at site $\mathbf{X}$ and $\mathbf{c}_\alpha$ are the velocities corresponding to the occupied links. At first sight, eqn. 5 may look similar to the expression for the site velocity in a Lorentz gas. The important difference is that for a Lorentz gas the site velocity is fixed by the (time-independent) distribution of random bonds. In contrast, for the LGCA model the site-velocity for a tagged particle changes with every time step. However, apart from this modification, we can use basically the same techniques that worked for the Lorentz gases, to compute the velocity ACF of a tagged particle in a lattice gas cellular automaton.

Consider a tagged particle that is moving at site $\mathbf{X}$ at time $t = 0$, moving with a velocity $\mathbf{c}_\alpha$. At time $t = 1$, the tagged particle will have collided at site $\mathbf{X} + \mathbf{c}_\alpha$ and its average post collisional velocity will be $\mathbf{v_{X+c_\alpha}}(t = 1)$. Clearly,

$$< \mathbf{v}(0).\mathbf{v}(1) > = \frac{1}{N_{tot}} \sum_{\alpha,\mathbf{X}} s_\alpha(\mathbf{X}, t = 0)\mathbf{c}_\alpha.\mathbf{v_{X+c_\alpha}}(t = 1) \tag{6}$$

where $s_\alpha(\mathbf{X}, t = 0) = 1(0)$ if link α at site $\mathbf{X}$ is occupied (empty) at $t = 0$.

To compute $< \mathbf{v}(0).\mathbf{v}(2) >$, we simply propagate the tagged particle that is at site $\mathbf{X} + \mathbf{c}_\alpha$ to all sites that will be reached from this site in the next time step. For all these sites (denoted by $\{\mathbf{X}(t = 2)\}$) we compute the average post collisional velocity, average it over the set $\{\mathbf{X}(t = 2)\}$ and this yields the average velocity of the tagged particle after two time-steps.

For the actual implementation of this technique the recipe is then the following:

1. compute the average post collisional velocity of the tagged particle at site $\mathbf{X}$, at time $-t$, for all possible $\mathbf{X}$ on the lattice;

2. propagate this average to all sites that will be reached in one cycle from site $\mathbf{X}$;

3. for every site thus reached, compute the average of all the averages that have been propagated to it.

4. iterate steps 2 and 3 until t iterations have been made in total.

Note that in order to compute the velocity correlation function of a tagged particle, use was made of all possible starting positions and trajectories that such a particle could have, compatible with the (deterministic) dynamics of the underlying 'uncolored' lattice gas. The only additional averaging is over all possible time origins and over independent initial conditions. The latter averaging would not be necessary if the lattice gas cellular automaton were strictly ergodic. But as the models studied in this paper are known to have spurious invariants [27], averaging over a number of independent initial conditions is advisable.

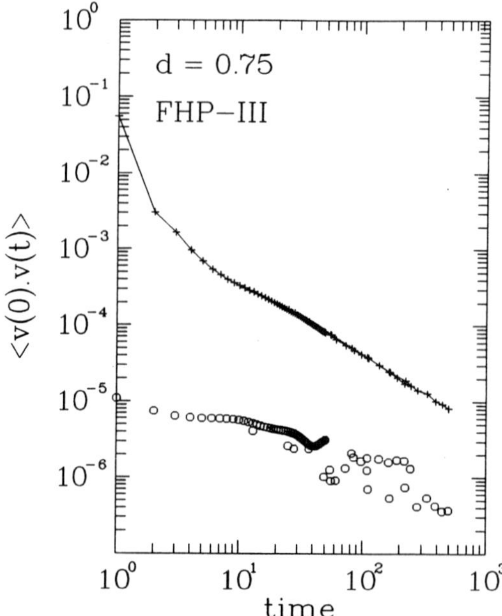

Figure 5: Example of the normalized velocity autocorrelation function of a tagged particle in the in a 2D lattice-gas cellular automaton (FHP-III) at a density $d=0.75$. Note that after an initial rapid decay (and overshoot), the velocity ACF approaches a power-law decay with an exponent -1. The estimated error (open circles) decreases with increasing t to a value of order 10^{-6}. In ref. [12] the statistical error in the long-time tail was of order 0.310^{-2}.

2.5 Models

Simulations of tagged particle diffusion in two-dimensional fluids were carried out for a lattice gas model ('FHP-III', 6 speed-1 particles, one rest-particle, defined in ref. [21]). System sizes of up to 500×500 lattice points were studied at densities varying from 5% to 75% occupancy. The simulations were either performed on a CYBER 205 vector-computer or, for the larger systems, on a NEC-SX2 super-computer. As a model for $3D$ fluids we used, the face-centered hyper-cubic(FCHC) lattice gas model [18,22]. Whereas FHP-III is discussed extensively elsewhere [18,21] we should explain the exact version of the 3D FCHC-model that we used in our simulations. In this lattice there are 24 possible velocities, so a collision would require a 2^{24}-word lookup table, which requires a very large shared memory [19]. In the algorithm used in the present paper the 24-bit state is split into two 12-bit sub-states [20], which requires only a small 12-bit lookup table. This splitting can be done in 6 different ways, one of which is choosen randomly at every collision.

The $3D$ simulations were carried out on systems of up to $60 \times 60 \times 60$ lattice points. In all cases correlations were only computed for time intervals less then the shortest time in which any particle could cross the periodic box. This is in contrast to corresponding simulations of long-time tails in atomic fluids [8] where time intervals up to 5 times the acoustic wave traversal time had to be used. In the present simulation the VACF is calculated for different densities varying from $d = 0.05$ to $d = 0.90$, where d is defined as the average number of particles per link per node ($d = \rho/b$). In order to estimate the statistical error of the VACF 5 to 10 independent simulations per density were performed.

2.6 Results

2.6.1 Three dimensional model

Let us first look at the results for one density $d = 0.1$. In figure 2 we show the VACF for this density. For shorter times we see that the decay is approximately exponential, and for longer times we clearly observe the algebraic decay, which appears in the $\log - \log$ plot as a straight line. A convenient way to present the results for a range of densities is to multiply the VACF with $t^{-3/2}$, see figure 3. If the decay is algebraic with the predicted exponent $-3/2$ these functions should reach a constant value in the limit $t \gg t_0$. This behavior is indeed observed. These plateau values should equal the amplitude of the algebraic tail and can be compared with the prediction of mode-coupling theory, see figure 4. This figure shows that there is essentially quantitative agreement between the simulated and the predicted amplitudes for all densities. We wish to stress that there are *no* adjustable parameters in the comparison of theory and simulation.

2.6.2 Two dimensional model

Simulations of tagged particle diffusion were carried out for a $2D$ lattice gas model ('FHP-III', 6 speed-1 particles, one rest-particle, with alternating collision rules for odd and even timesteps, see ref. [21]). System sizes of up to 500×500 lattice points were studied at densities d ranging from 0.05 to 0.75 (i.e. 5% to 75% occupancy per velocity state).

Figure 5 shows the velocity autocorrelation function of a tagged particle in the $2D$ lattice gas at a density $d = 0.75$. The velocity ACF has been normalized to 1 at $t = 0$. Here, and in all other cases shown, correlations were only computed for time intervals less than the shortest time in which any particle could cross the periodic box. In many of

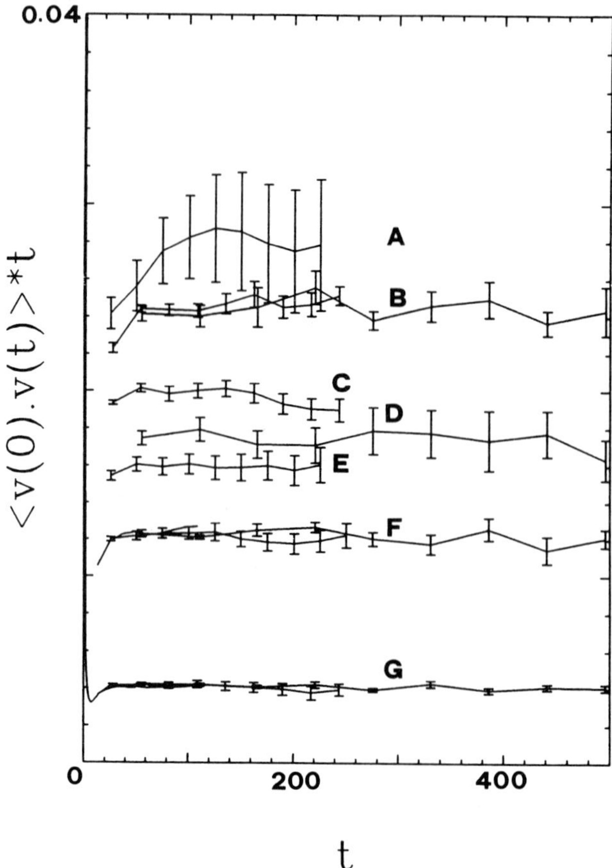

Figure 6: In order to detect possible deviations from the t^{-1} decay in the velocity ACF of a tagged particle in a $2D$ LGCA at long times, this figure shows $t < \mathbf{v}(0).\mathbf{v}(t) > / < v^2 >$ as a function of time t. The letters refer to the density: $A - G$ correspond to $d = 0.1, 0.2, 0.3, 0.35, 0.4, 0.5$ and 0.75 For some densities more than one simulation result is plotted. Within the statistical accuracy of the present calculations (the error bars have a length of 2 standard deviations) no systematic deviations from the t^{-1} decay can be detected. However, our error estimate is probably too conservative because a fit to the simulation data shows that for all 7 points with $d \geq 0.3$ the VACF decays with an effective exponent $\beta > 1$ (up to 3%).

our calculations the statistical error is of order 10^{-6}, which is about a factor 10^3 lower than has been achieved with conventional techniques. Note that such an error reduction corresponds to a gain of 10^6 in computer time. Initially the decay of the velocity ACF is approximately exponential. The characteristic decay time t_0 of this exponential ranges from 5.6 at $d = 0.05$ to 0.33 at $d = 0.75$. There is a surprisingly large time-interval where the decay is no longer exponential but not yet algebraic. However, after some 30 collision times the decay appears to become algebraic. In order to see this latter effect more clearly, fig. 6 shows the function $t < \mathbf{v}(0).\mathbf{v}(t) >$ for densities $0.1 \leq d \leq 0.75$. If the velocity ACF decays as t^{-1} then the curves in fig. 6 should approach a constant value as $t \to \infty$. Such behavior is indeed observed. This in itself is maybe not surprising, but it is reassuring as it has been argued that the hydrodynamic long-time tails observed in computer simulation on continuous systems may be due to a propagation of numerical errors [28]. In the present simulation the discrete dynamics of the lattice gas is solved exactly, hence propagation of numerical errors is ruled out as a factor affecting either the power-law tails or, for that matter, any corrections to the latter.

We have compared the measured amplitude of the t^{-1} tail with the predictions of mode-coupling theory [23] adapted to the LGCA. As it turns out, the expression for the amplitude of the t^{-1}-tail in a $2D$ LGCA is equal to eqn. 1 multiplied by a factor $(1 - d)$, where the density ρ for continuous fluids must be interpreted as $7d/v_0$, the number density per unit area for the FHP-III model (7 velocity states per site, volume of the unit cell of the triangular lattice $v_0 = \sqrt{3}/2$). The factor $(1 - d)$ is a consequence of the Fermi statistics and guarantees that the state occupied by the tagged particle contains no fluid particle. It should be stressed that the applicability of eqn. 1 to the FHP model is not self-evident because in the FHP-model there exist unphysical hydrodynamic modes (associated with the staggered momentum density, see ref. [27]). These modes can couple to the microscopic stress tensor, thereby affecting the amplitude of the long-time tail of the stress-stress ACF. However, to leading order in $1/t$ the same staggered modes do *not* couple to the tagged particle current.

In figure 7 we compare the simulation results for the amplitude of the d_0 with the predictions of mode coupling theory. As can be seen from the figure, the mode coupling predictions are very close to the simulation results. The remaining discrepancy of a few percent is comparable to that found by Kadanoff et al. [27].

Next, consider the behavior of $< \mathbf{v}(0).\mathbf{v}(t) >$ for $t/t_0 \gg 1$. Velocity correlations were studied for times up to $t = 500$, which corresponds to values of t/t_0 ranging from 10^2 to over 10^3. In this time regime, which has never before been studied numerically, we would expect to observe a $\log(t)/t$ correction to the $1/t$ tail similar to the one predicted by de Schepper and Ernst for the hard-disk model [9]. In continuous fluids, the relative importance of these logarithmic corrections grows linearly with the gas density. If we would assume that the expression for d_1 (see eqn. 2) given in ref. [9] also applies in the case of a lattice gas, then we find the following density dependence of the ratio d_1/d_0:

$$d_1/d_0 = -d_0\{(\nu_0 + D_0)^{-1} + (4\nu_0)^{-1} + (8(\nu_0 + \zeta_0))^{-1}\} = -0.247d \,, \tag{7}$$

where ζ_0 is the bulk viscosity and $\rho\zeta_0 = 1/14$ for $\rho = 0$ [21]. A ratio d_1/d_0 of this magnitude should be easily observable in the present simulations. However, as can be seen from figure 6 there is no clear evidence for faster than t^{-1} decay at any density. In fact, a $\log - \log$ fit to the long-time tail of the velocity ACF allows us to estimate the ratio d_1/d_0 as a function of density. If we assume a linear density dependence of d_1/d_0 we can put the

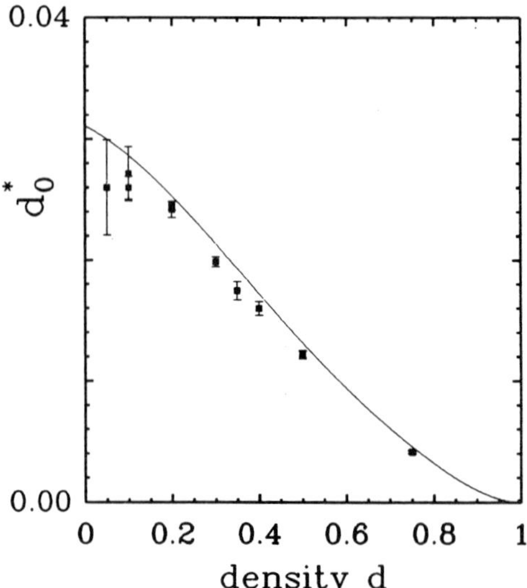

Figure 7: Density dependence of the amplitude of the t^{-1} algebraic tail of the tagged-particle VACF for the two-dimensional FHP-III model. Points: computer simulation results. Drawn curve: mode-coupling theory.

following bounds on the amplitude of this term:

$$d_1/d_0 = -0.02(3)d \,, \tag{8}$$

The kinetic theory estimate for continuous fluids exceeds this value by 7 standard deviations. This suggests that the expression for d_1/d_0 given in ref. [9] does not apply to lattice gases. It seems unlikely that this discrepancy is due to the fact that the simulations do not extend to long enough times. More likely, eqn. 7 is not correct for lattice gases, because the the density effects of the Fermi statistics have not been accounted for and the staggered momentum modes that do not contribute to d_0 are expected to affect d_1. It should be added that even though we compute $< \mathbf{v}(0).\mathbf{v}(t) >$ for $10^2 - 10^3$ mean-free times, we are still well removed from the asymptotic regime where the correlation functions is expected to decay as $1/(t\sqrt{\log t})$.

At present, the lattice-gas equivalent of the prediction for d_1/d_0 given in ref. [9] is still lacking. Clearly, such a theoretical result is highly desirable, as it would allow us to decide whether the suppression of significant corrections to the t^{-1} tail is peculiar to lattice gases or if it is indicative of the behavior of $2D$ fluids in general.

Acknowledgments

We would like to thank Matthieu Ernst and Tony Ladd for stimulating discussions and the steady exchange of information. This work is part of the research program of the Stichting voor Fundamenteel Onderzoek der Materie (Foundation for Fundamental Research on Matter) and was made possible by financial support from the Nederlandse Organisatie voor Zuiver Wetenschappelijk Onderzoek (Netherlands Organization for the Advancement of Research). Computer time on the NEC-SX2 at NLR was made available through a grant by the NFS(Nationaal Fonds Supercomputers).

References

[1] D.Frenkel in: **Liquids, Freezing and the Glass Transition**, J.P.Hansen and D.Levesque (editors), North-Holland, Amsterdam, 1990.

[2] B. J. Alder and T. E. Wainwright, Phys. Rev. **A1**:18(1970).

[3] J.R. Dorfman and E.G.D. Cohen, Phys. Rev. Lett. **25**:1257(1970), Phys. Rev. **A6**:776(1972),**A12**:292(1975).

[4] M.H. Ernst, E.H. Hauge and J.M.J. van Leeuwen, Phys. Rev. **A4**:2055(1971).

[5] Y. Pomeau and P. Résibois, Physics Reports **19**:63(1975).

[6] D. Levesque and W.T. Ashurst, Phys. Rev. Lett. **33**:277(1974).

[7] J.J. Erpenbeck and W.W. Wood, Phys. Rev. **A26**:1648(1982).

[8] J.J. Erpenbeck and W.W. Wood, Phys. Rev. **A32**:412(1985).

[9] I. de Schepper and M. H. Ernst, Physica **87A**:35(1977)

[10] D. Forster, D. R. Nelson and M. J. Stephen, Phys. Rev. **A16**:732(1977).

[11] U. Frisch, B. Hasslacher and Y. Pomeau, Phys. Rev. Lett. **56**:1505(1986).

[12] J.-P. Boon and A. Noullez in: Proceedings of workshop on **Discrete kinetic theory, lattice gas dynamics and foundations of hydrodynamics**, World Scientific, Singapore, 1989.

[13] P.M. Binder and D. d'Humières, Los Alamos preprint LA-UR-1341(1988) and in ref. [24].

[14] J. -P. Boon and A .Noullez, to be published.

[15] D.Frenkel in: Proceedings of workshop on **Cellular Automata and Modelling of Complex Physical Systems**, P. Manneville(editor), Springer, Berlin, 1989.

[16] D.Frenkel and M.H.Ernst, Phys. Rev. Lett. **63**:2165(1989).

[17] D. Frenkel, Phys. Lett. **A121**,385(1987)

[18] U. Frisch, D. d'Humières, B. Hasslacher, P. Lallemand, Y. Pomeau and J.-P. Rivet, Complex Systems **1**:649(1987).

[19] J.-P. Rivet, M. Hénon, U. Frisch and D. d'Humières, Europhys. Lett. **7**:231(1988).

[20] A.J.C. Ladd and D. Frenkel in: Proceedings of workshop on **Cellular Automata and Modelling of Complex Physical Systems**, P. Manneville(editor), Springer, Berlin, 1989.

[21] D. d'Humières and P. Lallemand, Complex Systems **1**:599(1987).

[22] D. d'Humières, P. Lallemand and U.Frisch, Europhys. Lett. **2**:291(1986).

[23] M.H. Ernst in: Proceedings of Les Houches Summerschool on **Liquids, Freezing and the Glass Transition**, J.P.Hansen and D.Levesque (editors), North-Holland, Amsterdam (1990).

[24] Proceedings of workshop on **Cellular Automata and Modelling of Complex Physical Systems**, P. Manneville(editor), Springer, Berlin, 1989.

[25] R. B. Pandey, D. Stauffer, A. Margolina and J. G. Zabolitzky, J. Stat. Phys. **34**:427(1984).

[26] M. E. Colvin, A. J. C. Ladd and B. J. Alder, Phys. Rev. Lett. **61**:381(1988)

[27] L. P. Kadanoff, G. R. McNamara and G. Zanetti, Phys. Rev. **A40**:4527(1989)

[28] R. F. Fox, Phys. Rev. **A27**,3216(1983).